C0-BOL-816

STUDIES OF HIGH TEMPERATURE SUPERCONDUCTORS

VOLUME 1

T. N

STUDIES OF HIGH TEMPERATURE SUPERCONDUCTORS

Advances in Research and Applications

Edited by Anant Narlikar
National Physical Laboratory
New Delhi

VOLUME 1

NOVA SCIENCE PUBLISHERS

Nova Science Publishers, Inc.
283 Commack Road
Suite 300
Commack, New York 11725

Library of Congress Cataloging-in-Publication Data
available upon request

ISBN 0-941743-54-3

Printed in the United States of America

Contributors to Volume 1

H. Adachi, Matsushita Electric Industries, Co.Ltd., Osaka, Japan.
S.K. Agarwal, National Phsyical Laboratory, New Delhi, India.
J.P. Carbotte, McMaster University, Hamilton, Canada.
K. Hirochi, Matsushita Electric Industries, Co.Ltd., Osaka, Japan.
G.H. Hyland, University of Warwick, Coventry, U.K.
Y. Ichikawa, Matsushita Electric Industries, Co.Ltd., Osaka, Japan.
Y. Iye, The University of Tokyo, Japan.
R.M. Iyer, Bhabha Atomic Research Centre, Bombay, India.
S.S. Jha, Tata Institute of Fundamental Research Centre, Bombay, India.
C.J. Jou, Lawrence Berkeley Laboratory, Berkeley, U.S.A.
U. Kabaswa, Osaka University, Osaka, Japan.
T. Kobayashi, Osaka University, Osaka, Japan.
K. Machida, Kyoto University, Kyoto, Japan.
F. Marsigilio, McMaster Universityy, Hamilton, Canada.
P.K. Mehta, Indain Institute of Technology, Bombay, India.
O. Meyer, Kernforschungszentrum, Karlsruhe, West Germany.
C.V. Narasimha Rao, National Physical Laboratory, New Delhi, India.
A.V. Narlikar, National Physical Laboratory, New Delhi, India.
B.D. Padalia, Indian Institute of Technology, Bombay, India.
J. Rammer, University of Bayreuth, Bayrueth, West Germany.
C.N.R. Rao, Indian Institute of Science, Bangalore, India.
K. Setsune, Matsushita Electric Industries, Co.Ltd., Osaka, Japan.
R. Srinivasan, Indian Institute of Technology, Madras, India.
K. Wasa, Matsushita Electric Industries, Co.Ltd., Osaka, Japan.
J. Washburn, Lawerence Berkeley Laboratory, Berkeley, U.S.A.
J.V. Yakmi, Bhabha Atomic Research Centre, Bombay, India.

ARTICLES PLANNED FOR FUTURE VOLUMES

Superconducting Weak Links in High Tc Thin Films *(M. Akinaga and L. Rinderer)*
Electron Microscopic Studies of High Tc Compounds *(S. Amelinckx)*
Low Dimensionality Aspects of High Temperature Superconductors *(S. N. Behra)*
A Review of the Preparation and Structure of Bi-Sr-Ca-Superconductors and Pb-Substituted Phases *(P. Bordet, J. J. Capponi, C. Chailout, J. Chenavas, A. W. Hewat, E. A. Hewat, J. J. Hodeau and M. Marezio)*
Mossbauer Spectroscopy of High Tc Superconductors *(P. Boolchand and D. McDaniel)*
Coprecipitation Technique for Preparation of High Tc Superconductors *(K. L. Chopra)*
Relation between Superconductivity and Magnetism with Special Attention to the Problem of Phase Diagram and Magnetic Structures *(G. Collin and D. Petigrand)*
Ginzburg-Landau Theory for High Tc Superconductors *(M. P. Das)*
Synthesis of High Temperature Superconducting Films - Processing Science and Issues *(C. Deshpandey and R. F. Bunshah)*
X-Ray Spectroscopic Study of High Temperature Superconductors *(K. B. Garg)*
Thermal Properties of High Temperature Superconductors *(E. Gmelin)*
Magnetic Relaxation of High Tc Superconductors *(R. Griessen and C. W. Hagen)*
High Pressure Studies on HTSC *(R. Griessen and R. J. Wijngaarden)*
High Resolution Electron Microscopy Study of High Tc Superconductors *(H. Hervieu, C. Michel and B. Raveau)*
Magnetic Pairing and Superconductivity *(K. Huang)*
The Materials Design for Very High Tc Superconductors *(H. Ihara)*
Atomic Imaging and Microstructure of Thallium-Based Superconductors *(Z. Iqbal and B.L. Ramakrishna)*
Anisotropic Superconducting and Normal State Transport Properties of High Temperature Superconducting Single Crystals *(Y. Iye)*
Multiple Hetero-epitaxial Growth of High Tc Superconductors *(T. Kobayashi)*
Normal and Superconducting Properties of the New High Tc Oxides and the Phonon-Plasmon Mechanism of Superconductivity *(V. Z. Kresin and H. Morawitz)*
Electronic Excitation Mechanisms of High Tc *(C. G. Kuper and J. Ashkenazi)*
Vibrational Spectra and Transport Properties of High Temperature Superconductors *(H. Kuzmany, E. Faulques, M. Matus and Pekker)*
Phonons in High Tc Superconductors *(S. Mase)*
Positron Annihilation Studies in High Temperature Superconductors *(T. Nagarajan)*
Twins in High Temperature Superconductors *(C. S. Pande and H. A. Hoff)*
Effect of Oxygen Ordering in YBCO Superconductors *(T. S. Radhaskrishnan)*
Field Ion Microscopic Studies of High Temperature Superconductors*(S. Ranganathan)*
Structural Chemistry of High Tc Superconductors *(B.Raveau, C. Michel and M. Hervieu)*

PREFACE

With Bednorz and Müller's pioneering discovery of high temperature superconductors in 1986 superconductivity has ceased to remain an area of mere academic curiosity and a preserve of a small community of low temperature physicists and cryogenists. Renouncing their cold confines freed from the grip of liquid helium, superconductors have stepped into the realm of high temperatures. The area has transformed into a rich field of intensive and highly competitive research, encompassing diverse disciplines such as structural chemistry, ceramic engineering, metallurgy, solid state electronics as well as experimental and theoretical condensed matter physics. The high temperature superconductors (HTSCs) discovered are from the family of ceramic oxides. Their large scale utilization in electrical utilities and in microelectronic devices are the frontal challenges which can perhaps be effectively met only through consolidated efforts and expertise of a multidisciplinary nature.

During the last two years the growth of the new field has occured on an international scale and perhaps has been more rapid than in most other fields. There has been an extraordinary rush of data and results which are continually being published as short texts dispersed in many excellent journals, some of which were started to ensure rapid publication exclusively in this field. As a result, the literature on HTSCs has indeed become so massive and unfortunately so diffuse that it is becoming increasingly difficult to keep abreast with the important and reliable facets of this fast-growing field. This provided the motivation to evolve a process whereby both professional investigators and students can have ready access to up-to-date in-depth accounts of major technical advances happening in this field. The present series "Studies of High Temperature Superconductors" has been launched to , at least in part, fulfill this need.

Each chapter in these volumes will comprise a detailed review or an extended paper focussing on one or more of the frontal aspects of research and applications, including the state-of-art technology pertaining to HTSCs. The contributors will be recognised authorities in the field. Starting with the present volume, a broad range of topics planned to be covered in the series will include theories of high temperature superconductivity, structural chemistry, superconducting and normal state properties, crystal structures and microstructural effects, experimental techniques of characterizing materials, thin film processing, device fabrication, production of wires and tapes for superconducting magnets and related applications, etc. While every effort will be made to avoid overlapping of the subject matter of the chapters, some repetition may inadvertently occur which is perhaps inescapable in a rapidly growing field. At the same time, although some of the chapters covering areas such as structural chemistry, microstructural aspects and high resolution electron microscopy may carry essentially similar broad chapter headings, the reader will hopefully find their contents, as prepared by various authorities in the field, to have expositions characteristically different and this is expected to pave the way for an improved understanding of the subject. The series is aimed at the professional scientist and engineer, as well as at graduate students in physics, chemistry, materials science, solid state electronics and engineering.

The present volume, i.e., Volume 1, contains 15 chapters covering some of the afore-said topics planned for the series. The chapters to be included in the forthcoming volumes are listed separately. In general, we will attempt to have each volume contain a range of different topics so as to make it useful to a wide cross-section of scientists and technologists working in the field.

The Editor would like to express his thanks to Mr. Frank Columbus, the President, Nova Science Publishers, for helpful and efficient cooperation. He thanks his colleagues, Dr. S.K. Agarwal and Dr. C.V. Narasimha Rao for their suggestions and help and he acknowledges the assistance received from his graduate students M/s V.P.S. Awana, P. Maruthi Kumar and V.N. Moorthy in the preparation of the Subject Index.

December 1988 — Anant Narlikar

CONTENTS

Contributors to Volume 1 v
Articles Planned for Future Volumes vii
Preface ix

CHAPTER-1: OXYGEN HOLE MECHANISM OF SUPERCONDUCTIVITY IN CUPRATES AND OTHER METAL OXIDES
C.N.R. Rao, Indian Institute of Science, Bangalore, India

Introduction 1
Experimental Results on Cuprates 1
Pairing of Oxygen Holes 6
Location of Oxygen Holes 7
Oxygen Holes in Other Oxides 8
References 9

CHAPTER-2: ELECTRONICALLY DRIVEN INSTABILITIES IN HIGH Tc SUPERCONDUCTORS AND RELATED MATERIALS
Kazushige Machida, Kyoto University, Kyoto, Japan

Overview 11
Heavy Fermion Superconductors 12
Copperless Cubic Oxides 22
Cu-Based Oxide Superconductors 34
References 38

CHAPTER-3: THE BCS PAIRING AND POSSIBLE MECHANISMS FOR HIGH Tc SUPERCONDUCTIVITY
Sudhanshu S. Jha, Tata Institute of Fundamental Research, Bombay, India

Introduction 41
The Generalized BCS Pairing Theory 44
The BCS Gap Equation for Layered Crystals 52
Some Properties of New High Tc Superconductors 55
Electrodynamic Response with Anisotropic Gap 56
Possible Pairing Mechanisms and Conclusion 58
References 62

CHAPTER-4: BOSON EXCHANGE MECHANISMS, BOUNDS AND ASYMPTOTIC LIMITS
J. P. Carbotte, McMaster University, Hamilton, Canada and
F. Marsiglio, University of California at San Diego, La Jolla, U.S.A.

Introduction 64
Formalism 66
Conventional Superconductors 69
Very Strong Coupling Regime 73
Functional Derivatives 79
Optimum Spectra 83
Asymptotic Limits 89
Specific to the Oxides 94
Combined Phonon-Exciton Mechanism 103
Conclusions 108

CHAPTER 5-A STRONG COUPLING APPROACH TO HIGH TEMPERATURE SUPERCONDUCTORS

Josef Rammer, University of Bayreuth, Bayreuth, West Germany

CHAPTER-6: ION BEAM MODIFICATION AND ANALYSIS OF THIN $YBa_2Cu_3O_7$ FILMS

O. Meyer, Kernforschungszentrum, Karlsruhe, West Germany.

CHAPTER-7: SUPERCONDUCTING PROPERTIES ASSOCIATED WITH SHORT COHERENCE LENGTH - FLUCTUATION EFFECT AND FLUX CREEP PHENOMENON IN HTSC

Yasuhiro Iye, the University of Tokyo, Tokyo, Japan

CHAPTER-8: BASIC THIN FILM PROCESSING FOR HIGH T_c SUPERCONDUCTORS

K. Wasa, H. Adachi, Y. Ichikawa, K. Setsune and K. Hirochi,
Matsushita Electric Industrial Co. Ltd., Moriguchi, Japan.

CHAPTER-9: THREE TERMINAL HIGH T_c SUPERCONDUCTING DEVICES WITH LARGE CURRENT GAIN
Takeshi Kobayashi and Uki Kabasawa, Osaka University, Toyonaka, Japan.

Introduction 211
Device Structure 212
Superconductor Thin Film Preparation 214
Device Fabrication 216
Three Terminal Device Characteristics 218
Possible Interpretation of Current Modulation Mechanism 223
References 228

CHAPTER-10: TWINS IN HIGH T_c $YBa_2Cu_3O_{7-\delta}$ SUPERCONDUCTORS
C. J. Jou and J. Washburn, Lawrence Berkeley Laboratory Berkeley, U.S.A.

Introduction 229
Formation of Coherent Twins 230
Models and Simulations of Oxygen Depleted Twin Boundaries 235
Effect of Oxygen Depleted Twin Boundaries on Superconducting Properties 237
Conclusion 241
References 242

CHAPTER-11: THE ROLE OF MOTT-INSULATION, NON-STOICHIOMETRY AND ALTERED VALENCE IN HIGH T_c SUPERCONDUCTIVITY
G. J. Hyland, University of Warwick, Coventry, U. K.

Introduction 244
Basic Properties of Mott-Insulators 246
Application to the Cuprate Superconductors 251
Resume and Outlook 263
References 265

CHAPTER-12: SOUND VELOCITY AND ELASTIC CONSTANTS IN OXIDE SUPERCONDUCTORS
R. Srinivasan, Indian Institute of Technology, Madras, India

Introduction 267
Doped Lanthanum Copper Compounds 270
Compounds Belonging to the Yttrium Barium Copper Oxide Family 273
References 281

CHAPTER-13: X-RAY PHOTOELECTRON SPECTROSCOPIC STUDIES OF HIGH T_c OXIDE SUPERCONDUCTORS
B. D. Padalia and P. K. Mehta, Indian Institute of Technology,Bombay, India

Introduction 283
Salient Features of XPS 285
Experimental 289
X-Ray Photoelectron Spectra 293
Conclusions 314

CHAPTER-14: SYNTHESIS OF HIGH Tc OXIDE SUPERCONDUCTORS IN Y-, Bi- AND T1- SYSTEMS: THE ROLE OF CHEMISTRY

R. M. Iyer and J. V. Yakhmi, Bhabha Atomic Research Center,Bombay, India.

CHAPTER-15: SUBSTITUTIONAL STUDIES ON HIGH TEMPERATURE SUPERCONDUCTORS

A. V. Narlikar, C. V. Narasimha Rao and S. K. Agarwal, National Physical Laboratory, New Delhi, India.

OXYGEN HOLE MECHANISM OF SUPERCONDUCTIVITY IN CUPRATES AND OTHER METAL OXIDES

C.N.R. Rao

Solid State and Structural Chemistry Unit
Indian Institute of Science
Bangalore-560012, India

INTRODUCTION

Several theoretical models have been proposed to explain high-temperature superconductivity in cuprates. An issue that is central to any model is the nature of copper and oxygen species in the cuprates since superconductivity clearly owes its origin to the Cu-O sheets universally present in all the cuprate families. Thus, the five families of cuprate superconductors, $La_{2-x}M_xCuO_4$ (M = Ca, Sr or Ba) of the K_2NiF_4 structure, $LnBa_2Cu_3O_{7-\delta}$ (Ln = Y or rare earth), $Bi_2(Ca, Sr)_{n+1}Cu_nO_{2n+4}$, $Tl_2(Ca,Ba)_{n+1}Cu_nO_{2n+4}$ and $Tl(Ca, Ba)_{n+1}Cu_nO_{2n+3}$, all contain two-dimensional Cu-O sheets. The Cu-O chains additionally present in the 123 compounds do not seem to play any crucial role. It has been generally believed that magnetic, superconducting and related properties of cuprates have some thing to do with the mixed valency of copper. For example, the resonating valence bond (RVB) model [1] requires the presence of holes on Cu sites (Cu^{3+} species). There are also a few models, however, based on the presence of holes on oxygen sites (O^- species); dimerization of oxygen holes has also been suggested to occur by a few workers. It is the purpose of this article to briefly present the available experimental evidence for the presence of oxygen holes and to discuss their role in high-temperature conductivity. It will be shown that these holes play a role in other oxide materials as well including the Cu-free $Ba_{1-x}K_xBiO_3$ superconductor.

EXPERIMENTAL RESULTS ON CUPRATES

In stoichiometric La_2CuO_4 which is an insulator copper is present only in the Cu^{2+} state (d^9 configuration or d^1 hole-state) while the oxygens are in the O^{2-} state (p^6 configuration or p^0 hole-state). The same would hold for the 123 compound $YBa_2Cu_3O_{6.5}$. When La in

La_2CuO_4 is partly substituted by Ba^{2+} or Sr^{2+} or when the oxygen stoichiometry of the 123 compound is $YBa_2Cu_3O_{7-\delta}$ ($\delta < 0.5$), as in the high T_c materials, we introduce holes in the Cu-O sheets. If the holes in these doped materials are on Cu sites, they would have the d^8 configuration (d^2 hole-state) as in Cu^{3+} ions. If, on the other hand, the holes are on oxygen sites, they would have the p^5 configuration (p^1 hole-state) as in O^{1-} species. We therefore have two possible alternatives (Cu^{3+} - O^{2-}) and (Cu^{2+} - O^{1-}). It is not as if that oxygen holes can only be present in such doped oxides. If the Cu-O bonds are highly covalent, it is quite conceivable that we can have species such as (Cu^{2+} - O^{1-}) and (Cu^{1+} - O^{1-}) instead of (Cu^{2+} - O^{2-}). Let us examine the available experimental evidence for the states of oxygen and copper in the cuprate superconductors.

Based on X-ray absorption (Cu K or L edge) spectroscopy, many workers initially came to the conclusion that Cu^{3+} was present in $YBa_2Cu_3O_{6.9}$ and $La_{1.85}Sr_{0.15}CuO_4$. Later measurements have, however, shown this to be incorrect [2,3]. The 21 eV feature in the Cu K-edge ($1s \longrightarrow 4p$ spectrum) earlier attributed to Cu^{3+}, is actually due to the poorly-screened satellites of the 15 eV feature which is its well-screened counterpart. In the absence of Cu^{3+}, the only possibility is that the holes reside on the oxygen orbitals. A $1s \longrightarrow 2p$ transition has indeed been identified in the oxygen K-edge spectrum, confirming the presence of holes in the oxygen band [4,5]. Electron energy loss spectroscopy also corroborates the presence of oxygen holes [6].

We have investigated the nature of holes in the superconducting cuprates for some time by employing photoelectron and Auger electron spectroscopies which provide direct information on the electronic structure [7-9]. Since the different valence states of copper would be associated with different binding energies, it becomes possible to determine the nature of the copper species by means of x-ray photoelectron spectroscopy (XPS) and Auger electron spectroscopy. The Cu(2p) spectra of $YBa_2Cu_3O_{6.9}$ and $La_{1.8}Sr_{0.2}CuO_4$ ($T_c \sim 36K$) show a 933 eV main feature with a weaker feature at 942 eV (Fig. 1). The 933 eV feature is due to the $d^{10}L^{-1} + d^{10}L^{-2}$ final states and the 942 eV feature due to the d^9L^{-1} final state. This means that there is some d^{10} (Cu^{1+}) species in the ground state. Furthermore, there was no evidence for a feature 23 eV away from the main peak suggesting that the amount of Cu^{3+} is negligible. Cu(LVV) Auger spectra show features due to a d^8 multiplet or ^1G state (Fig. 2) which also implies the presence of d^{10} species in the ground state.

Since these superconducting cuprates contain excess holes due to oxygen-excess stoichiometry, absence of Cu^{3+} necessitates that the excess holes be in the oxygen derived p-band. The O(1s) core level spectra of these cuprates show the presence of features with binding energies of around 529, 531 and 533 eV respectively (Fig.3). The normally expected oxide species, O^{2-}, with the filled $2p^6$ configuration is associated with the 529 eV feature. The O^{1-} species (corresponding to the presence of a hole in the 2p band) is expected to be associated with a slightly higher binding energy than O^{2-}; unfortunately CO_3^{2-} and OH^- which are inevitably present on such oxide surfaces show features around 531 eV. Scraping the sample surface decreases

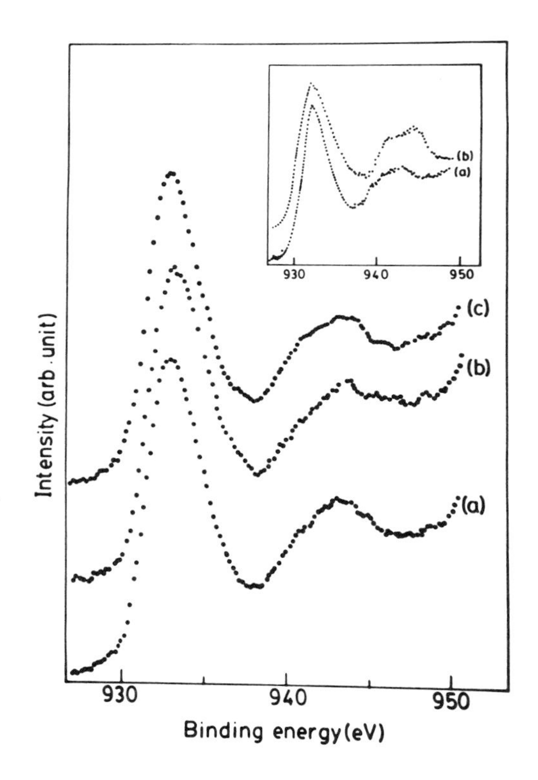

Fig. 1. Cu($2p_{3/2}$) region in the XPS of $YBa_2Cu_3O_{6.9}$: (a) at 300K; (b) after cooling to 80K; (c) after warming the sample (b) back to 300K. In the inset we show the Cu(2p) spectra of (a) $Tl_2Ca_2Ba_2Cu_3O_{10}$ and (b) $Bi_2(Ca, Sr)_3Cu_2O_8$ at 300K.

the intensity of the 531 eV feature or nearly eliminates it. We however find an oxygen species with a binding energy of ~ 533 eV which we consider to be due to the dimerization of O^{1-} holes. While $La_{1.8}Sr_{0.2}CuO_4$ showed only a small contribution due to the dimerized hole species at 300 in the O(1s) spectrum, the 533 eV signal due to this species is somewhat higher in $YBa_2Cu_3O_{6.9}$. We have studied the O(1s) spectra of a large number of samples of $YBa_2Cu_3O_7$ and all of them inevitably show the ~ 533 eV feature. This feature is not due to species arising from the interaction with H_2O since it is reversible and is found even well above 300K. Water disappears from such surfaces at 240K. The assignment of the 533 eV O(1s) feature to dimerized oxygen holes, O_2^{2-}, is based on the fact that ordinary peroxides and oxygen adsorbed on metal surfaces in the peroxo form

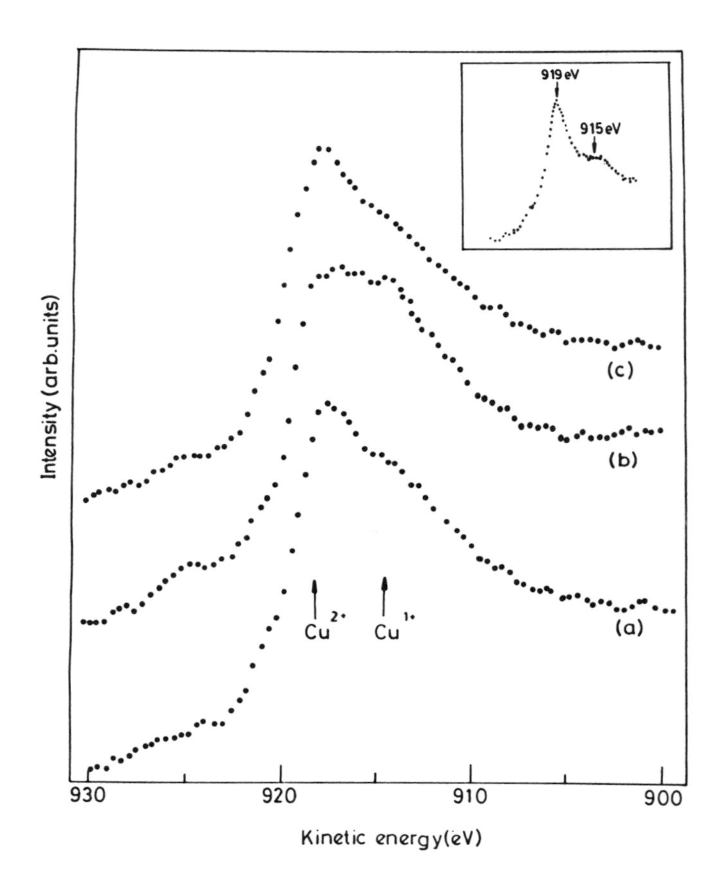

Fig. 2. Cu(L_3VV) Auger spectrum of $YBa_2Cu_3O_{6.9}$: (a) at 300K; (b) after cooling to 80K; (c) after warming the sample (b) back to 300K. In the inset, we show the Cu(L_3VV) spectrum of $Bi_2(Ca, Sr)_3Cu_2O_8$ at 300K.

show this feature. Recently, the 533 eV feature due to dimerized holes has been confirmed by other workers [10] although it was initially suspected to be an extrinsic feature by some workers.

The variation of the O(1s) signal intensity due to each species has been studied as a function of temperature in several samples. In Fig. 3, we show typical temperature-dependence of the 533 eV feature due to the dimerized hole species in the case of $YBa_2Cu_3O_{6.9}$. The proportion of dimerized holes increases with decreasing temperature and this variation is reversible. Accompanying such an increase of the hole dimers with lowering of temperature, the surface conductivity

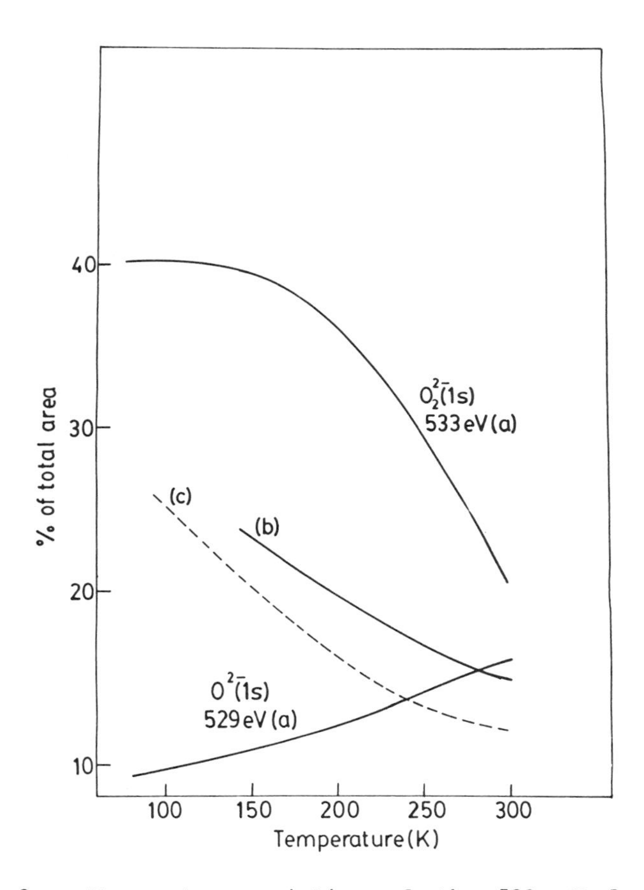

Fig. 3. Temperature-variation of the 533 eV feature due to the peroxide-like species in three independent measurements designated (a), (b) and (c). Temperature-variation of the 529 eV O^{2-} feature is shown in one case.

as measured by energy loss spectroscopy [11] increases. We feel that this dimer species plays an important role in the mechanism of superconductivity of cuprates. The preponderant presence of Cu^{1+} in the superconducting state is noteworthy and implies that a significant proportion of the oxygens is present in the hole state. The average electron occupation of the copper d-orbital would therefore be considerably less than 10.

Room-temperature Cu(2p) spectrum of $Tl_2Ca_2Ba_2Cu_3O_{10}$ ($T_c \sim 125K$) also shows features around 933 and 942 eV as shown in the inset of

(Fig. 1) [12]. The Cu(2p) spectrum of $Bi_2(CaSr)_3Cu_2O_{8+\delta}$ ($T_c \sim$ 90K) shows similar results (Fig. 1) [13). The Cu(LVV) spectrum of this oxide also shows the ^1G multiplet indicating the presence of Cu^{1+} (Fig. 2). The proportion of Cu^{1+} seems to increase on lowering the temperature in all these cuprates. Furthermore, we do not expect Cu^{3+} to be present significantly in the Bi and Tl cuprates based on stoichiometry considerations. Both Bi and Tl are present in the 3+ state in the cuprates, as found by x-ray photoemission and chemical means. Presence of Cu^{1+} and oxygen holes in the Bi and Tl cuprates is therefore of much significance. The O(1s) spectrum of $Tl_2CaBa_2Cu_2O_8$ also reveals that the 533 eV feature due to the oxygen hole dimers is present even at 400K and becomes more prominent on cooling, the changes being reversible.

Pairing of Oxygen Holes

In order to understand the nature of oxygen holes and the pairing of these holes, let us consider the square plane of 4 oxygens with a copper atom in the middle shown in Fig. 4. Of the four possible linear combination of the two p_x and two p_y orbitals only the following configuration can hybridize with the $Cu(d_{x^2-y^2})$ orbital:

$$p_{x,y} = p_x^1 - p_y^2 - p_x^3 + p_y^4 \tag{1}$$

Pairing of the oxygen holes in the $\underline{k}$-space can occur through the intermediacy of the Cu^{2+} spin, as pointed out by Emery[14] and Hirsch[15]. Can there also be real-space pairings? Chakraverty, Sarma and Rao [16] have examined this question in some detail. With two electrons per orbital of the oxygen band, the ground state is a filled non-bonding band of electrons unaffected by local lattice deformation. As soon as holes are added to the top of the electron band (the bottom

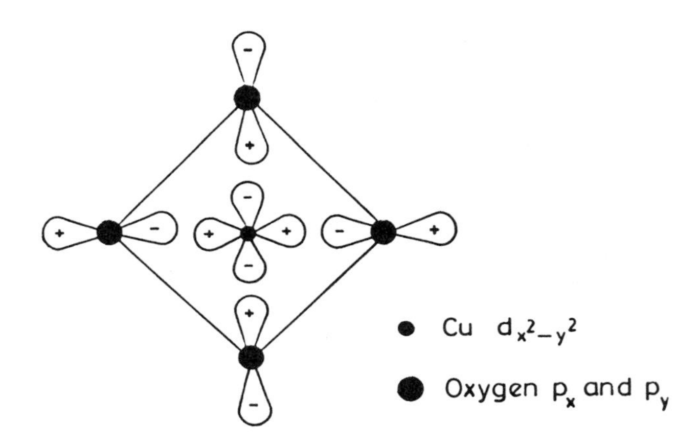

Fig. 4. Hybridization of Cu $|\underline{d}_{x^2-y^2}\rangle$ and oxygen p_x, p_y and (p_σ) orbitals.

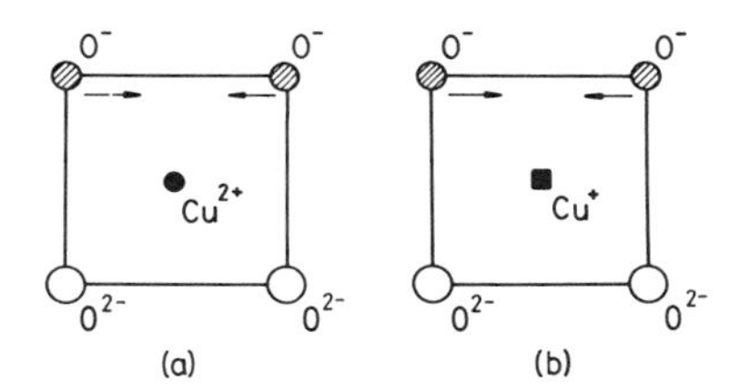

Fig. 5. Configurations of (a) bipolaron and (b) peroxiton; arrows indicate local deformation.

of the hole band), effective bond order gets added to the interatomic regions and the lattice responds by local bond contraction; this would pull out the empty state below the bottom of the hole band giving rise to the so-called a hole bipolaron ($O^- O^-$ dimer). Whether such a bound-state will actually occur or not depends on the intersite Coulomb repulsion terms between two near neighbour holes, V_{pp} and V_{pd}, depending on whether both are p-holes or one is a p (O^{1-}) the other a d-hole (Cu^{2+}). In Fig. 5 two possible configurations of (O^- - O^-) hole pairs are shown which we call bipolaron and peroxiton respectively. The net binding energy in the two configurations are:

$$E_{bipolaron} = -\lambda + 4V_{pd} + V_{pp}$$

$$E_{peroxiton} = -\lambda + 2V_{pd} + V_{pp} + \Delta$$

where λ is the electron-phonon coupling energy measured from the bottom of the oxygen-hole band. Because of the large inter-site Coulomb repulsion ($V_{pd} \sim 1$ eV), the bipolaron would have difficulty in forming a true bound state whereas the peroxiton (O^- - Cu^+ -O^-) pair would have much lower energy [16]. The peroxiton binding energy goes abruptly to zero, at some critical value of $\lambda/V_{pd} \sim 1$. At or above such a critical value, peroxiton formation will spontaneously occur. In a recent publication, Chakraverty et al [17] have developed the peroxiton hamiltonian in the small p-d hybridization limit and constructed a new ground state out of 'Squeezed Phonon Vacuum' such that the peroxitons will have an effective mass, $m^* \sim$ 4-10 m. These real-space hole pairs can now Bose-Condense at T_c. Making reasonable assumptions of the mass and concentration of peroxitons we get a T_c of 100K in cuprates [16].

LOCATION OF OXYGEN HOLES

We have proposed that the $d_{x^2-y^2}$ orbital of copper (in the Cu-O sheets) overlaps with the P_σ orbital of oxygen (formed by a combination of P_x and P_y orbitals) forming a broad Cu-O band consistent with high covalency of the Cu-O bands [16]. We also propose that the holes are in the P_σ state (within the Cu-O band) and that they are more favoured by the d^{10} state Cu^{1+} ions than by Cu^{2+} (d^9) ions. The presence of holes in P_σ also explains the absence of antiferromagnetism in the cuprates. Guo, Langlois and Goddard [18], based on cluster calculations, suggest that oxidation of Cu beyond Cu^{2+} creates oxygen P_π holes bridging two Cu^{2+} sites. The P_π holes are ferromagnetically coupled to adjacent Cu^{2+} d-electrons and hopping of the P_π holes in the Cu-O sheets from site to site is responsible for the conductivity. The P_π of these workers is not the P_z orbital but a $P_{x,y}$ orbital of oxygen. It is not clear that these will form narrow bands distinctly separated from the broad Cu-O band involving $d_{x^2-y^2}$(Cu) and oxygen P_σ orbitals. It is likely that this P_π will be within the broad Cu-O band. Cluster calculations may tend to overlocalize the holes; the high mobility required of the holes becomes possible when they are in the broad Cu-O band. It is, therefore, difficult to distinguish the P_π holes of Guo et al from that proposed from this laboratory.

If we consider that the $d_{x^2-y^2}$ (Cu) orbital overlaps with the P_x orbital of oxygen, then it is not at all certain that the P_y of oxygen will be well-separated from the Cu-O band. Holes in the P_π orbitals of oxygen have been proposed by other workers [19, 20] and on an examination of the models, it is found that they are actually P_σ in the BaO layers formed by overlap with the Cu(d_{z^2}) orbital. Such holes are likely to be localized and furthermore the oxygen atoms involved may be too far from the Cu atoms to really be involved in imparting conductivity to the oxide. Based on polarized x-ray absorption measurements, it had apparently been found that holes reside in the P_z orbitals. This does not, however, preclude the presence of holes in the $P_{x,y}$ orbitals, since partial hole occupation (in P_z) would suffice to yield such results. It should be noted that oxygen hole-pairing has been suggested by many theoreticians, but it is not entirely certain that the peroxo-type species observed in the O(1s) spectra correspond to the paired holes proposed by theoreticians.

We can relate the presence of O^{1-} type holes to the average charge, p of the $(Cu-O)^p$ bond since p is known to determine the superconducting properties of the cuprates [21]. A finite positive value of p (say, 0.2 which seems to delineate superconductors and insulators) can only result from $(Cu^{2+} - O^{1-})^{1+}$ in combination with $(Cu^{2+} - O^{2-})^0$ and $(Cu^{1+} - O^{1-})^0$ species since there is likely to be no $(Cu^{3+} - O^{2-})$ species. The relative importance of $(Cu^{2+} - O^{2-})^0$ and $(Cu^{1+} - O^{1-})^0$ determines whether a particular cuprate is a superconductor or an insulator.

OXYGEN HOLES IN OTHER OXIDES

Peroxides are not new to chemistry. Ordinary peroxides such as BaO_2 containing well-defined O-O bonds with a bond order of unity are insulators. The possibility of the existence of peroxide-type species in nominal oxide materials was proposed recently [22] to explain the nature of oxygen excess in $La_2NiO_{4+\delta}$. It was suggested that some of the oxygen in this oxide was present as O_2^{2-} instead of some of the nickel as Ni^{3+}. A study in this laboratory of transition of several metal oxides where the metal is expected to be in a high oxidation state (e.g. $LaCuO_3$, $Ba_2Cu_2O_5$, $LaNiO_3$) has shown the presence of holes in the oxygen 2p band giving rise to O^{1-} species, the metal being present in the normal low-oxidation (2+) state of the transition metal. Superconducting $Ba(Pb, Bi)O_3$ similarly seems to have Bi^{3+}, Pb^{2+} and oxygen holes. The discovery [23] of relatively high T_c (~ 30K) superconductivity in $Ba_{1-x}K_xBiO_3$ which has no copper is also explained on the basis of oxygen holes ($O^- - Bi^{3+} - O^-$). While the evidence for the presence of holes in the oxygen 2p bands of oxides may be considered to be of a recent origin, the occurrence of holes in sulphides, selenides and tellurides forming S-S, Se-Se and Te-Te bonds has been known for some time. Such species arising from hole dimerization are favoured in the presence of d^{10} ions such as Cu^{1+}. These holes impart unusual electrical properties to the chalcogenides. What we propose here in the case of the superconducting cuprates and other oxides therefore seems natural, although energetically the oxygen hole dimers would be less stable than the chalcogen hole dimers (O-O < S-S < Se-Se < Te-Te). Just as the oxygen holes give rise to superconductivity, they also act to hinder the antiferromagnetic state by progressively destroying the superexchange taking place through O^{2-}.

ACKNOWLEDGEMENTS

The author thanks the Department of Science and Technology and the University Grants Commission for support of this research.

REFERENCES

[1] P.W. Anderson, Science 235 (1987) 1196.
[2] P. Steiner, V. Kinsinger, I. Sander, B. Siegwart, S. Hufner, C. Politis, R. Hoppe and H.P. Müller, Z. Phys. B67 (1987) 497.
[3] A. Bianconi, M. De Sourtis, A.M. Flank, A. Fontaine, P. Lagarde, A. Marcelli, H.K. Yoshida and A. Kotani, Physica C, 153-155 (1988) 1760.
[4] J.A. Yarmoff, D.R. Clarke, W. Drube, U.O. Karlsson, A. Taleblbrahimi and F.J. Himpsel, Phys. Rev. B36 (1987) 3967.
[5] N. Nücker, J. Fink, B.Renker, D. Ewert, C. Politis, P.J.W. Weijs and J.C. Fuggle, Z. Phys. B67 (1987) 9.
[6] J.C. Fuggle, P.J.W. Weijs, R. Schoorl, G.A. Sawatzky, J. Fink, N. Nücker, P.J. Durham and W.M. Temmerman, Phys. Rev. B, B37 (1988) 123.
[7] D.D. Sarma, K. Sreedhar, P. Ganguly and C.N.R. Rao, Phys.Rev. B36 (1987) 2371.
[8] D.D. Sarma and C.N.R. Rao, Solid State Commun. 65 (1988) 47.
[9] C.N.R. Rao, D.D. Sarma, P. Ganguly and J. Gopalakrishnan, Mater. Res. Bull. 22 (1987) 1159.
[10] C.C. Chang, M.S. Hegde, J.M. Tarascon, T. Venkatesan, A. Inam, X.D. Wu and W.L. McLean, App. Phys. Lett. to be published.
[11] D.D. Sarma, K. Prabhakaran and C.N.R. Rao, Solid State Commun. 67 (1988) 263.

[12] A.K. Ganguli, K.S. Nanjunda Swamy, G.N. Subbanna, M.K. Rajumon, D.D. Sarma and C.N.R. Rao, Mod. Phys. Lett. B, in print.
[13] C.N.R. Rao, L. Ganapathi, R.Vijayaraghavan, G.R. Rao, K. Murthy and R.A. Mohan Ram, Physica C, in print.
[14] V.J. Emery, Phys. Rev. Lett. 58 (1987) 2794.
[15] J.E. Hirsch, Phys. Rev. Lett. 59 (1987) 228.
[16] B.K. Chakraverty, D.D. Sarma and C.N.R. Rao, Physica C, 156 (1988) 413; C.N.R. Rao, Mod. Phys. Lett. B. 2 (1988) 1217.
[17] B.K. Chakraverty, D. Feinberg, Z. Hang and M. Avignon, Solid St. Commun. 64 (1987) 1147.
[18] Y. Guo, J. Langlois and W.A. Godard III, Science 239 (1988) 896.
[19] H. Kamimura, to be published.
[20] A. Fujimori, to be published.
[21] Y . Takura, J.B. Torrance, T.C. Huang and A.I. Nazzal, to be published.
[22] C.N.R. Rao, P. Ganguly, M.S. Hegde and D.D. Sarma, J. Amer. Chem. Soc. 109 (1987) 6893.
[23] R.J. Cava, B. Batlogg, J.J. Krajewski, R. Farrow, L.W. Rupp, A.E. White, K. Short, W.F. Peck and T. Kometani, Nature 332 (1988) 814.

ELECTRONICALLY DRIVEN INSTABILITIES IN HIGH-Tc SUPERCONDUCTORS AND RELATED MATERIALS

Kazushige Machida*

Laboratory of Atomic and Solid State Physics
Clark Hall, Cornell University
Ithaca, New York 14853-2501

§1 OVERVIEW

It is not totally useless to learn the historical development if we assume continuous evolution in physics research, even in this high T_c (superconducting transition temperature) research. In the phenomenological level the high T_c oxide superconductors are very similar, in a broad sense, to the so-called magnetic superconductors.

Over a decade ago it was found that the magnetic ordering occurs out of the superconducting state in ternary compounds such as $(RE)Mo_6S_8$ and $(RE)Rh_4B_4$, where RE stands for a rare earth ion [1]. In the case of antiferromagnetic (AF) ordering, the superconductivity (SC) is not completely destroyed below the Nèel temperature (T_N), but is weakened by the onset of ordering. In these materials the two electron groups (one group being conduction electrons, coming mainly from 4d electrons, and the other being 4f localized electrons, which are responsible for the magnetism) are spatially separated because of their special crystal structures and interact weakly with each other. The pairing interaction here is ordinary phonon mediated attractive interaction, and the resulting pairing is the isotropic singlet of BCS theory.

Around 1980, two independent discoveries came at almost the same time: organic superconductors in Bechgaard salt and heavy Fermion superconductors (HFS). An example in the former system: $(TMTSF)_2PF_6$ exhibits a spin density wave (SDW) under ambient pressure which changes over the SC as the pressure increases [2]. Since there are no localized electrons to be responsible for the magnetism, the same conduction electron system plays two roles for SC and SDW. Thus the relationship of the two orderings becomes more delicate and subtle in these materials than in the ternary compounds.

In the latter system, HFS, close proximity between the SC and magnetism was observed. The 5f, or 4f electrons, which are localized at high temperatures, begin to move below the coherent temperature and to form a Fermi liquid state out of which both magnetism and SC emerge. It is therefore conceivable that the pairing interaction itself is greatly influenced by the presence of the magnetism. We will try to understand the high T_c mechanism in the recent copper-based oxides in this context, since close proximity between the magnetism and SC is also observed.

Another type of ordering which is driven electronically is charge density waves (CDW). The CDW is also discussed in conjunction with SC in various systems, such as A-15 compounds or $NbSe_3$ [3]. Here we consider the possibility that CDW can enhance T_c in copperless oxide high T_c materials: $Ba(Pb,Bi)O_3$ or $(Ba,K)BiO_3$. Through these studies we will try to understand a common empirical rule as to why the high T_c superconductors occur near the marginal instability of the metal-insulator transition irrespective of the nature of insulators.

The arrangement of this article is as follows: The heavy Fermion superconductors $CeCu_2Si_2$, UBe_{13}, UPt_3 and URu_2Si_2 are treated in §2. We discuss the copperless oxide superconductors $Ba(Pb,Bi)O_3$ and $(Ba,K)BiO_3$ in §3. The final section, §4, is devoted to reviewing $(La,A)_2CuO_4$ (A: Sr, Ba, Ca) and $YBa_2Cu_3O_7$, focusing on the common features shared with the above materials.

§2 HEAVY FERMION SUPERCONDUCTORS

All the four heavy Fermion superconductors (HFS) known to date are more or less related to magnetic phenomena, namely antiferromagnetism (AF) or spin density wave (SDW). As is seen from Fig. 2-1 where the schematic phase diagrams for HFS are drawn, the close proximity between the superconductivity and SDW is one of the most important key characteristics in understanding the nature of these extraordinary materials, especially the origin of the unconventional superconducting properties.

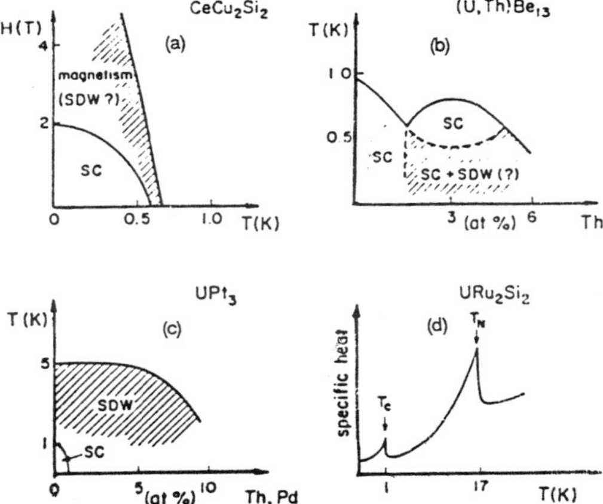

Fig. 2-1: Schematic phase diagrams of four heavy Fermion superconductors: (a) $CeCu_2Si_2$, (b) $(U,Th)Be_{13}$, (c) UPt_3, and (d) URu_2Si_2.

The purpose in this section is not to give a general survey of the HFS problem which was already treated by others [4], but to point out some common features in both HFS and the high T_c oxides, which might lead to a clue to understand the high T_c phenomena.

§2-1 Experimental Overview

It is now well established that the superconductivity in HFS is not an isotropic singlet pairing, but some anisotropic pairing. The SC gap has some node structure on the Fermi surface. This can be best seen from Fig. 2-2 where the nuclear relaxation rate T_1^{-1} is plotted for all four systems.

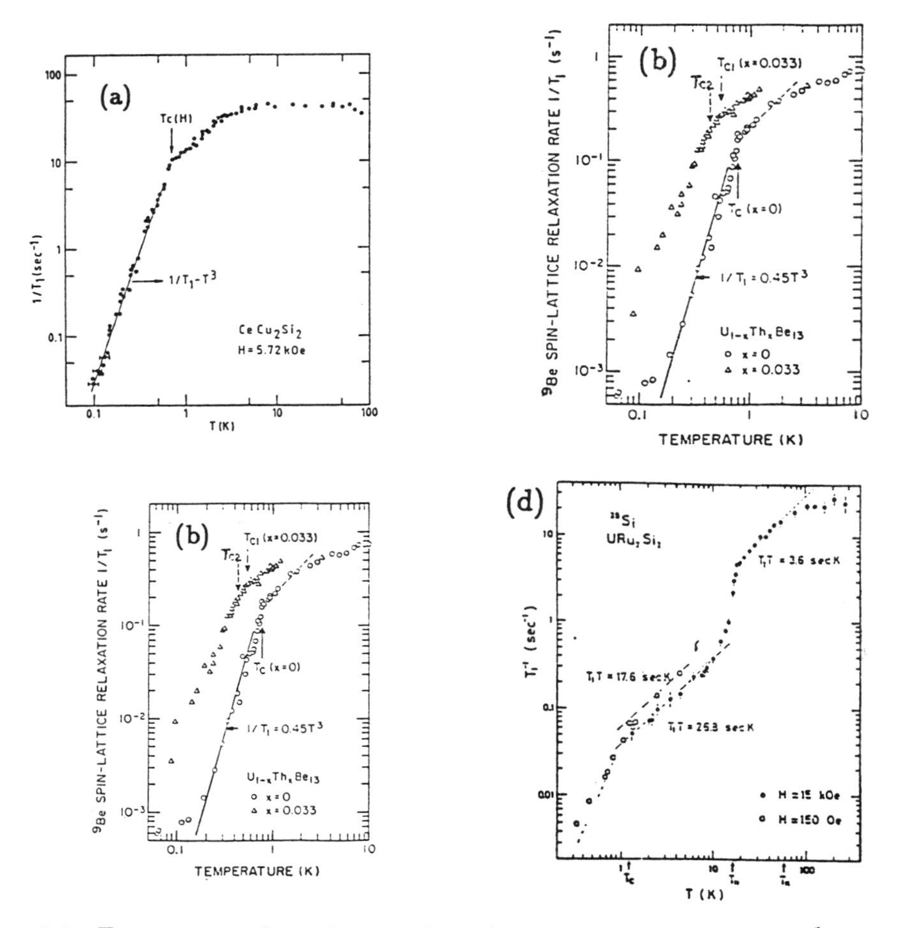

Fig. 2-2: Temperature dependences of the nuclear relaxation rate (T_1^{-1}) for (a) $CeCu_2Si_2$ [5], (b) $U_{1-x}Th_xBe_{13}$ [6], (c) UPt_3 [7], and (d) URu_2Si_2 [8].

If we look at Fig. 2-2 closely, we can recognize a common crossover behavior of T_1^{-1} except for UPt_3; that is, the relatively T-independent T_1^{-1} in high temperatures changes into the linear T-dependence, namely Korringa behavior just above T_c in the case of $CeCu_2Si_2$ and UBe_{13} and above T_N in URu_2Si_2. This indicates the change from the individual localized 5f (or 4f) moment regime to the Kondo coherent regime where the electrons begin to move, forming Fermi liquid quasi-particles with the renormalized mass. Although it is extremely difficult to describe this smooth transition from a microscopic model, which no one has succeeded yet, once the Kondo coherent state develops at lower temperatures, one can start out with a Fermi liquid state where the renormalized interaction between the quasi-particles with the renormalized heavy mass is expected to be relatively weak. This is one reason why the band calculations [9] can reproduce so well the topology of the observed Fermi surface of UPt_3 [10].

$CeCu_2Si_2$

This material is the first HFS found in 1979, which has been regarded as a typical superconductor. The recent finding of the static magnetic order by the zero field muon spin resonance (μSR) is a real surprise. The μSR experiment [11] done by using $CeCu_{2.1}Si_2$ is consistent with the other independent Si-NMR experiment [12] which also detects a sudden broadening of the distribution of the internal fields at a certain temperature T_N under an applied field. The zero field extrapolation of T_N in the NMR results smoothly comes to $T_N \sim 0.8$ K, indicating that both probes are seeing the same phenomenon, as sketched in Fig. 2-1. Therefore SC is covered by a magnetism, most probably, by antiferromagnetism or SDW in external field and temperature phase diagram.

The superconducting properties have been investigated extensively for a long time. There is little consensus as to the precise nature of the pairing function and the pairing mechanism although it is certainly unconventional. As T decreases the system becomes a Fermi liquid state around $T \lesssim 1 \sim 2$ K as seen from Fig. 2-2(a) below which T_1T = constant is obeyed. At T_c, T_1^{-1} does not exhibit enhancement characteristic in an ordinary BCS superconductor, which is indicative of unconventional pairing. At lower temperatures $T_1^{-1} \propto T^3$ and thermal conductivity ($\sim$ T) behaviors are consistent with line modes [13]. The system shows substantial Knight shift below T_c for both components parallel and perpendicular to an external field [14], implying that the parity of the pairing function is definitely even, not odd parity pinned to the crystalline axis.

$(U,Th)Be_{13}$

The thermodynamic properties of UBe_{13} such as the specific heat at lower temperatures below T_c are still much debated on what the power law is obeyed exactly (T^2, T^3 or other) [4]. On the other hand, the transport coefficients such as thermal conductivity ($\sim$ T), the nuclear relaxation rate ($\sim T^3$), the ultrasonic attenuation ($\sim T^2$) clearly indicate that the SC gap vanishes at lines on FS [13].

There is, however, no consensus on the parity of the pairing function because the two Knight shift experiments by μSR and NMR are mutually contradicted.

In $(U_{1-x}Th_x)Be_{13}$ as x increases T_c rapidly decreases while beyond $x \gtrsim x_{cr} = 0.017$ it stays rather constant and simultaneously there appears the second phase transition (T_{c2}) below T_c as sketched in Fig. 2-1(b). The specific heat [15] in this concentration region clearly shows a sharp successive phase transition as seen in Fig. 2-3, where the upper peak in the double porn structure corresponds to the SC transition. The second lower transition is also detected by the ultrasonic experiment. The μSR [16] observes the growth of the internal field below T_{c2},

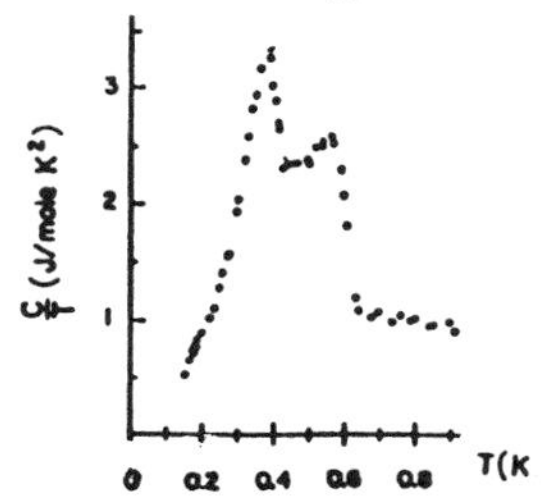

Fig. 2-3: The specific heat of $(U_{1-x}Th_x)$ Be_{13} ($x = 0.0308$) showing two successive transitions.

suggesting that some kind of magnetism appears. As we can see later, the unconventional pairing states are generally unstable against the SDW formation and some pairing states necessarily coexist with the SDW at low temperatures [17]. Others [18 ~ 21] regard T_{c2} as a phase transition between different classes of pairing states. Therefore the problem is still open at present. It is to be noted that the inelastic

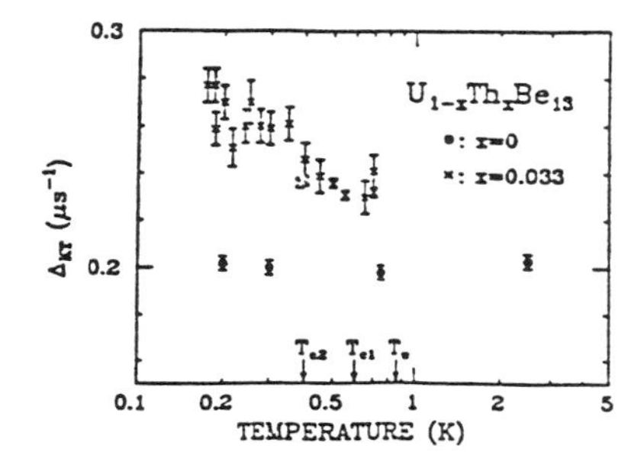

Fig. 2-4: Temperature dependence of the zero field relaxation rate Δ_{KT} of μSR in $(U_{1-x}Th_x)Be_{13}$ [16].

experiment on UBe_{13} by Mook *et al.* [22] observes the antiferromagnetic fluctuations at the zone boundary position in reciprocal space along [001] of the cubic crystal. On the other hand, according to the band structure calulation [23] there indeed exists a nesting feature between an object centered at Γ point and a pillbox at X point if we lower the Fermi level by $\sim$ 10 meV in their dispersion curve, coinciding with the observed antiferromagnetic correlation. Since Th-doping in (U,Th)Be_{13} can be regarded to act as decreasing the electron number as a first approximation, it is natural to consider that the doping induces the SDW with the wave number Q $= \frac{2\pi}{a}$ (0,0,1) which corresponds to Γ X vector in reciprocal space.

The lower critical field (B_{c1}) measurement [19] for $U_{0.97}Th_{0.03}Be_{13}$ indicates an interesting anomaly at T_{c2} (see Fig. 2-5), that is, B_{c1} is enhanced below T_{c2} while

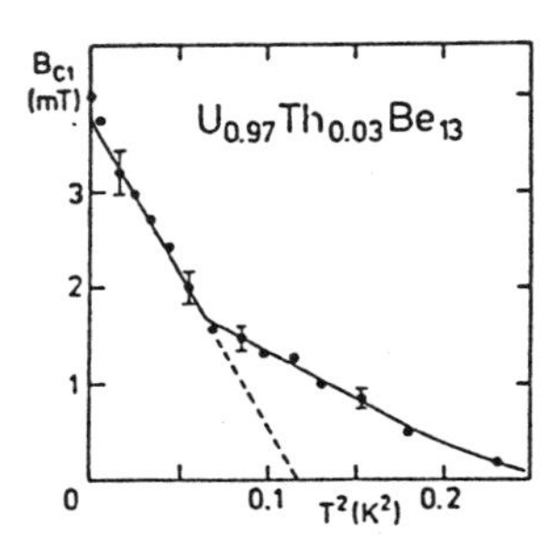

Fig. 2-5: Temperature dependence of the lower critical field B_{c1} for $(U_{1-x}Th_x)Be_{13}$ (x = 0.03) [19].

the upper critical field shows no anomaly, implying the enhancement of the SC condensation energy below T_{c2}. Naively we would expect the opposite, namely, the depression of the SC condensation because the coexisting SDW more or less weakens the SC. However, it is noted that the coexistence of the SDW and SC inevitably induces a different type of pairing function [24] whose condensation might compensate the loss due to the coexistence.

<u>UPt_3</u>

The power law temperature dependences in physical quantities such as thermal conductivity, ultrasonic attenuation and NMR relaxation rate (see Fig. 2-2) below $T_c \sim 0.5$K are consistently described by the line node of the SC gap on FS. The Knight shift experiment on Pt-NMR [7] shows essentially no change below T_c down to the lowest temperature ($\sim$60 mk) and cannot be explained by the even-parity state in the strong spin-orbit scattering limit. Thus the parity of the pairing function must be odd. The group-theoretic classifications in the strong spin-orbit coupling limit based on point group symmetry do not allow the odd-parity state with the line node.

Aeppli *et al.* [25] have made a totally unexpected discovery via the neutron diffraction that UPt_3 exhibits an antiferromagnetic transition at $T_N = 5$ K with the saturation moment ~ 0.02 μ_B/U-atom. The superconductivity coexists with the SDW below $T < T_c \sim 0.5$ K. The two orderings interfere with each other. As shown in Fig. 2-6, the SDW magnetization is depressed by the superconductivity. We notice that the SDW pattern is exactly the same as in the dilutely doped systems

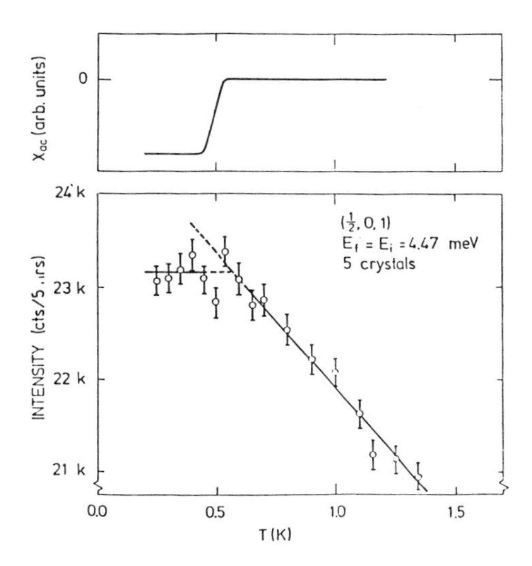

Fig. 2-6: The change of the amplitude of the SDW below T_c seen in neutron experiment [25]. The AC susceptibility χ_{ac} in the upper panel indicates the superconducting transition.

(U,Th)Pt_3 and U(Pt,Pd)$_3$ except that the moments are $\sim$ 0.6 μ_B/U-atom in the latter two materials [26 $\sim$ 28]. The dilute limit of T_N in these systems roughly coincides with $T_N \sim$ 5 K in UPt_3, giving rise to the phase diagram shown in Fig. 2-1. We must realize that in order to uncover the nature of the pairing state in UPt_3, we must take into account the existence of the SDW.

URu_2Si_2

This system exhibits SC at $T_c \sim$ 1.5 K below the SDW transition ($T_N \sim$ 17.5 K). As already mentioned, the localized-moment behavior evidenced by the Curie-Weiss law in the magnetic susceptibility or by NMR [8] changes into a Fermi liquid state as inferred from the Korringa like NMR relaxation around T $\sim$ 60 K (see Fig. 2-2(d)). The specific heat and NMR measurements indicate that the FS area removed by the SDW gap formation amounts to 30 $\sim$ 40% of the total. The SDW is commensurate with the underlying crystal lattice and the saturation moment is extremely small ($\sim$ 0.03 μ_B/U-atom) [29]. The small transition entropy and a BCS like specific heat jump at T_N are consistent with the itinerant electron magnetism picture.

§2-2 Model Consideration

To understand rich phenomena associated with SC and SDW in HFS, we must devise a model which contains essential features of this problem: An important interaction is the on-site Coulomb interaction between the renormalized quasi-particles. It is known that this repulsive interactions U is responsible for not only the SDW formation, but also anisotropic non-trivial pairing states which are more advantageous than the isotropic state because these can avoid U effectively. The origin of the attractive interaction $g(k,k')$ is not known so far in HFS. It is conceivable that a part of this comes from the SDW fluctuations [31] which are observed by neutron experiments [4] in HFS commonly. This suggests that U and $g(k,k')$ are not independent, but rather interconnected each other. However, at the present stage we regard these as independent parameters, yet we can understand some of the essences of the problem [11, 32], namely, within a mean field approximation in which we consider the SDW order parameter M and SC order parameter $\Delta(k)$, the total effective Hamiltonian is written as

$$\begin{aligned} H &= H_O + H_{SDW} + H_{SC} \qquad (2.1)\\ H_O &= \sum_{k\sigma} \varepsilon(k) C^+_{k\sigma} C_{k\sigma} \\ H_{SDW} &= -\sum_{k\sigma} \sigma(M C^+_{k+Q\sigma} C_{k\sigma} + h.c.) \\ H_{SC} &= -\sum_{k\sigma\sigma'} \Delta_{\sigma\sigma'}(k)(C^+_{k\sigma} C^+_{-k\sigma'} + h.c.) \end{aligned}$$

where for simplicity we consider a two-dimensional electron band given by

$$\begin{aligned} \varepsilon(k) &= \gamma_k + \delta_k \qquad (2.2)\\ \gamma_k &= -t(\cos k_x + \cos k_y) \\ \delta_k &= -t_1 \cos k_x \cos k_y \end{aligned}$$

t and t_1 are the nearest and next-nearest neighbor transfer integrals in a tight binding scheme respectively. The self-consistent equations for the SDW and the SC are given by

$$M = \frac{1}{2} U \sum_{k\sigma} \sigma < C^+_{k+Q\sigma} C_{k\sigma} > \qquad (2.3a)$$

$$\Delta_{\sigma\sigma'}(k) = \sum_{k'\mu\mu'} g_{\sigma\sigma'\mu\mu'}(k,k') < C^+_{k'\mu} C^+_{-k'\mu'} > \qquad (2.3b)$$

where the SDW nesting vector is fixed to $\vec{Q} = (\pi,\pi)$. We assume a separable attractive interaction form: $g_{\sigma\sigma'\mu\mu'}(k,k') = g\omega_{\sigma\sigma'}\omega_{\mu\mu'}\tau(k)\tau(k')$ where $\omega_{\sigma\sigma'}$ is a spin matrix, then the order parameter becomes $\Delta_{\sigma\sigma'}(k) = \Delta\omega_{\sigma\sigma'}\tau(k)$.

To diagonalize the Hamiltonian (2.1) and to obtain the self-consistent equations, there are two key symmetries of $\Delta(k)$, the parity and the translation by Q, namely,

$$\Delta(-k) = \pm\Delta(k) \tag{2.4}$$

$$\Delta(k+Q) = \pm\Delta(k) \tag{2.5}$$

respectively. Depending on the combined symmetries, we obtain two classes of a couple of the self-consistent equations for general $\Delta(k)$'s.

Class (I) (E, O), ($O_{||}$, O) and ($O_{\perp}$, E) case

$$\frac{1}{g} = T\sum_{\omega_n}\sum_{k}\frac{\tau^2(k)}{D(k,\omega_n)}\left[\omega_n^2+\gamma_k^2+\delta_k^2+M^2+\Delta^2\tau^2(k)\right] \tag{2.6a}$$

$$\frac{1}{U} = T\sum_{\omega_n}\sum_{k}\frac{1}{D(k,\omega_n)}\left[\omega_n^2+\gamma_k^2-\delta_k^2+M^2+\Delta^2\tau^2(k)\right] \tag{2.6b}$$

$$D(k,\omega_n) = \left[\omega_n^2+\gamma_k^2+\delta_k^2+M^2+\Delta^2\tau^2(k)\right]^2 - 4\delta_k^2(\gamma_k^2+M^2) \tag{2.6c}$$

Class (II) (E, E), ($O_{||}$, E) and ($O_{\perp}$, O) case

$$\frac{1}{g} = T\sum_{\omega_n}\sum_{k}\frac{\tau^2(k)}{D(k,\omega_n)}\left[\omega_n^2+\gamma_k^2+\delta_k^2-M^2+\Delta^2\tau^2(k)\right] \tag{2.7a}$$

$$\frac{1}{U} = T\sum_{\omega_n}\sum_{k}\frac{1}{D(k,\omega_n)}\left[\omega_n^2+\gamma_k^2-\delta_k^2+M^2-\Delta^2\tau^2(k)\right] \tag{2.7b}$$

$$\begin{aligned} D(k,\omega_n) = {} & \left[\omega_n^2+\gamma_k^2+\delta_k^2+M^2+\Delta^2\tau^2(k)\right]^2 \\ & - 4\delta_k^2(\gamma_k^2+M^2) - 4M^2\Delta^2\tau^2(k) \end{aligned} \tag{2.7c}$$

Here (x, y) means that the parity is x (Odd or Even) and the translation symmetry is y (Odd or Even), and $\perp$ or $||$ indicate that the spin quantization axis of triplet pairing is parallel or perpendicular to the magnetization M respectively. By using the self-consistent equations, we get the free energy for the above two classes

$$\delta F(\Delta, M) = \frac{\Delta^2}{g} + \frac{M^2}{U} - T\sum_{\omega_n}\sum_{k}\frac{1}{2}\ell n\frac{D(\Delta,M)}{D(O,O)} \tag{2.8}$$

where $D(\Delta, M)$ is given by eq. (2.6c) for class I and (2.7c) for class II.

From the structures of eqs. (2.6a) and (2.7a) it can be generally concluded that the pairing states in class II are more competitive than those in class I. This conclusion is independent of the dimensionality of the system, that is, holds for a three-dimensional system. Under a given SDW state the SC transition temperature

T_c in I is higher than that in II, that is, the state in I is more stable than that in II. It is proved that anisotropic pairing states in general have a tendency toward the SDW instability [11]. The states in I are easy to induce the SDW and coexist with it while the states in II tend to suppress the SDW instability, resulting in a first-order transition from the SC to the SDW for some states.

This close proximity between two orderings can be understood in terms of their intrinsic anisotropies: The SDW is anisotropic in the sense that the nesting portion connected by a vector Q is gapped on FS and the resulting SDW is anisotropic. The ungapped Fermi surface is available for the formation of the SC pairing whose energy gap should be anisotropic in order to maximally gain the attractive interaction. On the contrary, in the anisotropic pairing state there remain the normal electrons on the nodal portion of the Fermi surface. Because the residual net interaction is repulsive for these electrons, the system is unstable toward SDW. Here the built-in repulsive interaction plays double roles both for SC and SDW. This is quite contrasted with the conventional isotropic pairing state where the repulsive interaction is weak enough to be renormalized as the Coulomb pseudo-potential.

The SC transition temperature T_c is determined by letting $\Delta \to 0$ and $T \to T_c$ in eq. (2.6a) and (2.6b), that is

Class (I)

$$\frac{1}{g} = T_c \sum_{\omega_n} \sum_{k} \frac{\tau^2(k)}{D(k,\omega_n)} (\omega_n^2 + \gamma_k^2 + \delta_k^2 + M^2) \tag{2.9}$$

Class (II)

$$\frac{1}{g} = T_c \sum_{\omega_n} \sum_{k} \frac{\tau^2(k)}{D(k,\omega_n)} (\omega_n^2 + \gamma_k^2 + \delta_k^2 - M^2) \tag{2.10}$$

where

$$D(k,\omega_n) = (\omega_n^2 + \gamma_k^2 + \delta_k^2 + M^2)^2 - 4\delta_k^2(\gamma_k^2 + M^2). \tag{2.11}$$

The numerical results [33] are displayed in Fig. 2-7 for selected pairing states. We can see that T_c's in I are higher than those in II. The difference of T_c's among the same class is understandable from the relative position of the two energy gaps of SC and SDW on FS; the pairing state with higher T_c has larger (smaller) amplitude of $\Delta(k)$ on the portion of FS where the SDW is ungapped (gapped).

The self-consistent solutions give rise to two kinds of phase diagrams shown in Fig. 2-8 when $T_c < T_N$, depending on the pairing states I or II. The corresponding

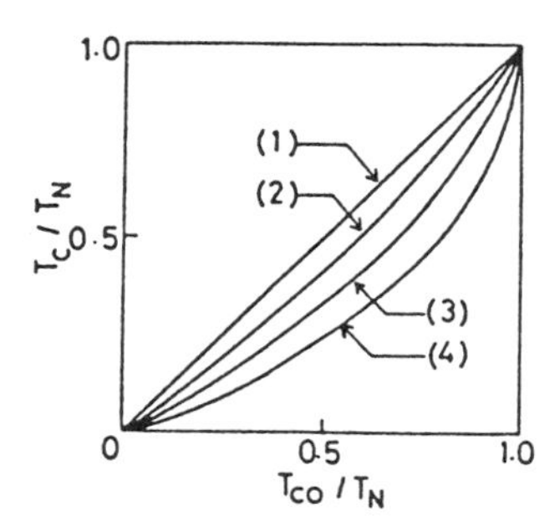

Fig. 2-7: Variations of the SC transition temperature T_c normalized by T_N as a function of T_{co}/T_N for various pairing states (1) (E, O), $\tau(k) = (\cos k_x - \cos k_y)/2$; (2) (E, E), $\tau(k) = 1$. (3) (O, O), $\tau(k) = (\sin k_x + \sin k_y)/2$. (4) (E, E), $\tau(k) = \sin k_x \sin k_y$ ($t/2\pi T_N = 100$, $t_1/2\pi\ T_N = 10$, $E_c/2\pi\ T_N = 30$ and $E_B/2\pi\ T_N = 50$ with E_c, E_B being the cutoff parameters).

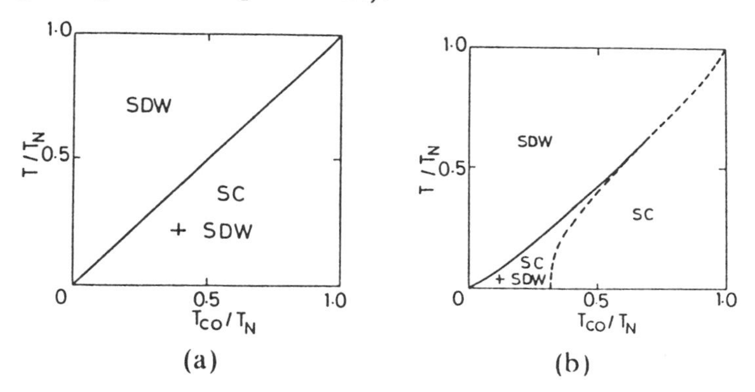

Fig. 2-8: Phase diagrams in T_c/T_N vs. T_{co}/T_N for (a) the d-wave like pairing $\tau(k) = (\cos k_x - \cos k_y)/2$ and (b) the isotropic s-wave pairing $\tau(k) = 1$ (the parameters are the same in Fig. 2-7).

T-dependences of the two order parameters are displayed in Figs. 2-9 and 2-10. The magnetization M is strongly suppressed below T_c for class II while M smoothly changes into the coexistence phase for class I. The latter feature is reminiscent of the T-dependence in UPt_3 (Fig. 2-6) where the growth of M ceases below the onset of the SC. This indicates that the SC pairing state in UPt_3 must be in class I, definitely excluding the isotropic singlet state. This is also true for the case of $(U,Th)Be_{13}$. The upper critical field H_{c2} exhibits no anomaly at T_{c2}, meaning that the SC order parameter does not change through T_{c2}. This is only possible when the SC state belongs to class I. The detailed identification of the pairing function for each materials by using group theoretical classification has been done in [34].

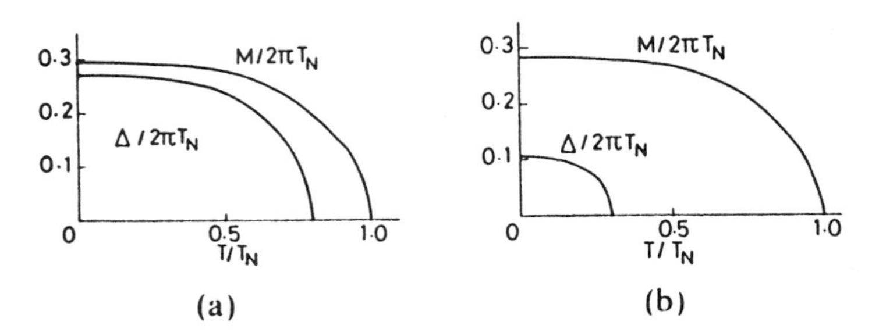

Fig. 2-9: Temperature dependence of the order parameters M and Δ for the d-wave like state $\tau(\mathrm{k}) = (\cos k_x - \cos k_y)/2$, (a) $\mathrm{T}_{c0}/\mathrm{T}_N = 0.800$, $\mathrm{T}_c/\mathrm{T}_N = 0.801$. (b) $\mathrm{T}_{c0}/\mathrm{T}_N = 0.300$ and $\mathrm{T}_c/\mathrm{T}_N = 0.302$.

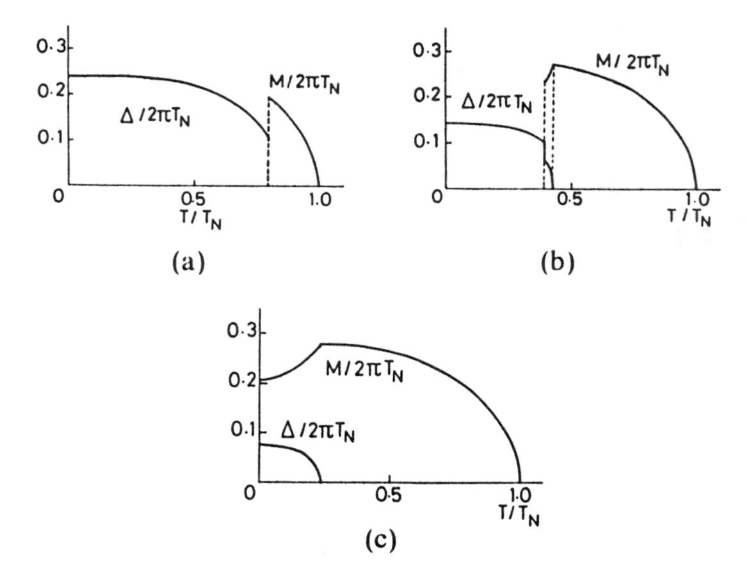

Fig. 2-10: Temperature dependence of the order parameters $M/2\pi T_N$ and $\Delta/2\pi\ \mathrm{T}_N$ for the isotropic s-wave pairing state $\tau(\mathrm{k}) = 1$. (a) SDW $\rightarrow$ SC: $\mathrm{T}_{c0}/\mathrm{T}_N = 0.856$ and $\mathrm{T}_{c1}/\mathrm{T}_N = 0.800$. (b) SDW $\rightarrow$ SDW + SC $\rightarrow$ SC: $\mathrm{T}_{c0}/\mathrm{T}_N = 0.500$, $\mathrm{T}_c/\mathrm{T}_N = 0.240$ and $\mathrm{T}_{c1}/\mathrm{T}_N = 0.397$. (c) SDW $\rightarrow$ SDW + SC: $\mathrm{T}_{c0}/\mathrm{T}_N = 0.300$ and $\mathrm{T}_c/\mathrm{T}_N = 0.240$. ($\mathrm{T}_{c1}$ is the first-order transition temperature.)

§3 COPPERLESS CUBIC OXIDES

In 1975 Sleight *et al.* [35] found that Ba(Pb,Bi)O_3 is a superconductor. We show its crystal structure in Fig. 3-1 at high temperatures. This is a cubic perovskite where the (Pb,Bi)O_6 octahedra are arranged three-dimensionally sharing a cornered

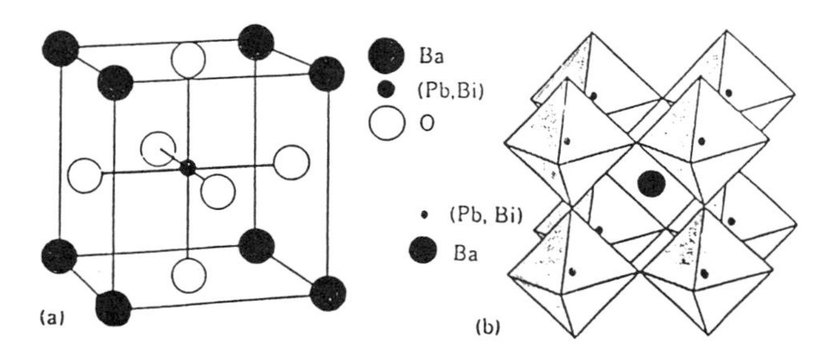

Fig. 3-1: (a) Cubic perovskite structure of $Ba(Pb,Bi)O_3$ at high temperatures. (b) Corner shared PbO_6 (or BiO_6) octahedra in cubic perovskite.

oxygen atom. Although the maximum $T_c \sim 13K$ is relatively low compared to that of the recent copper based oxides, it has been investigated extensively because of the peculiarity of its superconducting properties. Furthermore the recent finding [36, 37] of $T_c \sim 28$ K in this family $(Ba,K)BiO_3$ has boosted the investigation of these copperless materials.

In this section the present status of this material is reviewed by focusing on the similarity and dissimilarity of HFS in the previous section and of the copper oxides in the next section. The following items are concerned here:

(1) No d-electrons are involved near the Fermi level. The wide 6s and 2p derived bands play a role for superconductivity. Thus the electron correlation problem is not important.

(2) There is no experimental evidence to show that the superconductivity is unconventional.

(3) The attractive interaction for the Cooper pair formation comes from usual electron-phonon mechanism as in conventional superconductors.

Some of the topics are overlapped with the review article by Uchida *et al.* [38].

§3-1 Normal State Properties

A most prominent characteristic of this system lies in a sharp metal-insulator transition which occurs at x ~ 0.3 in $Ba(Pb_{1-x}Bi_x)O_3$. Shown in Fig. 3-2 is the phase diagram T vs x of this system together with the x-dependence of the optical energy gap. By doping Bi atoms in the end member $BaPbO_3$ which is a superconductor with $T_c = 0.45$ K [39], T_c dramatically rises and reaches $T_c \sim 13$ K just before the metal-insulator transition at x ~ 0.3 beyond which the system is an insulator, including the other end member $BaBiO_3$.

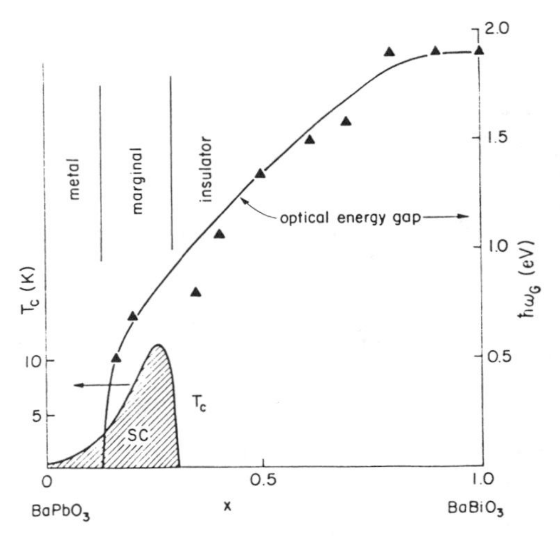

Fig. 3-2: Phase diagram of $Ba(Pb_{1-x}Bi_x)O_3$ where T_c is the superconducting transition temperature and the insulator gap $\hbar\omega_G$ is the optically measured energy gap.

A strikingly similar phase diagram in $(Ba_xK_{1-x})BiO_3$ is recently found by Hinks *et al.* [40] as shown in Fig. 3-3. Here again the semiconductivity in the $BaBiO_3$ side changes into the superconductivity quite abruptly at the Bi rich composition x ~ 0.75 where the maximum T_c ~ 28 K is attained.

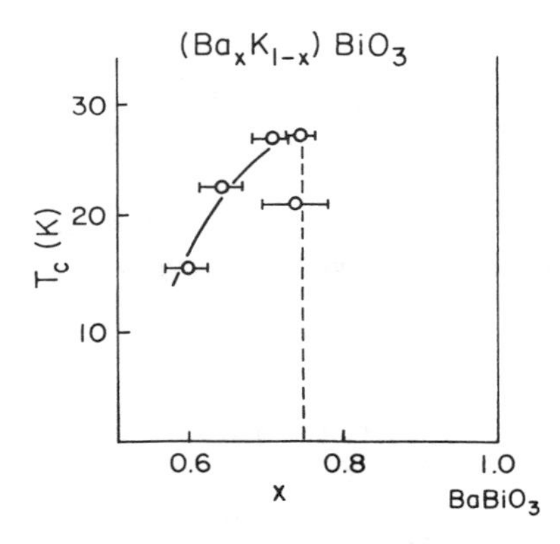

Fig. 3-3: The transition temperature T_c as a function of x in $(Ba_xK_{1-x})BiO_3$.

As is seen from Fig. 3-4, which exhibits the resistivity ρ(T) for various x in $Ba(Pb_{1-x}Bi_x)O_3$, the semimetallic behavior of ρ(T) in $BaPbO_3$ changes over the semiconducting one characterized by an activation type T-dependence [41] in Bi-

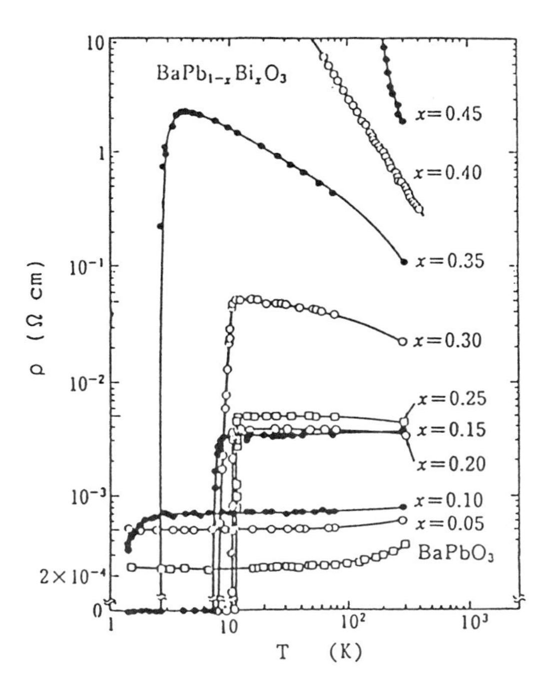

Fig. 3-4: Temperature dependence of the resistivity in $Ba(Pb_{1-x}Bi_x)O_3$ [41].

rich systems. The composition dependence of the activation energy determined from $\rho(T)$ is increased as x and tends to saturate to $\sim$ 0.3 eV at x = 1. This dependence agrees with that of the optical gap $\hbar\omega_G$ shown in Fig. 3-2. However, the absolute values do not coincide, because the transport measurement such as $\rho(T)$ is hampered by the morphology of actual samples. These transport measurements were conducted by using polycrystalline samples.

The optical conductivity measured by Tajima *et al.* [42, 43] for various x values is shown in Fig. 3-5. In the small x region (x = 0.05 and 0.17) we see a Drude-like behavior expected for a simple metal, showing a clear plasma edge. On the other hand, in the large x region the semiconducting behavior is clearly seen, that is, corresponding to the interband absorption across the semiconducting energy gap, the sharp gap edge structure can be identified. Inside the gap some phonon absorptions are seen whose structure is relatively unchanged with x.

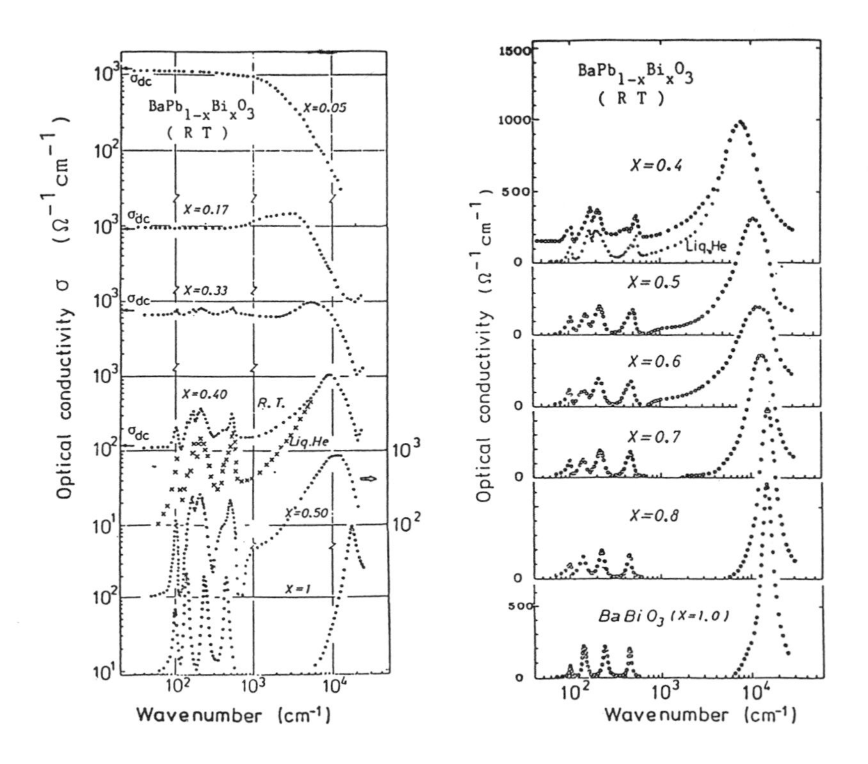

Fig. 3-5: Real part of the optical conductivity σ for various compositions in $Ba(Pb_{1-x}Bi_x)O_3$ [42, 43].

This important experiment is summarized as follows:

(1) The semiconducting gap $\hbar\omega_G$ is smoothly changed with x. No abrupt change occurs even passing through the metal-insulator transition at x $\cong$ 0.3. In Fig. 3-2 $\hbar\omega_G$ is shown.

(2) Even the superconducting system with x = 0.17 has an anomaly corresponding to the energy gap absorption. This means that this superconducting system has a semiconducting gap.

(3) The x-dependence of the plasma frequency ω_P behaves in a rigid band manner [42]. Since the Bi-doping donates one electron per Bi atom, the conduction band is progressively filled as x increases. This picture explains the experimental fact: $\omega_P^2 \propto$ n (n is the carrier number) which is expected from a simple metal theory.

The thermopower [44] and the spin susceptibility [45] experiments are also explained by this simple metal picture for at least small x regime (x $\lesssim$ 0.2). We note, however, that these results in larger x regime deviate from this simple picture.

The electronic band structure in Ba(Pb,Bi)O_3 is calculated by Mattheiss and Hamann [46]. In Fig. 3-6 we display their results. The overall band complex near

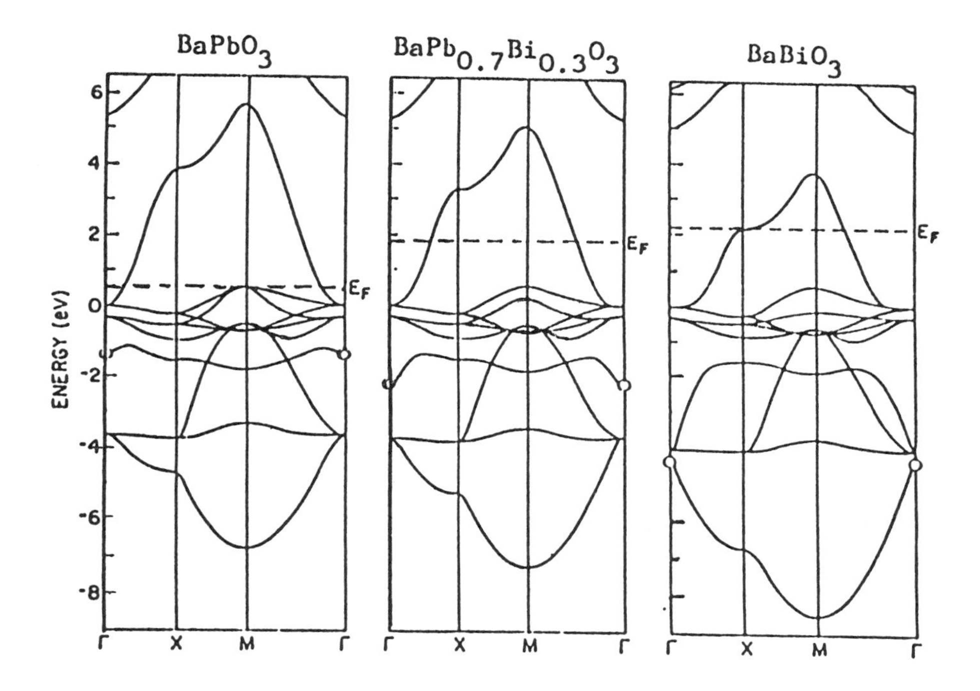

Fig. 3-6: Band structures for x = 0, 0.3 and 1.0 [46].

the Fermi level extends over 10 eV, forming a broad valence and conduction bands which consist of the 6s-2p bonding and antibonding bands in lower and higher energy regions, and of the non-bonding 2p in the middle. According to these results, the almost empty antibonding band in $BaPbO_3$, which explains its semimetallicity becomes filled progressively upon Bi-doping in a rigid manner, resulting in a half-filled band of this antibonding band in $BaBiO_3$. Therefore $BaBiO_3$ should be metallic.

It is easily recognized by closely looking at the antibonding band extended from 0 eV to 4 eV in $BaBiO_3$ (see Fig. 3-6) that this dispersion curve has a nice nesting feature characterized by the modulation vector $\vec{Q} = (\pi/a, \pi/a, \pi/a)$ where a is the lattice constant of the cubic crystal. If we approximate it as a tight binding Hamiltonian with only nearest neighbor hopping t: $\epsilon(k) = -t(\cos ak_x + \cos ak_y + \cos ak_z)$, then it is evident to see the perfect nesting: $\epsilon(k + Q) = -\epsilon(k)$, meaning that any perturbation connecting between these degenerate electron states triggers the electronic instability by opening up a gap. Indeed, the so-called breathing mode, which corresponds to the alternative contraction and expansion of the oxygen octahedra in the cubic perovskite, is known to give rise to a strong electron-phonon coupling. The resonance Raman experiment [47] clearly shows that the lattice distortion responsible for the gap formation or for the semiconductivity in $BaBiO_3$ is this breathing mode with the frequency $\sim 569\ cm^{-1}$.

Sleight [48] terms this distortion as the charge disproportionation where the valence of Bi atoms has two kinds of values: $2Bi^{+4} \rightarrow Bi^{+3} + Bi^{+5}$. This charge disproportionality in chemistry terms is responsible for the semiconductivity in $BaBiO_3$. In physics terms this is nothing but the charge density wave (CDW) of the electron-phonon coupled systems [49]. The strong electron-phonon (breathing mode) coupling gives rise to a large electronic energy gap which explains $\hbar\omega_G \sim 2$ eV. Thus the corresponding CDW transition temperature might be very high. The reported transition at around 800°C by Chaillout *et al.* [50] might be related to this.

As for the superconducting properties in $Ba(Pb,Bi)O_3$ except for the steep rise and then abrupt drop in T_c at $x \sim 0.3$ nothing is remarkable. (We notice, however, one experiment [51] which reports a T-linear term in the specific heat at lower temperatures. This result is neither conclusive, nor supported by others.) The system can be described by an ordinary type II superconductor with relatively large Ginzburg-Landau parameter $50 \sim 100$ and short mean free path ~ 10 Å. There is no reliable experiment to determine whether it is a weak or strong coupling superconductor. However, the tunneling experiment by Batlogg [52] concludes the former. The moderate oxygen isotope shift of T_c [53] ($T_c \sim M^{-\alpha}$ where $\alpha \sim 0.3$ and M is the oxygen atomic mass) renders a support to the phonon-mediated attractive pairing mechanism.

§3-2 $(Ba,K)BiO_3$

Since Mattheiss *et al.* [36] found $T_c > 20$ K in $(Ba,K)BiO_3$ and $(Ba,Rb)BiO_3$, the intensive research efforts are focused on these materials partly because:

(1) There is no copper atom in the system, which was thought to be a main ingredient to push T_c in the copper-based high T_c materials.

(2) Although $T_c \sim 28$ K, which was subsequently confirmed by Cava *et al.* [37], is relatively low (we note that $T_c \sim 28$ K is highest among the copperless superconductors), the physics on the high T_c mechanism in this system might contain some clue to understanding the common origin which is not necessarily identical, but may have some common features.

(3) There is an experimental challenge to further raise T_c comparable to $T_c \sim$ 100 K as in the copper-based oxides.

(4) These materials are cubic or at least more three dimensional than the copper oxides. The low dimensionality associated with the CuO_2 layered structure in the latter prevents wide practical applications. In this respect this family seems

promising for applications.

As shown in Fig. 3-3 substitution of K atom in $BaBiO_3$ abruptly raises T_c to $\sim$ 28 K. Further substitution only results in gradual decrease of T_c, being reminiscent of $Ba(Pb,Bi)O_3$ as is mentioned. A substantial oxygen isotope effect [54] comparable to ordinary superconductors is observed. It is confirmed that there is no trace of the magnetism, especially antiferromagnetism in this family ($BaBiO_3$, $(BaK)BiO_3$, and $Ba(Pb_{0.1}Bi_{0.9})O_3$) [55] as is anticipated. All lead to a common origin for the "high T_c mechanism" in this copperless family. In fact, according to the band calculation [56] the K-doping acts as lowering the chemical potential from the half-filling of the 6s-2p antibonding band in $BaBiO_3$, keeping the overall band structure intact. This is very similar to the $Ba(Pb,Bi)O_3$ situation.

It is believed that $T_c \lesssim 30$ K is explainable within the framework of the ordinary electron-phonon mechanism to produce the attractive interaction, which is consistent with the observed isotope shift in this material, although there is no rigorous proof.

§3-3 Model Considerations

It is now evident that in $BaPb_{1-x}Bi_xO_3$ the CDW at x = 1.0 persists all the way down to x $\sim$ 0.17 where the SC is stabilized. Usually the structural instability derived by CDW kills or weakens the superconductivity. Here the opposite seems to happen, that is, T_c seems to rise toward the marginal instability.

In order to resolve this puzzle we employ a model, which takes into account simultaneously the existence of the CDW and SC to see whether the former really helps to enhance T_c or not [57]. As we have already seen the relevant conduction band which comes from the 6s-2p antibonding band is approximated by a tight binding model:

$$\begin{aligned} H_0 &= -\sum_{k\sigma}(\epsilon(k)-\mu)C^+_{k\sigma}C_{k\sigma} \\ \epsilon(k) &= \gamma_k + \delta_k \\ \gamma_k &= -t(\cos k_x + \cos k_y + cos k_z) \\ \delta_k &= -t_1(\cos k_x \cos k_y + \cos k_y \cos k_z + \cos k_z \cos k_x). \end{aligned} \tag{3.1}$$

We have set the lattice constant a = 1 and introduced the (next) nearest neighbor integral $t(t_1)$. This model with $t_1/t = 0.3$ simulates the "canonical band" of the $Ba(Pb,Bi)O_3$ system which is obtained by averaging the two antibonding bands for $BaPbO_3$ and $BaBiO_3$ shown in Fig. 3-6. We compare the canonical band (upper panel) and the tight binding model (middle panel) along the symmetry lines in reciprocal space of the cubic crystal in Fig. 3-7. The Bi doping is to progressively raise the chemical potential μ to fill the band by donating one electron per Bi atom.

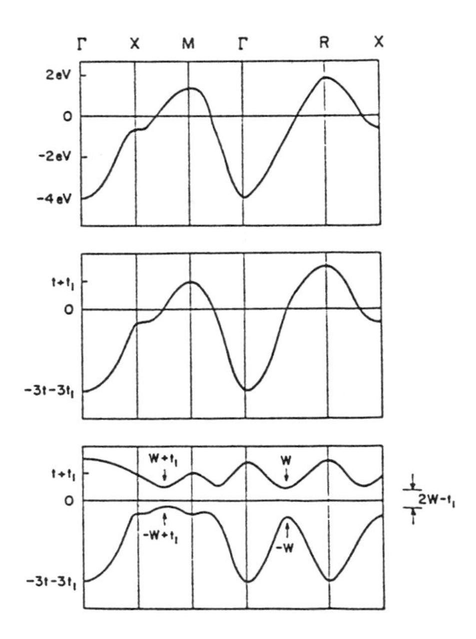

Fig. 3-7: Comparison of dispersion curves between the canonical band derived from band calculation (upper panel) and the tight binding model, eq. (3.1) (middle). The lowest panel shows the dispersion curve in the CDW state, eq. (3.8).

The charge density wave is characterized by the order parameter $\overline{W}$ and wave vector $\vec{Q}$ which denote the amplitude and wave length of the lattice distortion, respectively. Here we assume a commensurate CDW, that is, $\vec{Q} = (\pi, \pi, \pi)$ where the unit cell in the original cubic crystal is doubled, simulating the breathing mode distortion observed in $BaBiO_3$. The mean field Hamiltonian is given by

$$H_{CDW} = -W \sum_{k\sigma} (C^+_{k+Q\sigma} C_{k\sigma} + h.c.) \tag{3.2}$$

where W is proportional to lattice distortion.

We assume an attractive interaction g for the isotropic singlet pairing channel which could come from the ordinary electron-phonon coupling. It should be noted that the phonon modes responsible for this attractive interaction are not necessarily the breathing mode. Rather it is conceivable that the latter mode does not directly participate in the attractive interaction because the frequency of this mode is rather high (~ 569 cm^{-1}).

The mean-field Hamiltonian and self-consistent equation are written as

$$H_{SC} = -\Delta \sum_k (C^+_{k\uparrow} C_{-k\downarrow} + h.c.) \tag{3.3}$$

$$\Delta = g \sum_k < C^+_{k\uparrow} C^+_{-k\downarrow} > . \tag{3.4}$$

The diagonalization of the total Hamiltonian H = H_O + H_{CDW} + H_{SC} is easily done and yields the self-consistent equation for Δ:

$$1 = gT \sum_{\omega_n} \sum_k [\omega_n^2 + \gamma_k^2 + (\delta_k - \mu)^2 + \Delta^2 + W^2]/D(k, \omega_n) \tag{3.5}$$

$$D(k, \omega_n) = [\omega_n^2 + \gamma_k^2 + (\delta_k - \mu)^2 + \Delta^2 + W^2]^2 - 4(\delta_k - \mu)^2(\gamma_k^2 + W^2) \tag{3.6}$$

where $\omega_n = \pi T(2n+1)$. Taking $\Delta \to 0$ and $T \to T_c$ in eq. (3.5), we obtain the equation for T_c:

$$1 = gT_c \sum_{\omega_n} \sum_k [\omega_n^2 + \gamma_k^2 + (\delta_k - \mu)^2 + W^2] \times \left\{ [\omega_n^2 + (\delta_k - \mu + \sqrt{\gamma_k^2 + W^2})^2][\omega_n^2 + (\delta_k - \mu - \sqrt{\gamma_k^2 + W^2})^2] \right\}^{-1} . \tag{3.7}$$

Here we point out that the gap equation (3.5) has the same structure with eq. (2.6a) when $\tau(k) = 1$. This means that the class II pairing states now become favorable states under the CDW and class I states are more competitive ones. Therefore the isotropic singlet with $\tau(k) = 1$ is indeed enhanced by the CDW. This is due to the difference symmetry of the SDW and CDW under the time reversal.

The self-consistent equation (3.7) is solved under a given W whose x-dependence is taken from the experiment instead of determining W fully self-consistently. In Fig. 3-7 we plot the dispersion curve E_k, which is given by

$$E_k = \delta_k - \mu \pm \sqrt{\gamma_k^2 + W^2} \tag{3.8}$$

for CDW state. Note that the effective CDW gap is $2W$-t_1 rather than $2W$ because t_1 term destroys the perfect nesting. The corresponding density of states is displayed for various values of W in Fig. 3-8. It is seen that depending on the

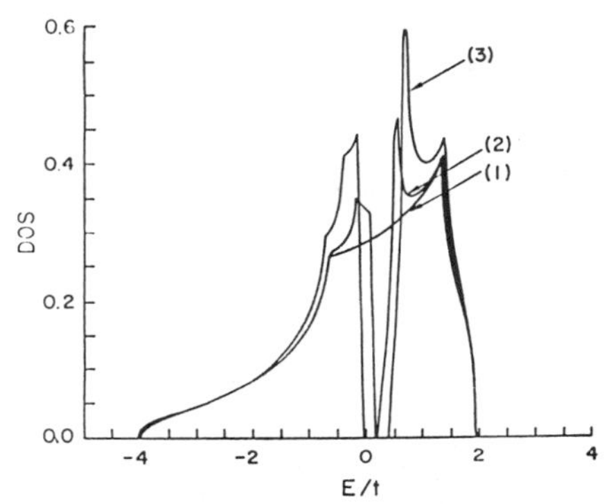

Fig. 3-8: Density of states for (1) the normal state ($W = 0$), (2) $W/t = 0.2$ and (3) $W/t = 0.4$ when $t_1/t = 0.35$. Note that the CDW gap edge singularities give rise to the large density of states.

electron filling, we have a partial gapping situation in the metallic state where a part of the Fermi surface is gapped and the remaining part is ungapped. The boundary region on the Fermi surface between these parts gives rise to a high density of states. When μ moves into the CDW gap region, the system becomes insulating and T_c goes to zero abruptly.

We have calculated T_c by solving eq. (3.7) numerically. The x-dependence of the order parameter shown in the lower panel in Fig. 3-9 is taken from the optical experiment shown in Fig. 3-2 where the effective energy gap is identified to $(2W - t_1)/t$. The results are displayed in Fig. 3-9 together with the experimental data for

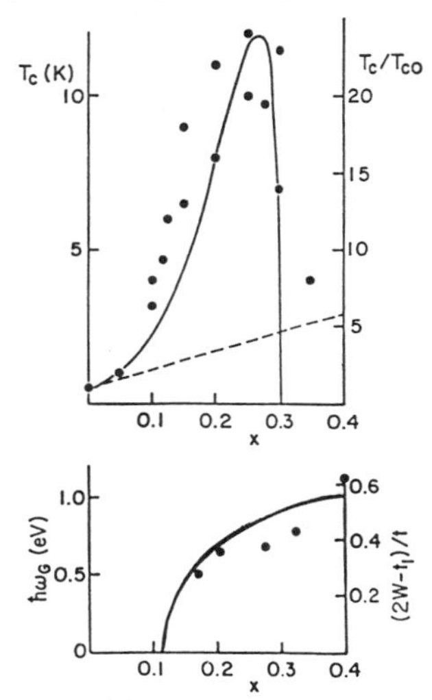

Fig. 3-9: The calculated T_c normalized by T_{c0} at x = 0 as a function of Bi-concentration x where $g/6t = 0.226$ and the Debye cutoff $\omega_D/t = 1.0$; the dots are experimental data. The dashed line is the T_c in the absence of the CDW (W

= 0).

T_c, where T_{c0} is the transition temperature at x = 0. The chemical potential μ is determined by accommodating the excess electrons upon doping Bi-atoms. A large T_c enhancement T_c (x = 0.25)/T_c (x = 0) $\simeq$ 24 is attained which is comparable to the observation.

We give rough estimates of the parameters used: If we put $t \cong$ O (0.5 eV), the band width $6t$ of the model band is $\sim$ 3 eV. Note that the antibonding band width is $\sim$ 3 eV for $BaBiO_3$ and $\sim$ 5 eV for $BaPbO_3$ according to the LAPW calculation. The effective band gap $(2W - t_1)/t = 0.5$ corresponds to the CDW gap $2W \sim$ 0.4 since $t_1 = 0.175$ eV ($t_1/t = 0.35$). The order of magnitude of this value is in agreement with the optical measurement.

So far we regard the CDW order parameter W as a given one taken from the experiment. Based on the above calculation, we can draw a realistic scenario of the successive phase changes as x varies, which a fully self-consistent calculation would give rise to: At x = 1 the antibonding band is exactly half-filled and thus its Fermi surface is almost perfectly nested with the CDW wave vector $\vec{Q} = (\pi, \pi, \pi)$. The stable CDW state with a large energy gap ($\sim$ 2 eV) is formed, making a mean field transition temperature $T_{CDW} \sim$ 6000 K. (We note that since T_{CDW} is high, the CDW coherence length ξ_{CDW} ($\sim \hbar v_F/k_B T_{CDW}$) is extremely short (a few lattice constants). This makes the real space pairing description of the CDW done by Rice's group [49] possible.) Therefore the CDW is preexisting in the system at higher temperatures. As x decreases and the Fermi level is lowered, the nesting is gradually deteriorated. The CDW stability is lowered, evidenced by the decrease of the CDW gap as shown in the lower panel in Fig. 3-9. The system, however, still remains semiconducting. Below x $\simeq$ 0.3, since the Fermi surface volume decreases substantially, there remains a little nesting feature in which $\vec{Q} = 2k_F$ (k_F is the Fermi wave number) differs from the commensurate value (π, π, π) and takes some incommensurate value to maximally stabilize the CDW state. In this imperfect nesting situation, only parts on FS connected by $\vec{Q}$ are gapped by the CDW. This situation is nothing but what Tanaka's group [38] calls a "pseudo-gap"; namely the optical reflectivity spectrum reveals the energy gap structure, yet the system is metallic. This is the case that we can expect a large T_c enhancement where the boundary between the CDW gapped and ungapped regions on FS always gives rise to the density of states (DOS) increase. In other words, there always exist such points on FS that the Fermi level exactly coincides with the CDW gap edge singularity, leading to the T_c enhancement. Upon further decreasing x (x $\lesssim$ 0.15) the FS nesting is completely destroyed even though the electron-phonon coupling relevant to the CDW is very strong. The CDW becomes unstable and the system is a normal metal (actually a semimetal). Here T_c becomes quite low because of the low DOS, reflecting a broad s-p derived band. (The T-linear coefficient γ of the specific heat $\gamma \sim$1 mJ/mole K^2.)

According to the above scenario, the even higher $T_c \sim$ 28 K in $(Ba,K)BiO_3$

than $Ba(Pb,Bi)O_3$ might be explained in the following ways: (1) Ba site substitution keeps the Bi-O conduction system intact in contrast to the octahedral site substitution in $Ba(Pb,Bi)O_3$. A similar situation occurs in $(La,A)_2CuO_4$.

(2) More importantly, fewer doping of K-atom relative to Pb-doping into $BaBiO_3$ implies that the CDW is still very stable. We would expect a sharper singularity in density of states, brought by the CDW, leading to a larger T_c enhancement. The existence of a large energy scale ($W \sim 2$ eV) is reminiscent of a large exchange constant ($J \sim 1200$ K) associated with the magnetism in copper based oxides.

§4 Cu-BASED OXIDE SUPERCONDUCTORS

Here we only consider $(La,A)_2CuO_4$ (A = Ba, Sr, Ca, $T_c \sim 40$ K) and $YBa_2Cu_3O_7$ ($T_c \sim 90$ K), focusing mainly on the magnetism related phenomena which are the keys to understanding the mechanism of the high T_c superconductivity.

It is now well established [58] that the strong antiferromagnetism is associated with these materials, namely La_2CuO_4 ($T_N \sim 230$ K), and $YBa_2Cu_3O_6$ ($T_N \sim 400$ K) with the moment $\sim 0.5\mu_B$/Cu-atom. The doping of holes by substituting +2 ions in La_2CuO_4 or by adding oxygens in $YBa_2Cu_3O_6$ results in a sharp drop of T_N, implying that the three-dimensional ordering is destroyed very quickly, but it does not necessarily mean that the two-dimensional antiferromagnetism associated with CuO_2 plane, which is indispensable for the high T_c mechanism, is also deteriorated simultaneously. In fact, the internal magnetic field seen by NMR and neutron only slowly decreases with the doping [59]. The in-plane exchange constant J between the Cu d-electron moment is estimated to be $J \sim 1200$ K within the Heisenberg model [60].

In Fig. 4-1 we display the phase diagram for $(La_{1-x}A_x)_2CuO_4$ (A = Sr and Ba). It is evident that the same marginal instability as in copperless oxides is also associated with this high T_c superconductor.

(1) $0 \leqq x < 0.01$; La_2CuO_4 is an antiferromagnetic insulator with the energy gap ~ 2 eV [61] and with $\gamma = 0$ (the T-linear specific coefficient) [62, 63]. We could argue this system in terms of both ways, localized picture or itinerant picture, which in principle should give the same answer at least for its ordered state. In particular, because of the lack of reliable measurements, we cannot solve this dichotomy.

(2) $0.01 \lesssim x \lesssim 0.04$; The three-dimensional (3D) transition temperature becomes very low ($\lesssim 10$ K) while the 2D magnetism does not change much. The linear specific heat γT appears ($\gamma \sim 4$mJ/mole K^2), implying that at the Fermi level free carriers must exist.

(3) $0.04 \lesssim x \lesssim 0.05$; $T_N^{(3d)}$ is almost zero and the 2D magnetism becomes deteriorated. Simultaneously the SC starts to appear and T_c sharply rises to ~ 20 K [64]. Some experiment such as μSR [65] reports that the magnetism coexists with the SC. The coefficient γ returns to zero.

(4) $0.05 \lesssim x \lesssim 0.15$; After passing through the maximum $T_c \sim 40$ K at around $x \sim 0.1$, the SC becomes again vanishing.

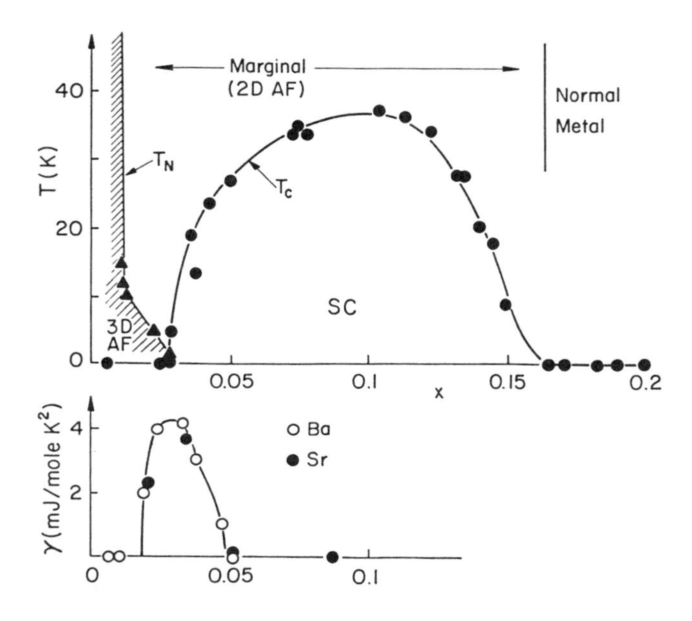

Fig. 4-1: Phase diagram of $(La_{1-x}A_x)_2CuO_4$ (A: Ba, Sr). The shaded area indicates the 3D antiferromagnetism. Variation of the linear coefficient γ in the specific heat.

(5) x $\gtrsim$ 0.15; The system becomes a normal metal with no trace of the SC (T_c < 5 K) [64].

This phase diagram is very similar to $Ba(Pb,Bi)O_3$ in §3, where La_2CuO_4 and $BaBiO_3$ are both insulators. The preexisting SDW in the former or CDW in the latter, whose transition temperatures are high and energy gaps are large, is weakened by doping. Because La_2CuO_4 is a layered structure, the doping affects mainly the interplanar coupling to reduce $T_N^{(3d)}$ while the 2D magnetism is rather intact as evidenced by NMR [59]. Therefore the doped carriers induce a metal-insulator transition around x $\sim$ 0.01. Further doping increases the density of states at the Fermi level and almost simultaneously induces the superconductivity. The absence of γ in x $\gtrsim$ 0.04 simply means the SC gap formation on the Fermi surface. If this scenario is correct, then a rapid T_c enhancement can be explained by the same way as in the previous section, provided that the SDW order or fluctuation preexists at higher temperatures.

§4-1 Model Considerations

It is rather easy to see that the class I pairing states is enhanced by the SDW order. Starting from eq. (2.9), we can calculate T_c for a given M, which could be either long-range order [66] or fluctuation [67]. As we show in Fig. 4-2, we can

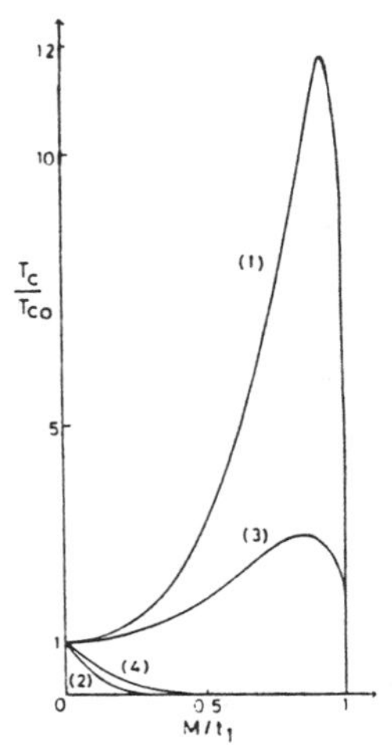

Fig. 4-2: Variations of T_c relative to T_{c0} as a function of M/t_1. (1) and (3): $\tau(k) = (\cos k_x\text{-}\cos k_y)/2$. (2) and (4): $\tau(k) = 1$. ((1) and (2): $t/2\pi T_N = 1000$, $t_1/2\pi T_N = 100$ and $E_c/2\pi T_N = 300$. (3) and (4): $t/2\pi T_N = 100$, $t_1/2\pi T_N = 10$ and $E_c/2\pi T_N = 30$.)

expect a large T_c enhancement compared to T_{c0} without SDW. It is to be noted that since the isotropic singlet state belongs to class II, there is no T_c enhancement for this state, while some anisotropic singlet state might be favorable. This is due to the intrinsic gap anisotropy of the SDW, meaning that the density of states right at the Fermi level is not increased by the SDW gap (or pseudo gap) formation.

The partial DOS at the particular points in reciprocal space in which the SC gap has a large magnitude should be enhanced. This is contrasted with the CDW case in the previous section where the total DOS at the Fermi level is enhanced to directly result in the T_c increase.

§4-2 Possible Scenario

Focusing on $(La,A)_2CuO_4$, we suggested a possible T_c enhancement due to the preexisting AF magnetism or SDW fluctuations. Let us summarize our scenario under doping:

The end-member La_2CuO_4 is a half-filled system. Due to the strong d-electron correlation the SDW is formed at a high temperature. The SDW gap opens at the

Fermi level, making the system insulator. By introducing excess holes upon A^{+2} doping the SDW becomes destabilized. The system tries to stabilize the SDW by changing the SDW vector $\vec{Q}$ away from the commensurate value $\vec{Q} = (\pi/a, \pi/a)$ in the undoped system. Upon further doping the nesting feature is lost and an imperfect nesting situation appears, simultaneously a large T_c enhancement also appears. If we identify La_2CuO_4 to $BaBiO_3$, the sequences of various phases are quite similar.

Possible objections on this scenario are the following:

(1) According to photoemission experiments [68], the valence of Cu-ion is Cu^{+2}: The d-electron is localized on the Cu-site. This is rather a quantitative matter, which only tells us that the degree of the 3d-2p mixing is small. This means immediately that the excess carriers upon doping enter mainly into the oxygen 2p orbital. This is the experimental fact. This does not contradict with our picture.

(2) As for the pairing symmetry, there are many conflicting experiments mainly concentrated on $YBa_2Cu_3O_7$. The most striking result is the ^{17}O NMR for 90 K material [69]. The Hebel-Slichter hump just below T_c observed in the nuclear relaxation rate could be explained both by the isotropic singlet state or by some anisotropic state. Note, however, that an extremely large hump observed in Cu-NMR of 60 K material [70] cannot be explained by any anisotropic state, only possible by the isotropic singlet state. But it should be checked whether the signal comes from the plane site which the authors claim or from the chain site. Since the Fermi surface associated with the chain electrons is one dimensional like, there is a good chance where all the Fermi surface (or Fermi point) is gapped even for anisotropic pairing states. Some of the experiments only see this chain state and conclude the isotropic singlet state, which confuses the situation.

As for the pairing mechanism we have said nothing. For 40 K superconductors even though it comes mainly from the magnetism-related attraction, there is still the possibility that the phonon mediated attraction plays some role. Probably it is a hard task to sort out. For 90 K superconductors we definitely need a new pairing mechanism although the T_c enhancement mechanism explained here can help to raise T_c.

*Permanent Address: Department of Physics, Kyoto University, Kyoto, 606 Japan

References

[1] For review, see *Superconductivity in Ternary Compounds I and II*, ed. M.B. Maple and O. Fischer (Springer-Verlag, New York, 1982); K. Machida, Applied Phys. **A35** (1984) 193.
[2] D. Jerome and H.J. Schulz, Adv. Phys. **31** (1982) 299.
[3] See, for example, *High-Temperature Superconductivity*, ed. V.L. Ginsburg and D.A. Kirzhritz (Consultants Bureau, New York, 1982).
[4] T.M. Rice, Jpn. J. Appl. Phys. **26** (1987), Suppl. 26-3, 1865. A.I. Goldman, *ibid.*, 1887. Z. Fisk, H.R. Ott and G. Aeppli, *ibid.*, 1882.
[5] Y. Kataoka, K. Ueda, T. Kohara, Y. Kohori and K. Asayama, in *Theoretical and Experimental Aspects of Valence Fluctuations and Heavy Fermions*, ed. L.C. Gupta and S.K. Malik (Plenum, New York, 1987).
[6] D.E. MacLaughlin, C. Tien, W.G. Clark, M.D. Lan, Z. Fisk, J.L. Smith and H.R. Ott, Phys. Rev. Lett. **53** (1984) 1853.
[7] Y. Kohori, T. Kohara, H. Shibai, Y. Oda, Y. Kitaoka and K. Asayama, J. Phys. Soc. Jpn. **57** (1988) 395.
[8] Y. Kitaoka, K. Ueda, Y. Kohori, T. Kohara and K. Asayama, in *Superconducting Materials*, Jpn. J. Appl. Phys. Series 1 (1988), pp. 104-107.
[9] C.S. Wang, M.R. Norman, R.C. Albers, A.M. Boring, W.E. Pickett, H. Krakauer and N.E. Christensen, Phys. Rev. B **35** (1987) 7260.
[10] L. Taillefer, R. Newbury, G.G. Lanzarich, Z. Fisk and J.L. Smith, J. Magn. Magn. Mat. **63 & 64** (1987) 372.
[11] Y. Uemura, W.J. Kossler, X.H. Yu, H.E. Schone, J.R. Kempton, C.E. Stronach, S. Barth, F.N. Gygax, B. Hitti, A. Schenck, C. Baines, W.F. Lankford, Y. Ōnuki and T. Komatsubara, preprint.
[12] Y. Kitaoka and K. Asayama, private communication.
[13] K. Miyake, J. Magn. Magn. Mat. **63 & 64** (1987) 411.
[14] K. Ueda, Y. Kitaoka, H. Yamada, Y. Kohori, T. Kohara and K. Asayama, J. Phys. Soc. Jpn. **56** (1987) 867.
[15] H.R. Ott, H. Rudigier, E. Felder, Z. Fisk and J.L. Smith, Phys. Rev. B **33** (1986) 126.
[16] R.H. Heffner, D.W. Cooke and D.E. MacLaughlin, in *Theoretical and Experimental Aspects of Valence Fluctuations and Heavy Fermions*, ed. L.C. Gupta and S.K. Malik (Plenum, New York, 1987).
[17] K. Machida and M. Kato, Phys. Rev. Lett. **58** (1987) 1986.
[18] U. Rauchschwalke, C.D. Bredel, F. Steglich, K. Maki and P. Fulde, Europhys. Lett. **3** (1987) 757.
[19] U. Rauchschwalke, F. Steglich, G.R. Stewart, A.L. Giorgi, P. Fulde and K. Maki, Europhys. Lett. **3** (1987) 751.
[20] P. Kumar and P. Wolfle, Phys. Rev. Lett. **59** (1987) 1954.

[21] M. Siegrist and T.M. Rice, preprint.
[22] H.A. Mook, B.D. Graulin, G. Aeppli, Z. Fisk and J.L. Smith, Bull. Amer. Phys. Soc. **32** (1987) 594.
[23] M.R. Norman, W.E. Pickett, H. Krakauer and C.S. Wang, Phys. Rev. B **36** (1987) 4058.
[24] K. Machida, K. Nokura and T. Matsubara, Phys. Rev. Lett. **44** (1980) 821.
[25] G. Aeppli, E. Bucher, C. Broholm, J.K. Kjems, J. Baumann and J. Hufnagl, Phys. Rev. Lett. **60** (1988) 615.
[26] A.I. Goldman, G. Shirane, G. Aeppli, B. Batlogg and E. Bucher, Phys. Rev. B **34** (1986) 6564. A.P. Ramirez, B. Batlogg, E. Bucher and A.S. Cooper, Phys. Rev. Lett. **57** (1986) 1072.
[27] P. Frings, B. Renker and C. Vettier, J. Magn. Magn. Mat. **63 & 64** (1987) 202.
[28] J.J.M. Franse, K. Kadowaki, A.A. Menovsky, M. Van Sprang and A. de Visser, J. Appl. Phys. **61** (1987) 3380.
[29] C. Broholm, J.K. Kjems, W.J.L. Buyers, P. Matthews, T.T.M. Palstra, A.A. Mydosh, Phys. Rev. Lett. **58** (1987) 1467.
[30] M.R. Norman, private communication.
[31] D.J. Scalapino, E. Loh, Jr. and J.E. Hirsch, Phys. Rev. B **34** (1986), 8190. K. Miyake, S. Schmitt-Rink and C.M. Varma, Phys. Rev. B **34** (1986) 6554.
[32] M. Kato and K. Machida, J. Phys. Soc. Jpn. **56** (1987) 2136.
[33] M. Kato and K. Machida, Phys. Rev. B **37** (1988) 1510.
[34] M. Ozaki and K. Machida, Phys. Rev. B (to be published).
[35] A.W. Sleight, J.L. Gillson and P.E. Bierstedt, Solid State Commun. **17** (1975) 27.
[36] L.F. Mattheiss, E.M. Gyorgy and D.W. Johnson, Jr., Phys. Rev. B **37** (1988) 3745.
[37] R.J. Cava, B. Batlogg, J.J. Krajewski, R. Farrow, L.W. Rupp, Jr., A.E. White, K. Short, W.F. Peck and T. Kometani, Nature, **332** (1988) 814.
[38] S. Uchida, K. Kitazawa and S. Tanaka, Phase Transitions **8** (1987) 95.
[39] V.V. Bogatko and Yu. N. Venevtsev, Sov. Phys. Solid State **22** (1980) 705.
[40] D.G. Hinks, B. Dabrowski, J.D. Jorgensen, A.W. Mitchell, D.R. Richards, Shiyou Pei and Donglu Shi, Nature, **333** (1988) 836.
[41] T.D. Thanh, A. Koma and S. Tanaka, Appl. Phys. **22** (1980) 205.
[42] S. Tajima, S. Uchida, A. Masaki, H. Takagi, K. Kitazawa, S. Tanaka and A. Katsui, Phys. Rev. B **32** (1985) 6302.
[43] S. Tajima, S. Uchida, A. Masaki, H. Takagi, K. Kitazawa, S. Tanaka and S. Sugai, Phys. Rev. B **35** (1987) 696.
[44] T. Tani, T. Itoh and S. Tanaka, J. Phys. Soc. Jpn. **49** (1980) Suppl. A. 309.
[45] S. Uchida, H. Hasegawa, K. Kitazawa and S. Tanaka, Physica C**156** (1988) 157.
[46] L.F. Mattheiss, Jpn. J. Appl. Phys. **24** (1985) Suppl. 24-2, 6. L.F. Mattheiss and D.R. Hamann, Phys. Rev. B **28** (1983) 4227.
[47] S. Sugai, S. Uchida, K. Kitazawa, S. Tanaka and A. Katsui, Phys. Rev. Lett.

55 (1985) 426, and Jpn. J. Appl. Phys. **24** (1985) Suppl. 24-2,13.

[48] A.W. Sleight, in *Superconductivity: Synthesis, Properties and Processing*, ed. W. Hatfield (Marcel Dekker, New York, 1988).

[49] E. Jurczek and T.M. Rice, Europhys. Lett. **1** (1986) 225. T.M. Rice and L. Sneddon, Phys. Rev. Lett. **47** (1981) 689.

[50] C. Chaillout, A. Santoro, J.P. Remeika, A.S. Cooper and G.P. Espinosa, Solid State Commun. **65** (1988) 1363. D.E. Cox and A.W. Sleight, Solid State Commun. **19** (1976) 969.

[51] C.E. Methfessel, G.R. Stewart, B.T. Matthias and C.K.N. Patel, Proc. Nat. Acad. Sci. USA **77** (1980) 6307.

[52] B. Batlogg, Physica. **126** B (1984) 275.

[53] H.-C. zur Loye, K.J. Leary, S.W. Keller, W.K. Ham, T.A. Faltens, J.N. Michaels and A.M. Stacy, Science **238** (1987) 1558.

[54] D.G. Hinks, private communication.

[55] Y.J. Uemura *et al.*, Nature **335** (1988) 151.

[56] L.F. Mattheiss and D.R. Hamann, Phys. Rev. Lett. **60** (1988) 2681.

[57] K. Machida, Physica C**156** (1988) 276. Also see K. Machida and M. Kato, Phys. Rev. B **35** (1987) 854.

[58] See for example, V.J. Emery, Nature **333** (1988) 14.

[59] Y. Kitaoka, S. Hiramatsu, K. Ishida, T. Kohara and K. Asayama, J. Phys. Soc. Jpn. **56** (1987) 3024.

[60] K.B. Lyons *et al.*, Phys. Rev. Lett. **60** (1988) 732.

[61] J.M. Ginder *et al.*, Phys. Rev. B **37** (1988) 7507.

[62] K. Kumagai *et al.*, Phys. Rev. Lett. **60** (1988) 724.

[63] M. Kato, Y. Maeno and T. Fujita, Physica C **152** (1988) 116.

[64] J.B. Torrance *et al.*, Phys. Rev. Lett. **61** (1988) 1127.

[65] H. Kitazawa, K. Katsumata, E. Torikai and K. Nagamine, preprint.

[66] K. Machida and M. Kato, Jpn. J. Appl. Phys. **26** (1987) L660.

[67] M. Kato *et al.*, J. Phys. Soc. Jpn. **57** (1988) 726.

[68] T. Takahashi *et al.*, Nature **334** (1988) 691.

[69] Y. Kitaoka *et al.*, preprint.

[70] H. Yasuoka *et al.*, preprint.

THE BCS PAIRING AND POSSIBLE MECHANISMS FOR HIGH-Tc SUPERCONDUCTIVITY

Sudhanshu S. Jha

Tata Institute of Fundamental Research, Bombay 400005, India

1. INTRODUCTION

Since there is no general physical principle which excludes the possibility of room temperature superconductivity, and because of its obvious importance in science and technology, there has been a steady experimental quest [1] to find superconducting materials with high enough transition temperature T_c. However, even after seventy five years of its discovery in 1911, the maximum observed T_c till early 1986 was no more that 23.2 K obtained in the case of Nb_3Ge alloy, with the so called A-15 structure. In this sense, the discovery of the new class of superconductors [2,3] in the late 1986, with T_c in the range of 30 - 40 K in $La_{2-x}M_xCuO_4$ (M = Ba, Sr, Ca), was indeed a great achievement and a pleasant surprise, particularly because of the subsequent discoveries of new oxides with T_c up to about 125 K. At present, there are at least four families of such superconducting mixed - valence copper oxides, with layered structures closely related to the perovskite. Besides the La-based compounds, these include the $YBa_2Cu_3O_{7-\delta}$ family ($T_c \simeq 90$ K, called 123 compound where Y can be replaced by many other trivalent rare-earth ion) [4], the bismuth copper-oxide family: $Bi_2Sr_2CaCu_2O_8$ ($T_c \simeq 85$ K - 100 K) [5], and the thallium copper-oxide family: $Tl_2Ba_2CaCu_2O_8$ ($T_c \simeq 108$ K, called 2212 compound) and $Tl_2Ba_2Ca_2Cu_3O_{10}$ ($T_c \simeq 125$ K, called 2223 compound) [6]. In all these materials, it seems remarkable that the copper-oxygen networks are always two-dimensional (layered), with or without additional one-dimensional chains [7]. The recent discovery of 30-K superconductivity in $Ba_{1-x}K_xBiO_3$ with Bi-O layers instead of the usual CuO layers may also be included in the above list.

While the experimental effort for obtaining high temperature superconductivity was progressing slowly, theoretical investigations to establish the existence conditions for such a possibility and novel ideas for fabricating and studying such systems were also being pursued for a long time. For a comprehensive review of this effort made till 1977, one should consult the book [8] by Ginzburg and Kirzhnits. The microscopic theory of Bardeen, Cooper and Schrieffer (BCS) predicts that superconductivity in a metal or alloy can arise below a transition temperature T_c, if the effective dynamic interaction between the conduction electrons occupying time-reversed states near the Fermi-surface is attractive. The original formulation of BCS was worked out by them only in the weak coupling case for the attractive interaction [9], but the theory was very soon extended by Gorkov [10] and Eliashberg [11] to the case of arbitrary coupling strength. The effective attractive interaction for superconductivity in most low-T_c materials is known to be provided by the exchange of lattice-phonons between the pair of electrons, for momentum and energy transfers of the order of the Fermi-momentum $\hbar k_F$ and the Debye-phonon energy $\hbar\omega_D$, respectively. The repulsive screened Coulomb interaction between the electrons, of course, should not dominate the attractive interaction due to the phonon exchange. The effective phonon-exchange coupling constant λ_{ph} is determined by

$$\lambda_{ph} \equiv 2 \int_o^\infty d\omega \, \frac{\alpha^2(\omega)\; F(\omega)}{\omega} \tag{1}$$

where, following the usual notations, $F(\omega)$ is the density of phonon states and $\alpha^2(\omega)$ is the square of the electron-phonon interaction strength, averaged over the possible polarizations of phonons. The range of the repulsive Coulomb coupling constant μ_c is of the order of the Fermi energy E_F, which is much larger than the range $\hbar\omega_D$ of the attractive phonon-exchange term, in almost all materials. Because of this, the Coulomb interaction parameter is considerably supressed,

$$\mu^* = \frac{\mu_c}{1 + \mu_c \ln(E_F/\hbar\omega_D)} \tag{2}$$

as compared to the attractive coupling parameter λ_{ph} in the actual expression for T_c,

$$T_c \simeq 1.13(\hbar\omega_D/k_B) \; \exp(-1/g); \; g \equiv \lambda_{ph} - \mu^* > o, \tag{3}$$

obtained as a solution to the weak-coupling BCS gap equation in the two-square model for the effective interaction. This implies that almost all metals can become superconductors at low enough temperatures provided $g > o$. However, the above simple expression for T_c is no longer valid in the strong-coupling limit [11], and also when carriers in more than one band are involved in the transition or when retardation effects in the energy-dependent interactions are crucial. In this sense, the general problem of predicting the maximum possible T_c even within the framework of the usual phonon-exchange mechanism has been difficult to tackle.

There is a common belief that the phonon-exchange mechanism in the BCS theory can not lead to T_c above 40 - 50 K, since the Debye temperature $\theta_D \equiv \hbar\omega_D/k_B$ in most metals is limited to 300 K to 400 K (except in metallic hydrogen under pressure) and $g \lesssim 0.5$. But this estimate can never be relied upon in the strong coupling case, when the phonon coupling constant λ_{ph} is much greater than 1, and for a given λ_{ph} the area

$$A \equiv \int_o^\infty d\omega\ \alpha^2(\omega) \tag{4}$$

is large [12]. Nevertheless, from the numerical solutions [13,14] of Eliashberg equations, it seemed almost certain that up to the intermediate coupling regime ($\lambda_{ph} \lesssim 2$), the attractive phonon-exchange mechanism alone may not give T_c beyond about 50 K. It was because of this that the possibilities of obtaining higher T_c due to the exchange of electronic excitations in the system, with a much larger energy range for the resulting attractive dynamic interaction between the electrons, were being explored for a long time. The early suggestions of Little [15] involving conducting organic molecules with highly polarizable side chains, and of Ginzburg [16] involving metal-semiconductor interface structures, together with many new systems, have been investigated for this purpose by many authors. The predictions of very high-T_c in metal-semiconductor sandwich structures made by Allender, Bray and Bardeen [17] were, however, shown to be an overestimate by Rangarajan [18] and Rangarajan and Jha [19]. It was shown [18-20] that such structures may give high-T_c only if the metal thickness or radius (in a granular medium, with metal-semiconductor matrix) is less that 2-3 Å. But many other possibilities [8] for obtaining high T_c, including our prediction [21] of T_c in the range of 100 K in a two-band electron-hole system with carrier densities of about 10^{21} cm^{-3} and their effective mass ratios greater than 7, arising mainly due to the acoustic-plasmon exchange mechanism, and the possible enhancement of T_c due to spin fluctuations in the temperature region in which the inverse magnetic susceptibility $\chi_m^{-1}(q,T)$ becomes quite small before a possible magnetic ordering in systems with very large Coulomb correlations [22,23], have already been considered. Against this background, the important question which we must try to answer is whether the superconductivity in the new families of oxide superconductors, with T_c ranging between 40 K to 125 K, still arises from the phonon-exchange mechanism or whether the exchange of some kind of electronic-excitations is involved in these materials, within the framework of the BCS pairing idea. In fact, many scientists are suggesting that the superconducting transition in these new oxide superconductors, because of their many peculiar superconducting and normal state properties, including a very short superconducting coherence length ξ of the order of 10^{-7} cm or less, may not have anything to do with the BCS pairing at all [24,25]. Because of the conflicting experimental data on various crucial physical parameters due to uncontrolled but extremely important variations in the preparation of the new ceramic superconducting materials with varying stoichiometry, oxygen deficiencies and unintended inhomogeneities or impurity phases, it has not been possible still to answer these questions unambiguously. Perhaps, the extensive single-crystal results on the rare-earth-based materials as well as on the new more stable compounds (as far as oxygen stoichiometry is concerned) involving Bi or Tl may give more convincing clues in these directions, in the near future. At present, it is enough to say that we do not find any overwhelming reason to abandon our search for finding the correct mechanism for superconductivity in the new oxide materials, within the conventional framework of the generalized BCS pairing theory. It is in this spirit that this chapter has been written. We will not discuss here the alternative proposals, e.g. superconductivity due to Bose-condensation of the so called charged "holons" in the resonating valence bond (RVB) theory of Anderson. The idea is to present a brief exposition of the essential points of the BCS pairing theory in order to guard against various misconceptions in the current literature, in connection with the comparison of experimental results with the BCS predictions, and the confusion between the results of the generalized BCS pairing theory and the extremely simplified one-parameter (square-well) weak-coupling results of the so-called "BCS-model".

In what follows, we first review the main points of the generalized BCS pairing theory in Sec.2. The weak coupling limit as well as the strong coupling limit are considered. Since the new class of oxide-superconductors has highly anisotropic normal as well as superconducting state properties, a recent formulation of the theory to deal with a crystal with layered structure [26] is presented in Sec.3. Since, the dielectric formulation for the effective interaction has been used in these discussions, it allows one to consider different types of attractive exchange mechanisms, within the BCS framework. Some important physical properties of the new oxide superconductors will be discussed in Sec.4, and the nature of possible electrodynamic response of such superconductors with anisotropic energy gap will be considered in Sec.5. Based on these expositions, the present status for establishing probable pairing mechanisms in the high-T_c superconductors will be discussed in the concluding Sec.6.

2. THE GENERALIZED BCS PAIRING THEORY

The most important concept in the generalized BCS pairing due to mutual dynamic interaction of electrons in time-reversed states near the Fermi-surface in a metallic system is the introduction of the corresponding effective dynamic interaction function V between the electrons, in the presence of moving ions and other conduction electrons in the system. If one ignores the spin-orbit interaction, these time-reversed states are nothing but the states of opposite spins $\uparrow$ and $\downarrow$, implying the special case of spin-singlet pairing. The formulation for the spin-triplet pairing follows similar lines, but we omit that in our explicit consideration here. In an approximate way [8], it is possible to write V in terms of the bare Coulomb interaction and an appropriate inverse dielectric function of the system. In the normal state of the metal, it is assumed that the electronic properties are characterized by a well-defined Fermi-surface and weakly interacting quasi-particle excitation spectrum without any energy gap. The quasi-particle spectrum is determined by the poles of the temporal Fourier transform of the single-particle Green's function

$$G_{\alpha\beta}(1,2) \equiv -i\langle T(\Psi_\alpha(1)\ \Psi^+_\beta(2))\rangle\ \delta_{\alpha\beta} \qquad (5)$$

defined as the fixed-chemical potential thermal average of the time-ordered electron field operators in the Heisenberg representation. Here, 1 and 2 refer to space-time points $\vec{r}_1$, t_1 and $\vec{r}_2$, t_2, respectively, and α, β refer to spin-$\uparrow$ or spin $\downarrow$ states. In the absence of any magnetic field, $G_{\uparrow\uparrow} = G_{\downarrow\downarrow} = G$. For the superconducting state, the generalized BCS pairing of opposite-spin states (more correctly, time-reversed states) is introduced through the nonvanishing of the anomalous Green's function

$$\mathfrak{F}_{\uparrow\downarrow}(1,2) \equiv -i\langle T(\Psi_\uparrow(1)\ \Psi_\downarrow(2))\rangle = -\mathfrak{F}_{\downarrow\uparrow}(2,1) \equiv \mathfrak{F} \neq o,\ T < T_c \qquad (6)$$

and its complex conjugate. In the normal state, $\mathfrak{F} = o$, so that it acts as an order parameter for the superconducting state.

With the total effective Lagarangian for the electronic system determined by the electronic kinetic energy T and the effective electron-electron interaction V, one can obtain the rate of change of the field operators with time, in the form

$$(i\hbar \frac{\partial}{\partial t_1} - T(\vec{r}_1) + \mu)\ \Psi_\alpha(1) = \left\{\sum_\beta \int d2\ V(2,1)\ \Psi^+_\beta(2)\ \Psi_\beta(2)\right\} \Psi_\alpha(1) \qquad (7)$$

Using this, one can obtain the equations for the rate of change of G and $\mathfrak{F}$ with time. In the time-dependent Hartree-Fock approximation, these reduce to coupled integro-differential equations [8,27], for determining these quantities. The general procedure for taking account of higher-order terms in V, at least perturbatively, are much more complicated, even though contributions corresponding to bubble polarization diagrams have not to be included in such a calculation since they are already contained in the effective V. Even the coupled equations for G and $\mathfrak{F}$ obtained in the time-dependent H.F. approximation, are not very easy to handle. As discussed in Ref.8, in what follows we first consider the simpler case of the weak coupling approximation for V, before taking up the strong coupling case.

The Weak Coupling Formalism

In the lowest-order weak coupling approximation, the equation for $\mathfrak{F}$ simplifies considerably. In such a case, one finds

$$\mathfrak{F}(1,2) = i \int d3 \int d4\ G(1,3)\ V(3,4)\,\mathfrak{F}(3,4)\ G(4,2) \qquad (8)$$

The so-called "energy-gap" function is defined in terms of $\mathfrak{F}$ as,

$$\Delta(1,2) \equiv iV(1,2)\,\mathfrak{F}(1,2) \qquad (9)$$

For a homogeneous equilibrium system, all these quantities depend only on the difference 1-2. For the simplest case of carriers in a single band with Bloch energies

$$\varepsilon_{\vec{k}} - \mu = \xi_{\vec{k}} \tag{10}$$

as measured from the Fermi-energy μ, one can then approximately derive the usual weak-coupling BCS gap equation in the $\vec{k}$-space [8]:

$$\Delta(\vec{k}) = -\sum_{\vec{k}'} V^{sup}_{\vec{k},\vec{k}'} \frac{\Delta(\vec{k}') \tanh(E_{\vec{k}'}/2k_B T)}{2E_{\vec{k}'}} \tag{11}$$

where,

$$E_{\vec{k}} = + \left[\xi^2_{\vec{k}} + \Delta^2(\vec{k})\right]^{\frac{1}{2}}, \quad \Delta(\vec{k}) \equiv - \sum_{\vec{k}'} V^{sup}_{\vec{k},\vec{k}'} \langle C_{\vec{k}'\uparrow} C_{-\vec{k}'\downarrow} \rangle \tag{12}$$

and where $C_{\vec{k}\sigma}$ ($C^{+}_{\vec{k}\sigma}$) is the electron annihilation (creation) operator in the $\vec{k}$ - space. For an effective homogeneous system under consideration, the effective dynamic interaction can be assumed to be of the form

$$V(\vec{q},\omega) = \frac{4\pi e^2}{q^2} \varepsilon^{-1}(\vec{q},\omega) \tag{13}$$

in terms of which one has the explicit expression for V^{sup} in the BCS gap equation:

$$V^{sup}_{\vec{k},\vec{k}'} = V^{sup}(\vec{q} = \vec{k} - \vec{k}', \xi_{\vec{k}}, \xi_{\vec{k}'}),$$

$$V^{sup}(\vec{q},\xi,\xi') = \frac{4\pi e^2}{q^2}\left[1 + \frac{2}{\pi}\int_0^\infty d\omega' \frac{\mathrm{Im}\ \varepsilon^{-1}(\vec{q},\omega')}{(\omega' + |\xi| + |\xi'|)}\right] \tag{14}$$

The excitation spectrum in the superconducting state is determined in this approach by

$$H_{BCS} = W_o + \sum_{\vec{k}} E_{\vec{k}}(\gamma^{+}_{ko}\gamma_{ko} + \gamma^{+}_{k1}\gamma_{k1}) \tag{15}$$

where

$$\gamma_{\vec{k}o} = u_{\vec{k}} C_{\vec{k}\uparrow} - v_k C^{+}_{-\vec{k}\downarrow}\ ;\ \gamma_{\vec{k}1} = v_{\vec{k}} C^{+}_{\vec{k}\uparrow} + u_{\vec{k}} C_{-\vec{k}\downarrow} \tag{16}$$

$$u_{\vec{k}}^2 = \tfrac{1}{2}(1 + \xi_{\vec{k}}/E_{\vec{k}})\,; \quad v_{\vec{k}}^2 = \tfrac{1}{2}(1 - \xi_{\vec{k}}/E_{\vec{k}}) \tag{17}$$

In the three-dimensional isotropic case, after averaging over the angles between $\vec{k}$ and $\vec{k}'$, the BCS gap equation (11) can be rewritten in the form

$$\Delta(\xi) = - \int_{-\infty}^{\infty} d\xi\ K(\xi,\xi')\ \Delta(\xi') \left[\tanh(E'/2k_B T)\right]/2E' \tag{18}$$

with the kernel

$$K(\xi,\xi') = N(\xi')\ \overline{V}(\xi,\xi') = \frac{N(\xi')}{2k(\xi)\ k(\xi')} \int_{|k(\xi) - k(\xi')|}^{k(\xi) + k(\xi')} dq\ q\ V^{sup}(q,\xi,\xi') \tag{19}$$

where $N(\xi)$ is the density of the electronic states (of one spin only) and $E' = E(\xi') = [\Delta^2(\xi') + \xi'^2]^{\frac{1}{2}}$. Close to a spherical Fermi surface, both $k(\xi)$ and $k(\xi')$ can be approximated by the Fermi-wave vector k_F, and $N(\xi')$ by $N(o)$.

As $T \to T_c$, $\Delta \to o$, so that the transition temperature T_c can be obtained from the nontrivial solution of the linearized integral equation

$$\Delta(\vec{k}) = - \sum_{\vec{k}'} V^{sup}_{\vec{k},\vec{k}'} \frac{\Delta(\vec{k}')\ \tanh(\xi_{k'}/2k_B T_c)}{2\xi_{k'}} \tag{20}$$

which for the three-dimensional isotropic case reduces to

$$\Delta(\xi) = - \int_{-\infty}^{\infty} d\xi'\ K(\xi,\xi')\ \Delta(\xi') \left[\tanh(\xi'/2k_B T_c)\right] (2\xi')^{-1} \tag{21}$$

where the interaction kernel is defined by Eqs. (19) and (14) in terms of the inverse dielectric function $\varepsilon^{-1}(q,\omega)$. Thus, in this approximation the main input relating to the metal which determines T_c is the dynamic effective interaction $V(q,\omega)$ or equivalently $\varepsilon^{-1}(q,\omega)$, and the nature of the density of states $N(\xi)$ in the normal phase. The kernel $K(\xi,\xi')$ in Eq. (21) for obtaining T_c is then fixed uniquely. If to the first approximation $\varepsilon^{-1}(q,\omega)$ for such a system is identified with the inverse linear dielectric function of the system, it can be split up as a sum of terms arising from different longitudinal excitation modes of the normal state, corresponding to the exchange of phonons, plasmons, spin-fluctuations, excitons, etc., having poles at the respective mode frequencies. Thus one may write

$$\frac{1}{\varepsilon(q,\omega)} = \sum_i \varepsilon_i^{-1}(q,\omega)$$

$$= \frac{1}{\varepsilon_{coul.}} + \frac{1}{\varepsilon_{phonon}} + \frac{1}{\varepsilon_{exciton}} + \frac{1}{\varepsilon_{plasmon}} + \frac{1}{\varepsilon_{spin-fluc.}} + \ldots. \quad (22)$$

so that the net attractive interaction may not be provided by the phonons alone. The splitting of $K(\xi,\xi')$, defined by Eqs.(14) and (19), corresponding to the splitting (22) of $\varepsilon^{-1}(q,\omega)$, then defines the corresponding coupling constants

$$\lambda_i \equiv -K_i(o,o) \quad (23)$$

which are attractive for negative $K_i(o,o)$. In terms of the more familiar notation for the coupling in the case of phonon exchange,

$$\alpha^2_{ph}(\omega)\, F_{ph}(\omega) \simeq -\frac{N(o)}{2\pi k_F^2}\int_o^{2k_F} dq\, q\left(\frac{4\pi e^2}{q^2}\right) \mathrm{Im}\, \varepsilon^{-1}_{ph}(q,\omega) \quad (24)$$

$$K_{ph}(\xi,\xi') = -2\int_o^{\infty} d\omega' \frac{\alpha^2_{ph}(\omega')\, F_{ph}(\omega')}{\omega' + |\xi| + |\xi'|} \quad (25)$$

$$\lambda_{ph} \equiv -K_{ph}(o,o) = 2\int_o^{\infty} d\omega \frac{\alpha^2_{ph}(\omega)\, F_{ph}(\omega)}{\omega} \quad (26)$$

The one-parameter BCS model is defined by Eqs.(18) and (21) in which the kernel K is replaced by $-\lambda_{ph}$, for $|\xi|$, $|\xi'| < \hbar\omega_D$, and zero otherwise. The usual two-square model in which the repulsive Coulomb kernel is added to the phonon kernel in the form of an additional contribution $-\lambda_c = \mu$ to K in the range $|\xi|$, $|\xi'| < E_F$, leads to the wellknown weak coupling equations (3) and (2), discussed earlier. However, it should be emphasized here that these parametrizations of the actual kernel $K(\xi,\xi')$ in Eqs.(18) and (21) to determine Δ as a function of temperature, and T_c, respectively, are extreme simplifications, even for the case of phonon-exchange only. Any general prediction based on the so-called one-parameter BCS model, namely, the ratio $2\Delta(T=o)/k_BT_c = 3.5$, the specific heat jump $[\Delta C]_{T_c} = 1.43\, C_{normal}(T_c)$ at the transition temperature, the critical magnetic field $H_c(o) = [4\pi N(o)]^{\frac{1}{2}}\,\Delta(o)$ at zero temperature, etc., can only be an approximate one. Only for negligible μ^* in Eq.(3), T_c varies with isotopic mass in elemental superconductors as $M^{-\alpha}$, with $\alpha = 0.5$. In general, even in the weak coupling theory, the above "BCS-model" predictions must be modified depending on the detailed nature of $K(\xi,\xi')$. It must be noted once again that contrary to the assumptions of various recent authors, these predictions are not the exact predictions of the

BCS pairing theory, even in the weak coupling limit. In fact, it is amazing that the above predictions of the simple equivalent one-parameter BCS model can explain so well so many properties of many superconducting materials. However, these are not the gospels of the BCS theory; only results of an extremely simplified model in the theory, which can be solved easily. Moreover, for actual crystals homogeneous and isotropic approximations are not exactly valid, and sometimes there may be carriers occupying more than one band. In such a case, the weak coupling equations (20) and (14) for determining T_c must be generalized to

$$\Delta_n = - \sum_{n'} \frac{V^{sup}_{nn'} \Delta_{n'} \tanh(\xi_{n'}/2k_BT_c)}{2\xi_{n'}} \tag{27}$$

$$V_{nn'}(\vec{q} = \vec{k}-\vec{k}', \xi_n, \xi_{n'}) = \sum_{\vec{G}} \sum_{\vec{G}'} \langle n'| e^{-i(\vec{q}+\vec{G}).\vec{r}} |n\rangle\langle n| e^{i(\vec{q}+\vec{G}').\vec{r}} |n'\rangle$$

$$\times \frac{4\pi e^2}{|\vec{q}+\vec{G}|^2}\left[\delta_{\vec{G},\vec{G}'} + \frac{2}{\pi}\int d\omega' \, \frac{\mathrm{Im}\,\varepsilon^{-1}(\vec{q}+\vec{G}, \vec{q}+\vec{G}', \omega')}{\omega' + |\xi_n| + |\xi_{n'}|}\right] \tag{28}$$

in which n stands for the band index b and wavevector $\vec{k}$, and $\vec{G}$, $\vec{G}'$ are reciprocal lattice wave vectors of the system with $\varepsilon^{-1}(\vec{q}+\vec{G}, \vec{q}+\vec{G}', \omega)$ as the purely longitudinal component of the reciprocal dielectric tensor. This can again change the simple predictions of one-parameter "BCS-model" completely. For example, in the two-band model, one gets two coupled equations for the gap functions, involving interaction parameters V_{11}, V_{22} and V_{12}. Although, one gets a single T_c, $2\Delta(o)/k_BT_c$ is no longer 3.5. Before considering possible simplifications of the above equations (27) and (28) for the case of layered materials in the next section, we will briefly discuss the BCS pairing theory when the coupling is strong.

The Strong Coupling Formalism

It has already been emphasized earlier that the reduction of the equations for $\mathfrak{F}$ and G to the form (11) is possible only if one keeps terms to the lowest-order in V, e.g., the equation (8) for $\mathfrak{F}$. No vertex correction and renormalization of energies have been included in the lowest-order diagram. However, for the case of electron-phonon interaction, Migdal [28] has shown that one can tackle this problem for the case of normal phase for arbitrary strong coupling, since there is a small parameter $(m/M)^{\frac{1}{2}}$,the square root of the ratio of the electronic to ionic mass, in the problem. To the lowest-order in this parameter, it is possible to obtain the electronic vertex function Γ and the self-energy Σ (in the electronic Green's function) for an orbitary electron-phonon coupling strength. For electronic states close to the Fermi surface, the self-energy $\Sigma(\vec{k},\omega)$ varies appreciably with ω, although its overall magnitude may be small compared to the Fermi energy E_F. This affects the density of states at the Fermi surface, e.g. the mass renormalization

$$\frac{m^*}{m} = Z(k_F) = 1 + \lambda_{ph} \tag{29}$$

in the expression for the low-temperature specific heat in the normal phase, where as before λ_{ph} is given by Eq. (26). In the superconducting phase, there are two self-energies Σ_1 and Σ_2, corresponding to the Green's functions G and $\mathfrak{F}$, both of which can be calculated to the same accuracy as in the normal case [11] for any arbitary electron-phonon coupling. In a 2 x 2 matrix notation [29], if one uses Nambu's Green's function matrix $\bar{G}$ and the self-energy matrix $\bar{\Sigma}$, such that

$$\bar{G} = \begin{bmatrix} G & \mathfrak{F} \\ \mathfrak{F}^* & G \end{bmatrix}, \quad \bar{\Sigma} = \begin{bmatrix} \Sigma_1 & \Sigma_2 \\ \Sigma_2 & \Sigma_1 \end{bmatrix} \tag{30}$$

one can obtain the matrix Dyson equation

$$\bar{G} = \bar{G}_o + \bar{G}_o \, \bar{\Sigma} \, \bar{G} \tag{31}$$

where $\bar{G}_o$ is the Green's function in the absence of the interaction. For a homogeneous and isotropic system, one finds the renormalization function $Z(\omega)$ and the gap function $\Delta(\omega)$ through the relations

$$\Sigma_1(\omega) = \omega \, Z(\omega) - \omega, \quad \Sigma_2(\omega) = Z(\omega) \, \Delta(\omega) \tag{32}$$

both of which can be complex.

Ginzburg and Kirzhnits [8] have shown that to the lowest-order in $(m/M)^{1/2}$, one can indeed obtain the strong-coupling equations of Eliashberg [11] which are nothing but the coupled nonlinear integral equations for $Z(\omega)$ and $\Delta(\omega)$, starting from the general equations for G and $\mathfrak{F}$, discussed earlier. If the Coulomb repulsion is introduced through the square-well coupling constant μ_c, these equations are

$$Z(\omega)\,\Delta(\omega) = \int_{-\infty}^{\infty} d\omega' \, \mathrm{Re}\left[\frac{\Delta(\omega')}{[\omega'^2 - \Delta^2(\omega')]^{1/2}}\right]$$

$$\times \int_o^{\infty} d\Omega \, \alpha^2_{ph}(\Omega) \, F_{ph}(\Omega) \left[\frac{f(-\omega') + N(\Omega)}{\omega' + \Omega - \omega} + \frac{f(\omega') + N(\Omega)}{\omega' - \Omega - \omega}\right]$$

$$- \mu_c \int_o^{E_F} d\omega' \, \mathrm{Re}\left[\frac{\Delta(\omega')}{[\omega' - \Delta^2(\omega')]^{1/2}}\right] \tanh(\omega'/2k_BT) \tag{33}$$

$$[1 - Z(\omega)]\omega = \int_o^\infty d\omega' \; \mathrm{Re}\left\{\frac{\omega'}{[\omega'^2 - \Delta^2(\omega')]^{\frac{1}{2}}}\right\}$$

$$\times \int_o^\infty d\Omega \alpha^2_{ph}(\Omega) \; F_{ph}(\Omega) \left[\frac{f(-\omega') + N(\Omega)}{\omega' + \Omega + \omega} - \frac{f(-\omega') + N(\Omega)}{\omega' + \Omega - \omega}\right.$$

$$\left. + \frac{f(\omega') + N(\Omega)}{-\omega' + \Omega + \omega} - \frac{f(\omega') + N(\Omega)}{-\omega' + \Omega - \omega}\right] \tag{34}$$

where $f(\omega)$ and $N(\Omega)$ are the usual Fermi and Bose distribution functions $[\exp(\omega/k_BT) + 1]^{-1}$ and $[\exp(\Omega/k_BT) - 1]^{-1}$, respectively. The transition temperature T_c is of course obtained from the above equations by putting $\Delta^2(\omega') \to o$ in all the denominators on the right hand sides of these equations. It can be shown that in the weak coupling limit, these equations reduce to the corresponding BCS gap equations. The solutions of Eliashberg equations have been used for many strong coupling superconductors. In fact, an analysis of experimental tunneling data [30] can give Re $\Delta(\omega)$ and Im $\Delta(\omega)$, from which it is possible to reconstruct $\alpha^2(\omega)$ $F(\omega)$ of Eliashberg equations. In all superconducting materials discovered till 1985, where this procedure has been followed through, one finds that the resulting $\alpha^2(\omega)$ $F(\omega)$ resembles phonon interaction spectrum of the metal, with reasonable values of μ^*(0.1 to 0.2). Numerical integrations of Eliashberg equations for λ_{ph} very large, particularly for $\lambda_{ph} \geqq 10$, have not been very accurate mainly because of large oscillations in $\Delta(\omega)$ even if one performs the calculations in much more convenient "temperature"-technique, with discrete Matsubara imaginary frequency representation. However, for $\lambda_{ph} \lesssim 2.5$, numerically [14] one finds that one can obtain T_c using the expression

$$T_c \simeq \frac{\hbar \bar{\omega}_{log}}{1.43 \; k_B} \exp\left\{- \frac{1 + \lambda_{ph}}{\lambda_{ph} - \mu^*(1 + \alpha)}\right\} \tag{35}$$

where

$$\alpha \simeq \frac{0.5 \; \lambda_{ph}}{1 + \lambda_{ph}} \; , \; \mu^* = \mu_c\left[1 + \mu \; \ln(E_F/\hbar\bar{\omega}_{log})\right]^{-1} \tag{36}$$

$$\ln \bar{\omega}_{log} = \frac{2}{\lambda_{ph}} \int_o^\infty \frac{d\omega}{\omega} \; [\ln\omega] \; \alpha^2_{ph}(\omega) \; F_{ph}(\omega) \tag{37}$$

Many of the ratios predicted in the weak-coupling one-parameter BCS model are no longer valid in the strong coupling case [12]. For example, $2\Delta(o)/k_BT_c$ is always greater than 3.5, etc., in the isotropic case.

It should be emphasized here that no generalization of Eliashberg equations for the electronic-exchange mechanisms has yet been obtained from the first principles. This is obviously because of the lack of a small parameter like $(m/M)^{\frac{1}{2}}$ in the case of phonon-exchange. It is, however, possible to consider individual electronic exchange processes to generalize them to the strong-coupling case. However, if one is interested only up to the intermediate coupling regime, Eliashberg equations (33) and (34) and their approximate solutions (35) - (37) can also be used for non-phonon exchange mechanisms, provided one replaces $\alpha^2_{ph}(\omega)\ F_{ph}(\omega)$, λ_{ph}, etc., by appropriate $\alpha^2_i(\omega)\ F_i(\omega)$, λ_i, etc., given in terms of the corresponding $\varepsilon_i^{-1}(q,\omega)$ and $K_i^{\cdot}(\xi,\xi')$ in Eqs.(24)-(26), for the boson-exchange being considered. Thus, once one can set up the appropriate $\varepsilon_i^{-1}(q,\omega)$ for the relevant attractive exchange mechanism under consideration, one is able to formulate the corresponding Eliashberg equations completely, which may be valid at least up to the intermediate coupling limit.

3. THE BCS GAP EQUATION FOR LAYERED CRYSTALS

It has already been noted in the last section that in real crystals the knowledge effective electron-electron interaction and the study of the resulting superconducting phase transition require the expression for the complete longitudinal inverse dielectric matrix $\varepsilon^{-1}(\vec{q}+\vec{G},\ \vec{q}+\vec{G}',\ \omega)$ in the reciprocal lattice-vector space. In actual calculations, one usually replaces the dielectric matrix by an average local $\varepsilon^{-1}(q,\omega)$ for an approximately equivalent homogeneous and isotropic medium. It has been argued by us recently [26] that this approximation is too crude for the new class of oxide-superconductors. In all these new materials, the motion of the charged carriers is mostly confined in two-dimensional layers with a very weak coupling between the layers. Perhaps, the most important layers in the unit cell responsible for superconductivity are the Cu-O layers and the layers adjacent to them, which may be either metallic or insulating. The higher transition temperatures in Bi-based and Tl-based oxides suggest that T_c can be enhanced if there are more Cu-O layers in the unit cell. In view of this, it is extremely important to avoid the use of 3-dimensional homogeneous-isotropic approximation for such crystals. It may, however, be used for the 2-dimensional motion within each layer, at least approximately, but not for the motion in the direction perpendicular to the layers.

Superconductivity in a crystal with equally-spaced alternate metallic and insulating molecular layers was first investigated by Bulaevskii and Kukharenko [31,8]. The calculation of the effective electron-electron interaction and the expression for the superconducting gap equation for a layered crystal, with an arbitrary number of layers at different distances between them in any unit cell, have been given by us recently [26,31]. If the layers in any unit cell are taken to be approximately in the (x-y)-plane, with total number of layers m in the cell, one can obtain the relevant effective interaction $V(q_t,\omega,z,z')$ and $\varepsilon^{-1}(q_t,\omega,z,z')$ as $m\times m$ matrices whose elements j,j' give the interaction of electrons in the layers j and j'. Here, q_t is the magnitude of the two-dimensional wave vector in the x-y plane of the layers. If all the different layers in the crystal perpendicular to the z-axis is labelled by the running index n, different layers in any unit cell N can be labelled by two numbers N,j, such that $n = mN+j$, $j=0,1,2.......m$;

$N = 0,1,2 \ldots$, where layers N,m and $N+1,0$ are same. If the length L is the periodicity of the crystal in the Z-direction (length of the unit cell along Z), the z-coordinate of the nth layer is given $z(N,j) = NL + R_j$, where R_j is the distance of the jth layer to the 0th-layer in the unit cell (with $R_o = o$). Using the discrete Fourier transform with respect to the variables $(N - N')$, to go over the variables θ,

$$g(N - N') \equiv \int_o^{2\pi} \frac{d\theta}{2\pi} g(\theta) \exp\left[i(N - N')\theta\right] \tag{38}$$

and a simple diagrammatic perturbation theory [26,31], for the $m \times m$ matrices $V(q_t,\omega,\theta)$ and $\varepsilon^{-1}(q_t,\omega,\theta)$, one finds

$$V(q_t,\omega,\theta) = \varepsilon^{-1}(q_t,\omega,\theta)\, V^{(o)}(q_t,\theta) \tag{39}$$

$$\varepsilon^{-1}(q_t,\omega,\theta) = I + V^{(o)}(q_t,\theta)\, \Pi(q_t,\omega) \tag{40}$$

where $\Pi_{jj'}(q_t,\omega) = \delta_{jj'}\, \pi_j(q_t,\omega) = \delta_{jj'}\, \pi_{j+m}(q_t,\omega)$ are the two-dimensional polarization functions of the layer j in the unit cell, and the Fourier transforms of the matrix elements of the bare Coulomb interaction $V^{(o)}(q_t,\omega,z_n,z_{n'}) \equiv V^{(o)}_{jj'}(q_t,\omega,N,N') = (2\pi e^2/q_t)\exp[-q_t|(N-N')L + R_{jj'}|]$, are given by

$$V^{(o)}_{jj'}(q_t,\theta) = \begin{cases} \dfrac{2\pi e^2}{q_t}\left[P(\theta)\exp(-q_t R_{jj'}) + N(\theta)\exp(q_t R_{jj'})\right]; & j > j' \\[2ex] \dfrac{2\pi e^2}{q_t}\left[P(-\theta)\exp(q_t R_{jj'}) + N(-\theta)\exp(-q_t R_{jj'})\right]; & j < j \end{cases} \tag{41}$$

with

$$\begin{aligned} P(\theta) &= \exp(i\theta)\left[\exp(i\theta) - \exp(-q_t L)\right]^{-1} \\ N(\theta) &= \exp(-q_t L)\left[\exp(-i\theta) - \exp(-q_t L)\right]^{-1} \end{aligned} \tag{42}$$

In Eq.(40), I is the unit $m \times m$ matrix. Note that $\varepsilon^{-1}(q_t,\omega,\theta)$ is a non-diagonal matrix because of which the effective electron-electron interaction V_{jj}, confined within a given layer j, depends not only on the polarizability function π_j of that layer but also on the polarizability functions of other layers.

In general, the BCS gap equation is now a matrix equation in the unit cell layer space for the gap functions $\Delta_{jj'}(k_t,\theta)$. However, for simplicity if we try to find the superconducting transition temperature

T_c arising from the motion of electrons or holes within any given conducting layer j only (i.e., $N=N'$, $j=j'$), we have to find the nontrivial solution of the reduced BCS gap equation

$$\Delta_{jj}(\vec{k}_t) \simeq -\sum_{\vec{k}'_t} \int_0^{2\pi} \frac{d\theta}{2\pi} V_{jj}(q_t = |\vec{k}_t - \vec{k}'_t|, \omega = \xi_j - \xi'_j, \theta) \Delta_{jj}(\vec{k}'_t) \tanh(\xi'_j/2k_B T_c)/(2\xi'_j) \tag{43}$$

where ξ_j and ξ'_j are single-particle energies corresponding to 2-dimensional in-plane wave vectors $\vec{k}_t$ and $\vec{k}'_t$, respectively, measured from the Fermi-energy of the carriers. As in the case of 3-dimensional isotropic case, $V_{jj}(|\vec{k}_t - \vec{k}'_t|, \omega)$ can be averaged over angles between $\vec{k}_t$ and $\vec{k}'_t$, to find

$$\Delta_{jj}(\xi) = - \int d\xi' \, N_j(\xi') \, V_{jj}(\xi,\xi') \, \Delta_{jj}(\xi') \left[\tanh(\xi'/2k_B T_c)\right] (2\xi')^{-1} \tag{44}$$

where

$$\bar{V}_{jj}(\xi,\xi') = \frac{1}{2\pi} \int_0^{2\pi} d\phi \, V_{jj}(q_t = \sqrt{k_t^2 + k_t'^2 - 2k_t k'_t \cos\phi}, \omega = \xi - \xi', N = N')$$

$$= \frac{2}{\pi} \int_{|k_t - k'_t|}^{(k_t + k'_t)} dq_t \int_0^{2\pi} \frac{d\theta}{2\pi} \frac{V_{jj}(q_t, \xi - \xi', \theta)}{[2k_t^2 + 2k_t'^2 - q_t^2 - (k_t^2 - k_t'^2)^2/q_t^2]^{\frac{1}{2}}} \tag{45}$$

and where $N_j(\xi') = d^2k'_t/(4\pi^2 \, d\xi')$ is the 2-dimensional density of states for one type of spins in the layer j.

The main input in the above approach is the specification of the polarizability functions $\pi_i(q_t,\omega)$ for different layers. Once this is known, one can determine $V_{jj}(q_t,\omega)$ which can be split further into contributions from the exchange of different possible excitations. In the random-phase approximation, it can be shown that the nature of the expression $(2\pi e^2/q_t)\,(q_t L/2)^{-1}\,\pi_j(q_t,\omega)$ is similar to the corresponding 3-dimensional susceptibility times 4π. Their properties for conducting and insulating layers have been discussed explicitly in Ref.26. Further, it is shown there that the interlayer contribution to the net attractive interaction within a given conducting layer can be appreciable if $2k_F d \sim 1$, where d is the separation between these layers and k_F is the 2-d Fermi wave vector of the conducting layer under consideration. It has been argued that the exchange of electronic excitations of an adjacent, e.g., insulating layer by carriers in a conducting layer can give rise to a large enhancement of T_c arising from the conducting layer alone, if $2k_F d \sim 1$. For low layer-carrier densities of the order of 10^{14} cm^{-2} (volume density $n = n_s/L \sim 10^{21}$ cm^{-3}), as reported in most of these new oxide superconductors, this is quite favourable for interlayer separations d of the order of 2Å. For many conducting layers in any unit cell, well separated by insulating layers, T_c may be enhanced further. In the more general case of appreciable coupling in between these conducting layers, there may be finite non vanishing off-diagonal elements of the gap, and variation of the gap functions with respect to θ/L, the equivalent wave vector perpendicular to $\vec{k}_t$.

4. SOME PROPERTIES OF NEW HIGH-T_c SUPERCONDUCTORS

One of the most severe difficulties one faces in the La-based as well as Y-based superconducting materials is the control of oxygen stoichiometry in these ceramics. These materials with oxygen deficient pervoskite structures are quite unstable with respect to oxygen contents, and one has to prepare them with extra care in oxygen atmosphere. Even when one tries to obtain a single-phase compound in the form of a single-crystal instead of a pellet compacted from powder samples, one finds many voids and twins inside. Because of these uncontrolled but crucial variations, we still have conflicting experimental data regarding many of the crucial physical parameters of the new superconductors. Even in Bi-based and Tl-based superconductors, there is no real consensus about various experimental results obtained so far, although single-crystal work in these materials have yet to be done on a larger scale. Since various physical properties of all these materials, including their structure, thermodynamic properties, magnetic and electrodynamical properties, energy gaps, tunneling characteristics, etc., are being discussed in great detail in other chapters by various experts, our aim here is only to highlight some of the results which may be important from the point of view of establishing a theoretical understanding of superconductivity in these new materials.

Thermodynamically, the onset of the traditional superconducting state is characterized as a second-order phase transition in the absence of any magnetic field, with a jump ΔC in the electronic specific heat at $T = T_c$. However, at relatively high values of T_c in the new materials there is a large lattice-heat capacity contribution to the total specific heat of the material at $T = T_c$. This masks the discontinuity in the measured total specific heat, apart from smearing due to inhomogeneities, etc.. However, careful measurements and analysis [32,33] of data have shown that $\Delta C/\gamma T_c$ ranges from 1 to 3, in most of the new materials. Here γT_c is the normal state electronic specific heat at $T = T_c$, and in the simple one-parameter weak-coupling BCS model $\Delta C/\gamma T_c = 1.43$. Many workers [33] also find a linear-T term in the electronic specific heat at low temperatures in the superconducting state. Actually, the presence of an energy gap in the BCS excitation spectrum leads to the usual exponential temperature dependance in the superconducting state. In a gapless superconductor, the linear-T term is possible. However, many experimentalists feel that there may be other reasons for the linear term, particularly because the volume of the sample showing the Meissner effect is never 100% in most of these measurements. Also, the role of the possible anisotropy in the energy gap has also to be considered in this connection. Although, the origin of this linear term in specific heat is still controversial, many workers have taken this seriously to use this as a basis for an alternative theory, like the RVB theory [24], as an alternative to the BCS pairing theory.

As far as the so called isotope effect is concerned, the replacement of ^{16}O by ^{18}O in these new materials has shown that in general the exponent α, in $T_c \sim M^{-\alpha}$, is smaller than 0.5, and can be almost zero in the Y-based 1-2-3 compounds. The superconducting gap has been measured in these materials by different techniques. Although the interpretation of experimental data has not been so easy, in general various tunneling data have given larger values of $2\Delta(o)/k_BT_c$ as compared to infrared absorption data. These range from as low as 1.5 to 8, whereas the weak coupling BCS model value is 3.5. On the other hand, careful flux-trapping experiments [35] have clearly established that the magnetic flux is quartized in the units of hc/2e in these superconductors, indicating a pairing of two electrons of charge e.

The normal state resistivity in most of the new materials is different in the direction perpendicular ($\perp$) to the (a-b)-plane of the layers than in the direction parallel ($||$) to the a-b plane. The carriers are more mobile in the a-b plane. There is a general agreement that the superconductivity in these layered materials is anisotropic, with different values of the coherence lengths $\xi_{a,b}$ and ξ_c, and penetration depths $\lambda_{b,a}$ and λ_c for the magnetic field H parallel to the C-axis or perpendicular to it. This implies that the upper critical field $H_{c2}(T)$ in these type-II superconductors is also anisotropic, with $H_{c2}(o)$ in the C-direction much smaller than $H_{c2}(o)$ in the a,b plane. To get an order of magnitude for these parameters, one can quote, $\xi_{a,b} \sim 15$ to 30 Å, $\xi_c \sim 4$-7 Å, $\lambda^{H\perp C}_{b,a} \sim 8000$ Å, $\lambda^{H||C}_{b,a} \sim 1000$ Å, $H^c_{c2}(o) \sim 500$ kilogauss, $H^{a,b}_{c2}(o) \sim 2000$ kilogauss. The carriers are probably holes with densities of the order of 10^{21} cm^{-3}, the effective-mass $m^* \sim 4$ to $7m$, and Fermi energy $E_F \sim 0.5$ to 0.8 eV. The estimates for the lower critical field H_{c1} varies considerably in the literature, and may be as low as a few gauss [36] if one uses the criterion of strict reversible thermodynamic state below $H < H_{c1}$ in the H-T plane. Whereas the theoretical limits for the critical current densities T_c in these materials are as high as 10^8 A/cm^2, in reality transport critical currents less than about 10^6 A/cm^2 have been obtained so far, at low temperatures.

5. ELECTRODYNAMIC RESPONSE WITH ANISOTROPIC GAP

The most essential properties of a super conductor which distinguish it from a normal conductor in a crucial way are the electrodynamic response of the material. The infinite d.c. conductivity, perfect diamagnetism below $H < H_{c1}$, the absence of high-frequency electromagnetic absorption for $\hbar\omega < 2\Delta$ at low temperatures, etc. are well known for the case of usual superconductors. All these properties can be explained in terms of the BCS pairing theory in which the quasi-particle excitations are determined by H_{BCS} of Eq.(15), in the superconducting state, with

$$E_{\vec{k}} = \left[(\varepsilon_{\vec{k}} - \mu)^2 + \Delta^2(\vec{k})\right]^{\frac{1}{2}}, \quad \varepsilon_{\vec{k}} - \mu \equiv \xi_{\vec{k}} \tag{46}$$

where $\xi_{\vec{k}}$ is the single-particle Bloch energy in the normal state as measured from the Fermi-Energy (chemical potential μ). Mattis and Bardeen [37] were the first to work out in detail the high-frequency electromagnetic response of a superconductor with an isotropic (in $\vec{k}$-space) energy-gap as well as $\varepsilon_{\vec{k}}$, with explicit expressions for the limiting case of the extreme anomalous limit for which the penetration of the field in the sample is small compared to the coherence length ξ. Later on Abrikosov [38] and Abrikosov and Gorkov [39] obtained explicit expressions for some more general limiting cases including the case in which the normal state meanfree path $\ell << \xi_o$. However, they also assumed the gap function to be isotropic in the $\vec{k}$-space. Inspite of the fact that in reality, in most cases the normal state is anisotropic with the possibility of a multiply - connected Fermi surface, and that all the known properties of the superconductors can not be described by the idealized model of an isotropic energy gap that does not depend on $\vec{k}$, with an isotropic singly-connected Fermi-surface [40], there are only a few realistic studies of electromagnetic response of a superconductor with anisotropic gap and $\xi_{\vec{k}}$. The available studies are usually based on a perturbative approach (the anisotropy in gap is small compared to the isotropic part) with small anisotropies, and limiting cases of $\vec{v}_F\cdot\vec{q}/2\Delta >>1$ and $<< 1$, where q is the wavevector of the electromagnetic wave [41,42]. These approximations are not valid in the case of new superconductors with large anisotropies in $\xi_{\vec{k}}$ and also in $\Delta(\vec{k})$.

As already emphasized in Sec.3, the properties of new class of high-T_c oxide superconductors are highly anisotropic. It would be extremely crude to assume $\xi_{\vec{k}}$ to be isotropic. Similarly, because of the possibility of even a zero-gap or at least only a small gap $\Delta(\vec{k}_t = o,\ k_z)$ in the direction perpendicular to the a-b plane, one must consider the electromagnetic response of a BCS superconductor with general $\xi_{\vec{k}}$ and $\Delta(\vec{k})$, before comparing experimental results with the BCS pairing. It is extremely misleading to find such comparisons in the literature, with theoretical expressions derived for the ideal case of an isotropic superconductor.

The general expression for the linear electromagnetic response function, the complex conductivity $\sigma_{\mu\nu}(\vec{q},\omega)$ or the dielectric function $\varepsilon_{\mu\nu}(\vec{q},\omega) = \delta_{\mu\nu} - (4\pi/i\omega)\sigma_{\mu\nu}(\vec{q},\omega)$, for the system described by the BCS Hamiltonian (15), with anisotropic $\xi_{\vec{k}}$ and $\Delta(\vec{k})$, can be obtained relatively easily if the normal state mean free pth $\ell \to \infty$ (pure material). For example, one can find an expression of this type in Ref.43, for the case of parabolic $\xi_{\vec{k}}$. For the more interesting case of an anisotropic superconductor, in which the normal state mean free-path ℓ is large compared to the coherence length but small compared to the penetration depth of e.m. wave, (1/q), an expression has been derived recently by us [44]. This is the most relevant expression to be used to analyse experimental results on single-crystals. One finds that the physical properties of an anisotropic superconductor are quite different than the usual results obtained for the case of an isotropic superconductor. For example, even in the limit of pure superconducting material, the infra-red absorption threshold is determined approximately by the expression

$$\omega^2 \geqq (\vec{q}.\vec{v}_F)^2 + 4\Delta^2(\hat{k})/\hbar^2 \qquad (47)$$

where for simplicity $\Delta(\vec{k})$ is assumed to depend only on the direction of $\hat{k}$. In other words, if the electromagnetic wave is incident in the z-direction, perpendicular to the a-b plane, and if $\Delta(\hat{k}_z)$ is small compared to $\Delta(\hat{k}_t)$, there will be a significant electromagnetic absorption below the maximum gap $2\Delta(\hat{k}_t)$, assuming qv_{Fz} is small. In general, the absorption structure and the infra-red reflectivity will have a complicated structure below the in-plane gap $2\Delta(\hat{k}_t)$, particularly if the role of the finite mean free path is also included in the analysis.

6. POSSIBLE PAIRING MECHANISMS AND CONCLUSION

In preceding sections, we have reviewed the general framework of the spin-singlet BCS pairing theory, indicating various approximate formalisms and equations which have to be considered and solved in different simplifying conditions. Of course, there are situations in which spin-triplet pairing becomes important, mostly due to the major role played by spin-fluctuations, e.g. in ^3He, but that kind of situation may not be important in the case of new oxide-superconductors. However, in these new materials what seems to be important is the significant anisotropy in the pairing gap $\Delta(\hat{k})$, so that the symmetry and structure of the unit cell in the plane of the layers and perpendicular to it have to be taken into account explicitly. In other words, it is no longer relevant to talk about pairing in the orbital-angular momentum space, e.g. the usual ℓ = o,s-wave pairing for an isotropic case. As explained earlier, in view of the large anisotropy in $\Delta(\hat{k})$, and its vanishing (or smallness) in certain directions, it is possible to obtain completely different behaviour of various physical quantities in these superconductors than in the conventional superconductors. For example, it is possible to obtain a power-law temperature-dependence in heat capacity, spin-susceptibility, etc., instead of the usual exponential behaviour $\exp(-\Delta/k_BT)$ in superconductors with an isotropic gap parameter Δ. In reality, there is always some anisotropy in the gap parameter in conventional superconductors also. However, in most of those materials, either because of the smallness of the mean free path ℓ as compared to the coherence length ξ (Anderson theorem, dity limit) or due the smallness of the field penetration depth compared to ξ (anomalous limit in electromagnetic experiments), the anisotropy in Δ becomes unimportant while interpreting the experimental results. This situation may not be valid in the new oxide - superconductors where $\xi << \ell << \lambda$, particularly if anisotropy in Δ is very large. The question whether the superconductivity in these materials is essentially 2-dimensional with weak Josephson - like coupling between superconducting Cu-O planes or it is 3-dimensional in nature with highly anisotropic $\Delta(\hat{k})$ is not yet completely settled. Nevertheless, it seems quite clear that for a proper understanding of this pehenomena, any theoretical approach must distinguish the motion of the carriers in the a-b plane from their motion perpendicular to it. Eventually, in the weak coupling case, it perhaps does not matter whether one uses the complete multi-band, reciprocal-space matrix formulation given by Eqs.(27) and (28) or the simplified coupled multi-layer formulation of Sec.3. Unfortunately, most of the discussions in the literature, proposing possible mechanisms for the new high-T_c superconductivity either uses a single-layer 2-dim picture or an isotropic 3-dimensional picture. Obviously, such a simplification can, at best, give only crude qualitative results. In the multi-layer approach, as already suggested, the polarizabilities of each layer can be obtained from the quantities

$$4\pi X_j (q_t,\omega) = \frac{2\pi e^2}{q_t} \cdot \frac{1}{(q_t L/2)} \cdot \pi_j(q_t,\omega) \tag{48}$$

which have the same structure in the $q_t - \omega$ space as the corresponding three-dimensional susceptibility $4\pi\ \chi(q,\omega)$. For example, for an insulating layer, $4\pi\ X_j(q_t,\omega)$ has poles at the transverse exciton frequencies. In general, all X_j will have poles at transverse phonon frequencies and at other transverse modes of the system, e.g., transverse anti-paramagnons etc. For a conducting plane $4\pi\ X_j(q_t,\omega) \to 1/q_t^2$ as $\omega \to o$, and behaves as $-\omega_{pj}^2/\omega^2$, for $\omega \to \infty$, $q_t \to o$, where ω_{pj} is the plasma frequency. As usual, the longitudinal modes, including the plasmons, are determined by the zeros of $1 + 4\pi\ X_j(q_t,\omega)$.

There have been many efforts [45] to formulate the BCS pairing theory involving different exchange-mechanisms to explain superconductivity in new oxides. However, it is quite fair to say that non of these calculations are complete or at the stage where they can be directly compared with experiments. Besides, there still exist great uncertainties in interpreting available experimental results. Among the various possibilities, the role of phonon-exchange mechanism (ionic excitations), exciton and plasmon-exchange mechanisms (electronic charge-density excitations) and spin-fluctuations (electronic spin-density excitations) have been considered seriously.

In one of the earliest calculations [46,47], Weber has argued that at least in 40K La-based oxides, superconductivity can be explained in terms of the usual phonon-exchange mechanism. Because of the light oxygen mass, the phonon-coupling constant λ_{ph} was shown to be greater than 2. Since in these compounds, a small isotope effect arising from the replacement of oxygen is observable, the phonon-mechanism may indeed be responsible for superconductivity, at least partially. For obtaining T_c in 90K range and beyond, it is not clear whether even a strong-coupling phonon-exchange mechanism is enough [12] for that purpose, due to the possibility of structural instability with large λ_{ph}. However, serious attempts have been made by many to go beyond the perturbation theory and invoke structural transition or Jahn-Teller effect to explain high-T_c [48] with the phonon-exchange mechanism. Some of these are being discussed in another chapter in this volume. The effect of multi-band BCS coupling [49] on various parameters like T_c and $2\Delta(o)/k_B T_c$ has been considered by Yamaji and Abe to show that it is possible to get very high values of these parameters, quite different from the simple one parameter BCS model.

Verma et al [50] have considered the exchange of pure electronic intra-cell charge-transfer excitons in Cu-O layers. They have argued that charge fluctuations between Cu^{++} and O^{--} sites are unscreened because of low carrier densities in these materials, although the direct Coulomb interaction among the carriers is screened. Superconductivity is shown to arise from the exchange of these charge-transfer excitons. In such a model, no exchange of magnetic fluctuations is involved. Because of the recent discovery [51] of 30K superconductivity in $Ba_{1-x}K_xBiO_3$, which contains Bi-O layers instead of the usual Cu-O layers, and where it is hard to imagine any role for magnetic fluctuations, Nunez Regueiro and Aligia [52] have considered these excitons in both CuO_2 and BiO_3, to substantiate the role of exciton-mechanism. Ruvalds, and Z. Kresin, have also considered the role of 2-dim plasmons [53] which may be well-defined even for large wavevectors $q_t \sim k_F$, in contrast to the 3-dim case where such plasmons are heavily damped. As one knows, attractive interaction for large wave vector transfers of the order of k_F is crucial for superconductivity.

At least in the case of La-based and Y-based copper-oxide compounds, antiferromagnetic spin-order has been observed close to the superconducting transition in the phase diagram. In $La_{2-x}M_xCuO_4$, there seems to be no superconductivity for x greater than 0.3 (with Maxm.T_c for $x = .15$, when $M = Sr$). But in the absence of doping, in La_2CuO_{4+y} one finds an anti-ferromegnetic ordering with Neél temperature $T_N \simeq 220K$. Similarly, for $YBa_2Cu_3O_{7-\delta}$, superconductivity is observed for $0 \leqq \delta \leqq 0.5$, but for $\delta > 0.5$ it undergoes antiferromagnetic ordering ($T_N \simeq 400K$ for $\delta \simeq 0.85$). In both these cases, the structure is orthorhombic ($a \neq b \neq c$) near the superconducting transition. Although, this may not be quite relevant in Bi-based and Tl-based compounds, proponents of the spin-fluctuation mechanisms argue that antiferromagnetic and superconducting phase transitions may be related. In fact, as already stated earlier, in the resonating Valence-bond (RVB) theory [24], one gives up the BCS pairing idea altogether to construct a theory based on only repulsive interactions. As summarized recently by Schrieffer et al. [54], it is important to decide whether the motion of electrons in these materials is best described by band states or by localized states. If the effective Hubbard on-site repulsive interaction U is large compared to the band width 8t for the nearest neighbour hopping in an equivalent square-lattice of Cu-sites, it may indeed be useful to start with the Hubbard model Hamiltonian theory. In such a case, starting with the 2-dim Hubbard model, it is shown that the charge and spin degrees of freedom of the holes created at Cu-sites due to doping of La_2CuO_4 by Sr, are decoupled. These give rise to two kinds of quasi-particles: holons (bosons of charge e) and spinons (neutral spin-½ fermions). The pairing condensation of holon gives superconductivity below T_c, but the normal specific heat (linear T-dependence) at low temperatures is due to spinons. Although, for large U/8t, the usual starting point from the side of Fermi-liquid theory is indeed not very good, it is not clear that in the materials under consideration U/8t is very different than 1. Also, in the two-dimensional RVB theory, the magnetic flux quantization is in units of hc/e, contrary to the experimental observation (units hc/2e, as in all other conventional superconductors). Moreover, there does not seem to be any real energy gap in the quasi-particle excitation spectrum in this theory. That in a fully 3-dim. RVB theory, these problems will be resolved, has still not been fully realized. In fact, it seems that holons repel each other in any given plane; the pairing condensate is indeed stablized by interplanar hopping. Even in the 2-dimensional case, it is not clear whether the RVB state is really a lower energy state compared to an antiferromagnetic phase. However, the role of spin-fluctuations has been considered in the conventional BCS pairing aproach also, notable by Emery [55], Scalapino [56] and Schrieffer [57]. Schrieffer has shown the possibility of obtaining the usual s-wave BCS pairing gap due to an effective attractive pairing potential between the holes in a so-called spin-bag created by anti-ferromagnetic fluctuations in the system. However, at present it is not at all clear whether spin-fluctuations have any major role to play in determining high T_c, particularly due to recent discoveries of superconductivity in many other kinds of oxides where spin-fluctuations may not be very important. Nevertheless, it should be pointed out here that even in the conventional band picture, in the presence of very large Coulomb correlation energy, the effective dynamic electron-electron interaction is known to become attractive [23] in the temperature region in which the magnetic susceptibility $\chi_m^{-1}(q,T)$ becomes small, before a possible magnetic ordering. If this happens for wavevectors $q \sim k_F$, superconductivity with very high-T_c is indeed possible.

In conclusion, it is fair to say that a lot of theoretical work still remains to be done to explore superconductivity in the new oxide materials, using more realistic models and mechanisms for the generalized BCS pairing. Even in the case of usual conventional superconductors, one must realize that it has not been easy to obtain T_c from a realistic first-principle calculation. With the added uncertainties of experimental data, which also must get resolved soon, and the complications due to the composition and structure of these new materials, together with the role of electronic-exchange mechanisms, strong coupling limit, anisotropy, multi-band formulation, etc., it is not surprising that we are still far away from the proper theoretical understanding of high-T_c superconductivity. Although, it may take a long time to establish the actual exchange-mechanisms, one should avoid the prevailing temptation to abandon the generalized BCS pairing idea, based on the misleading comparisons of existing experimental results with extremely simple and unrealistic models of the theory. Whereas some recent results based on the measurement of reflectivity and conductivity in $YaBa_2Cu_3O_7$ seem to show that [58] a strong-coupling pairing theory is enough to explain high T_c in such a material, it is not our intention to say that the issue is settled. Apart from very careful single crystal experiments to get better data, one must obtain theoretical expressions for various physical quantities in the superconducting state, with a more general anisotropic gap structure and normal state Fermi surface. It is only then that one can even consider a realistic formulation of the theory based on specific exchange-mechanisms in specific materials, and calculate their properties for comparison with experiments. At that stage only, we may be in a position to tell whether there is a need to go beyond the generalized BCS pairing theory to describe superconductivity in the new high-T_c materials.

REFERENCES

[1] B.T. Matthias and P.R. Stein, Physics of Modern Materials, Vol.2, Int. Atomic Energy Agency, Vienna, (1980) p.212.

[2] J.B. Bednorz and K.A. Müller, Z. Phys. B64, (1986) 189.

[3] R.J. Cava, R.B. vanDover, B. Batlogg and E.A. Rietman, Phys. Rev. Lett. 58, (1987) 408.

[4] M.K. Wu, J.R. Ashburn, C.J. Torng, P.H. Hor, R.L. Meng, L. Gao, Z.J. Huang, Y.Q. Wang, C.W. Chu, Phys. Rev. Lett. 58, (1987) 908.

[5] H. Maeda, Y. Tanaka, M. Fukotomi and T. Asano, Japn. J. Appl. Phys. 27, (1988) L 209; C.W. Chu, J. Bechtold, L. Gao, P.H. Hor, Z.J. Huang, R.L. Meng, Y.Y. Sun, Y.Q. Wang and Y.Y. Xue, Phys. Rev. Lett. 60, (1988) 941.

[6] Z.Z. Sheng, A.M. Herman, E. El Ali, C. Almasan, J.E. Estrada, T. Datta and R.J. Matsui, Phys. Rev. Lett. 60, (1988) 937.

[7] See, e.g., B. Raveau, C. Michel, M. Hervieu and J. Provost, Revs. Solid State Sc., Vol.2, Progress in High Temperature Superconductivity - Vol.16, World Scientific, Singapore, (1988) 115.

[8] V.L. Ginzburg and D.A. Kirzhnits, High Temperature Superconductivity, Consultants Bureau, N.Y., (1982).

[9] J. Bardeen, L.N. Cooper and J.R. Schrieffer, Phys. Rev. 108, (1957) 1175.

[10] L.P. Gorkov, Sov. Phys. JETP 7, (1958) 505.

[11] G.M. Eliashberg, Sov. Phys. JETP 11, (1960) 696; N.N. Bogoliubov, Sov. Phys. JETP 7, (1958) 41.

[12] F. Marsiglio, R. Akis and J.P. Carbotte, Phys. Rev. B36, (1987) 5245.

[13] W.L. McMillan, Phys. Rev. 167, (1968) 331.

[14] P.B. Allen and R.C. Dynes, Phys. Rev. B12, (1975) 905.

[15] W.A. Little, Phys. Rev. A134, (1964) 1416.

[16] V.L. Ginzburg, Phys. Lett. 13, (1964) 101.

[17] D. Allender, J. Bray and J. Bardeen, Phys. Rev. B7, (1973) 1020; ibid B8, (1973) 4433.

[18] S. Rangarajan, Phys. Rev. B10, (1974) 872.

[19] S. Rangarajan and S.S. Jha, Pramana - J. Phys. 6, (1976) 161.

[20] S. Srinivasan and S.S. Jha, Phys. Rev. B18, (1978) 2169.

[21] P. Bhattacharyya and S.S. Jha, J. Phys. C11, (1978) L805; S. Srinivasan, P. Bhattacharyya and S.S. Jha, Pramana - J. Phys. 13, (1979) 131.

[22] Y.A. Uspenskii, Sov. Phys. JETP 49, (1979) 822.

[23] R. Chaudhury and S.S. Jha, Pramana - J. Phys. 22, (1984) 431.

[24] P.W. Anderson, G. Baskaran, G. Zou and T. Hsu, Phys. Rev. Lett. 58, (1987) 2790; P.W. Anderson, Science 235, (1987) 1196.

[25] B.K. Chakravarty, Revs. Solid State Sc., Vol.2, Progress in High-Temp. Superconductivity - Vol.16, World Scientific, Singapore, (1988) 297.

[26] S.S. Jha, Pramana - J. Phys. 29, (1987) L615.

[27] S.S. Jha, Proc. International Bose Conference on Frontiers of Theoretical Physics (Eds. F.C. Auluck et al.), Indian Nat. Sc. Academy, New Delhi, (1978) p.453.

[28] A.B. Migdal, Sov. Phys. JETP 7, (1958) 996.

[29] Y. Nambu, Phys. Rev. 117, (1960) 648.

[30] J.M. Rowell, P.W. Anderson and D.E. Thomas, Phys. Rev. Lett. 10, (1963) 334.
[31] S.S. Jha, Revs. Solid State Sc., Vol.2, Progress in High-Temp Superconductivity - Vol.16, World Scientific, Singapore, (1988) 307.
[32] B. Batlogg, A.P. Ramirez, R.J. Cava, R.B. vanDover and E.A. Rietman, Phys. Rev. B35, (1987) 5340; M.V. Nevitt, G.W. Grabtree and T.E. Klippert, Phys. Rev. B36, (1987) 2398.
[33] R. Srinivasan and V. Sankaranarayanan, Revs. Solid State Sc., Vol.2, Progress in High - Temp Superconductivity - Vol.16, World Scientific, Singapore, (1987) 177.
[34] See, e.g., K.E. Gray, M.E. Hawley and E.R. Moog, Novel Mechanisms of Superconductivity, Plenum, N.Y., (1987) p.611.
[35] C.E. Gough, M.S. Colclough, E.M. Forgan, G.R. Jordan, M. Keene, C.M. Muirhead, J.M. Rae, N. Thomas, J.S. Abell and S. Sutton, Nature 326, (1987) 855.
[36] A.K. Grover, C. Radhakrishnamurthy, P. Chaddah, G. Ravikumar and G.V. Subbarao, Pramana-J Phys. 30, (1988) 569.
[37] D.C. Martis and J. Bardeen, Phys. Rev. 111, (1958) 412.
[38] A.A. Abrikosov, Sov. Phys. JETP 8, (1959) 182.
[39] A.A. Abrikosov and L.P. Gorkov, Sov. Phys. JETP 8, (1959) 1090.
[40] A.G. Shepelev, Sov. Phys. JETP 11, (1969) 690.
[41] V.L. Pokorovskii and M.S. Ryvkin, Sov. Phys. JETP 13, (1961) 1306.
[42] A.J. Bennet, Phys. Rev. 140, (1965) A1902.
[43] G. Rickayzen, Theory of Superconductivity, John Wiley, N.Y. (1965).
[44] S.S. Jha, to be published.
[45] See, e.g., Proc. Int. Conf. High Temp. Superconductors and Materials and Mechanisms of Superconductivity, Interlaken, Physica C 153-155, (1988).
[46] W. Weber, Phys. Rev. Lett.. 58, (1987) 1371.
[47] L.F. Mattheiss, Phys. Rev. Lett. 58, (1987) 1028.
[48] H. Kamimura, Jap. J. Appl. Phys. Suppl. 26-3, (1987) 1092.
[49] K. Yamaji and S. Abe, Physica C 153-155, (1988) 1209.
[50] C.M. Verma, S. Schmitt-Rink and E. Abrahams, Solid State Commun. 62, (1987) 681.
[51] L.F. Matheiss, E.M. Gyorgy and D.W. Johnson, Phys. Rev. B37, (1988) 3745.
[52] M.D. Nunez Regueiro and A.A. Aligia, Phys. Rev. Lett. 61, (1988) 1889.
[53] J. Ruvalds, Phys. Rev. B35, (1987) 8868; Z. Kresin, Phys. Rev. B35, (1987) 8716.
[54] J.R. Schrieffer, X.G. Wen and S.C. Zhang, Physica C 153-155, (1988) 21.
[55] V.J. Emery, Phys. Rev. Lett. 58, (1987) 2794.
[56] D.J. Scalapino and E. Loh, Phys. Rev. B35, (1987) 6694.
[57] J.R. Schrieffer, Phys. Rev. Lett. 60, (1988) 944.
[58] R.T. Collins, Z. Schlesinger, F. Holtzberg, P. Chaudhuri and C. Field, to be published.

BOSON EXCHANGE MECHANISMS, BOUNDS AND ASYMPTOTIC LIMITS

J.P. Carbotte and F. Marsiglio†

Physics Department, McMaster University, Hamilton, Ontario, Canada L8S 4M1

I - Introduction

BCS theory[1] predicts universal laws for the normalized thermodynamic and other properties of superconductors. The only parameter in the theory, which models the pairing interaction, can be fit to the measured size of the critical temperature and everything else follows. For example, the ratio of twice the gap edge Δ_o to the critical temperature is 3.54 while the specific heat jump at T_c ($\Delta C(T_c)$) normalized to the normal state value of the electronic specific heat $\gamma_o T_c$ (with γ_o the Sommerfeld constant) is 1.43. Other universal numbers also apply. In real materials, deviations from BCS laws are observed[2-6]. which carry information on details of the microscopic parameters involved that cannot be modeled well by a constant pairing potential approximation. Eliashberg[7] theory, which is an extension of BCS theory in which the full details of the electron-phonon interaction are properly treated, has been found to describe quantitatively the observed deviation from BCS in all conventional superconductors.[3]

The Eliashberg equations deal with a gap and renormalization function from which thermodynamic properties follow as well as the gap edge (Δ_o) at zero temperature. The kernels in these equations are an electron-phonon spectral density $\alpha^2F(\omega)$[8] which is a function of phonon energy and a Coulomb pseudopotential μ^*. The electron-phonon kernel contains details about the underlying normal state electronic structure, phonon spectrum and electron-phonon interaction and does not involve any approximations. Basically, it is a phonon frequency distribution in which each mode has been properly weighted by the strength of the electron-phonon interaction. To calculate this spectral density is complicated and requires an underlying band calculation in addition to a calculation[9-14] of the electron-phonon matrix elements and a knowledge of the phonon spectrum. In many cases, it can be measured directly from inversion of current (I) voltage (V) characteristics of tunnel junctions.[15] Once $\alpha^2F(\omega)$ is known the thermodynamics and all other properties, can be computed

† Now at the University of California at San Diego, La Jolla, California 92093, U.S.A.

from numerical solutions of the Eliashberg equations[3] with coulomb pseudopotential μ^* adjusted to get the measured T_c value.

The Eliashberg equations themselves[3,5–6] are two nonlinear coupled equations which represent the sum of all perturbation theory diagrams in the many body expansion except for the vertex corrections which are left out. This approximation is justified for the electron-phonon vertex because of Migdal's[16] theorem which states that such corrections are of order $\sqrt{\omega_D/E_F}$ as compared with 1; here ω_D is the Debye energy and E_F is the Fermi energy.

For other boson exchange mechanisms such as excitons, plasmons, paramagnons,Ref. 17 etc., the equivalent of Migdal's theorem has not yet been proven. Such a proof is beyond the scope of the present work. Here we will assume that the same Eliashberg equations can still be used, at least as a first approximation, as done by Allender et al.[18] for excitons. It can be hoped that vertex corrections, if important, can be partially included in effective kernels without modification of the form of the basic equations. Since the kernels will be left arbitrary in much of what follows, this is not a serious limitation. In any case, it is important to know the range of the thermodynamic properties which are possible within Eliashberg theory for arbitrary kernels. This is particularly relevant with the advent of superconductivity in the oxides[19–23] with T_c up to 125K. Should the properties of these superconductors fall outside the range of possible properties for boson exchange superconductivity this could be taken as strong evidence that a new theory is required or, at the very least, that important modifications to the conventional equations are required (modifications which go beyond the basic kernels). In a word, we are interested in this chapter in calculating the full range of predictions that is possible for thermodynamic properties within Eliashberg theory, irrespective of the shape, size, and origin of the kernels.

Besides this brief introduction, there are nine other sections involved in this chapter with the last being a short conclusion. In section II, we introduce the necessary formalism without derivation as the isotropic Eliashberg equations and associated free energy formula are well known. Section III is devoted to a discussion of conventional superconductors which can be described very well in terms of the Eliashberg equations with tunneling derived kernels. Approximate formulas for various dimensionless thermodynamic ratios are derived with adjustable parameters fit to give the observed overall trends for thermodynamic properties in a large number of phonon superconductors. In the process of establishing these useful formulas, it is shown that the strong coupling index $T_c/\omega_{\ln}$, with $\omega_{\ln}$[24] a well defined characteristic boson exchange energy, is very useful in defining qualitative trends. It is the simple most important parameter characteristic of the spectral density other than the mass renormalization λ. In section IV, we extend the work to the very strong coupling regime which is characterized by values of $T_c/\omega_{\ln}$ that fall beyond the conventional electron-phonon range of $0 \leq T_c/\omega_{\ln} \leq 0.25$. In as much as unconstrained values of the kernels are used, we are not limiting ourselves to phonons and so, our results should apply to any boson exchange mechanism treated within the Eliashberg theory. As the relationship between superconducting properties and the kernels in the Eliashberg equations is highly nonlinear, it is very useful in obtaining limited information on this relationship to calculate functional derivatives. These give the response of the property of interest to an infinitesimal change in the spectral density $\alpha^2F(\omega)$ at a particular frequency ω. Such functional derivatives are discussed in section V. They are used extensively in section VI where we discuss optimum spectra and establish definite limits for the thermodynamic properties of an Eliashberg superconductor which apply to any value of the kernels independent of their size, shape, or origin. In section VII, we establish asymptotic limits for the gap to critical temperature ratio which hold when $\lambda \to \infty$. This completes the formal part of the chapter which is followed by a short discussion (section VIII) specific to the new oxide superconductors. We review some of their properties and compare with the theory developed in the previous sections. It is argued that a combined phonon plus exciton (really any boson exchange mechanism of electronic origin) is indicated as a possibility at the present time. Such a combined mechanism is the subject of section IX which is followed by a brief conclusion (section X).

II - Formalism

The isotropic Eliashberg[7] equations which form the basis of the work presented here, were derived for an electron-phonon system and based on Migdal's[16] theorem which allows the neglect of vertex corrections. The resulting equations have been tested against experiment in conventional superconductors and found to be accurate at the few percent level.[3] Here, we will use these equations, as a first approximation, even for other boson exchange mechanisms such as excitons, plasmons, spin fluctuations, etc., even though, in these cases, the issue of vertex corrections has not been addressed to date. Such corrections go beyond the scope of the present work. While we cannot expect the Eliashberg equations to be as accurate for the more exotic boson exchange mechanisms compared with phonons, we can hope that they represent a first good approximation and that they capture the essential physics. Some of the complications inherent in these higher energy mechanisms can probably be absorbed in effective kernels with the form of the equations left unaltered. Since the kernels, which involve an electron-boson exchange spectral density $\alpha^2F(\omega)$ and Coulomb pseudopotential μ^* will be left arbitrary, this is not an important limitation.

The Eliashberg equations have been derived in many places.[2,5] In particular, details can be found in the recent review of Allen and Mitrović[25] to which the reader is referred. They deal with two non linear coupled equations. In Matsubara representation they are[3,6]

$$\Delta(i\omega_n)Z(i\omega_n) = \pi T \sum_{m=-\infty}^{\infty} [\lambda(i\omega_n - i\omega_m) - \mu^*(\omega_c)\theta(\omega_c - |\omega_m|)] \frac{\Delta(i\omega_m)}{\sqrt{\omega_m^2 + \Delta^2(i\omega_m)}} \tag{2.1}$$

and

$$Z(i\omega_n) = 1 + \frac{\pi T}{\omega_n} \sum_{m=-\infty}^{\infty} \lambda(i\omega_m - i\omega_n) \frac{\omega_m}{\sqrt{\omega_m^2 + \Delta^2(i\omega_m)}} \; . \tag{2.2}$$

The $\Delta(i\omega_n)$ are the Matsubara gaps defined at the Matsubara frequencies, $i\omega_n \equiv i\pi T(2n-1)$, $n = 0, \pm1, \pm2, \ldots$. Similarly, $Z(i\omega_n)$ are the renormalization factors. The electron-phonon spectral density, $\alpha^2F(\nu)$, appears through the relation,

$$\lambda(z) \equiv \int_0^\infty \frac{2\nu \, d\nu \, \alpha^2F(\nu)}{\nu^2 - z^2} \tag{2.3}$$

and $\mu^*(\omega_c)$ is the Coulomb pseudopotential with cutoff ω_c. The value of this cutoff is typically 5–10 times the maximum phonon frequency. Physically, the cutoff for the Coulomb interaction should be of order of the Fermi energy, E_F, but as Morel and

Anderson[26] first showed, this cutoff can be scaled down to ω_c, with a result that the potential is renormalized:

$$\mu^*(\omega_c) = \frac{\mu(E_F)}{1 + \mu(E_F)\ln(\frac{E_F}{\omega_c})} \,. \tag{2.4}$$

The above equations have been written in the isotropic limit, often referred to as the "dirty" limit in which all anisotropy effects are neglected[3]. Equation (2.1)-(2.3) are the Eliashberg equations written on the imaginary frequency axis. Most of the computations in this paper were done using this formulation. Note that all quantities involved are real, and simple summations are required for iteration. This will contrast remarkably with the "real axis" Eliashberg equations, about which we will have more to say later. Many quantities follow from the imaginary axis formulation and in particular the difference in free energy between superconducting (F^S) and normal (F^N) state. The Bardeen-Stephen[27] formula for this difference $\Delta F \equiv F^S - F^N$ is

$$\begin{aligned} \frac{\Delta F}{N(0)} = &- \pi T \sum_m \left[\sqrt{\omega_n^2 + \Delta^2(i\omega_n)} - |\omega_n|\right]\Big[Z^S(i\omega_n) \\ &- Z^N(i\omega_n)\frac{|\omega_n|}{\sqrt{w_n^2 + \Delta^2(i\omega_n)}}\Big] \end{aligned} \tag{2.5}$$

where $Z^N(i\omega_n)$ is the renormalization function in the normal state. It is given by equation (2.2) with $\Delta(i\omega_n)$ set equal to zero on the right hand side. Finally, $Z^S(i\omega_n)$ stands for $Z(i\omega_n)$ in the superconducting state which is given by equation (2.2) and $N(0)$ is the single spin electronic density of states at the Fermi surface.

From the free energy, we can calculate the specific heat difference between superconducting and normal state using the thermodynamic formula

$$\Delta C(T) = -\frac{T d^2 \Delta F}{dT^2} \tag{2.6}$$

and the critical magnetic field $H_c(T)$ at the temperature T from

$$H_c(T) = [-8\pi\Delta F]^{1/2} \tag{2.7}$$

In discussions of the temperature variation of the critical magnetic field it has become conventional to introduce a deviation function $D(t)$ defined by[2]

$$D(t) = \frac{H_c(T)}{H_c(0)} - (1 - t^2) \tag{2.8}$$

with $t \equiv T/T_c$, the reduced temperature.

For other applications, such as calculating the gap edge Δ_o in the quasiparticle excitation spectrum, it is necessary to go to the real axis formulation in which the gap and renormalization $\Delta(\omega)$ and $Z(\omega)$ become complex functions of a real frequency ω. The standard form of these equations is[28,2,5]

$$\begin{aligned} \Delta(\omega)Z(\omega) = &- \int_{-\infty}^{\infty} d\omega' \int_0^{\infty} d\nu\, \alpha^2 F(\nu) I(\omega + i\delta, \nu, \omega')\, \Re e \left(\frac{\Delta(\omega')}{\sqrt{\omega'^2 - \Delta^2(\omega')}}\right) \\ &- 2T\mu^*(\omega_c) \int_{-\infty}^{\infty} d\omega'\, \Re e \left(\frac{\Delta(\omega')}{\sqrt{\omega'^2 - \Delta^2(\omega')}}\right) \sum_{n=1}^{\infty} \frac{\omega'\theta(\omega_c - \omega_n)}{\omega'^2 + \omega_n^2} \end{aligned} \tag{2.9}$$

and

$$Z(\omega) = 1 - \frac{1}{\omega}\int_{-\infty}^{\infty} d\omega' \int_0^{\infty} d\nu\, \alpha^2 F(\nu) I(\omega + i\delta, \nu, \omega')\, \Re e\left(\frac{\Delta(\omega')}{\sqrt{\omega'^2 - \Delta^2(\omega')}}\right) \quad (2.10)$$

where

$$I(\omega + i\delta, \nu, \omega') = \frac{N(\nu) + 1 - f(\omega')}{\omega + i\delta - \nu - \omega'} + \frac{N(\nu) + f(\omega')}{\omega + i\delta + \nu - \omega'} \,. \quad (2.11)$$

It is to be understood that "ω" always has a small positive imaginary part ($i\delta$). The branch of the square roots must be chosen such that the real part has the same sign as ω. Note that we have adopted a model for the Coulomb self-energy which has a sharp cut-off on the imaginary axis, so that, following Leavens and Fenton,[29] we write the Coulomb part in Eq. (2.9) in a non-standard way. The functions in (2.11) are the standard Bose ($N(\nu)$) and Fermi ($f(\omega)$) functions. Note that gap and renormalization parameters, $\Delta(\omega)$ and $Z(\omega)$, respectively, are complex, and principal value integrals are required. This makes this method of solution somewhat cumbersome, numerically.

A new method for analytical continuation to the real frequency axis which avoids much of the above mentioned complication was recently given by Marsiglio et al.[30] Their final equations, which involve both real axis quantities $\Delta(\omega)$ and $Z(\omega)$ and Matsubara functions $\Delta_m \equiv \Delta(i\omega_m)$, are[30]

$$\begin{aligned}\Delta(\omega)Z(\omega) = \pi T \sum_{m=-\infty}^{\infty} [\lambda(\omega - i\omega_m) - \mu^*(\omega_c)\theta(\omega_c - |\omega_m|)] \frac{\Delta_m}{\sqrt{\omega_m^2 + \Delta_m^2}} \\ + i\pi \int_0^{\infty} d\nu\, \alpha^2 F(\nu) \Big\{ [N(\nu) + f(\nu - \omega)] \frac{\Delta(\omega - \nu)}{\sqrt{(\omega - \nu)^2 - \Delta^2(\omega - \nu)}} \\ + [N(\nu) + f(\nu + \omega)] \frac{\Delta(\omega + \nu)}{\sqrt{(\omega + \nu)^2 - \Delta^2(\omega + \nu)}} \Big\}\end{aligned} \quad (2.12)$$

and

$$\begin{aligned}Z(\omega) = 1 + \frac{i\pi T}{\omega} \sum_{m=-\infty}^{\infty} \lambda(\omega - i\omega_m) \frac{\omega_m}{\sqrt{\omega_m^2 + \Delta_m^2}} \\ + \frac{i\pi}{\omega} \int_0^{\infty} d\nu\, \alpha^2 F(\nu) \Big\{ [N(\nu) + f(\nu - \omega)] \frac{(\omega - \nu)}{\sqrt{(\omega - \nu)^2 - \Delta^2(\omega - \nu)}} \\ + [N(\nu) + f(\nu + \omega)] \frac{(\omega + \nu)}{\sqrt{(\omega + \nu)^2 - \Delta^2(\omega + \nu)}} \Big\} \,.\end{aligned} \quad (2.13)$$

Note that ω can now be interpreted as the real axis variable, "$\omega + i\delta$", or as being anywhere in the upper half of the complex plane. These equations will be used to obtain the real frequency gap $\Delta(\omega)$ and, in particular, the gap edge Δ_o at zero temperature ($T = 0$) which follows from the equation

$$\Re e\, \Delta(\omega = \Delta_o) = \Delta_o \,. \quad (2.14)$$

Before leaving these equations, we note that on setting $\omega = i\omega_n$ in equations (2.12) and (2.13), we recover our Matsubara representation equations (2.1) and (2.2) because $f(\nu - i\omega_n) = -N(\nu)$.

All the numerical results and analytic expression to be given in the following sections follow from the equations given above. Our main interest will be in values of the gap edge and of the thermodynamics for arbitrary values of the kernels $\alpha^2F(\omega)$ and μ^* which define the normal state parameters characteristic of a particular material.

III - Conventional Superconductors

The most compelling evidence in favour of the Eliashberg theory of superconductivity is perhaps the inversion accomplished through the measurement of the current-voltage (I-V) characteristic in a tunnelling junction.[15] The idea is the following. A trial $\alpha^2F(\omega)$ is assumed, and the quasiparticle density of states $N(\omega)/N(0)$ calculated. This requires a knowledge of the real frequency solution of the Eliashberg equations, namely (2.9) and (2.10) of the previous section. It is given by[15]

$$\frac{N(\omega)}{N(0)} = \Re e\left\{\frac{\omega}{\sqrt{\omega^2 - \Delta^2(\omega)}}\right\} \tag{3.1}$$

where $N(0)$ is, as before, the electronic density of states at the Fermi energy. In general, the calculated quasi-particle density of states will disagree with that measured in the I-V. Hence, $\alpha^2F(\omega)$ is adjusted in an iterative fashion until agreement is reached. This procedure is usually a very reliable determination of the microscopic function, $\alpha^2F(\omega)$. The Coulomb pseudopotential μ^* is then obtained by some requirement, for instance, that the measured gap edge is obtained theoretically. In this fashion the microscopic parameters from which all superconducting properties can be calculated are known. This procedure may seem circular, except that the experimentally measured (I-V) characteristic is used only up to a frequency which is the gap plus the maximum phonon frequency ($\omega_{\max}$) in the spectral function. However, once $\alpha^2F(\omega)$ has been determined in this way, $N(\omega)/N(0)$ can be calculated beyond $\omega_{\max}$ and compared with experiment. The agreement for Pb, for instance, is excellent.[15] That it is phonons which are mediating the interaction is supported upon comparison of $\alpha^2F(\omega)$ with measured phonon density of states obtained through neutron scattering experiments. We should also add that theoretical calculations[9-14] of $\alpha^2F(\omega)$ are possible, and these generally agree well with the measured functions, where comparison is possible.

The tunneling information on $\alpha^2F(\omega)$ which we will use in this section is summarized in appendix A. In all cases, we have carried out full numerical solutions of the Eliashberg equations to determine the thermodynamics and the gap edges for these materials. Our results can be found in tabular form in the thesis by Marsiglio.[31] We will have occasion to present them graphically in this section.

As a guide to the detailed numerical results obtained, we begin with approximate but analytical solutions to the full Eliashberg equations (2.1) and (2.2) and associated free energy formula (2.5). Following Allen and Dynes,[24] we define a single characteristic boson energy for a given spectral density $\alpha^2F(\omega)$ by

$$\omega_{\ln} = exp\left\{\frac{2}{\lambda}\int_0^\infty \frac{\alpha^2F(\omega)}{\omega}\ln(\omega)d\omega\right\}. \tag{3.2}$$

This quantity plays an important role in our final formulas. First, we ignore μ^* from the start. (It can be kept, as in Ref. 32, and retained as an extra parameter; however the benefits are minimal, at the cost of an extra parameter. Note, moreover, that in the strong coupling parameter, $T_c/\omega_{\ln}$ which will enter our final formulae, the T_c is to be regarded as coming from experiment. It contains, therefore, some effect of μ^* already.) The model we use is a step-function approximation on the imaginary axis:[32]

$$\Delta(\omega_n) = \begin{cases} \Delta_o(T) & |\omega_n| < \omega_0 \\ 0 & |\omega_n| > \omega_0 \end{cases} \tag{3.3a}$$

$$Z(\omega_n) = \begin{cases} Z_0(T) & |\omega_n| < \omega_0 \\ 1 & |\omega_n| > \omega_0 \end{cases} \ . \tag{3.3b}$$

Here, ω_0 represents roughly a few times the maximum phonon frequency in the system. Note that, for self-consistency we would require $\lambda(i\omega_n - i\omega_m)$ to be independent of n. This would reduce to the so called $\lambda^{\theta\theta}$ approximation[25] and lead to BCS theory. Instead, we evaluate Eqs. (2.1) and (2.2) at $n = 1$. Thus, Δ_1 is the constant gap in Eq. (3.3a). The procedure is outlined in the thesis of F. Marsiglio[31]. Essential to the approximations used is the requirement, $T_c/\omega_{\ln} \ll 1$. We have also assumed that ω_0 is sufficiently large that $\omega_{\ln}/\omega_0 \ll 1$. Expansions near T_c and at $T = 0$ are required.

Results for the specific heat ($\Delta C(T)$) near T_c are

$$\frac{\Delta C(T)}{\gamma_0 T_c} = f - (1-t)g \ , \tag{3.4}$$

with γ_0 the normal state electronic specific heat Sommerfeld constant and $t = T/T_c$ the reduced temperature. Here

$$f = \frac{\Delta C(T_c)}{\gamma_o T_c} = 1.43\left[1 + 53\left(\frac{T_c}{\omega_{\ln}}\right)^2 \ln\left(\frac{\omega_{\ln}}{3T_c}\right)\right] \tag{3.5}$$

and

$$g = 3.77\left[1 + 117\left(\frac{T_c}{\omega_{\ln}}\right)^2 \ln\left(\frac{\omega_{\ln}}{2.9T_c}\right)\right] \ . \tag{3.6}$$

In the limit $T_c/\omega_{\ln} \to 0$ we recover the famous BCS numbers 1.43 and 3.77 for f and g respectively. The form of these expressions has been derived and is similar to that used previously.[33–35] In addition, we have specified an average phonon frequency, $\omega_{\ln}$, and fitted coefficients to numerical data. The derived expressions are plotted in Fig. (3.1) and (3.2) (solid line) along with the numerical data (solid dots). Experimental data have been omitted, although the Eliashberg theory results are generally accurate to within 10%. The origin of the spectral functions used in these calculations is described in Appendix A where references are given. Note that there is some scatter, especially amongst the A15 compounds and Hg. In the case of Hg, there is a very low frequency peak in the $\alpha^2F(\nu)$ spectrum, so that the assumption $\nu \gg T_c$ for all important ν has broken down. Also note that the results from amorphous materials have not been plotted. They are not well described by these formulas at all. Their values for $T_c/\omega_{\ln}$ are near 0.3 . This value is beyond the limit of validity of formulas (3.5) and (3.6) (and others to be presented later). More importantly however, the spectral shapes of $\alpha^2F(\nu)$ for these materials is such that they are not well described

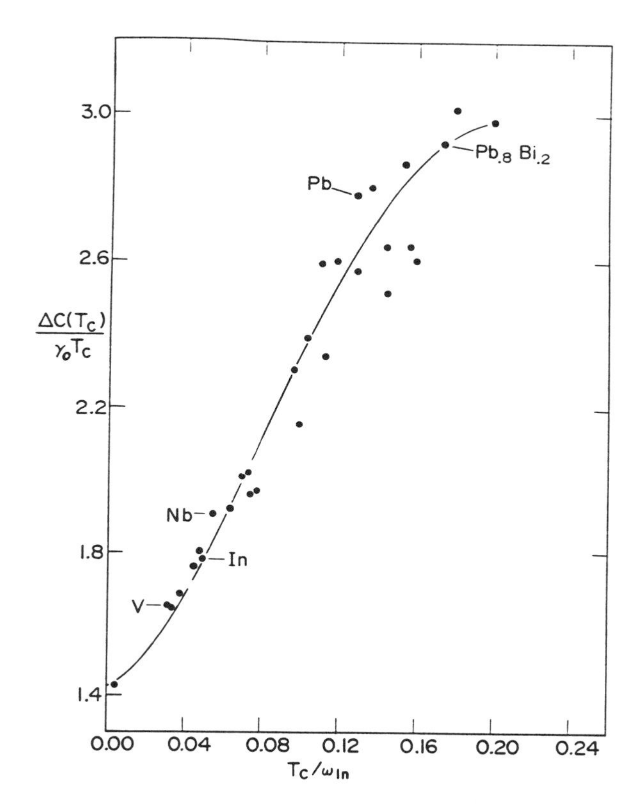

Figure 3.1 Specific heat jump ratio $f \equiv \Delta C(T_c)/\gamma_o T_c$ vs. $T_c/\omega_{\ln}$. The dots represent the accurate results from the full numerical solutions of the Eliashberg equations. Experiment tends to agree to within 10%. In increasing order of $T_c/\omega_{\ln}$, the dots correspond to the following systems: Al, V, Ta, Sn, Tl, $Tl_{0.9}Bi_{0.1}$, In, Nb (Butler), Nb (Arnold), V_3Si_1, V_3Si (Kihl.), Nb (Rowell), Mo, $Pb_{0.4}Tl_{0.6}$, La, V_3Ga, Nb_3Al (2), Nb_3Ge (2), $Pb_{0.6}Tl_{0.4}$, Pb, Nb_3Al (3), $Pb_{0.8}Tl_{0.2}$, Hg, Nb_3Sn, $Pb_{0.9}Bi_{0.1}$, Nb_3Al (1), Nb_3Ge (1), $Pb_{0.8}Bi_{0.2}$, $Pb_{0.7}Bi_{0.3}$, and $Pb_{0.65}Bi_{0.35}$. The drawn curve corresponds to $f = 1.43(1 + 53(T_c/\omega_{\ln})^2 \ln(\omega_{\ln}/3T_c))$.

by an Einstein spectrum with frequency $\nu_E = \omega_{\ln}$ of the material. The opposite tends to be true for the crystalline materials. Coombes and Carbotte[49,50] have analysed shape dependence in detail. Some of the scatter is also due to variations in μ^*. In any event, the point of these expressions is to describe the general trend of superconducting properties as a function of strong coupling. The result is a continual increase as $T_c/\omega_{\ln}$ varies from 0 (BCS) to 0.20 . Note that there are signs of saturation, and in fact, later it will be seen that as the coupling is increased further, $\Delta C(T_c)/\gamma_o T_c$ will decrease to values below 1.43 . The physical reason behind the increase in $\Delta C(T_c)/\gamma_o T_c$ can be traced to the gap opening up more rapidly just below T_c as the coupling strength is increased. The specific heat jump, which is a measure of steepness of the ascent of the gap, will increase as well.

A calculation of other thermodynamic properties requires a knowledge of strong coupling corrections at zero temperature. The procedure is similar to that near T_c,

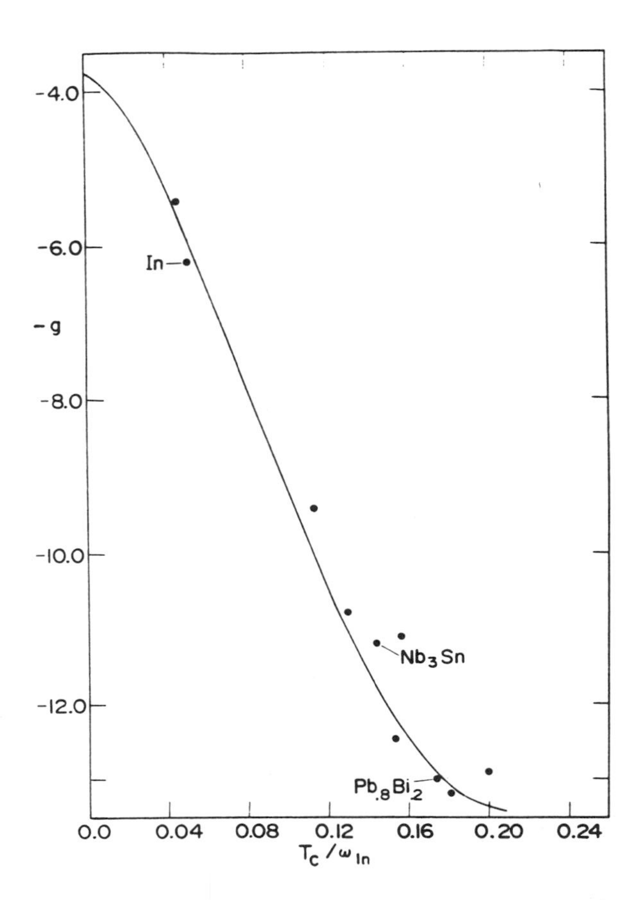

Figure 3.2 Plot of g (see Eq. (2.29) in text) vs. $T_c/\omega_{\ln}$ for a selected number of systems. Dots correspond to the results extracted from numerical solutions for $D(t) \equiv H_c(T)/H_c(0) - (1-t^2)$ vs. t, using the Eliashberg equations. In increasing order of $T_c/\omega_{\ln}$, the dots correspond to Tl, In, Nb_3Al (2), Nb_3Al (3), Nb_3Sn, $Pb_{0.9}Bi_{0.1}$, Nb_3Al (1), $Pb_{0.8}Bi_{0.2}$, $Pb_{0.7}Bi_{0.3}$, and $Pb_{0.65}Bi_{0.35}$. The drawn curve corresponds to $-g = -3.77\{1 + 117(T_c/\omega_{\ln})^2 \ln(\omega_{\ln}/2.9T_c)\}$. The fit is remarkably good, considering the constraints on the coefficients.

and is also outlined in the thesis of F. Marsiglio[31]. The result for the gap ratio is readily obtained:

$$\frac{2\Delta_o}{k_B T_c} = 3.53\left[1 + 12.5\left(\frac{T_c}{\omega_{\ln}}\right)^2 \ln\left(\frac{\omega_{\ln}}{2T_c}\right)\right] . \tag{3.7}$$

This result was first obtained by Mitrović *et al*[51] using the real axis equations. We have used the same fit as Mitrović *et al.*[51] This result is plotted in Fig. (3.3) (solid line), along with numerical data (solid dots). Note that data from more A15 compounds have been included, and has increased the amount of scatter slightly. However, overall, the fit is very good and the trend is well described by Eq. (3.7). The shape dependence of $2\Delta_o/k_B T_c$ has been studied by Coombes and Carbotte[49] and more recently in

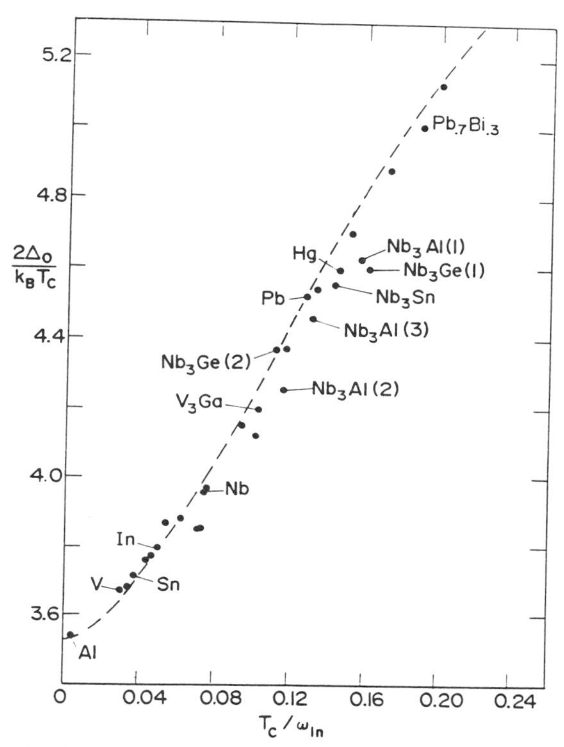

Figure 3.3 The ratio $2\Delta_o/k_BT_c$ vs. $T_c/\omega_{\ln}$. Most of the points have been reproduced from Ref. 51. We have also included some numerical solutions for A15 compounds. See Fig. 3.1 for identification of materials. The curve corresponds to $2\Delta_o/k_BT_c = 3.53\left[1 + 12.5(T_c/\omega_{\ln})^2 \ln(\omega_{\ln}/2T_c)\right]$.

Ref. 52 . In Ref. 50 it was noted that shape dependence was much more prominent for $\Delta C/\gamma_o T_c$ than for $2\Delta_o/k_BT_c$. This conclusion is supported by the increased amount of scatter in Fig. (3.1) relative to Fig. (3.3). Also note that the trend shows no sign of saturation, as was the case for $\Delta C/\gamma_o T_c$. It will be seen later, in fact, that $2\Delta_o/k_BT_c$ saturates only in the limit of $T_c/\omega_{\ln} \to \infty$.[53] We should also add that amorphous compounds have once again been excluded, as they are not well described by Eq. 3.7 for the same reasons given earlier. The enhancement of $2\Delta_o/k_BT_c$ with increased coupling can be understood by the following simple argument: as the coupling strength increases both T_c and Δ_o increase. However, the detrimental effect of thermal phonons is also felt more strongly by T_c, whereas Δ_o is unaffected since it is a zero temperature property. The result is a larger increase of Δ_o compared to T_c[5].

IV - Very Strong Coupling Regime

In the previous section, we have seen that, for the select properties considered, $T_c/\omega_{\ln}$ is a good first characterization of the electron boson spectral density and that it is meaningful to plot properties against $T_c/\omega_{\ln}$. The calculations were limited,

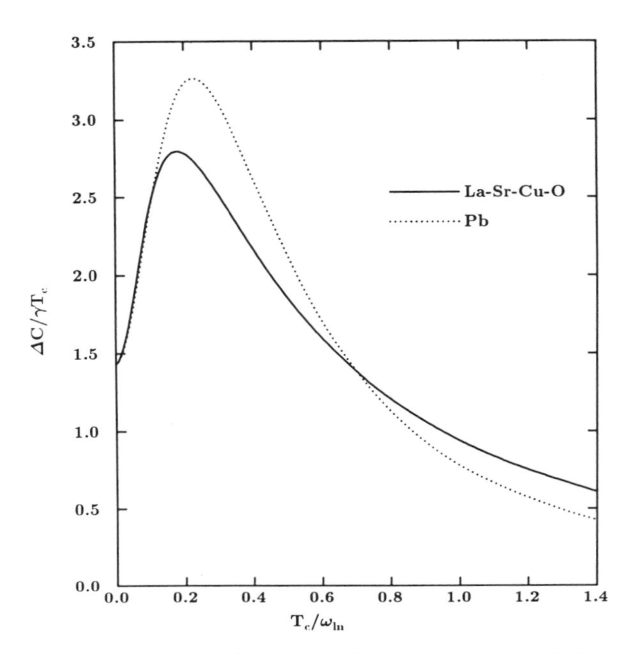

Figure 4.1 Plot of $\Delta C(T_c)/\gamma_o T_c$ vs. $T_c/\omega_{\ln}$. As $T_c/\omega_{\ln}$ increases beyond the conventional regime ($T_c/\omega_{\ln} \sim 0.25$), the normalized jump decreases to values lower than the BCS value.

however, to the conventional electron-phonon regime for which $T_c/\omega_{\ln} \leq 0.25$. With the discovery of superconductivity in the oxides[19–23] well beyond 100K, it is of some interest to extend the calculations of the previous section to the case when T_c is of the same magnitude or even bigger than the characteristic boson energy $\omega_{\ln}$. Here we will refer to this regime as the very strong coupling region[4] so as to differentiate it from the conventional region.

Since it is not practical to carry out numerical calculations for many electron-boson spectral shapes, it is necessary to make a definite choice for this function. To make sure that the results obtained are not qualitatively dependent on the exact shape of spectral density we will use, in what follows, two quite different spectra. Since W. Weber[14] has calculated an electron-phonon spectral density $\alpha^2F(\omega)$ for $La_{1.85}Sr_{0.15}CuO_4$(LSCO) it seems reasonable to use this function as a basis for the oxides, although we are not by this choice really wanting to commit ourselves to a phonon mechanism and other choices could have been made. To make sure the results are not qualitatively dependent on this choice, we will also use the $\alpha^2F(\omega)$ for Pb in a second calculation which serves as a comparison.

In terms of the model spectral densities $\alpha_M^2F(\omega)$ for Pb and LSCO, the kernel used in this work is

$$\alpha^2F(\omega) = B\alpha_M^2F(b\omega) \tag{4.1}$$

with B and b scaling factors to be varied at will. From the definition of $\omega_{\ln}$ (formula

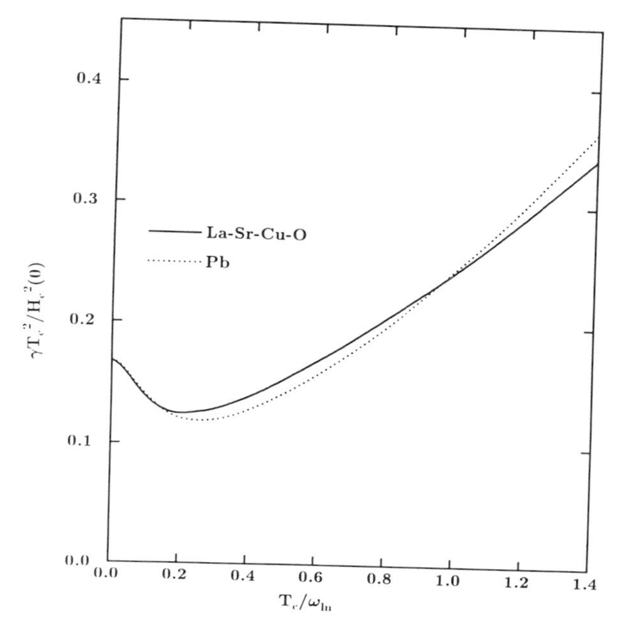

Figure 4.2 Plot of $\gamma_o T_c^2/H_c^2(0)$ vs. $T_c/\omega_{\ln}$ in very strong coupling regime. The trend has reversed, as it did for the jump, and values above that of BCS are found in this regime.

(3.2)), it is easily verified that the scaling B drops out of this quantity and that

$$\omega_{\ln} = \frac{1}{b}\omega_{\ln}^{M} \tag{4.2}$$

where $\omega_{\ln}^{M}$ is characteristic of our model spectral density and is 13.9 meV for LSCO and 4.8 meV for Pb. Any desired value of $\omega_{\ln}$ can be achieved through an appropriate choice of b. Further, for a fixed $\omega_{\ln}$, any value of T_c and hence of $T_c/\omega_{\ln}$ can be achieved by exploiting the freedom provided by the second scaling factor B. To be definite, we will choose $T_c = 96$K but this is not an important restriction because T_c scales with b in exactly the same way as $\omega_{\ln}$. Consequently, the ratio $T_c/\omega_{\ln}$ is fixed for systems in which μ^* is not an important player. This can be verified upon substitution of (4.1) into equations (2.1) and (2.2). With simple algebra, these equations can be rewritten as[49,50,54]

$$\bar{\Delta}(i\bar{\omega}_n)Z(i\bar{\omega}_n) = \pi\bar{T}\sum_m\left[B\lambda^M(i\bar{\omega}_n - i\bar{\omega}_m) - \mu^*\theta\left(\omega_c - \frac{|\bar{\omega}_m|}{b}\right)\right]\frac{\bar{\Delta}(i\bar{\omega}_m)}{\sqrt{\bar{\omega}_m^2 + \bar{\Delta}^2(i\bar{\omega}_m)}} \tag{4.3}$$

and

$$Z(i\bar{\omega}_n) = 1 + \frac{\pi\bar{T}}{\bar{\omega}_n}\sum_m B\lambda^M(i\bar{\omega}_n - i\bar{\omega}_m)\frac{\bar{\omega}_m}{\sqrt{\bar{\omega}_m^2 + \bar{\Delta}^2(i\bar{\omega}_m)}} \tag{4.4}$$

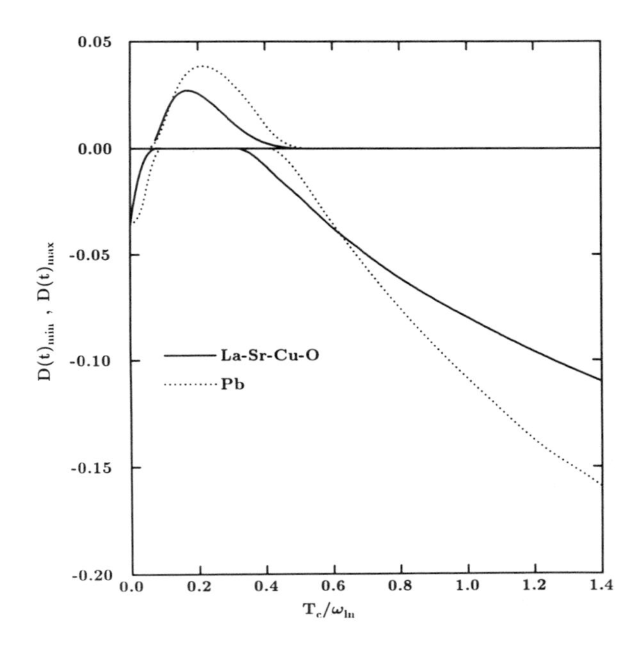

Figure 4.3 Plot of the maximum or minimum (or both when it is S-shaped) of the critical magnetic field deviation function vs. $T_c/\omega_{\ln}$. In the very strong coupling regime the curve becomes negative definite with minimum values which exceed the BCS value in absolute terms.

in which $\bar{\Delta} = b\Delta$, $\bar{\omega} = b\omega$ and $\bar{T} = bT$. The factor b has dropped out of these equations provided it is ignored in the cut off in the Matsubara sum in (4.3). We will make this approximation. If we take $B = 1$, equations (4.3) and (4.4) apply to a superconductor defined by the microscopic parameters $\alpha_M^2 F(\omega)$ and μ^*. Denoting these solutions by $\Delta_M(i\bar{\omega}_n)$ and $Z_M(i\bar{\omega}_n)$, we conclude that the solutions for the scaled spectrum are related to these by

$$\Delta(i\omega_n) = \frac{1}{b}\Delta_M(i\bar{\omega}_n) \tag{4.5}$$

$$Z(i\omega_n) = Z_M(i\bar{\omega}_n)\ . \tag{4.6}$$

It is clear then that, $T_c^M = bT_c$ and so $T_c/\omega_{\ln} = T_c^M/\omega_{\ln}^M$ since each quantity in this ratio has been shifted by $1/b$ compared to the case $b = 1$. Reference to the free energy formula (2.5) gives a relationship between the free energy at temperature T of the scaled system ($\Delta F(T)$) with that for the model system ΔF^M, namely[49,50,54]

$$\Delta F(T) = \frac{1}{b^2}\Delta F^M(\bar{T}) = \frac{1}{b^2}\Delta F^M(bT)\ . \tag{4.7}$$

The above results imply that the scaling b does not affect ratios such as $T_c/\omega_{\ln}$, $2\Delta_\circ/k_BT_c$, $\Delta C(T_c)/\gamma_\circ T_c$, $\gamma_\circ T_c^2/H_c^2(0)$ and $D(t)$.

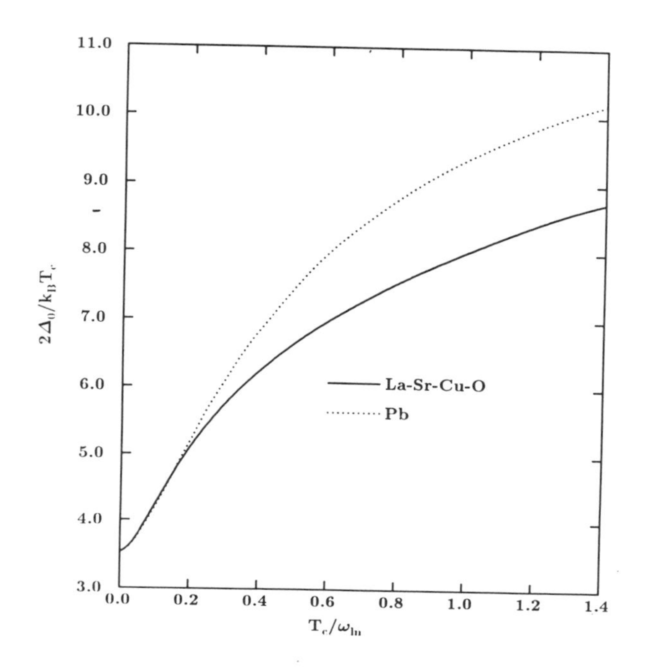

Figure 4.4 The gap ratio, $2\Delta_o/k_BT_c$ vs. $T_c/\omega_{\rm ln}$ for two spectral function shapes, that of Pb and LSCO. Note that in the very strong coupling regime, $2\Delta_o/k_BT_c$ continues to rise above the BCS value and that the shape dependence is more pronounced in this regime; however, the qualitative feature of increasing $2\Delta_o/k_BT_c$ seems to be shape-independent.

From this point onward, the calculations need to proceed numerically and results are presented in Figs. 4.1 to 4.4 . In Fig. 4.1, we show the normalized specific heat jump at T_c, $\Delta C(T_c)/\gamma_o T_c$ as a function of the strong coupling ratio $T_c/\omega_{\rm ln}$. The solid curve is based on a spectral density for the boson exchange modeled after LSCO while the dotted curve is modeled on Pb and is included for comparison. We see that both curves show the same qualitative trend, although there are quantitative differences particularly in the size of the maximum, it does occur at nearly the same value of $T_c/\omega_{\rm ln}$ in both cases. Both curves start at the well known result of 1.43 in the BCS limit $T_c/\omega_{\rm ln} \to 0$ and then increase rapidly through the conventional strong coupling regime which ends around the maximum for $T_c/\omega_{\rm ln} \sim 0.25$. The curves then drop to lower values and can fall below BCS. For T_c equal to $\omega_{\rm ln}$, $\Delta C(T_c)/\gamma_o T_c$ has already fallen below 1.0. These results were unexpected and can be taken to represent a clear signature of the very strong coupling regime.

Besides the normalized specific-heat jump at T_c, the dimensionless ratio $\gamma_o T_c^2/H_c^2(0)$ is also often discussed. Here $H_c(0)$ is the zero-temperature thermodynamic critical field. Results are shown in Fig. 4.2 . The solid line applies, as before, to the La-Sr-Cu- O base spectrum, while the dotted line is for a Pb base. In this case, the results of the two models are closer than found for the specific-heat jump. As $T_c/\omega_{\rm ln} \to 0$ the value of $\gamma_o T_c^2/H_c^2(0) \to 0.168$ (the BCS value). As $T_c/\omega_{\rm ln}$ increases the ratio decreases, has a minimum near $T_c/\omega_{\rm ln} \approx 0.2$ to 0.3, and then starts rising towards values that can be as large as 0.35 for $T_c/\omega_{\rm ln}$=1.4. Again we note that all the

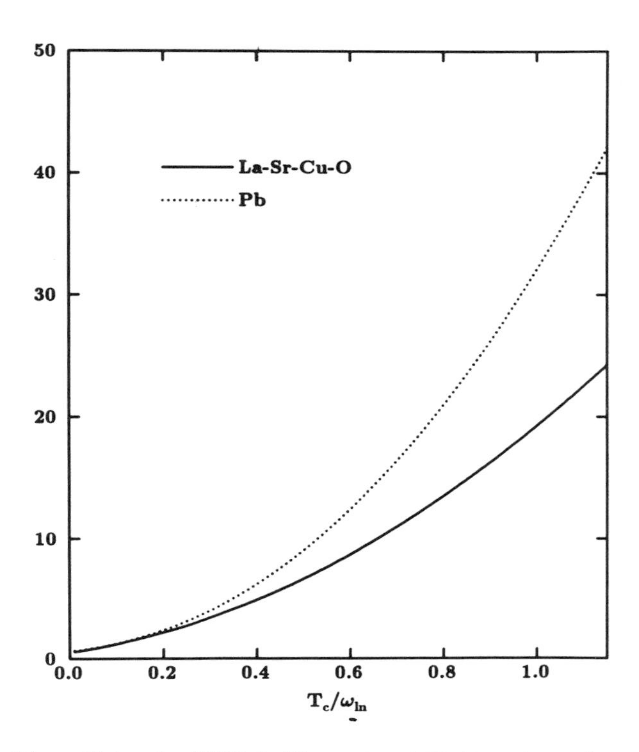

Figure 4.5 Plot of λ vs. $T_c/\omega_{\ln}$ for LSCO (solid curve) and Pb (dotted curve) shaped spectra, with $\mu^* = 0.1$.

conventional superconductors discussed in the previous section fall in the small region before the minimum. We stress that in the very strong coupling regime, Eliashberg theory predicts a behaviour for this ratio which is quite different from what is found in the conventional case. This should serve as a clear signature of a large electron-boson parameter $T_c/\omega_{\ln}$.

In addition to the zero temperature critical magnetic field, we can consider its finite temperature counterpart described by the deviation function $D(t)$, given by formula (2.8), with $t = T/T_c$. In Fig. 4.3, we show results for the maximum or minimum (or both) value of this function versus $T_c/\omega_{\ln}$. In BCS $D(t)$ is negative definite with a minimum of -0.036 achieved around the middle range for the reduced temperature $t = T/T_c$. As the coupling is increased, the curve takes on an S shape and exhibits both minimum and maximum. At yet higher values of $T_c/\omega_{\ln}$, a positive definite curve can result, but as $T_c/\omega_{\ln}$ is increased beyond the conventional strong coupling regime, the maximum in $D(t)$ peaks and then begins to drop. Eventually both a maximum and a minimum are present [S shape curve for $D(t)$] and finally there is only a minimum with $D(t)$ negative definite. Note that in the very-strong-coupling limit, the minimum of $D(t)$ can be very much smaller than the BCS value of -0.036. This is again a definite prediction that could help confirm or rule out the very strong coupling regime as the mechanism for superconductivity in the high-T_c oxides.

In Fig. 4.4, we turn to our results for the ratio of the energy gap Δ_o to T_c. To calculate this quantity from our imaginary- frequency Eliashberg solutions we need to perform an analytic continuation to real frequencies as described in section II. As reviewed by Scalapino[5] inelastic phonon scattering leads to quasiparticle damping which should reduce both Δ_o and T_c. However, the reduction in T_c should be larger because of thermal phonons and so $2\Delta_o/k_BT_c$ is larger than in BCS. Our results show that $2\Delta_o/k_BT_c$ simply keeps rising as T_c/ω_{ln} increases. This was also observed in the work of Carbotte, Marsiglio, and Mitrović.[54] These authors were the first to point out that a maximum value exists for this ratio in Eliashberg theory. Similar curves to our Pb results are to be found in our reference (52) which was directed mainly towards $La_{1.85}Sr_{0.15}CuO_4$ so that the range of values for T_c/ω_{ln} considered was much more restricted than that considered here.

The results of Fig. 4.4 are in striking contrast to those of Figs. 4.1 to 4.3 . In the very-strong-coupling regime Eliashberg theory predicts a large $2\Delta_o/k_BT_c$ of the order of 10 or more while at the same time $\Delta C(T_c)/\gamma(0)T_c$ is much smaller than the BCS value of 1.43 and $\gamma(0)T_c^2/H_c^2(0)$ much larger than 0.168. Also, $D(t)$ is negative definite with a minimum value much less than -0.036. These predictions are very different from BCS theory and also from the pattern of behaviour predicted and observed in conventional strong coupling systems. They can be taken to be a clear signature of the very strong coupling regime for boson exchange mechanisms.

Finally, it is of interest to know the value of λ that is implied for a given value of T_c/ω_{ln}. This is not a unique relationship as the results depend on the shape of the spectral density used. Results for LSCO and for Pb base spectra are shown in Fig. 4.5 where λ is plotted against T_c/ω_{ln}. It is clear that in both cases λ increases steadily with increasing T_c/ω_{ln} and that for $T_c/\omega_{\text{ln}} \cong 1.0$, λ can get quite large. Certainly, such values are an order of magnitude larger than the ones observed in conventional cases. It is not at all clear that such large values are consistent with lattice stability and this can be taken to be an argument against the very strong coupling regime.

V - Functional Derivatives

The relationship between a given superconducting property such as the free energy difference between superconducting and normal state and microscopic parameters is nonlinear and complicated. In such cases, functional derivative techniques can be very useful in obtaining limited information on this relationship. The question that is asked and answered is how does the property Q response to an infinitesimal change in the electron-boson spectral density $\alpha^2F(\omega)$ at a particular frequency Ω. More specifically [55–65]

$$\frac{\delta Q}{\delta\alpha^2F(\Omega)} = \lim_{\epsilon\to 0}\frac{Q[\alpha^2F(\omega)+\epsilon\delta(\omega-\Omega)]-Q[\alpha^2F(\omega)]}{\epsilon}\,. \tag{5.1}$$

This gives information on how a particular boson frequency Ω affects Q. Many interesting questions can be answered from such quantities as we will see later.

For the remainder of this section, it is convenient to rewrite the Eliashberg equations (2.1) and (2.2) in a slightly different form. We introduce $\tilde{\omega}(i\omega_n) = \omega_n Z(i\omega_n)$ and $\tilde{\Delta}_n(i\omega_n) = \Delta(i\omega_n)\tilde{\omega}(i\omega_n)/\omega_n$. The equations are now

$$\tilde{\Delta}(i\omega_n) = \pi T\sum_m[\lambda(i\omega_n - i\omega_n) - \mu^*]\frac{\tilde{\Delta}(i\omega_m)}{\sqrt{\tilde{\omega}^2(i\omega_m)+\tilde{\Delta}^2(i\omega_m)}} \tag{5.2}$$

and

$$\tilde{\omega}(i\omega_n) = \omega_n + \pi T \sum_m \lambda(i\omega_n - i\omega_m) \frac{\tilde{\omega}(i\omega_m)}{\sqrt{\tilde{\omega}^2(i\omega_m) + \tilde{\Delta}^2(i\omega_m)}} . \tag{5.3}$$

Bergman and Rainer[55] first discussed the functional derivative of T_c with respect to $\alpha^2F(\omega)$ on the basis of the linearized form of equations (5.2) and (5.3) which apply near T_c. Introducing $\bar{\Delta}_n = \tilde{\Delta}(i\omega_n)/(|\tilde{\omega}(i\omega_n) + \rho)|$ where ρ is a pair breaking parameter to be set equal to zero and substituting (5.3) into (5.2) yield the equation[55–56]

$$\rho\bar{\Delta}_n = \pi T_c \sum_m \left(\lambda(i\omega_n - i\omega_m) - \mu^* - \delta_{n,m} \frac{|\tilde{\omega}(i\omega_n)|}{\pi T_c} \right) \bar{\Delta}_m \tag{5.4}$$

which is an eigenvalue equation with kernel

$$K_{n,m} = \pi T_c \left[\lambda(i\omega_n - i\omega_m) - \mu^* - \delta_{n,m} \frac{|\tilde{\omega}(i\omega_n)|}{\pi T_c} \right] . \tag{5.5}$$

A variation $\delta K_{n,m}$ in the kernel leads to a change in ρ of $\delta\rho$ which can be written in the form

$$\delta\rho = \frac{\sum\limits_{n,m} \bar{\Delta}_n \delta K_{n,m} \bar{\Delta}_m}{\sum_n \bar{\Delta}_n^2} \tag{5.6}$$

with $\bar{\Delta}_n$ in this last equation the eigenvector for $\rho = 0$ and $T = T_c$.

The functional derivative of the critical temperature is given by[55–56]

$$\frac{\delta T_c}{\delta\alpha^2F(\omega)} = -\frac{\frac{\delta\rho}{\delta\alpha^2F(\omega)}}{\left(\frac{\partial\rho}{\partial T}\right)_{T_c}} \tag{5.7}$$

where $\partial\rho/\partial\alpha^2F(\omega)$ means the variation in ρ due to the explicit dependence of $K_{n,m}$ only on $\alpha^2F(\omega)$.

Results for the functional derivative of T_c in the case of Pb and Tl are shown in Fig. 5.1 where we note that both display the same shape. The curves go to zero smoothly as $\omega \to 0$ and also as $\omega \to \infty$ displaying a simple maximum in between. The existence of a maximum in these curves indicates that an optimum frequency exists for T_c. Low frequency and high frequency phonons are not effective in T_c while intermediate ones are, over a broad range around the maximum at about $7k_bT_c$. We will take up the idea of an optimum frequency in more detail later on.

To obtain the functional derivative of the critical field deviation function or of the specific heat, we need the functional derivative of the free energy difference between superconducting and normal state. To accomplish this, the use of the Bardeen-Stephen[27] formula is not very convenient because it is not zero under variations of either $\Delta(i\omega_n)$ or $Z(i\omega_m)$ and so these variations would need to be taken. Rainer and Bergmann[6] noted that this difficulty could be circumvented by using instead the Wada[66] formula given by:

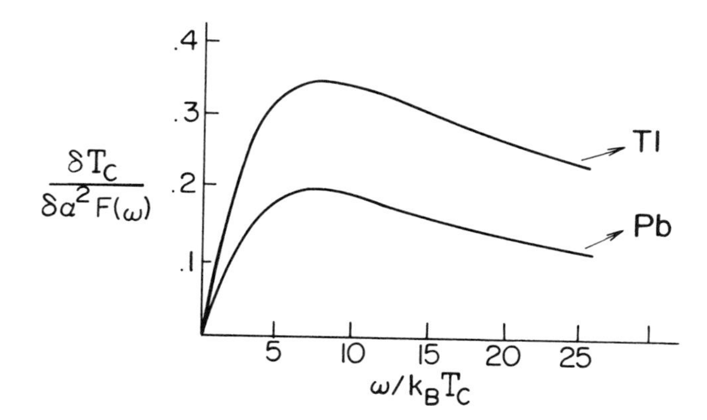

Figure 5.1 Functional derivative of T_c as a function of ω/T_c

$$
\begin{aligned}
\frac{\Delta F}{N(0)} = &- 2\pi T \sum_n \omega_n \Big[\frac{\tilde{\omega}(i\omega_n)}{\sqrt{\tilde{\omega}^2(i\omega_n) + \tilde{\Delta}^2(i\omega_n)}} - \mathrm{sgn}(\omega_n)\Big] \\
&- (\pi T)^2 \sum_{n,m} \Big\{ \Big[\frac{\tilde{\omega}(i\omega_n)}{\sqrt{\tilde{\omega}^2(i\omega_n) + \tilde{\Delta}^2(i\omega_n)}} \frac{\tilde{\omega}(i\omega_m)}{\sqrt{\tilde{\omega}^2(i\omega_m) + \tilde{\Delta}^2(i\omega_m)}} \\
&- \mathrm{sgn}(\omega_n \omega_m)\Big] \lambda(i\omega_n - i\omega_m) \\
&+ \frac{\tilde{\Delta}(i\omega_n)}{\sqrt{\tilde{\omega}^2(i\omega_n) + \tilde{\Delta}^2(i\omega_n)}} \frac{\tilde{\Delta}(i\omega_m)}{\sqrt{\tilde{\omega}^2(i\omega_m) + \tilde{\Delta}^2(i\omega_m)}} \\
&\times \Big[\lambda(i\omega_n - i\omega_m) - \mu^* \theta(\omega_c - |\omega_m|)\theta(\omega_c - |\omega_n|)\Big] \Big\}
\end{aligned} \tag{5.8}
$$

where $\tilde{\Delta}(i\omega_n) = \Delta(i\omega_n)\tilde{\omega}(i\omega_n)/\omega_n$ and $\tilde{\omega}(i\omega_n) = \omega_n Z(i\omega_n)$. Taking variations with respect to $\tilde{\Delta}$ or $\tilde{\omega}$ and setting each equal to zero gives respectively the Eliashberg equations (5.2) and (5.3).

In the functional derivative of the specific heat and we note that t is to be kept constant when taking variations with $\alpha^2 F(\omega)$ so we have[65]

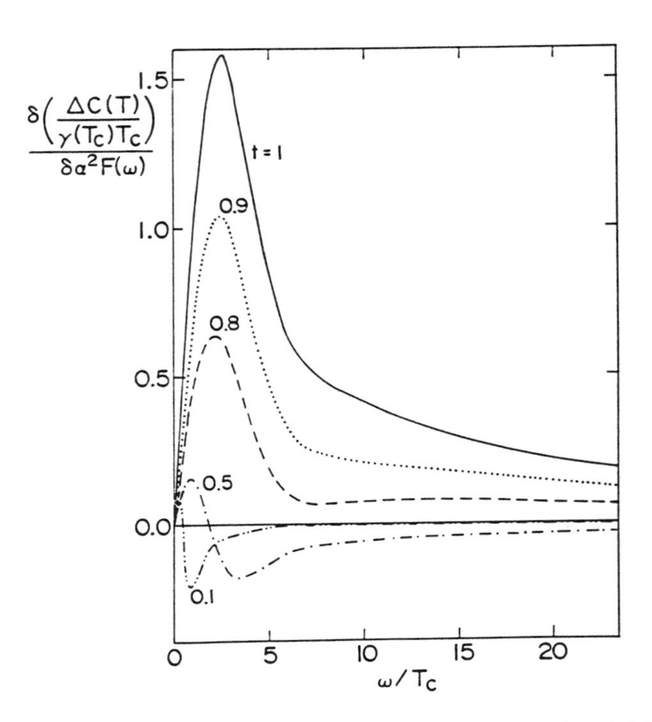

Figure 5.2 Plot of the functional derivative of the specific heat difference, $\delta[\Delta C(T)/\gamma(T_c)T_c]/\delta\alpha^2F(\omega)$ vs. ω/T_c for Pb, for various reduced temperatures. Note that as $\omega \to 0$, the derivative approaches zero, in contrast to the situation when γ_0 is used instead of $\gamma(T_c)$ as will be seen in section VI.

$$\frac{1}{\gamma}\frac{\delta[\Delta C(T)/T_c]}{\delta\alpha^2F(\omega)} = \frac{N(0)}{\gamma_0}\frac{T}{T_c}\frac{d^2}{dT^2}\left\{(\pi T)^2\sum_{n,m}\left[\frac{\tilde{\omega}(i\omega_n)}{\sqrt{\tilde{\omega}^2(i\omega_n)+\tilde{\Delta}^2(i\omega_n)}}\right.\right.$$

$$\times\frac{\tilde{\omega}(i\omega_m)}{\sqrt{\tilde{\omega}^2(i\omega_m)+\tilde{\Delta}^2(i\omega_m)}} - \mathrm{sgn}(\omega_n\omega_m)$$

$$\left.+\frac{\tilde{\Delta}(i\omega_n)}{\sqrt{\tilde{\omega}^2(i\omega_n)+\tilde{\Delta}^2(i\omega_n)}}\frac{\tilde{\Delta}(i\omega_m)}{\sqrt{\tilde{\omega}^2(i\omega_m)+\tilde{\Delta}^2(i\omega_m)}}\right]$$

$$\times\left[\frac{2\omega}{\omega^2+(\omega_n-\omega_m)^2} - \frac{2}{T_c}(\omega_n-\omega_m)^2\right.$$

$$\left.\left.\times\frac{\delta T_c}{\delta\alpha^2F(\omega)}\int\frac{2\omega'\alpha^2F(\omega')d\omega'}{[\omega'^2+(\omega_n-\omega_m)^2]^2}\right]\right\}. \tag{5.9}$$

The functional derivative indicated in (5.9) is not the only one of interest. n this equation, we are normalizing to γ_o which is $\frac{2}{3}\pi^2 k_B^2 N(0)(1+\lambda)$. This is, trictly speaking, only valid at low T. When the electron-boson interaction is large, ve should be using a temperature dependent Sommerfeld constant $\gamma(T)$ as described y Grimvall.[67] It is given by

$$\gamma(T) = \gamma_o\left\{1 + \frac{1}{1+\lambda}\int_0^\infty 2\frac{d\omega}{\omega}\alpha^2(\omega)F(\omega)\left[Z\left(\frac{T}{\omega}\right) - 1\right]\right\} \tag{5.10}$$

vith $\gamma_o \equiv \gamma(0) = \frac{2}{3}\pi^2 k_B^2 N(0)(1+\lambda)$. The kernel $Z(x)$ can be written in terms of a niversal function introduced by Grimvall.[67] If we designate $\gamma \equiv \gamma(T_c)$

$$I(T) = \frac{1}{\gamma}\frac{\delta[\Delta C(T)/T_c]}{\delta\alpha^2 F(\omega)} \tag{5.11}$$

$$\frac{1}{\gamma T_c}\frac{\delta \Delta C(t)}{\delta\alpha^2 F(\omega)} = I(T) + \frac{1}{T_c}\frac{\Delta C(T)}{\gamma T_c}\frac{\delta T_c}{\delta\alpha^2 F(\omega)} \tag{5.12}$$

nd

$$\frac{\delta}{\delta\alpha^2 F(\omega)}\left(\frac{\Delta C(t)}{\gamma(T_c)T_c}\right) = I(T) - \frac{1}{\gamma}\frac{\Delta C(t)}{\gamma T_c}\left[\frac{\delta\gamma(T_c)}{\delta\alpha^2 F(\omega)} + \frac{\partial\gamma(T)}{\partial T}\bigg|_{T_c}\frac{\delta T_c}{\delta\alpha^2 F(\omega)}\right] . \tag{5.13}$$

Results for $\delta(\Delta C(t)/\gamma(T_c)T_c)/\delta\alpha^2 F(\omega)$ are given in Fig. 5.2[65] for several values f the reduced temperature t. At $t = 1$, the functional derivative is positive definite nd exhibits a single peak at an optimum frequency and goes to zero for both $\omega \to 0$ nd $\omega \to \infty$. As the temperature t is lowered, the peak shifts but remains and t sufficiently low t part of the curve becomes negative and a much more complex ehaviour results, particularly in the small ω region. In what follows we will be nterested only in the $t = 1$ results.

VI - Optimum Spectra

In this section, we follow up on results obtained, or at least hinted at in the revious section, and employ functional derivative techniques to prove the existence within Eliashberg theory) of extrema in superconducting properties. As a first example, we consider the value of the critical temperature itself. We have already seen that he functional derivative of T_c with $\alpha^2 F(\omega)$ displays a maximum around $7k_B T_c$, goes o zero like ω as $\omega \to 0$ and like $1/\omega$ as $\omega \to \infty$. This result suggests the existence of n optimum boson energy and an optimum spectral shape. We can argue that taking nfinitesimal weight out of the base $\alpha^2 F(\omega)$ at a frequency away from the maximum n its functional derivative and placing the same amount of weight at the position of he maximum (where the phonons are more effective in T_c) should increase the critical emperature. This procedure does not change the total spectral weight available, i.e. he area under $\alpha^2 F(\omega)$ which is[68–70]

$$A = \int_0^\infty \alpha^2 F(\omega) d\omega . \tag{6.1}$$

If, for any base function, the functional derivative retains its shape, i.e. displays a maximum, then we should be able to repeat the procedure step after step and conclud that, for a given spectral weight A, the best shape for $\alpha^2F(\omega)$ in order to maximiz T_c is a delta function at some optimum Einstein frequency ω_E. This frequency shoul also be exactly equal to the frequency defining the maximum in its own functiona derivative.[68]

The above concept can be quantified by first considering a delta function spec trum of the form $\alpha^2F(\omega) = A\delta(\omega - \omega_E)$ with ω_E a general Einstein frequency. Sub stitution of this function into the Eliashberg equations (2.1) and (2.2) leads, afte simple algebra, to the equations[68–71]

$$\bar{\Delta}(i\bar{\omega}_n)Z(i\bar{\omega}_n) = \bar{T}\pi\sum_m\Big[\frac{2\bar{\omega}_E}{\bar{\omega}_E^2 + (\bar{\omega}_n - \bar{\omega}_m)^2} - \mu^*\theta(\omega_c - |\bar{\omega}_m|A)\Big]\frac{\bar{\Delta}(i\bar{\omega}_n)}{\sqrt{\bar{\omega}_m^2 + \Delta^2(i\bar{\omega}_m)}} \tag{6.2}$$

and

$$Z(i\bar{\omega}_n) = 1 + \frac{\pi\bar{T}}{\bar{\omega}_n}\sum_m\frac{2\bar{\omega}_E}{\bar{\omega}_E^2 + (\bar{\omega}_n - \bar{\omega}_m)^2}\frac{\bar{\omega}_m}{\sqrt{\bar{\omega}_m^2 + \bar{\Delta}^2(i\bar{\omega}_m)}} \tag{6.3}$$

where $\bar{\Delta}(i\bar{\omega}_n) = \Delta(i\omega_n)/A$, $\bar{T} = T/A$, $\bar{\omega}_E = \omega/A$ and $\bar{\omega}_n = \omega_n/A$. Equations (6.2 and (6.3) are clearly independent of A and depend only on the material paramete ω_E and on μ^* provided we follow Leavens[69] and neglect the A dependence in th cut off which is needed to make the Coulomb repulsion term converge. For man superconductors, μ^* is not an important parameter and the above approximation i justified. It follows from our scaling law that $\bar{T}_c = T_c/A$ with $\bar{T}_c$ a function only c ω_E and μ^*. Denoting this function, that follows from the numerical solution of th linearized version of (6.2) and (6.3), by $f(\omega_E, \mu^*)$ we have the important relationship[7

$$T_c = Af(\bar{\omega}_E, \mu^*)\ . \tag{6.4}$$

Equation (6.4) applies for any delta function spectrum with $f(\bar{\omega}_E, \mu^*)$ a univer sal curve for fixed Coulomb repulsion μ^*. In Fig. 6.1, we plot this universal functio for three different values of μ^* namely μ^*=0.0, 0.1 and 0.2. We see that as a functio of $\bar{\omega}_E = \omega_E/A$ all three curves have the same shape, each exhibiting a maximum a some intermediate value of $\bar{\omega}_E$ which decreases with increasing μ^*. On either side c the broad maximum, the curves drop smoothly towards lower values and go to zer at $\bar{\omega}_E = 0$ and $\bar{\omega}_E = \infty$. It is clear that for a delta function spectral density ther is a unique optimum frequency which makes T_c largest for a fixed value of A. If w denote the normalized frequency $\bar{\omega}_E$ where the maximum occurs by $\bar{\omega}_E^*(\mu^*)$ we wi find that at this frequency[68]

$$\frac{T_c}{A} = c(\mu^*) \qquad \text{for} \quad \bar{\omega}_E = \bar{\omega}_E^*(\mu^*) \equiv d(\mu^*) \tag{6.5}$$

where c and d are universal functions of μ^*.

To demonstrate that (6.5) represents a local maximum not just for delta func tion shape but indeed, for any $\alpha^2F(\omega)$ with fixed A, we show in the lower part of Fig 6.1 the functional derivative of T_c/A for the base spectrum

$$\alpha^2F(\omega) = A\delta(\omega - \omega_E^*) \tag{6.6}$$

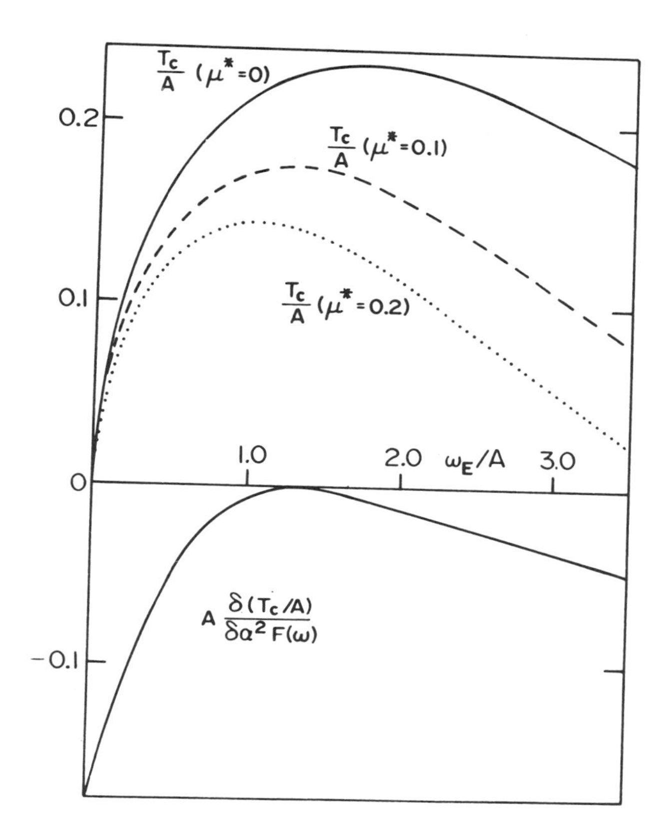

Figure 6.1 Plot of T_c/A vs. ω_E/A for Einstein spectra located at frequency ω_E with weight A, for various μ^*. The curves are universal and all exhibit a maximum, indicating that an optimum spectrum exists. Also drawn is the functional derivative of T_c/A for the optimum Einstein spectrum for μ^*=0.1, with $\bar{\omega}_E^*$=1.3. Note that it is negative definite, with a maximum value of 0.0 at the location of the optimum spectrum ($\bar{\omega}_E^* = 1.3$. Clearly, T_c/A cannot be enhanced.

with $\omega_E^* = 1.3A$ for μ^*=0.1. What is plotted is $A\delta(T_c/A)/\delta\alpha^2F(\omega)$ which is found to be negative definite and zero exactly at $\bar{\omega}_E$=1.3≡ $\bar{\omega}_E^*$. That is, the functional derivative is zero exactly at the frequency of the base delta function. This means that adding weight to $\alpha^2F(\omega)$ at this frequency leaves T_c/A unchanged and adding weight anywhere else reduces this ratio. Thus we have maximized T_c/A with our optimum spectrum (6.6) for arbitrary A. Therefore, the following inequality holds[68]

$$T_c \leq Ac(\mu^*) \tag{6.7}$$

where the equality applies for a delta function and the strict less than for any other shape, a result first established by Leavens[68] using somewhat different arguments. The relationship (6.7) is tested against conventional superconductors in Fig. 6.2 where we show $c(\mu^*)$ as a function of μ^* (solid line with open circles). On the same figure, we have also plotted the ratio T_c/A (solid dots) for a large number of the

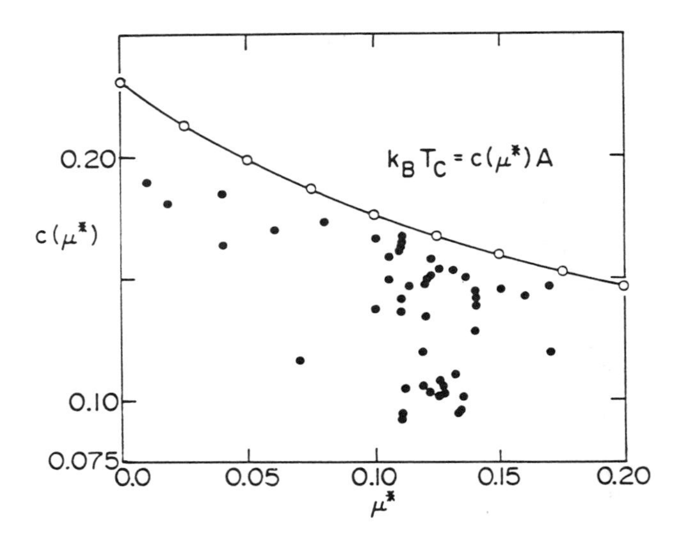

Figure 6.2 Plot of T_c/A as a function of μ^*. The solid line represents the equality $k_B T_c = c(\mu^*)A$ and is a maximum. The solid dots are results for various conventional superconductors.

conventional superconductors discussed in a previous section. It is clear that several fall close to our theoretical maximum indicating that in nature some systems have boson modes that fall near the broad maximum around the optimum frequency. In these cases, T_c is nearly as large as it can be made for a given A. Also, the equality in equation (6.7) indicates that there is no limit on T_c imposed by the Eliashberg equations themselves. As the area under $\alpha^2 F(\omega)$ is increased indefinitely, T_c also increases indefinitely. For example, for μ^*=0.1, $c \cong$0.175 so that T_c=0.175A with ω_E=1.3A which corresponds to a value of λ=1.53. To achieve a T_c of 16.0 meV, we need a value of $A \cong$90 meV which implies an ω_E value of approximately 120 meV. This is, however, an electronic energy rather than a phonon energy. Nevertheless, Eliashberg theory includes large values of T_c even for modest values of λ.

We next turn to a discussion of the thermodynamics.[71] We have already seen, in the previous section on very strong coupling, that for a LSCO or Pb base spectrum, the normalized specific heat jump at T_c displays a maximum at intermediate values of $T_c/\omega_{\ln}$. It is interesting to know how dependent this maximum is on the shape of the spectral density $\alpha^2 F(\omega)$ and if some optimum shape will maximize its absolute value. That this is likely is indicated by the shape of the functional derivative of $\Delta C(T_c)/\gamma(0)T_c$. In Fig. 6.3,[71] we show results for

$$(1+\lambda)T_c \frac{\delta}{\delta\alpha^2 F(\omega)}\left(\frac{\Delta C(T_c)}{\gamma(0)T_c}\right) \tag{6.8}$$

in the case of Pb (solid curve). It is seen that this functional derivative diverges toward $-\infty$ at low ω. This divergence can be traced to our use of $\gamma(0)$ in equation (6.8) since $\delta\gamma(0)/\delta\alpha^2 F(\omega) \sim 1/\omega$. If instead $\gamma(T_c)$ had been used, the functional derivative would go smoothly to zero as seen in the previous section. At higher frequencies, around $\omega/T_c \cong 5$, a positive maximum occurs and then the functional derivative

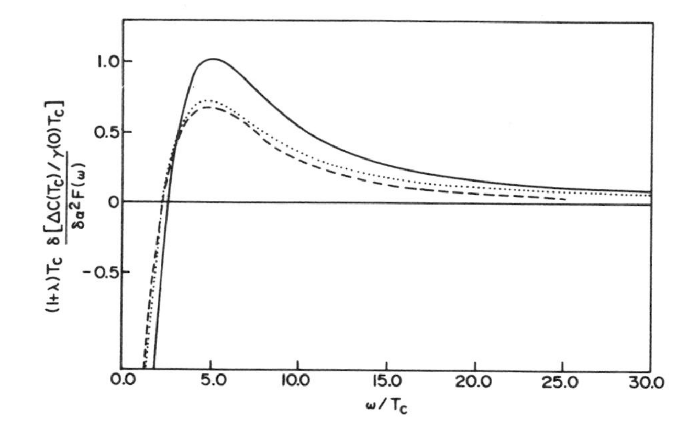

Figure 6.3 The functional derivative of the specific heat jump at T_c normalized to $\gamma(0)T_c$, where $\gamma(0)$ is the Sommerfeld constant. What is given is $(1+\lambda)T_c\delta[\Delta C(T_c)/\gamma(0)T_c]\delta\alpha^2F(\omega)$ vs. ω/T_c. The solid curve is for Pb, the dotted line for Al, and the dashed line for any weak coupling superconductor in a two square well model.

slowly decays towards zero as $\omega \to \infty$. We have tried many other realistic tunneling-derived spectra instead of Pb, and find, in all cases, the same shape for the functional derivative. For example, the dotted line was calculated using the Al spectrum given by Leung, Carbotte, Taylor, and Leavens.[10] It is very close to the dashed line which was calculated using a two-square-well model for $\lambda(n-m)$. Details of the two-square-well model results can be found in the work of Marsiglio and Carbotte.[64] In this simplified model, no assumption is made about the shape of the spectral density, except that all important phonon frequencies should be much greater than several k_BT_c's. This is the usual BCS limit, and applies to all weak coupling systems.

From the above, we conclude that the shape of $\delta[\Delta C(T_c)/\gamma(0)T_c]/\delta\alpha^2F(\omega)$ is fairly universal for realistic values of $\alpha^2F(\omega)$ (i.e., actual measured shapes for real materials). This leads to the suggestion that in order to increase $\Delta C(T_c)/\gamma(0)T_c$ for a given $\alpha^2F(\omega)$, we should take weight from some frequency where the functional derivative is smaller than its value at maximum, and transfer it to the optimum frequency, keeping the total area under $\alpha^2F(\omega)$ constant. This suggests that to maximize $\Delta C(T_c)/\gamma(0)T_c$, for a given value of $A = \int \alpha^2F(\omega)d\omega$, we should use a delta function with all its weight placed at the same Einstein frequency ω_E.

For a delta function spectrum, the scaled equations (6.2) and (6.3) hold from which we can conclude that $\Delta(i\omega_n)/A$ is a function only of $\bar{\omega}_E$ and of $\bar{T}$ as is $Z(i\omega_n)$. Reference to the free energy formula yields immediately that

$$\Delta F = N(0)A^2g(\bar{\omega}_E,\bar{T}) \tag{6.9}$$

where g is an appropriate function related to formula (2.5). Thus, the specific heat difference between superconducting and normal state is

$$\Delta C(T) = T\frac{d^2\Delta F(T)}{dT^2} = \bar{T}AN(0)\frac{d^2g(\bar{\omega}_E,\bar{T})}{d\bar{T}^2} \tag{6.10}$$

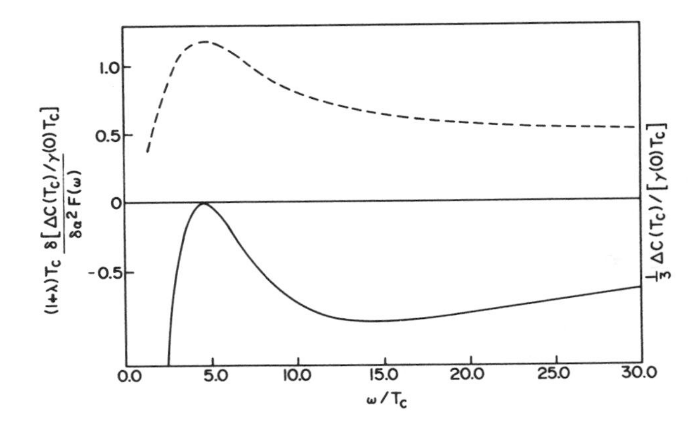

Figure 6.4 The value of $\frac{1}{3}\Delta C(T_c)/[\gamma(0)T_c]$ (right hand label) for an Einstein spectrum as a function of ω_E/T_c (dashed curve), where ω_E is the position of the phonon frequency. The results are independent of the value used for the area (A) under the delta function. The μ^* value was 0.051. The functional derivative $\delta[\Delta C(T_c)/\langle\gamma(0)T_c\rangle]/\delta\alpha^2F(\omega)$ multiplied by $(1+\lambda)T_c$ (left hand label) for the case of a delta function spectrum $\alpha^2F(\omega) = A\delta(\omega-\omega_E^*)$, where ω_E^* is the frequency of the maximum in the dashed curve.

and, therefore,

$$\frac{\Delta C(T)}{\gamma(0)T_c} = \frac{3}{2\pi^2 k_B^2(1+\frac{2}{\bar{\omega}_E})}\frac{d^2 g(\bar{\omega}_E,\bar{T})}{d\bar{T}^2} \ . \tag{6.11}$$

But $T_c = Af(\bar{\omega}_E,\mu^*)$ so that we can finally write

$$\frac{\Delta C(T)}{\gamma(0)T_c} \equiv \mathcal{G}(\omega_E/T_c,\mu^*,t) \tag{6.12}$$

with t the reduced temperature $t = T/T_c$ which is independent of A. In particular, the specific heat jump depends only on ω_E/T_c for a specific choice of Coulomb pseudopotential μ^*.

In Fig. 6.4, we show results of calculations for the normalized jump as a function of ω_E/T_c (dashed curve) for μ^*=0.051. What is plotted is $\frac{1}{3}\Delta C(T_c)/\gamma(0)T_c$. We see that on lowering the position of the delta function ω_E/T_c the normalized jump increases until a maximum of 3.57 is reached for $\bar{\omega}_E^*$=4.55 after which it drops rather rapidly towards very small values as ω_E goes toward zero. This curve then proves that, for a delta function, there is an optimum frequency for the specific heat jump and that the maximum in this quantity is 3.57 (for μ^*=0.051) independent of A. We can use functional derivatives to prove that any other shape will lower $\Delta C(T_c)/\gamma(0)T_c$. Returning to Fig. 6.4, the solid curve is $(1+\lambda)T_c\delta[\Delta C(T_c)/\langle\gamma(0)T_c\rangle]\delta\alpha^2F(\omega)$ (left-hand label) for a model delta function $\alpha^2F(\omega) = A\delta(\omega-\omega_E^*)$ with ω_E^* the frequency giving the maximum of the dashed curve. It is clear from the figure that the functional derivative is now very different from those found for realistic $\alpha^2F(\omega)$ spectra. It is negative definite with value zero right at ω_E^*. This proves that a delta function at ω_E^* gives a local maximum. Removal of some weight from the delta function at ω_E^* and

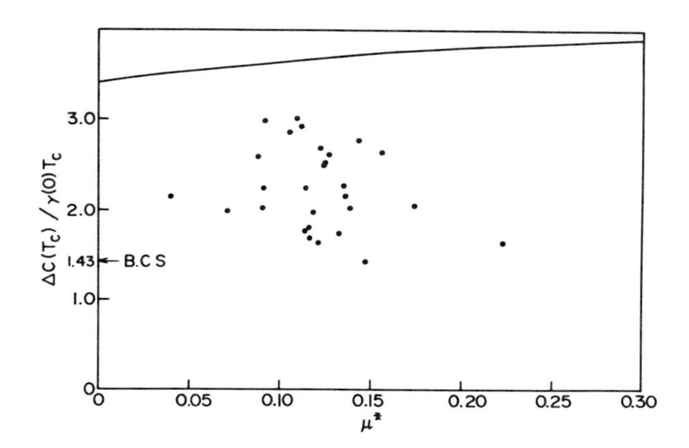

Figure 6.5 The maximum possible value for $\Delta C(T_c)/[\gamma(0)T_c]$ as a function of μ^*. The solid dots represent theoretical values for the following materials in order of decreasing value of $\Delta C(T_c)/[\gamma(0)T_c]$: $Pb_{0.7}Bi_{0.3}$, $Pb_{0.65}Bi_{0.35}$, $Pb_{0.8}Bi_{0.2}$, $Pb_{0.9}Bi_{0.1}$, Pb, $Pb_{0.8}Tl_{0.2}$, Nb_3Sn, Nb_3Al, Nb_3Ge, $Pb_{0.6}$ $Tl_{0.4}$, Hg, $Pb_{0.75}Bi_{0.25}$, $Pb_{0.4}Tl_{0.6}$, V_3Ga, $Pb_{0.5}Bi_{0.5}$, La, Ga, (amorphous), Bi (amorphous), V_3Si, Mo (amorphous), Nb, In, $Tl_{0.9}Bi_{0.1}$, Tl, Sn, Ta, V, Al (BCS).

placing it at any other frequency reduces the specific-heat jump. Our experience with realistic spectra would lead us to believe that this is, in fact, an absolute maximum, although we have not found a rigorous mathematical proof. Figure 6.5 sheds more light on this point. In this figure we show as the solid line, the maximum value of $\Delta C(T_c)/\gamma(0)T_c$, obtained for different choice of μ^*. On the same figure, we have shown (dark points) values for the same quantity obtained in realistic cases. They all fall below our theoretical local maximum indicating that for physical systems, the solid line is, indeed, an absolute maximum.

VII - Asymptotic Limits

In section IV, we considered the very strong coupling regime and give results for values of the strong coupling index, $T_c/\omega_{\ln}$, ranging up to 1.2 in some cases. We saw that, while the corresponding electron-boson mass enhancement factor (λ) value depends somewhat on the shape of the electron-boson spectral density which is used, $T_c/\omega_{\ln}$=1.2 probably already corresponds to unrealistically large values of λ. Nevertheless, it is of some interest to know how the thermodynamic properties behave at even larger values of λ[73] Instead of simply continuing the numerical work of section IV to larger values of $T_c/\omega_{\ln}$, it seems more reasonable to attempt to derive analytic expressions for the $\lambda \to \infty$ limit. Here we treat only the gap to critical temperature ratio.

Throughout the work described in this section, it will be convenient to use a delta function spectral density of weight A centered at the Einstein boson energy ω_E. That is, we will take

$$\alpha^2F(\omega) = A\delta(\omega - \omega_E)\ . \tag{7.1}$$

In this case $\lambda = 2A/\omega_E$ and, if we keep A fixed as we will, the $\lambda \to \infty$ limit corresponds to letting $\omega_E \to 0$. For convenience, we also set μ^*=0. Other calculations could be done at finite μ^* but no qualitative changes are expected. In the above model the electron-boson factor $\lambda(i\omega_n - i\omega_m)$ simplifies:

$$\lambda(i\omega_n - i\omega_m) = \frac{2A\omega_E}{\omega_E^2 + (\omega_n - \omega_m)^2} \, . \tag{7.2}$$

As we are only interested in the $\omega_E \to 0$ limit, we can ignore the ω_E^2 factor in the denominator of equation (7.2) provided that $n \neq m$ and that we assume $\omega_E \ll 2\pi T$. This inequality will become important later. With the above approximation, we have for $n \neq m$

$$\lambda(i\omega_n - i\omega_m) = \frac{1}{2\pi^2 \bar{T}^2 (n-m)^2} \tag{7.3a}$$

and for $n = m$

$$\lambda(i\omega_n - i\omega_m) = \frac{2A}{\omega_E} \equiv \lambda \, . \tag{7.3b}$$

In (7.3a), we have introduced the scaled quantity $\bar{T} = T/\sqrt{A\omega_E}$ and the form (7.3a) is now completely independent of any material parameter and suggests that we introduce the same scaling as above into the gap equations themselves. Indeed (2.1) and (2.2) can be rewritten in the form

$$\bar{\Delta}(i\bar{\omega}_n) Z(i\bar{\omega}_n) = \pi \bar{T} \sum_{m \neq n} \frac{1}{2\pi^2 \bar{T}^2 (m-n)^2} \frac{\bar{\Delta}(i\bar{\omega}_m)}{\sqrt{\bar{\omega}_m^2 + \bar{\Delta}^2(i\bar{\omega}_m)}} + \bar{T}\pi\lambda \frac{\bar{\Delta}(i\bar{\omega}_n)}{\sqrt{\bar{\omega}_n^2 + \bar{\Delta}^2(i\bar{\omega}_n)}} \tag{7.4}$$

and

$$Z(i\bar{\omega}_n) = 1 + \frac{\pi \bar{T}}{\bar{\omega}_n} \sum_{m \neq n} \frac{1}{2\pi^2 \bar{T}^2 (m-n)^2} \frac{\bar{\omega}_m}{\sqrt{\bar{\omega}_m^2 + \bar{\Delta}^2(i\bar{\omega}_m)}} + \frac{\bar{T}\pi}{\bar{\omega}_n} \lambda \frac{\bar{\omega}_n}{\sqrt{\bar{\omega}_n^2 + \bar{\Delta}^2(i\bar{\omega}_n)}} \, . \tag{7.5}$$

where we have used the notation $\bar{Q} \equiv Q/\sqrt{A\omega_E}$ for any quantity Q. While these equations seem to be dependent on material parameters through the appearance of the λ factor in the last term of both equation (7.4) and (7.5), these terms cancel when (7.5) is substituted into (7.4) and we get a gap equation which is universal and applies to the asymptotic limit independent of A or ω_E. The equation is

$$\bar{\Delta}(i\bar{\omega}_n) = \bar{T}\pi \sum_{m \neq n} \frac{1}{2\pi^2 \bar{T}^2 (n-m)^2} \left\{ \frac{\bar{\Delta}(i\bar{\omega}_m)}{\sqrt{\bar{\omega}_m^2 + \bar{\Delta}^2(i\bar{\omega}_m)}} - \frac{\bar{\omega}_m}{\bar{\omega}_n} \frac{\bar{\Delta}(i\bar{\omega}_n)}{\sqrt{\bar{\omega}_n^2 + \bar{\Delta}^2(i\bar{\omega}_n)}} \right\} \tag{7.6}$$

which shows quite explicitly that in this limit $\bar{\Delta}(i\bar{\omega}_n)$ is only a function of the reduced temperature t i.e. we can write

$$\bar{\Delta}(i\bar{\omega}_n) \equiv f(t, i\bar{\omega}_n) \, , \tag{7.7}$$

where $f(t, i\bar{\omega}_n)$ which can be determined from (7.6) through a numerical solution, is universal, and depends only on the form of the Eliashberg equations and not on any kernel (i.e. material parameter). It follows immediately from these remarks that $\bar{T}_c$

(the scaled critical temperature) is a unique number. A few iterations of the linearized form of (7.6) gives

$$\bar{T}_c = .2584 \,, \qquad \text{or} \qquad \bar{T}_c = .2584\sqrt{A\omega_E} = .183\omega_E\sqrt{\lambda}\,. \tag{7.8}$$

This last equation was first obtained by Allen and Dynes.[24] An approximate value for the proportionality factor .183 in (7.8) can easily be obtained from equation (7.6) if we assume that in the very strong coupling limit only $\bar{\Delta}(i2\pi\bar{T})$ and $\bar{\Delta}(-i2\pi\bar{T})$ are important with higher Matsubara gaps negligible. This gives

$$\bar{\Delta}(i2\pi\bar{T}) = \pi\bar{T}\frac{1}{\pi^2\bar{T}^2}\frac{\bar{\Delta}(i2\pi\bar{T})}{\sqrt{(2\pi\bar{T})^2 + \bar{\Delta}^2(i2\pi\bar{T})}} \tag{7.9}$$

where we have used the symmetry $\Delta(i\omega_{-n}) = \Delta(i\omega_{n+1})$. For $\bar{\Delta} \to 0$ we obtain

$$\bar{T}_c = \frac{1}{\pi\sqrt{2}} \qquad \text{or} \qquad \bar{T}_c = .225\sqrt{A\omega_E} \tag{7.10}$$

which gives the same dependence of T_c on material parameters as the exact relation (7.8) but with a different coefficient.

Returning now to equation (7.7) for the gap $\bar{\Delta}(i\bar{\omega}_n)$ at any reduced temperature, an analytic continuation $i\bar{\omega}_n \to \omega$ implies that $\Delta(\omega)$ satisfies

$$\Delta(t,\omega) = \sqrt{A\omega_E}\, f\Big(t, \frac{\omega}{\sqrt{A\omega_E}}\Big)\,. \tag{7.11}$$

Defining the temperature gap edge through

$$\Re e\,[\Delta(t, \Delta_0(t))] = \Delta_0(t) \tag{7.12}$$

it then follows that $\Delta_o(t) = \sqrt{A\omega_E}\, h(t)$, where $h(t)$ is some universal function of t. Hence, combining this result with equation (7.8) for T_c implies $2\Delta_o(t)/k_BT_c =$ constant, provided that $T \gg 2\pi\omega_E$. How can zero temperature be achieved with this last restriction? The point is that we want to achieve zero temperature *behaviour*, which does not necessarily entail solving $T = 0$ equations. This is standard procedure for conventional materials, where, utilizing the imaginary axis equations, we actually solve at $t = 0.1$, and call this our zero temperature solution. This is justified by the fact that the gap edge, for example, does not change (to $\sim 6-7$ significant digits) when we use $t = 0.05$, or $t = 0.025$. The reason for this is of course, well understood: the presence of a gap causes exponential behaviour. Simple Boltzmann factors like $e^{-\beta\epsilon}$ become $e^{-\beta E}$ in BCS theory, where $E^2 = \epsilon^2 + \Delta^2$. Hence, even as $\epsilon \to 0$, $e^{-\Delta_o/T}$ remains. For example, in BCS theory, the low temperature behaviour of the gap is given by

$$\Delta(T) \approx \Delta_o - (2\pi\Delta_oT)^{1/2}\, e^{-\Delta_o/T}\,, \tag{7.13}$$

where Δ_o is the zero temperature gap. Thus, when $T \ll \Delta_o$, the second term in Eq. (7.13) switches off very quickly, and zero temperature behaviour is achieved. Within BCS, $2\Delta_o/k_BT_c = 3.53$ so that the term is governed by the factor $e^{-1.75/t}$, to be compared with unity (the first term). At $t = 0.1$, this is already eight orders of magnitude smaller than unity, so zero temperature behaviour has been achieved at even higher temperature.

The analysis in Ref. 54 assumed this would be the case in the strong coupling limit as well, without proof. If it is, then $T = 0$ behaviour will be achieved for the case when $e^{-a/t}$ is small, or $t \approx 0.1a$, where a is some *number*. Then the conditions $T \gg \omega_E$ and yet $T \approx 0$ (behaviour-wise) can be simultaneously achieved, so that the proof will be applicable to the zero temperature ratio, $2\Delta_\circ/k_BT_c$. Clearly then, the important issue is the low temperature behaviour of the universal function $f(t, i\bar{\omega}_n)$. We have proceeded numerically to investigate this function.

For low t, Eq. (7.6) was solved for $\bar{\Delta}_n$, for $n = 1, 2, \ldots N_c$ with N_c some large cutoff. As t is lowered, the number of Matsubara frequencies increases since the temperature mesh becomes finer, and the $\bar{\Delta}_n$ tend to increase with decreasing temperature, as is expected, since the $\bar{\Delta}_n$ are order parameters. We found that at $t \approx 0.01$, the size of the gaps is no longer changing. The effect of lowering the temperature, then, is to fill in gap values at more Matsubara frequencies. Hence, we are assured that a good approximation to zero temperature behaviour has been achieved, as far as the imaginary axis calculations go, where now the solution is a continuous curve ($i\omega_n \to i\omega$).

It remains to analytically continue $\Delta(i\omega)$ and $Z(i\omega)$ to the real axis. Here we can distinguish between very low temperatures and zero temperature, which is necessary, since there can be a qualitative difference in the solutions (say, at $\omega = 0$). Inspection of Eqs. (2.12) and (2.13) reveals that this occurs in two places. The first is in the summations, which are integrals at $T = 0$. The behaviour for $\omega \lesssim \pi T$ will depend strongly on whether $T = 0$ or $T > 0$. However, this frequency region is unimportant. The second is in the Bose and Fermi factors. In particular, for $T = 0$, $N(\nu) = 0$. However, for $T > 0$, in the asymptotic limit, $\lim_{\nu \to 0} N(\nu) = T/\nu$ and hence we get an infinity. Since we are interested in zero temperature, we put $N(\nu) = 0$, and $f(x) = \theta(-x)$, the unit step function. These conditions represent the freezing out of thermal phonons, and the sharpening of the occupation of electron states, respectively. Eqs. (2.12) and (2.13) are then scaled according to

$$\begin{aligned}\bar{\Delta}(\bar{\omega})Z(\bar{\omega}) =& \pi\bar{T} \sum_{m=-\infty}^{\infty} \frac{2}{(\bar{\omega} - i\bar{\omega}_m)^2} \frac{\bar{\Delta}_m}{\sqrt{\bar{\omega}_m^2 + \bar{\Delta}_m^2}} \\ &+ i\pi \sqrt{\frac{A}{\nu_E}} \frac{\bar{\Delta}(\bar{\omega} - \bar{\omega}_E)}{\sqrt{(\bar{\omega} - \bar{\omega}_E)^2 - \bar{\Delta}^2(\bar{\omega} - \bar{\omega}_E)}}\end{aligned} \tag{7.14}$$

$$\begin{aligned}Z(\bar{\omega}) = 1 +& i\frac{\pi\bar{T}}{\bar{\omega}} \sum_{m=-\infty}^{\infty} \frac{2}{(\bar{\omega} - i\bar{\omega}_m)^2} \frac{\bar{\omega}_m}{\sqrt{\bar{\omega}_m^2 + \bar{\Delta}_m^2}} \\ &+ i\pi \sqrt{\frac{A}{\nu_E}} \frac{\bar{\omega} - \bar{\omega}_E}{\sqrt{(\bar{\omega} - \bar{\omega}_E)^2 - \bar{\Delta}^2(\bar{\omega} - \bar{\omega}_E)}}\end{aligned} \tag{7.15}$$

These equations do not obey exact scaling laws. However in the limit $\omega_E \to 0$ the material dependence drops out. Hence, we have solved Eqs. (7.14) and (7.15) numerically, using progressively smaller values of $\bar{\omega}_E$. The dependence on $\bar{\omega}_E$ is linear at small $\bar{\omega}_E$, and hence the $\omega_E \to 0$ behaviour can be extrapolated.

In this manner, we find

$$\left. \frac{2\Delta_\circ}{k_BT_c} \right|_{max} \cong 13.0 \tag{7.16}$$

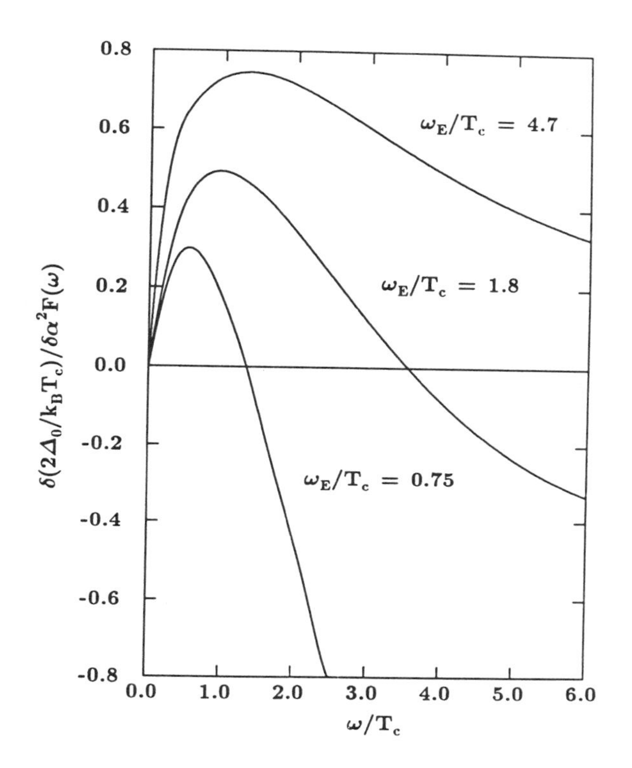

Figure 7.1 The functional derivative of the gap Δ_o to critical temperature ratio, $2\Delta_o/k_bT_c$, for three delta function based spectra labelled by normalized Eliashberg frequency ω_E/T_c. The upper curve for which $\omega_E/T_c = 4.7$ is characteristic of results found for real materials. As the Einstein frequency of the base spectrum is lowered towards zero, however, the curves distort in shape and become negative for much of the frequency range except for a positive peak at ever lower values of ω/T_c. The progression is for the functional derivative curve to become negative definite for $\omega_E/T_c = 0$ with maximum exactly at $\omega/T_c = 0$.

which is remarkably close to the result found by Kresin.[74]. However the approximate form of the gap as a function of frequency used by him is not qualitatively correct. Contrary to what was found for the case of the normalized specific heat jump in section VI the maximum in the gap to critical temperature ratio occurs in the limit $\omega_E \to 0$ rather than for a finite frequency. This means that maximum and asymptotic limit coincide. It is of interest to investigate this maximum a little more closely using functional derivatives. We have established (7.16) on the basis of a delta function spectral density. We would like to show that any infinitesimal change of shape of $\alpha^2F(\omega)$ will decrease $2\Delta_o/k_BT_c$. To see that this is the case, we have calculated the functional derivative

$$\frac{\delta\left(\frac{2\Delta_o}{k_BT_c}\right)}{\delta\alpha^2F(\omega)} \tag{7.17}$$

for a series of base delta functions. Remember that the area under $\alpha^2F(\omega)$ drops out and only the position of ω_E is relevant. Upon lowering the frequency of the Einstein mode in the base function, towards the frequency of the maximum of its own functional derivative, we found, upon calculating the functional derivative for the new spectrum, that the peak in this new functional derivative had shifted to a lower frequency. This continues to occur for ever smaller values of ω_E and, in fact, one must approach the limit $\lambda \to \infty$ in order to find the optimum $2\Delta_o/k_BT_c$. But this is also the asymptotic limit. The situation is illustrated in Fig. 7.1 where we show a sequence of functional derivatives for several ever decreasing values of ω_E in the base delta function. We note the shift of the peak frequency in the functional derivative to ever smaller values and that this peak frequency is always lower than that of the base delta function. The magnitude of the functional derivative is also decreasing and it becomes more negative at the higher frequencies as ω_E is lowered. It is only in the limit of $\omega_E \to 0$ that the peak moves to $\omega = 0$ and that the functional derivative is negative definite with zero exactly at $\omega = 0$. Adding weight to the spectral density in this case at $\omega = 0$ leaves $2\Delta_o/k_BT_c$ unchanged while adding it at any finite $\omega \neq 0$ will reduce $2\Delta_o/k_BT_c$. Therefore, we have achieved a maximum in the asymptotic limit.

VIII Specific to the Oxides

Many theoretical proposals[17] have been put forth to explain the superconductivity in the high T_c oxides. The numerous proposals can probably be classified under two general headings: BCS-like and non-BCS-like. The first category includes any theory in which the superconducting electrons are thought to pair via the exchange of some virtual boson; this boson has been identified in conventional materials to be the phonon. We have regarded the proposals that come under this heading as involving different mechanisms, in the sense that the boson which mediates the electron-electron attraction differs from mechanism to mechanism. Possibilities included plasmons, spin fluctuations, excitons, demons, *etc.* The second category is of course quite vague in name, but is perhaps best exemplified by the resonating valence bond (RVB) theory first proposed by Anderson[75–77] in 1973. In these theories electron-electron correlations are deemed to be so important in the normal state, that a BCS description based on quasiparticles in the normal state is inappropriate. In the RVB, for example, it is thought that a superconducting condensate occurs through a Bose condensation of charge carriers, which are bosons because of the environment they find themselves in (resonating valence bonds). The RVB theory has only recently received much attention, and so suffers the disadvantage of being in its infancy, whereas theories of the first type have been relatively well studied. Here we will be concerned only with theories of the BCS-type but of course not necessarily phonon mediated[17].

We stress once again that Eliashberg theory is based on the calculation of the electron self-energy due to the emission and absorption of some boson. Hence, Eliashberg theory in principle is applicable to all BCS-type theories. However, the infinite set of diagrams summed in the self-energy expansion is only a subset of all possible diagrams. The neglect of all other diagrams ("vertex corrections") is justified by Migdal's theorem which ensures that contributions from these corrections are $O(m/M)^{1/2}$ (or $O(\omega_D/E_F)$) where m is the electron mass and M is the ion mass. This approximation is extremely accurate in the case of phonon mechanism, since $\omega_D \ll E_F$. However, other boson mechanisms include "electronic" bosons have a characteristic frequency of order $O(E_F)$ and hence the use of Migdal's theorem will be suspect. One can nonetheless proceed to investigate the consequences[18] of Eliashberg theory in the same spirit that resulted in BCS theory. For the time being, experiment may be able to decide on the correctness of this approach.

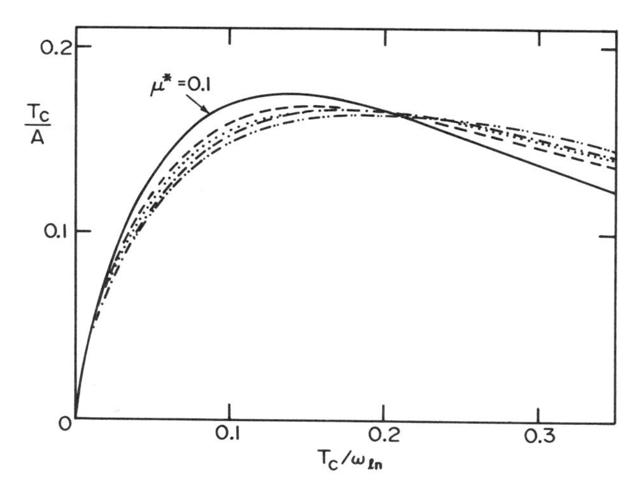

Figure 8.1 Plot of T_c/A vs. $T_c/\omega_{\ln}$. The solid curve represents an Einstein spectrum with $\mu^* = 0.1$. Also illustrated are curves for various real spectra: Pb (- - - -), Nb (····), V(– · · – ·) and Nb_3Sn and Nb_3Al (– ·· – ··). Note that the curves describing spectra corresponding to realistic shapes practically form a universal curve.

What is the need to introduce another mechanism? As has already been mentioned, Eliashberg theory imposes no limit on the maximum attainable T_c. However, crude estimates based on the phonon mechanism have been proposed in the past[78,79] yielding $T_c^{\max} \lesssim 35\,\mathrm{K}$ due to lattice instability. Hence, observation of a high T_c alone justifies the search for an alternative mechanism. At the same time one must be wary of maximum T_c arguments. They have been wrong before and they will possibly be wrong again.

The standard argument in favour of some (usually) higher frequency electronic boson mechanism uses an approximate equation for T_c. In a version of Eliashberg that reduces to BCS theory the critical temperature is

$$T_c = 1.13\omega_c e^{-\frac{1+\lambda}{\lambda-\mu^*}} \tag{8.1}$$

where ω_c is usually the Debye cutoff, ω_D. It is clear that for the same value of λ, an increase in ω_c to $O(E_F)$ from ω_D will result in a substantial increase in T_c. Again, one must be wary with this naive argument since (8.1) is a very approximate form for T_c.

With knowledge of the value of T_c only, there is little one can determine. Fig. 8.1 illustrates this point clearly. What is plotted is T_c/A against $T_c/\omega_{\ln}$ for an Einstein spectrum as well as for Pb, Nb, V, NB_3Sn and Nb_3Al based spectra. While such curves depend somewhat on the underlying spectral shape, the general trend does not. It is clear that one can obtain T_c of any magnitude simply by adjusting $\omega_{\ln}$ and A independently. A system with a given T_c could then be described by parameters corresponding to any point along the universal trend curve in Fig. 8.1 . However, Fig. 8.2 illustrates the gap ratio $2\Delta_o/k_BT_c$ vs. $T_c/\omega_{\ln}$ for a variety of spectral shapes and μ^*. This figure emphasizes what was already discussed before; $2\Delta_o/k_BT_c$ is an almost universal function of $T_c/\omega_{\ln}$, regardless of shape. Given $2\Delta_o/k_BT_c$, Figures

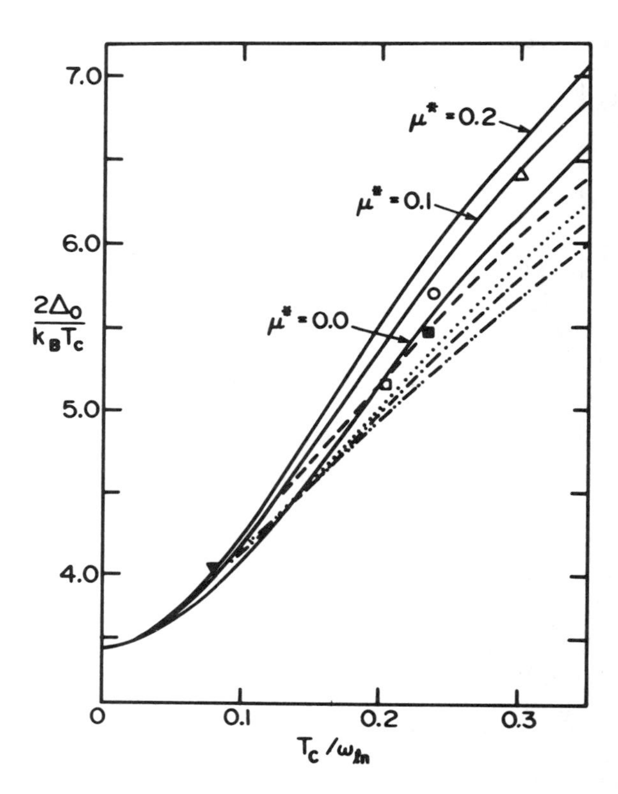

Figure 8.2 Plot of $2\Delta_o/k_BT_c$ vs. $T_c/\omega_{\ln}$. The solid curves are for Einstein spectra with $\mu^* = 0.0$, 0.1, and 0.2. These curves are universal, and indicate that, for given $T_c/\omega_{\ln}$, μ^* causes $2\Delta_o/k_BT_c$ to increase. Also shown are various scaled real spectra, with $\mu^* = 0.1$. They are universal in the sense that they represent spectra which have been scaled by an arbitrary amount along the frequency axis. Note that they form a rather tight band as a function of $T_c/\omega_{\ln}$. The materials are Pb(— — —), Nb(· · · ·), V(— · · —·), and Nb_3Sn and Nb_3Al(— · · · — ··). The various symbols represent the δ—function spectra described in Table 8.1 as follows: ▲, $\lambda = .75$; ▼, $\lambda = 1.0$; ■, $\lambda = 2.6$; ○, $\lambda = 2.3$; ◻, $\lambda = 2.9$; △, $\lambda = 4.4$.

(8.1) and (8.2) together provide us with the data required to determine $\omega_{\ln}$ and A, the fundamental parameters characterizing $\alpha^2F(\nu)$ (recall $\lambda \sim 2A/\omega_{\ln}$). With knowledge only of A and $\omega_{\ln}$ it is natural to think of a delta function spectrum, although this is not essential as is illustrated in Table 8.1 where we have included results for several model spectra consisting of Einstein spectra and the scaled Pb spectrum. Several possibilities are illustrated; the first entry, for example, would be representative of an $\alpha^2F(\nu)$ which is comprised largely of phonons which represent the oxygen breathing modes in the oxides. On the assumption that they fall around 60 meV, the strong coupling parameter, $T_c/\omega_{\ln}$ is then very small, and $2\Delta_o/k_BT_c$ (and other superconducting ratios) attains a value close to its BCS value. Model spectra with even higher modes will yield BCS results, and would be representative of say, an exciton model. Models with lower frequency phonon modes are representative of $\alpha^2F(\nu)$'s with strongly

Table 8.1
Summary of model spectra used in Fig. (8.2).

Freq. of δ–fns.	Area (meV)	λ	$T_c/\omega_{\ln}$	$\frac{2\Delta_o}{k_B T_c}$
60	23.1	.75	.05	3.7
30,60	20.0	1.0	.08	4.0
10,60	22.7	2.6	.23	5.5
10,20,30	18.4	2.3	.20	5.2
10,20	19.0	2.9	.24	5.6
10	21.9	4.4	.30	6.4
Pb($\gamma = 1$)	32.3	12.4	.62	8.0
Pb($\frac{1}{3}$)	18.3	2.3	.21	5.2
Pb($\frac{1}{5}$)	18.1	1.4	.12	4.4

renormalized oxygen breathing modes, as well as spectra with acoustic modes in a more conventional frequency regime. These are illustrated in Table 8.1 by models with one, two or even three delta functions of equal weights and varying positions for the Einstein energies. One immediate consequence of these models is that $2\Delta_o/k_BT_c$ attains values ≈ 5 or higher. It is seen how an accurate measurement of $2\Delta_o/k_BT_c$ can decide between these two possibilities. Moreover, the area required remains more or less constant at $A \sim 20\,\text{meV}$. This value is not so unreasonable, though it is unprecedented; the A15 compounds have values of $A \lesssim 13$. Similar remarks hold if a Pb spectrum is used with $b = 1/3$ or $1/5$. Thus, one sees that the electron-phonon coupling must be very large. The above results are all for the 35 K superconductor. The effects become even more pronounced for the 95 K superconductor. It can be seen for example, on the basis of Fig. 8.2 that a BCS gap measurement rules out the phonon mechanism completely, since phonon modes generally cannot exceed 100 meV. A phonon mechanism would invariably require large values of $2\Delta_o/k_BT_c$, and an area $A \gtrsim 50\,\text{meV}$ would be required. The possibility of a lattice instability would have to be seriously considered.

The amount of information one can glean without a knowledge of the underlying spectral function is somewhat limited, although if one is fortunate, it may be possible to pin down the characteristic boson frequency, which is of obvious significance. Next we will delineate the quantitative predictions based on Weber's calculated $\alpha^2F(\nu)$ spectrum[14] and then return to further considerations based on model spectra.

Based on the first-principles energy band results of L. Mattheiss,[80] W. Weber[14] has calculated an $\alpha^2F(\nu)$ spectrum for $La_{2-x}Sr_xCuO_4$ (LSCO) in the framework of the non-orthogonal tight binding theory of lattice dynamics developed by himself and C. M. Varma.[13] This framework is highly sophisticated and had already proven to be quite successful in describing the A15 compounds. In LSCO he found that the material bordered on a lattice instability (a fact which is not observed), achieving, with μ^* ($\omega_c = 540\,\text{meV}$) $= 0.13$, a $T_c \sim 35 - 40\,\text{K}$. The spectrum is illustrated in

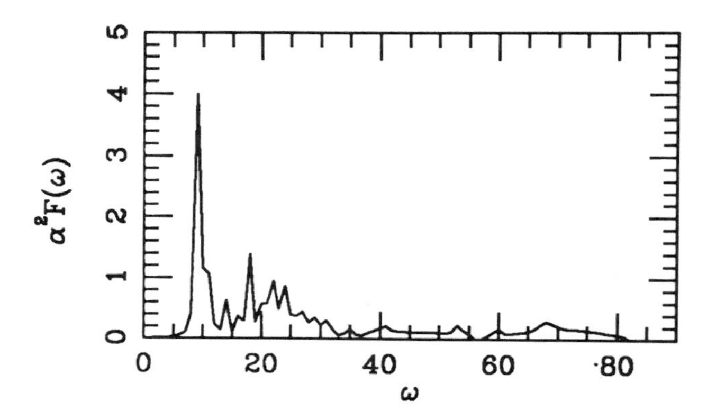

Figure 8.3 The electron-phonon spectral density $\alpha^2F(\omega)$ for LSCO calculated by Weber.

Fig. 8.3. Prominent features are (i) a very large peak at $\omega \sim 10\,\text{meV}$, and (ii) coupled phonon modes extending out to beyond 80 meV. The high frequency range is due to the low oxygen mass. Using this spectrum we have calculated many superconducting properties, based on the isotropic Eliashberg equations.[81] For most properties no adjustable parameters are involved and, although the free energy itself simply scales with $N(0)$, the reduced properties remain unaffected. Calculated ratios are presented in Table 8.2 (see last column). There is nothing unusual about the calculated values. The various properties have values characteristic of the strong coupling regime as described in section III. Unfortunately, nature has conspired to be somewhat secretive about the high T_c oxides. Properties involving γ_o have not been measured because γ_o is difficult to obtain. The critical temperature T_c is sufficiently high that the normal state specific heat at this temperature is dominated by the phonon contribution. Furthermore, the critical magnetic fields are so large that at low temperatures it is impossible to force the material to become normal, and hence measure γ_o. Finally, the gap ratio,[82] has been measured by many groups (see Table 8.3). Unfortunately, there remains considerable controversy about the experimental value of the gap in $La_{1.85}Sr_{0.15}CuO_4$. Far-infrared measurements tend to give values of the gap ratio close to the BCS value (with one exception) whereas tunneling measurements give values which are rather higher, although some measurements now give values closer to BCS. Both types of measurements display somewhat anomalous features: the far-infrared measurements give considerable absorption below twice the gap edge where there should be none. Moreover, the presence of phonons obscures the analysis somewhat. The tunneling measurements tend to be quite erratic, even from measurements at different spots on the same sample. Anomalous structure is also seen in the I-V characteristic. The theoretical value of 5.3 obtained from Weber's spectrum would certainly be compatible with some of the tunneling measurements. Unfortunately no definitive conclusion can be reached at this time.

The specific heat jump, ΔC, has also been measured by several groups,[82] and as is apparent from Table 8.3, there is considerable disagreement for this measurement as well. It is clear that there is considerable sample dependence. This is also exemplified by the fact that the normal state resistivity is (anomalously) linear in temperature down to T_c although in some measurements there is an upturn before the sample goes superconducting. It should be kept in mind that almost all experimental results

Table 8.2

Summary of theoretical results obtained for $\delta = 1.0$ (pure exciton or BCS), $\delta = 0.0$ (pure phonon) and $\delta = 0.25, 0.5$ (combined phonon-exciton mechanism to be discussed in the next section).

Property	BCS	$\delta^b = 0.50$	$\delta = 0.25$	$\delta = 0.0$
$\frac{2\Delta_0}{k_B T_c}$	3.53	4.0	4.4	5.3
β_{ox}	0.0	0.14	0.24	0.28
$\frac{\Delta C(T_c)}{\gamma T_c}$	1.43	1.7	2.1	2.8
$\frac{\gamma T_c^2}{H_c^2(0)}$	0.168	0.153	0.140	0.124
$h_c(0)$	0.576	0.55	0.52	0.48

$^b\delta \equiv \lambda_{ex}/\lambda_{tot}$

reported here were obtained from measurements on polycrystalline samples. There is also increasing evidence that while the superconductivity is bulk, it is by no means 100 % bulk. There seem to be both metallic and insulating components present. Moreover, the granularity of the samples varies from sample to sample, and this feature is known to affect many superconducting properties.[86–88] Nonetheless, we can proceed to investigate the consistency of the data in the following fashion. On the basis of the results quoted in Table 8.3, we choose a value of $\Delta C(T_c)/T_c = 17$ mJ/mole K^2. Using the calculated value $\Delta C/\gamma_o T_c = 2.8$ (see Table 8.2), we find $\gamma_o = 6.1$ mJ/ mole K^2, from which we can extract the band density of states. A useful formula is:

$$N(0)\left[\frac{\text{states}}{\text{eV}-\text{f.u.}-\text{spin}}\right] = \frac{0.212}{1+\lambda}\gamma_o\left[\frac{\text{mJ}}{\text{mole K}^2}\right] . \qquad (8.2)$$

We find $N(0) = 0.36\frac{\text{states}}{(\text{eV}-\text{f.u.}-\text{spin})}$. This value should correspond to the band structure density of states. We have assumed that the important electron-electron correlation effects have been included in the band structure calculation. The band structure density of states for La_2CuO_4 has been obtained by several groups.[85] There is general agreement that $N(0) \approx 0.65\frac{\text{states}}{(\text{eV}-\text{f.u.}-\text{spin})}$. Freeman *et al.*[85b] and Papaconstantopoulis *et al.*[85e] have used a rigid band model to include the effect of doping. They find for $La_{1.85}Sr_{0.15}CuO_4$, $N(0) = 0.9\frac{\text{states}}{(\text{eV}-\text{f.u.}-\text{spin})}$ and $N(0) = 1.1\frac{\text{states}}{(\text{eV}-\text{f.u.}-\text{spin})}$ respectively. Hence there is a discrepancy of a factor of 3 for $N(0)$. It may be argued that this rules out such a large value of λ for $La_{1.85}Sr_{0.15}CuO_4$. However, as has been alluded to already, the measured value of $\Delta C/T_c$ may be lower than the "ideal" value due to effects of granularity, anisotropy, and a significant non-superconducting fraction. We note that Phillips *et al.*[82g] have attempted to account for this latter effect.

The upper critical magnetic field has been measured near T_c by many groups.[84] Again as Table 8.3 indicates there is significant variation in the slope at T_c ($H'_{c2}(T_c)$) not only from sample to sample, but also depending on whether the onset or midpoint

Table 8.3
Summary of data for LSCO

$\frac{\Delta C(T_c)}{T_c}$ $\frac{mJ}{mole K^2}$	$\frac{2\Delta_o}{k_B T_c}$	$\mid H'_{c2}(T_c) \mid \frac{T}{K}$	$N_B(E_F)^*$
7.6 ± 1.8[a]	2.4[a]	$2.2 - 5.0$[a]	0.66[a]
20 ± 5[b]	$1.6 - 2.7$[b]	1.51[b]	0.6 $(0.95^{\flat})$[b]
16.8[c]	2.5[c]	1.7[c]	0.65[c]
8.8[d]	$2.9 - 4.5$[d]	$2.7 - 6.0$[d]	0.82[d]
$22 - 26$[e]	$5.2 - 9.1$[e]	$0.3 \parallel, 4 \perp$[e]	0.62 $(1.08^{\dagger})$[e]
11[f]	< 4.5[f]	1.8[f]	0.83[f]
9.9[g]	$5 - 8.7$[g]	2.13[g]	1.03[g]
10 ± 2[h]	$4.07 - 4.78$[h]	$1.3 - 4.0$[h]	
6.5[i]	$4.5 - 5.8$[i]	2[i]	
	$3 - 6$[j]	2.1 ± 0.1[j]	
	$3.5 - 4.0$[k]	2.2[k]	
	$0.7 - 2.7$[l]	$2.0 - 3.7$[l]	
	$8 - 18$[m]		
	2.6[n]		
	1.3 ± 0.2[o]		
	5[p]		
	4.7[q]		
	7 ± 2[r]		
	$2(c - axis)$[s]		

See Refs. (82,83,84,85) for columns (1-4), respectively.

$*$ units are: $\frac{\text{states}}{\text{ev}-\text{Cu}-\text{atom}-\text{spin}}$

$\flat$ with doping: $x = 0.15$

$\dagger$ with doping: $x = 0.15$

of the resistivity drop is used. Strictly speaking the onset point should be used, but the polycrystalline samples and the possible fluctuations above T_c complicate this simple prescription. Moreover, anisotropy[89] now plays a more significant role, as recent measurements[88e] on single crystals indicate. We proceed nonetheless with the isotropic theory, noting that simple modifications are required to treat an anisotropic electron gas.

The measured slope can also be used to give an independent estimate of the Sommerfeld γ_o, through the relation[90]

$$H'_{c2}(T_c) = -4.48 \times 10^4 \gamma_o \, \rho_{\Omega\,\text{cm}} \, \eta_{H_{c2}}(T_c) \quad [\text{OeK}^{-1}] \tag{8.3}$$

$\rho_{\Omega\,\mathrm{cm}}$ is the residual resistivity in Ω cm. The factor $\eta_{H_{c2}}(T_c)$, first introduced by Bergmann and Rainer,[55] takes into account strong coupling effects. It remains within 30% of unity for a moderate strong coupling material and so is not an important factor in this regime. The problematic factor is $\rho_{\Omega\,\mathrm{cm}}$. Normally, the residual resistivity is found by measuring the zero temperature normal state resistance; in this manner, the phonon-assisted resistance is removed. This is impossible, however, in he oxides; moreover, extrapolation is difficult—the linear resistivity behaviour is not well understood, and sometimes there is a sharp rise just before T_c. In some resistivity measurements, the extrapolation would cause $\rho(T=0)$ to be very near to zero, which is not understood. Kwok *et al.*[84b] used $\rho(T_c)$ in their analysis and obtained $\gamma_o \approx 4.9 \frac{\mathrm{mJ}}{\mathrm{mole{-}K}^2}$. A linear extrapolation to zero temperature gives a resistivity of just more than half that at T_c, so that we get $\gamma_o \approx 8 \frac{\mathrm{mJ}}{\mathrm{mole{-}K}}^2$ ($\eta_{H_{c2}} \approx 1$ for LSCO[14]). This is in reasonable agreement with our previous estimate for γ_o. It is clear however hat considerable uncertainty remains in such analyses.

Before closing this section, we should also mention the important isotope effect measurements which have taken place.[91,92] One group[91] finds $\beta_{ox} = 0.16 \pm 0.02$, where $\beta_i = -d(\ln T_c)/d(\ln M_i)$, and M_i is the ith element, in this case oxygen. The other group[92] finds a similar result, but with a much larger error margin. The naive isotope effect coefficient one expects is $\beta = 0.5$. It is normally argued that this can be reduced due to finite μ^*. In fact, when large values of λ are present to produce such high T_c's, he reduction in β due to μ^* is very small. Another important point is that here a partial isotope effect is involved, *i.e.* only oxygen atoms are replaced by O^{18} isotopes. Hence the simple fact that $\omega_{\mathrm{ph}} \sim M^{-\frac{1}{2}}$ no longer holds since phonon modes in general involve all the atoms in the unit cell. Rainer and Culetto[93] have analysed in detail partial isotope effects and Ashauer *et al.* [94] have found that under the assumption hat phonon modes that are due to the oxygen atoms can be isolated from the rest at high frequency, a small isotope effect is not unexpected. The most rigorous manner in which to compute a theoretical estimate for β_{ox} is for Weber to recompute $\alpha^2F(\omega)$, using O^{18} in place of O^{16} to compute the phonon dynamics for LSCO. He has done his,[91] and finds $\beta_{ox} \approx 0.30$. This is quite a bit larger than the measured value. This ould be used as an argument for a joint mechanism: phonon plus some excitonic ontribution. It is fair to say however at this early stage that it is difficult to say or sure that the measured isotope effect rules out the phonon mechanism. We note hat in Nb_3Sn, a well established electron-phonon superconductor, $\beta \approx 0.08$,[95] in significant disagreement with that expected from simple arguments.

It is clear that the experimental data is sufficiently vague and the isotope effect sufficiently small that few firm conclusions can be reached about the validity of Weber's $\alpha^2F(\nu)$. Most important in this regard will be a repeat of experiments on single crystals, along with a theoretical analysis which includes anisotropy.

We turn next to the case of $YBa_2Cu_3O_{7-y}$. The tunneling experiments (see Tables 8.3 and 8.4) on both LSCO and $YBa_2Cu_3O_{7-y}$ (YBCO)[96] indicate that the gap ratio $2\Delta_o/k_BT_c$ may be very large. As was mentioned previously, $2\Delta_o/k_BT_c$ increases as λ or $T_c/\omega_{\ln}$ increases. Hence it is of interest to study the regime $T_c/\omega_{\ln} \approx 1$ or the high T_c oxides. Further motivation comes from the simple observation that $T_c/\omega_{\ln} \approx 0.22$ for LSCO. In the absence of a calculated $\alpha^2F(\nu)$ spectrum for YBCO (Weber[97] has since produced a calculation which yields only a very low T_c) one can assume a spectrum having the same shape as the LSCO compound, but of course with

Table 8.4
Summary of Data for YBCO

$\frac{\Delta C(T_c)}{T_c}$ $\frac{\text{mJ}}{\text{moleK}^2}$	$\frac{2\Delta_o}{k_B T_c}$	$\lvert H'_{c2}(T_c) \rvert \frac{T}{K}$	$\text{N}_B(E_F)^\ddagger$
18^a	3.2 ± 0.3^a	$1.3(50\%)^a$	1.5^a
15.5^b	3.5 ± 0.3^b	$3(10\%)^a$	$0.56(0.43^*)(0.26^\dagger)^b$
13^c	2.0^c	$2.5(50\%)^b$	$1.1(0.92 - 0.97^\flat)^c$
1.32^d	3.5^d	$1.2(?)^c$	
16^e	$3.7 - 5.6^e$	$2.4(?)^c$	
7^f	$1.6 - 3.4^e$	$2.3 \parallel, 0.46 - 0.71 \perp^d$	
11 ± 2^g	$2.3 - 3.5^f$	$4.6(10\%)^e$	
23 ± 5^h	$2.5 - 4.2^g$	$0.6(0\%)^e$	
13^i	1.3 ± 0.2^h	$1.25(?)^f$	
11.3^j	3.3^i	$3.8(50\%)^g$	
20^k	3.2^j	$1.0(1\%)^g$	
	11^k	$2.35(50\%)^h$	
	$4.5 - 6.0 \perp^l$	$0.37 \perp (50\%)^i$	
	$3.9 - 4.8 \parallel^l$	$1.95 \parallel (50\%)^i$	
	4.8^m	$2.2(50\%)^j$	
	$3.8 - 4.5^n$	$1.8(50\%)^k$	
	10°	$2.2 - 3.6(50\%)^l$	
	13^p	$1.75(50\%)^m$	
	3.9^q	$5.3(10\%)^m$	
	3.2 ± 0.4^r	$3 \parallel, 0.9 \perp (100\%)^n$	
	$7 - 13^s$	$1.27(50\%)^o$	
	3.5^t	$4.7(90\%)^p$	
	5 ± 0.2^u	$1.9 \pm 0.2(?)^q$	
	3.4 ± 1.5^v	$1.3(50\%), 5(onset)^r$	
	4.8 ± 0.5^w	$1.9(50\%)^s$	
	$\sim 8^x$	$2.9(50\%)^t$	
	7.5^y	$5.3(90\%)^t$	
	3.9^z	$3 \pm 0.3(?)^u$	
	3.8^{aa}		

See Refs. (96,98,100,99) for columns (1-4), respectively.

$\ddagger$ units are: $\frac{\text{states}}{\text{ev}-\text{Cu}-\text{atom}-\text{spin}}$

$*$ $y = 0.1$, $\dagger$ $y = 0.2$, $\flat$ $y = 0.5$.

:oupling strength scaled to give $T_c \sim 90\,\mathrm{K}$. This immediately gives $T_c/\omega_{\ln} \approx 0.6$, and, n fact mode softening is expected, due to renormalization from increased electron-›honon coupling, so that even larger $T_c/\omega_{\ln}$ is expected. Of course, we have assumed hat a lattice instability has not occurred first, which at present is in conflict with Weber's findings.

The very strong coupling regime has been studied in section IV. Fig. 4.4 illus-rates the variation of $2\Delta_o/k_b T_c$ with increasing $T_c/\omega_{\ln}$. For $T_c/\omega_{\ln} \approx 0.6$, $2\Delta_o/k_B T_c$ hould be of the order of 7.0 . Reference to Table 8.4 shows that this is consistent vith some experiments but in conflict with others so that no firm conclusion can be nade. If $2\Delta_o/k_B T_c$ is BCS-like as is indicated in some other experiments, $T_c/\omega_{\ln}$ will ›e small and some high energy boson exchange mechanism is involved. As we now how the data on specific heat jump does not help in sorting out which possibility is orrect i.e. $T_c/\omega_{\ln} \approx 0.6$ or $T_c/\omega_{\ln} \approx 0.0$.

Many specific heat measurements[98] have also been performed, the results of vhich are listed in Table 8.4. Let us consider as given an average measured value f $\Delta C(T_c)/T_c = 15\,\mathrm{mJ/mole-Cu-K^2}$. Then, using the theoretically obtained $f \equiv \Delta C(T_c)/\gamma_o T_c$ values we can calculate the electron density of states at the Fermi urface, $N(0)$. With $f = 1.43$, $N(0)(1+\lambda) = 2.2\,\frac{\text{states}}{(\text{eV}-\text{Cu}-\text{atom}-\text{spin})}$. Table 8.4 lso lists some values of $N(0)$ calculated from band structure. With $\lambda \approx 0.5$ we vould obtain $N(0) \sim 1.5\,\frac{\text{states}}{(\text{eV}-\text{Cu}-\text{atom}-\text{spin})}$ in agreement with the upper limit of he calculated values (see Table 8.4) . Higher values of λ would violate the BCS ssumption. With $f = 2.8$, $N(0)(1+\lambda) = 1.1\,\frac{\text{states}}{(\text{eV}-\text{Cu}-\text{atom}-\text{spin})}$. Choosing $\lambda \approx$1.5–.0 gives $N(0) \sim$0.28–0.45 $\frac{\text{states}}{(\text{eV}-\text{Cu}-\text{atom}-\text{spin})}$, which agrees with Massidda *et al.*'s[99b] alues including doping. Finally, to represent the very strong coupling limit, we use ıe $\lambda = 30$ spectrum. Then, $f \approx 0.6$, so that $N(0)(1+\lambda) = 5.3\,\frac{\text{states}}{(\text{eV}-\text{Cu}-\text{atom}-\text{spin})}$, nd hence $N(0) \approx 0.2\,\frac{\text{states}}{(\text{eV}-\text{Cu}-\text{atom}-\text{spin})}$. This is just below the range indicated by Iassidda *et al.*'s[99b] calculations with doping.

On the basis of the existing data, then, it is not possible to favour one or ıe other of these possibilities. Many uncertainties exist in the analysis itself. The ulk specific heat capacity should be used, whereas most of the measurements were erformed on polcrystalline samples with as low as 25% Meissner effects at low tem-eratures. This might indicate that not all the material was superconducting and ıus $\frac{1}{4}f$ might be used in the analysis. Moreover, Deutscher *et al.*[87] have noted that ·anularity in a sample tends to reduce the specific heat jump. Finally, anisotropy, is ıown to reduce the jump as well. All of these effects point toward larger γ_o values ıan is obtained in this analysis, and hence larger mass renormalization, which would vour the very strong coupling limit. However, Fermi liquid effects have been ig-›red, and these may also play a large role in contributing to a mass renormalization. inally, the disagreement in values for $N(0)$ produced by band structure theorists also disturbing, and prevents this sort of analysis from being quantitative, at this age.

IX Combined Phonon-Exciton Mechanism

Exciton superconductivity was first suggested as a theoretical possibility in)64 by Little[101] and Ginzburg.[102] The idea is very similar to that involving phonons, ccept that the polarization is not due to a movement of the ions themselves, but rather to movements of electrons (say on the ion cores). This polarization will result in a net positively charged region of space to which a conduction electron will be attracted. Little[101] envisioned a one-dimensional geometry with conduction chains along side "exciton" chains. The two types of electrons (conduction and those that produce the excitons) are then physically separated from one another. Ginzburg considered a two-dimensional geometry with a metallic layer sandwiched between two dielectric layers. This geometry was later investigated in detail by Allender, Bray and Bardeen,[18] within the Eliashberg formalism. There are many objections to the idea of excitonic superconductivity, which have been neatly summarized by Little.[103] We will not reiterate them here, but simply note that many can be overcome under the proper conditions. Perhaps the most serious problem is that vertex corrections may become important,[104] since the ratio of the electron mass to exciton mass is no longer small. We will nonetheless proceed, following Allender *et al.*,[18] and ignore this complication.

The most compelling reason for considering excitonic superconductivity is associated with this last problem, in that the mediating boson (exciton) now has an electronic energy scale. Hence, in the simple BCS picture, T_c will be significantly enhanced because the prefactor is now $\omega_{\mathrm{ex}} \gg \omega_D$. Theoretical work[105–107] has also suggested the possibility of excitonic superconductivity in the high T_c oxides. Several experiments have indicated that excitons may be playing a role in these materials. Far-infrared optical measurements[108–110] reveal an increased absorption at high frequencies (0.44 eV in YBCO and 0.37 eV in LSCO), which may be due to an exciton mode being present. Moreover, the presence of the increased absorption has been shown[110,111] to be strongly correlated with the presence of superconductivity in the material. (The 'peak' has since disappeared in single crystal measurements.[112a] However, the significance of the anomalous "Drude-like" absorption is still controversial.[113]) Meanwhile, most far-infrared and some tunneling measurements indicated a BCS value for the gap ratio (see Tables 8.3 and 8.4), which, as we have argued earlier, implies that the mediating boson is electronic in origin. Finally, measurements of the isotope effect have suggested that the superconductivity is not phononic in origin. Measurements were first performed[114,115] for YBCO samples with ^{16}O replaced by ^{18}O. No isotope effect was observed. A subsequent measurement[116] reported a small isotope effect of $\beta_{\mathrm{ox}} \approx 0.05$, where $\beta_{\mathrm{ox}} \equiv -d[\ln T_c]/d[\ln M_{\mathrm{ox}}]$. The subscript "*ox*" indicates that only oxygen atoms are replaced. As previously discussed measurements on LSCO samples revealed much higher isotope effects with one group reporting[91] $\beta_{\mathrm{ox}} = 0.16 \pm 0.02$ and the other reporting[92] $0.15 < \beta < 0.35$. On the basis of Weber's calculation,[91] the expected isotope effect is $\beta_{\mathrm{ox}} = 0.30$, so that a discrepancy exists with the first result. Note, however, as we have already mentioned, this theoretical estimate can be significantly reduced[94] if more higher frequency oxygen phonon modes are participating in the electron-phonon interaction. In the meantime, the isotope effect in BaPbBiO has been measured[117] to be $\beta_{\mathrm{ox}} \approx 0.22$. One would like to think that the same mechanism is causing superconductivity in all three of these compounds, so we are led to investigate[118,119] a combined phonon-exciton mechanism. The case of a 96 K superconductor has been treated in Ref. 118, using the model:

$$\alpha^2F(\nu) = \frac{1}{2}\lambda_{\mathrm{ph}}\,\omega_{\mathrm{ph}}\,\delta(\nu-\omega_{\mathrm{ph}}) + \frac{1}{2}\lambda_{\mathrm{ex}}\,\omega_{\mathrm{ex}}\,\delta(\nu-\omega_{\mathrm{ex}}) \tag{9.1}$$

where $\lambda_{\mathrm{ph}}\,(\lambda_{\mathrm{ex}})$ is the electron-phonon (electron-exciton) electronic mass renormal-

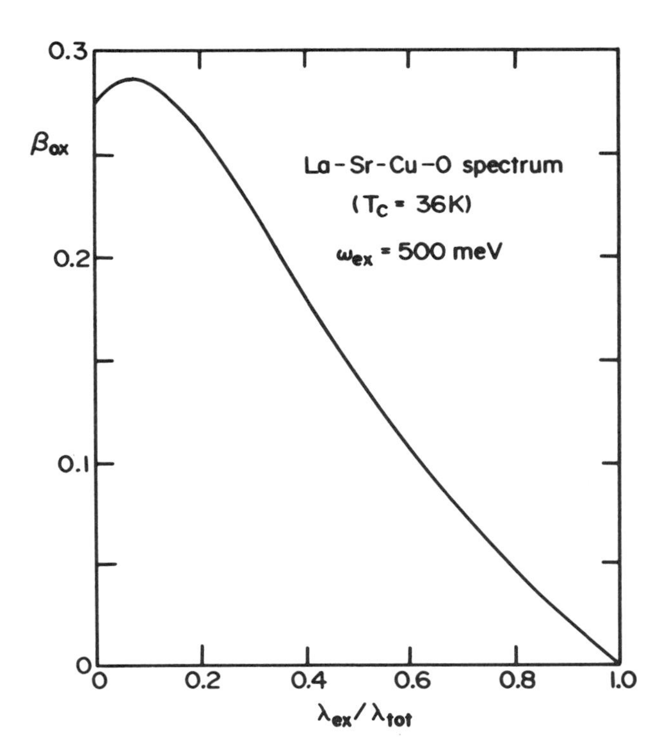

'igure 9.1 The isotope effect coefficient β_{ox} vs. $\lambda_{ex}/\lambda_{tot}$, with $\mu^* = 0.15$ for the combined phonon nd exciton spectrum. This calculation uses Weber's spectral functions calculated with and without xygen replacement in the Cu-O planes. Note that the experimentally measured value $\beta_{ox} = 0.16 \pm 0.02$ ives rather stringent constraints on the possible value of $\lambda_{ex}/\lambda_{tot}$ within this model.

zation. The two Einstein modes have been placed at frequencies $\omega_{ph} = 8\,\text{meV}$ (to imulate the phonon spectrum) and $\omega_{ex} = 500\,\text{meV}$ (to simulate the exciton specrum). The coupling strengths were then varied such that $\lambda_{ex}/\lambda_{tot}$ spanned the range rom 0 to 1 ($\lambda_{tot} = \lambda_{ex} + \lambda_{ph}$). We will simply report briefly on the results found here. The full isotope effect, gap ratio, specific heat jump and critical magnetic field roperties were calculated as a function of $\lambda_{ex}/\lambda_{tot}$. It was found that an isotope ffect of $0 < \beta < 0.05$ was not very restrictive, *i.e.* a large range of $\lambda_{ex}/\lambda_{tot}$ was ossible. Furthermore calculation of the various superconducting properties resulted n values very close to BCS values, and hence, would make a combined mechanism lifficult to be distinguished from a pure exciton mechanism through experiments. It s noteworthy, however, that in all the properties calculated, the trend as a function f decreasing $\lambda_{ex}/\lambda_{tot}$ (from the "BCS" value of 1) was qualitatively similar to the rends observed in the very strong coupling regime. This indicated that while most f the T_c (about 90 K) was caused by the "exciton" peak at $\omega_{ex} = 500\,\text{meV}$, the thernodynamic and critical magnetic field properties were largely affected by the phonon eak at $\omega_{ph} = 8\,\text{meV}$, which, considered on its own, gives the ratio $T_c/\omega_{ln} \approx 1$, vhich is in the very strong coupling regime. Note, however, that $2\Delta_o/k_BT_c$ remained elatively BCS-like for the range of $\lambda_{ex}/\lambda_{tot}$ for which $\beta \leq 0.05$.

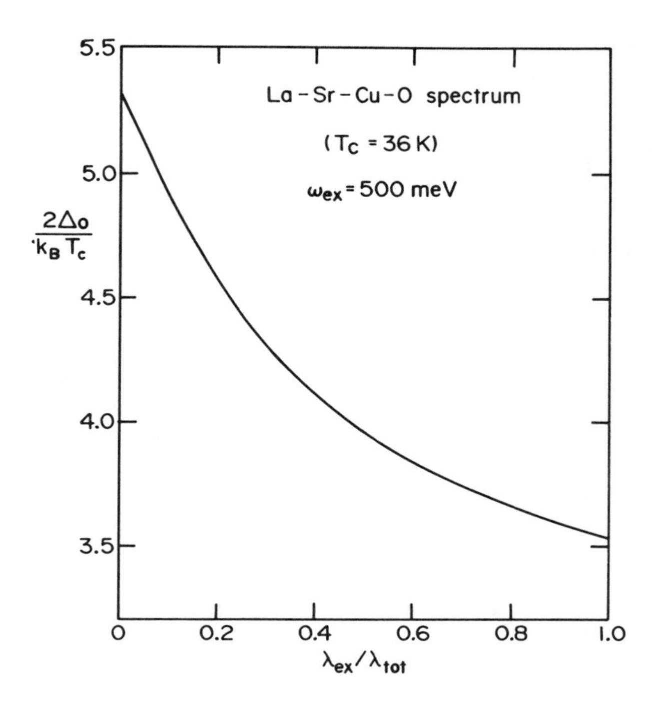

Figure 9.2 Plot of $2\Delta_o/k_BT_c$ vs. $\lambda_{ex}/\lambda_{tot}$ for the combined phonon-exciton model considered in the text. A choice of $\lambda_{ex}/\lambda_{tot} \approx 0.5$ implies $2\Delta_o/k_BT_c \approx 4$ which is in better agreement with far-infrared measurements of the gap edge.

Because of the relatively large isotope effect observed in the LSCO compound, the effects of a combined exciton-phonon mechanism are much more interesting to study. Moreover, the detailed calculation of $\alpha^2F(\nu)$ by Weber[14] allows for a more detailed calculation, although we follow the previous example and model the exciton mode by the same Einstein spectrum located at $\omega_{\text{ex}} = 500\,\text{meV}$. The relative weightings of the phonon and exciton contributions are then adjusted keeping $T_c = 36\,\text{K}$. We have used $\mu^*\,(\omega_c = 1.53\,\text{eV}) = 0.15$. The results for the isotope effect coefficient, β_{ox} are illustrated in Fig. (9.1). Note that when only phonons are present $\left(\lambda_{\text{ex}}/\lambda_{\text{tot}} = 0\right)$, β_{ox} is near 0.275, whereas, when only excitons are present, the isotope coefficient approaches zero, as is expected. We have used Weber's calculated spectrum with ^{16}O replaced by ^{18}O in the planes. Also note that had the isotope coefficient been measured to be near zero, the phonons could not have played a significant role, since the increase in β_{ox} from zero is relatively steep (compared to that for the 96 K superconductor—see Ref. 118). Note that the value for the pure phonon case would be slightly higher, except that there is a μ^* present. The small maximum near $\lambda_{\text{ex}}/\lambda_{\text{tot}} = 0$ is indicative of the fact that as λ_{ex} increases from zero it initially has the effect of representing a negative μ^* and hence tends to cancel some of the effect of the existing μ^*. The manner in which the isotope effect was calculated followed the method used by Rainer and Culleto.[93] They kept the cutoff in the Δ-channel fixed, and simply shifted the frequencies in the $\alpha^2F(\nu)$ spectrum downwards

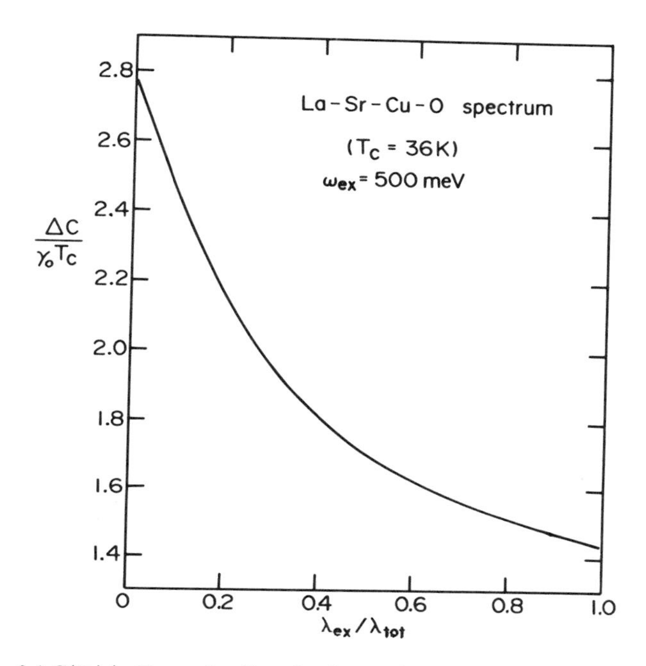

igure 9.3 Plot of $\Delta C(T_c)/\gamma_o T_c$ vs. $\lambda_{ex}/\lambda_{tot}$ for the combined phonon-exciton model considered in e text.

ı inverse proportion to the square root of the mass change. This method contrasts ith the more complicated procedure[26,120,121] of referring the cutoff to the phonon ectrum so that μ^* acquires an artificial mass dependence through its cutoff. We ave the added advantage of having Weber's more accurate calculation of the phonon equency shifts at our disposal as well.

Fig. (9.1) allows us to decide on the ratio of λ_{ph} to λ_{ex} in LSCO on the basis the isotope effect measurement[91] alone. Taking into account the axial site oxygen otope effect we require $\beta_{ox} \approx 0.14$ in Fig. (9.1), which determines $\lambda_{ex}/\lambda_{tot} \approx 0.5$. his choice will depend somewhat on our choice for μ^* and ω_{ex}, as well as the manner which we simply scaled Weber's $\alpha^2F(\nu)$ spectrum. However, for definiteness and r purposes of illustration we fix $\lambda_{ex}/\lambda_{tot} = 0.5$ and investigate other properties.

Fig. (9.2) displays the gap ratio, $2\Delta_o/k_BT_c$, as a function of $\lambda_{ex}/\lambda_{tot}$. Again the pure phonon case, we have simply Weber's spectrum, so that the results of the evious section apply, and $2\Delta_o/k_BT_c = 5.3$. In the other extreme a purely excitonic echanism implies that $T_c/\omega_{ln} \approx 0.006$, so that a BCS result will be achieved. Our oice of $\lambda_{ex}/\lambda_{tot}$ implies that $2\Delta_o/k_BT_c \approx 4$ (see also Table 8.2), which is certainly in e thick of things, as far as experiments go (see Table 8.3). Note that this and ensuing sults will differ slightly from those of Ref. 119, where the choice of $\lambda_{ex}/\lambda_{tot} \approx 0.4$ was ade based on the calculated *full* isotope effect. The normalized specific heat jump, $C(T_c)/\gamma_o T_c$, is illustrated in Fig. (9.3). $\lambda_{ex}/\lambda_{tot} = 0.5$ implies $\Delta C(T_c)/\gamma_o T_c = 1.7$ ee also Table 8.2). Again, here we cannot check this against experiment, but instead ly on it to determine γ_o. We use once again the experimental value $\Delta C(T_c)/T_c =$

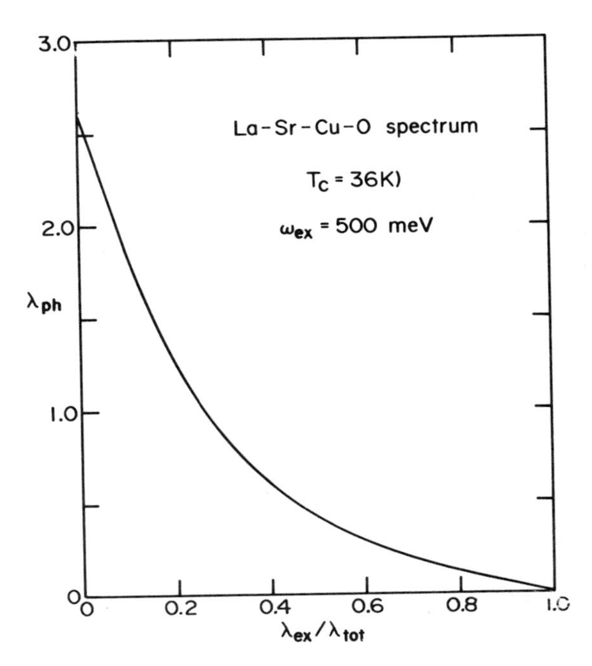

Figure 9.4 Plot of $\lambda_{\rm ph}$ vs. $\lambda_{\rm ex}/\lambda_{\rm tot}$ for the combined phonon-exciton model considered in the text.

$17\,\frac{\rm mJ}{\rm mole\,K^2}$ (see Table 8.3) so that $\gamma_o \approx 10\,\frac{\rm mJ}{\rm mole\,K^2}$. To determine the density of states we need to know $\lambda_{\rm tot}$. Fig. (9.4) illustrates $\lambda_{\rm ph}$ vs. $\lambda_{\rm ex}/\lambda_{\rm tot}$, from which $\lambda_{\rm tot}$ can be determined for $\lambda_{\rm ex}/\lambda_{\rm tot} = 0.5$. We find $\lambda_{\rm tot} \approx 0.9$. Using Eq. (8.2), we find $N(0) = 1.12\,\frac{\rm states}{{\rm eV}-fu-{\rm spin}}$. This value is far more consistent with the values 0.95[85b] and 1.08[85e] determined through band structure calculations than the value determined directly from Weber's spectrum (0.36 in the same units), and hence lends support to the combined phonon-exciton model for LSCO. The degree of agreement can be varied, however, by adjustment of choice of $\lambda_{\rm ex}/\lambda_{\rm tot}$ and by variation of μ^*, not to mention the choice of the experimental value of $\Delta C(T_c)/T_c$. What is clear, however, is that significant improvement has been achieved over the pure phonon model.

While the possibility of a combined phonon-exciton model looks very promising the existing experimental information is not sufficient to rule out either the pure phonon or pure "exciton" (BCS) case. In particular Schossmann *et al.* [81] have examined the former. Arberg *et al.* [94] have considered the latter case and find consistency with their experimental results.

X Conclusions

In the first part of this review, we derived strong coupling corrections to BCS universal constants. It was found that a semiphenomenological approach worked quite well, so that, rather than evaluate integrals requiring detailed information about $\alpha^2F(\nu)$ for each material, averages of the spectral moment were defined, in terms of single parameter, $T_c/\omega_{\ln}$. Coefficients were fit on the basis of realistic spectral hapes available. A more sophisticated approach could have been used; however, his would have defeated our purpose. We wished simply to describe the trends of roperties semi-quantitatively as a function of strong coupling. We found that all the orrections were of the form,

$$bx^2 \ln(1/cx) \tag{10.1}$$

vhere $x \equiv T_c/\omega_{\ln}$.

The expression (10.1) signifies the fact that a single parameter could describe uite well strong coupling properties. As an immediate consequence, this implies hat various properties are simply related to one another. Hence, our analysis allows stimates of the gap edge based on thermodynamic measurements, for example.

All known conventional materials, with the exception of amorphous materials, re found to fall in the regime where the strong coupling expansion used is valid. With he discovery of the high-T_c oxides, however, the possibility remains that their $T_c/\omega_{\ln}$ alues fall outside the regime described by expressions of type (10.1). On the basis of spectral function calculated by W. Weber, we have tried to evaluate the degree of greement with experiment. Inspection of Table 8.3, however, exhibits the variation n the measured properties. Moreover, the Sommerfeld constant γ_o is critical to our nalysis, and, at the same time, very difficult to measure. Disagreement by a factor f about three has been found for the density of electron states, although this can nly be considered serious after some of the measurements have settled down. Single rystal measurements may provide more uniformity, although our calculations should e extended to include anisotropy. Large anisotropy in the electron gas alone could ause a significant non-constant density of states at the Fermi surface. No doubt n anisotropic electron gas will also manifest itself in an anisotropic electron-phonon nteraction so that superconducting properties will also be affected by the resulting nisotropic gap.

The high T_c's observed in these materials have motivated us to explore the ery strong coupling regime, in which the parameter $T_c/\omega_{\ln}$ can take on values of bout unity. Many signatures of this regime have been found. The gap ratio will be ery large, while, at the same time, the normalized specific heat jump will be *less* han BCS. The critical thermodynamic magnetic field also behaves in a distinctive ay as does the universal ratio $\gamma_o T_c^2/H_c^2(0)$ with γ_o the Sommerfeld constant, T_c the ritical temperature and $H_c(0)$ the zero temperature critical magnetic field.

Besides the very-strong coupling regime which obtains for large but finite values f λ we have been investigating the $\lambda \to \infty$ regime. We found that the gap (Δ_o) to T_c ratio $2\Delta_o/k_BT_c$ saturates at a value of about 13. Limits on other thermodynamic atios were also established although they occur for finite values of λ. As an example he normalized specific heat jump $\Delta C(T_c)/\gamma_o T_c$ saturates at a value less than 4.0 with ome variations with μ^*. Known conventional superconductors are found to exhibit alues below this limit. Should a superconductor fall above this bound we could onclude directly that its thermodynamics cannot be described within the framework f isotropic Eliashberg theory whatever the size, shape and origin of the kernel used. imilar important bounds were established for other properties and should be of ome use when considering experiments. In the course of establishing these bounds unctional derivatives were introduced, calculated and found to be very useful.

In light of the various measured isotope effects, we have also investigated a ombined phonon-exciton mechanism, where by "exciton" here, we mean some high equency electronic excitation which mediates as electron-electron attraction. In oth LSCO and YBCO, this possibility would be difficult to detect on the basis of thermodynamic and magnetic measurements alone. In the former, the effect of an added exciton mechanism is to mimic a pure phonon mechanism with a higher characteristic frequency. In the latter, the isotope effect is almost nil, so even if phonons are partially responsible, the predicted properties would be very BCS like. Nonetheless, the interesting result that values of $\Delta C(T_c)/\gamma_o T_c$ below BCS could be obtained with the addition of a few low frequency phonons is surprising. It indicates that while the exciton component is most responsible for the high T_c, the phonon component is drastically affecting the thermodynamic properties. In the case of LSCO, the precision of the isotope effect measurement looks promising, but the other measurements have to settle down before a quantitative analysis as we have suggested can be taken seriously. Still a combined mechanism is fairly consistent with the data presently available.

Acknowledgements

The research reported here was supported in part by the Natural Sciences and Engineering Research Council of Canada (NSERC) and the Ontario Centre for Materials Research (OCMR). J. P. Carbotte is a fellow of the Canadian Institute for Advanced Research (CIAR).

Appendix A - Spectral Function Sources

We tabulate here the sources for the $\alpha^2F(\nu)$ spectra used in our calculations. Most of the spectra come from a tabulation of Rowell, McMillan and Dynes.

1.) Al comes from a theoretical calculation of Leung *et al.*[10]

2.) Several sources are available for Nb. Nb(R) comes from Robinson and Rowell[36]. Nb(A) comes from G.B. Arnold *et al.*[37] These have been measured through tunneling. A theoretical calculation (Nb(B)) comes from Butler *et al.*[12]

3.) V has been obtained through tunneling by Zasadzinski *et al.*[38]

4.) Amorphous Bi and Ga have been obtained through tunneling by Chen *et al.*[39]

5.) La is from Lou and Tomasch[40] (tunneling).

6.) Mo is from Kimhi and Geballe[41] (tunneling).

7.) Nb_3Sn has been obtained from Shen[42] through tunneling.

8.) V_3Si (Kihl.) has been obtained through tunneling by Kihlstrom.[43] We have scaled it to give $\lambda = 1$ and used $T_c = 16.4$ K (rather than 15.4 K as measured by Kihlstrom), which is in better agreement with the single crystal T_c value.

9.) V_3Si_1 was obtained from scaling $G(\Omega)$ obtained through inelastic neutron scattering, such that $\lambda = 1$, by Schweiss *et al.* .[44]

10.) Nb_3Al has three possibilities; two of them ((1) and (2)) were obtained from tunneling measurements by Kwo and Geballe;[45] the third (3) is a phonon spectrum obtained by Schweiss *et al.* [44] and scaled to give $\lambda = 1.7$.

11.) Nb_3Ge (1) was obtained through tunneling by Geerk *et al.* [46] whereas Nb_3Ge (2) was from neutron scattering data obtained by Müller *et al.* [47] to give $\lambda = 1.6$. μ^* has been fitted to give $T_c = 20$ K.

12.) V_3Ga has been obtained by tunneling by Zasadzinski *et al.* [48]

13.) Finally, the spectra by Weber[14] have been obtained by theoretical calculation.

References

[1] J. Bardeen, L.N. Cooper, J.R. Schreiffer, *Phys. Rev.* **108**, (1957) 1175.

[2] J.R. Schreiffer, Theory of Superconductivity, Benjamin, New York, 1964.

[3] J. M. Daams and J.P. Carbotte, *J. Low Temp. Phys.* **43**, (1981) 263.

[4] See R. Meservey and B.B. Schwartz, in Superconductivity, edited by R.D. Parks (Marcel Dekker, Inc., New York, 1969), Vol. 1, p.117.

[5] D. J. Scalapino, in Superconductivity, edited by R.D. Parks (Marcel Dekker, Inc., New York, 1969), Vol. 1, p.449.

[6] D. Rainer and G. Bergmann, *J. Low Temp. Phys.* **14**, (1974) 501.

[7] G.M. Eliashberg, *Zh. Eksp. Teor. Fiz.* **38**, (1960) 966; *Sov. Phys.-JETP* **11**, (1960) 696.

[8] G. Grimvall, The Electron-Phonon Interaction in Metals, North-Holland, New York, 1981.

[9] J.P. Carbotte and R.C. Dynes, *Phys. Rev.* **172**, (1968) 476.

[10] H.K. Leung, J.P. Carbotte, D.W. Taylor and C.R. Leavens, *J. Low Temp. Phys.* **24**, (1976) 2534.

[11] P.G. Tomlinson and J.P. Carbotte, *Can. J. Phys.* **55**, (1977) 751.

[12] W.H. Butler, H.G. Smith and N. Wakabayashi, *Phys. Rev. Lett.* **39**, (1977) 1004.

[13] C.M. Varma and W. Weber, *Phys. Rev. B* **19**, (1979) 6142.

[14] W. Weber, *Phys. Rev. Lett.* **58**, (1987) 1371. See also W. Weber, *Phys. Rev. Lett.* **58**, (1987) 2154(E).

[15] W.L. McMillan and J.M. Rowell, in Superconductivity, edited by R.D. Parks (Marcel Dekker, Inc., New York, 1969), Vol. 1, p.561. See also E.L. Wolf and G.B. Arnold, *Phys. Reports* **91**, (1982) 32.

[16] A.B. Migdal, *Zh. Eksp. Teor. Fiz.* **34**, (1958) 1438; *Sov. Phys.-JETP* **34**, (1958) 996.

[17] See, for example, Novel Superconductivity, edited by S.A. Wolf and V.Z. Kresin [Plenum, New York, 1987].

[18] D. Allender, J. Bray and J. Bardeen, *Phys. Rev. B* **7**, (1973) 1020.

[19] J.G. Bednorz and K.A. Müller, *Z. Phys. B* **64**, (1986) 189.

[20] C.W. Chu, P.H. Hor, R.L. Meng, L. Gao, Z.J. Huang, and Y.Q. Wang, *Phys. Rev. Lett.* **58**, (1987) 405.

[21] R.J. Cava, R.B. Van Dover, B. Batlogg, and E.A. Rietman, *Phys. Rev. Lett.* **58**, (1987) 408.

[22] S. Uchida, H. Takagi, K. Kitazawa, and S. Tanaka, *Jpn. J. Appl. Phys.* **26**, (1987) L1.

[23] M.K. Wu, J.R. Ashburn, C.J. Torng, P.H. Hor, R.L. Meng, L. Gao, Z.J. Huang, Y.Q. Wang, and C.W. Chu, *Phys. Rev. Lett.* **58**, (1987) 908.

[24] P.B. Allen and R.C. Dynes, *Phys. Rev. B* **12**, (1975) 905.

[25] P.B. Allen and B. Mitrović, in Solid State Physics, edited by H. Ehrenreich, F Seitz, and D. Turnbull (Academic, New York, 1982), Vol. 37, p.1.

[26] P. Morel and P.W. Anderson, *Phys. Rev.* **125**, (1962) 1263.

[27] J. Bardeen and M. Stephen, *Phys. Rev.* **136**, (1964) A1485.

[28] J.R. Schreiffer, D.J. Scalapino, and J.W. Wilkins, *Phys. Rev. Lett.* **10**, (1963) 336.

[29] C.R. Leavens and E.W. Fenton, *Solid State Comm.* **33**, (1980) 597.

[30] F. Marsiglio, M. Schossmann and J.P. Carbotte, *Phys. Rev. B* **37**, (1988) 4965.

[31] F. Marsiglio, Ph.D. Thesis, McMaster University (1988) unpublished.

[32] F. Marsiglio and J.P. Carbotte, *Phys. Rev. B* **33**, (1986) 6141.

[33] B.T. Geilikman and V.Z. Kresin, *Fiz. Tverd. Tela (Leningrad)* **1**, (1965) 3294. [*Sov. Phys.-Solid State* **7**, (1966) 2659.]

[34] N.F. Masharov, *Fiz. Tverd. Tela* **16**, (1974) 2342. [*Sov. Phys.-Solid State* **16** (1975) 1524.]

[35] V.Z. Kresin and V.P. Parkhomenko, *Fiz. Tverd. Tela (Leningrad)* **16**, (1974) 3363. [*Sov. Phys.-Solid State* **16**, (1975) 2180.].

[36] R. Robinson and J. Rowell (private communication)

[37] G.B. Arnold, J. Zasadzinski, J.W. Osmun and E.L. Wolf, *J. Low Temp. Phys.* **40** (1980) 225.

[38] J. Zasadzinski, D.M. Burnell, E.L. Wolf and G.B. Arnold, *Phys. Rev. B* **25** (1982) 1662.

[39] T.T. Chen, J.T. Chen, J.D. Leslie, and H.J.T. Smith, *Phys. Rev. Lett.* **22**, (1969 526.

[40] L.F. Lou and W.J. Tomasch, in Low Temperature Physics-LT 13, edited by K.D Timmerhaus, W.J. O'Sullivan, and E.F. Hammel (Plenum Press, New York, 1972) p. 599.

[41] D.B. Kimhi and T.H. Geballe, *Phys. Rev. Lett.* **45**, (1980) 1039.

[42] L.Y. Shen, *Phys. Rev. Lett.* **29**, (1972) 1082.

[43] K.E. Kihlstrom, *Phys. Rev. B* **32**, (1985) 2891.

[44] B.P. Schweiss, B. Renker, E. Schneider and W. Reichardt, in Superconductivity in d- and f-Band Metals, edited by D.H. Douglass (AIP, New York, 1972).

[45] J. Kwo and T.H. Geballe, *Phys. Rev. B* **23**, (1981) 3230.

[46] J. Geerk, J.M. Rowell, P.H. Schmidt, F. Wuchner and W. Schaver, in Superconductivity in d- and f- Band Metals, edited by W. Buckel and W. Weber (Kernforschungszentrum Karlsruhe, Karlsruhe, West Germany, 1982).

[47] P. Muller, N. Nucker, W. Reichardt and A. Muller in Superconductivity in d and f- Band Metals, edited by W. Buckel and W. Weber (Kernforschungszentrum Karlsruhe, Karlsruhe, West Germany, 1982).

[48] J. Zasadzinski, W.K. Schubert, E.L. Wolf and G.B. Arnold in Superconductivity in d- and f- Band Metals, edited by H. Suhl and M.B. Maple (Academic Press New York, 1980).

[49] J.M. Coombes and J.P. Carbotte, *J. Low Temp. Phys.* **63**, (1986) 431.

50] J.M. Coombes and J.P. Carbotte, *Phys. Rev. B* (in press).

51] B. Mitrović, H.G. Zarate, and J.P. Carbotte, *Phys. Rev. B* **29**, (1984) 184.

52] F. Marsiglio and J.P. Carbotte, *Solid State Comm.* **63**, (1987) 419.

53] J.P. Carbotte, F. Marsiglio and B. Mitrović, *Phys. Rev. B* **33**, (1986) 6135.

54] F. Marsiglio, R. Akis and J. P. Carbotte, *Phys. Rev. B* **36**, (1987) 5245.

55] G. Bergmann and D. Rainer, *Z. Physik* **263**, (1973) 59.

56] J.M. Daams and J.P. Carbotte, *Can. J. Phys.* **56**, (1978) 1248.

57] B. Mitrović, C.R. Leavens and J.P. Carbotte, *Phys. Rev. B* **21**, (1980) 5048.

58] E. Schachinger, J.M. Daams and J.P. Carbotte, *Phys. Rev. B* **22**, (1980) 3194.

59] J.M. Daams and J.P. Carbotte, *Solid State Comm.* **33**, (1980) 585.

60] J.M. Daams, E. Schachinger and J.P. Carbotte, *J. Low Temp. Phys.* **42**, (1981) 69.

61] B. Mitrović and J.P. Carbotte, *Solid State Comm.* **37**, (1981) 1009.

62] M. Ashraf and J.P. Carbotte, *Solid State Comm.* **46**, (1983) 63.

63] M. Whitmore, *J. Low Temp. Phys.* **56**, (1984) 129.

64] F. Marsiglio and J.P. Carbotte, *Phys. Rev. B* **31**, (1985) 4192.

65] F. Marsiglio, J.P. Carbotte and E. Schachinger, *J. Low Temp. Phys.* **65**, (1986) 305.

66] Y. Wada, *Phys. Rev.* **135**, (1964) A1481.

67] G. Grimvall, *J. Phys. Chem. Solids* **29**, (1968) 1221.; *Phys. Kondens. Mater.* **9**, (1969) 283.. See also V.Z. Kresin and G.O. Vaitsev, *Zh. Eksp. Teor. Fiz.* **74**, (1978) 1886; *Sov. Phys.-JETP* **47**, (1978) 983.

68] C.R. Leavens, *Solid State Comm.* **17**, (1975) 1499.

69] C.R. Leavens, *J. Phys. F* **7**, (1977) 1911.

70] C.R. Leavens and J.P. Carbotte, *J. Low Temp. Phys.* **14**, (1974) 195.

71] J.W. Blezius and J.P. Carbotte, *Phys. Rev. B* **36**, (1987) 3622.

72] J.W. Blezius and J.P. Carbotte,*Jour. Low Temp. Phys.* (in press).

73] F. Marsiglio, P. Williams and J.P. Carbotte (unpublished)

74] V.Z. Kresin, *Solid State Comm.* **63**, (1987) 725.

75] P.W. Anderson, *Mater. Res. Bull.* **8**, (1973) 153.

76] P.W. Anderson, *Science* **235**, (1987) 1196.

77] P.W. Anderson, G. Baskaran, Z. Zou, and T. Hsu, *Phys. Rev. Lett.* **58**, (1987) 2790.

78] C.M. Varma, in Superconductivity in d- and f- Band Metals, edited by W. Buckel and W. Weber (Kernforschungszentrum Karlsruhe, Karlsruhe, West Germany, 1982) p.603.

79] M. Cohen and P.W. Anderson, in Superconductivity in d- and f-Band Metals, edited by D.H. Douglass (AIP, New York, 1972) p.17.

80] L.F. Mattheiss, *Phys. Rev. Lett.* **58**, (1987) 1028.

[81] M. Schossmann, F. Marsiglio and J.P. Carbotte, *Phys. Rev. B* **36**, (1987) 3627.

[82] For the references containing measured values of the specific heat jump for LSCO see reference 31.

[83] For the references containing measured values of the gap ratio for LSCO see reference 31.

[84] For the references containing measured values of the slope of the upper critical field near T_c for LSCO see reference 31.

[85] For the references containing band structure calculations for LSCO and also contain estimates for the electron density of states at the Fermi level see reference 31.

[86] R.L. Filler, P. Lindenfeld, T. Worthington and G. Deutscher, *Phys. Rev. B* **21** (1980) 5031.

[87] G. Deutscher, O. Entin-Wohlman, S. Fishman and Y. Shapira, *Phys. Rev. B* **21** (1980) 5041.

[88] C. Ebner and D. Stroud, *Phys. Rev. B* **31**, (1985) 165.

[89] H. Teichler in Anisotropy Effects in Superconductors, edited by H.W. Weber (Plenum, New York, 1977) p.7.

[90] N.R. Werthamer, E. Helfand and P.C. Hohenberg, *Phys. Rev.* **147**, (1966) 295.

[91] B. Batlogg, G. Kourouklis, W. Weber, R.J. Cava, A. Jayaraman, A.E. White K.T. Short, L.W. Rupp, and E.A. Rietman, *Phys. Rev. Lett.* **59**, (1987) 912.

[92] T.A. Faltens, W.K. Ham, S.W. Keller, K.J. Leary, J.N. Michaels, A.M. Stacy, H zur Loye, D.E. Morris, T.W. Barbee,III, L.C. Bourne, M.L. Cohen, S. Hoen, and A. Zettl, *Phys. Rev. Lett.* **59**, (1987) 915.

[93] D. Rainer and F.J. Culetto, *Phys. Rev. B* **19**, (1979) 2540.

[94] B. Ashauer, W. Lee, D. Rainer, and J. Rammer (preprint).

[95] C. Kittel, Introduction to Solid State Physics, John Wiley and Sons, Inc. Toronto, 1986.

[96] For the references containing measurements of the gap ratio for YBCO see reference 31.

[97] W. Weber, private communication.

[98] For the references containing measurements of the specific heat jump for YBCO see reference 31.

[99] For the references containing band structure calculations for YBCO and also contain estimates for the electron density of states at the Fermi level see reference 31.

[100] For the references containing measurements of the slope of the upper critical magnetic field near T_c for YBCO see reference 31.

[101] W.A. Little, *Phys. Rev.* **134**, (1964) A1416.

[102] V.L. Ginzburg, *Zh. Eksp. Teor. Fiz.* **47**, (1964) 2318. *Sov. Phys.-JETP* **20** (1965) 1549.

[103] W.A. Little, in Novel Superconductivity, edited by S.A. Wolf and V.Z. Kresin [Plenum, New York, 1987], p. 341.

[104] M. Grabowski and L.J. Sham, *Phys. Rev. B* **29**, (1984) 6132.

[105] C.M. Varma, S. Schmitt-Rink and E. Abrahams, *Solid State Comm.* **3**, (1987) 3.

[106] J. Yu, S. Massidda, A.J. Freeman and D.D. Koeling, *Phys. Lett.* **A122**, (1987) 203.

[107] C.F. Gallo, L.R. Whitney and P.J. Walsh, in Novel Superconductivity, edited by S.A. Wolf and V.Z. Kresin [Plenum, New York, 1987], p. 385.

[108] J. Orenstein *et al.* , *Phys. Rev. B* **36**, (1987) 729.

[109] S.L. Herr *et al.* , *Phys. Rev. B* **36**, (1987) 733.

[110] K. Kamarás *et al.* , *Phys. Rev. Lett.* **59**, (1987) 919.

[111] S. Etemad, D.E. Aspnes, M.K. Kelly, R. Thompson, J.M. Tarascon and G.W. Hull, *Phys. Rev. B* **37**, (1987) 3396.

[112] I. Bozovic *et al.* , *Phys. Rev. Lett.* **59**, (1987) 2219.

[113] S. Etemad, T. Timusk *et al.* (preprint). It was definitively established at the Materials and Mechanisms conference (March, 1988) that a mid-infrared absorption does indeed exist in the oxides.

[114] B. Batlogg, R.J. Cava, A. Jayaraman, R.B. van Dover, G.A. Kourouklis, S. Sunshine, D.W. Murphy, L.W. Rupp, H.S. Chen, A. White, K.T. Short, A.M. Mujsce, and E.A. Reitman, *Phys. Rev. Lett.* **58**, (1987) 2333.

[115] L.C. Bourne, M.F. Crommie, A. Zettl, H. zur Loye, S.W. Keller, K.L. Leary, A.M. Stacy, K.J. Chang, M.L. Cohen and D.E. Morris, *Phys. Rev. Lett.* **58**, (1987) 2337.

[116] K.J. Leary, H. zur Loye, S.W. Keller, T.A. Faltens, W.K. Ham, J.N. Michaels and A.M. Stacy, *Phys. Rev. Lett.* **59**, (1987) 1236.

[117] B. Batlogg, private communication.

[118] F. Marsiglio and J.P. Carbotte, *Phys. Rev. B* **36**, (1987) 3937.

[119] F. Marsiglio, R. Akis and J.P. Carbotte, *Solid State Comm.* **64**, (1987) 905.

[120] W.L. McMillan, *Phys. Rev.* **167**, (1968) 331.

[121] C.R. Leavens, *Solid State Comm.* **15**, (1974) 1329.

A STRONG-COUPLING APPROACH TO HIGH TEMPERATURE SUPERCONDUCTORS

Josef Rammer

Physikalisches Institut der Universitat Bayreuth
D-8580 Bayreuth, West Germany

1. INTRODUCTION

In their classic paper on high temperature oxide superconductors Bednorz and Müller [1] also considered the eventual origin o superconductivity in the system they were studying. According to Bednor and Müller,.their search for superconductivity in the Ba-La-Cu-O syste was motivated by indications of strong electron-phonon interactions i related compounds such as Nb-doped $SrTiO_3$ (T_c=0.7K) and $BaPb_{1-x}Bi_xO_3$ (T_c=13K) [1,2]. The unusually high ratio $T_c/N(0)$ (N(0) is the electroni density of states at the Fermi energy) in both superconductors implie (using conventional Bardeen-Cooper-Schrieffer (BCS) theory[1] [3]) correspondingly strong electron-phonon interaction. From these result Bednorz and Müller concluded that "within the BCS mechanism, one may fin still higher T_c's in perovskite-type or related metallic oxides, if th electron-phonon interactions and the carrier densities at the Fermi leve can be enhanced further" [1].

The guiding idea in the search for superconductivity in metalli oxides in [1] was based on polaron formation due to the Jahn-Telle effect. This mechanism should provide a strong electron-phono interaction, and leads to the concept of a superconductivity (metallic) t bipolaronic (insulator) transition phase diagram as proposed b Chakraverty [4] and Alexandrov and Ranninger [5]. According to thes ideas, high superconducting transition temperatures are expected just a the borderline of the metal-insulator transition. Indeed, it turned ou

[1] In this article the term "BCS theory" or "BCS model" denotes all pairing theories which involve the BCS Cooper-pair groundstate; the notion "conventional BCS theory" refers to the phonon-mediated pairing mechanism

that whether the system becomes superconducting or insulating depends critically on the stoichiometry, and even the superconducting compounds often show a characteristic upturn in the resistivity with decreasing temperature (reminiscent of the onset of localization), before the curve eventually drops to zero. This feature is common to all known high temperature oxide superconductors of the perovskite type.

The present situation concerning the superconductive mechanism in the high temperature oxide superconductors $BaPb_{1-x}Bi_xO_3$ (T_c~13K) [6], $La_{2-x}Ba_xCuO_4$ (T_c~30K) [1] , $YBa_2Cu_3O_{7-x}$ (T_c~90K) [7], $Tl_2Ca_2Ba_2Cu_3O_x$ (T_c~120K) [8], and related compounds is highly controversial. In particular, it is not clear whether the above concept represents a relevant approach to understanding why these materials are (high temperature) superconductors. Moreover, there is also no consensus about the role of the phonons, i.e. whether or not a (more or less) conventional pairing mechanism driven by the electron-phonon interaction is responsible for the superconducting condensation.

Despite the controversial status characterizing the present discussion, various familiar features (as compared with conventional superconductors) have been established experimentally: the Josephson effects have been measured and it has been demonstrated that pairs of electrons are present in the superconducting state [9-12]; the same result has been obtained from the direct observation of the flux-line lattice (which has hexagonal symmetry) in $YBa_2Cu_3O_{7-x}$ by means of a decoration technique [13]; successful Andreev-scattering experiments have been performed, and they are consistent with the BCS Cooper-pair ($k\uparrow,-k\downarrow$) groundstate [14,15]; the value of the gap ratio $2\Delta_0/k_BT_c$, although still controversial, is in the range expected from BCS theory [16]. All these experiments (and others which are not cited) are consistent with the BCS model, according to which the superconducting transition occurs due to a Bose-like condensation of Cooper pairs at T_c. For this reason, most of the theoretical work nowadays is based on the BCS theory, as well as extensions or modifications thereof.

In this article I do not intend to propose or favor a microscopic mechanism for the high temperature superconductors. Due to the as yet inconclusive situation, this question is open to me, and I believe an answer can only emerge from a careful comparison between experiments on well-characterized samples and the various theories to be tested. The scope of this article is to discuss selected aspects of BCS-like theories, in particular the potential role of strong-coupling effects on superconducting properties, as studied in the literature. After a very brief summary of some basic concepts of strong-coupling theory in section 2 (which follows closely Rainer's review article [17]), I introduce in section 3 the Gorkov-Eilenberger equations, which determine magnetic and thermodynamic properties of inhomogeneous superconductors in the framework of the BCS theory. Section 4 is devoted to experimental and theoretical studies of phonon properties of high temperature superconductors; such investigations are crucial for assessing the role of the phonons in the pairing mechanism. In section 5 I discuss various approaches based on phonon as well as non-phonon pairing mechanisms, and their experimental consequences. Conclusions will be given in section 6.

2. ELEMENTS OF STRONG-COUPLING THEORY

If the high temperature oxide superconductors are of the BCS type, the question arises about the nature of the Cooper-pairing mechanism, i.e. which excitation in the phononic or electronic system mediates the net attractive interaction between the electronic quasiparticles forming the Cooper pair. It is well known that the energy scale of this excitation (characteristic frequency ω_o) critically determines the properties of the superconducting state; if $\hbar\omega_0$ is much larger than k_BT_c, the superconductor will show *universal* weak-coupling BCS behaviour, whatever the origin of ω_0 (here it is assumed that all other relevant normal-state energies are much larger than k_BT_c as well). On the other hand, if $\hbar\omega_0$ (or other normal-state energies like the Fermi energy E_F or the band width D) is of the order of k_BT_c, superconducting properties will be modified in a characteristic way.

In the above sense, I define strong-coupling effects as *deviations from weak-coupling BCS theory that are due to low-lying normal-state excitations* (e.g. ω_0) [17]. It is evident that strong-coupling effects, if they occur, provide some insight into the internal dynamics of the interaction mediating the Cooper-pair formation. This fact has in the past been extensively exploited in obtaining the coupling characteristics of conventional (phonon) strong-coupling superconductors in tunneling measurements[2] [18]. In this way, information about the microscopic superconductive mechanism can in principle be obtained. However, if the superconductor is of the weak-coupling type, measurements of superconducting properties will not allow one to determine whether a high-frequency phonon, exciton, plasmon, etc mechanism is active, since all these mechanisms would give rise to the same universal weak-coupling behaviour[3].

At the present there is no comprehensive theory that would allow one to calculate strong-coupling effects as defined above. Fortunately, however, in most superconductors the relevant energy scales are well separated; the characteristic phonon energy $\hbar\omega_{ph}$ (ω_{ph} stands for ω_0 in conventional superconductors) is usually much smaller than typical electronic normal-state energies. This fact is crucial for formulating a tractable strong-coupling theory. Consequently, Eliashberg's strong-coupling theory [17,20,21] (to be called alternatively "strong-coupling theory" in the following) is based on the smallness of certain expansion parameters associated with the above energy scales. These expansion parameters are the dimensionless ratios k_BT_c/E_F, $\hbar\omega_{ph}/E_F$ and $1/k_F\ell$ ($1/k_F$ is a characteristic electronic wavelength, and ℓ is the mean free path of conduction electrons). Eliashberg's strong-coupling theory is correct to leading order in these fundamental expansion parameters, since higher corrections are neglected. It should be noted that Eliashberg's theory

---

[2] For instance the tunneling conductance of typical strong-coupling superconductors (such as Pb) shows phonon features which can be used to determine the electron-phonon coupling function $\alpha^2F(\omega)$.

[3] In the case of phonon-mediated pairing, however, an isotope effect on T_c is expected (see section 5.3).

does not need a small $k_BT_c/\hbar\omega_{ph}$; in this limit, one obtains the weak-coupling theory.

The standard method for formulating many-body problems in solid-state physics is the Green's function technique [22,23]. One defines the electron Green's function $\hat{G}(x,x';\epsilon_n)$ and the phonon Green's function $D(n\nu,m\mu;\omega_m)$, which obey Dyson's equations:

$$\int d^3y\Big[(i\epsilon_n - \hat{\tau}_3(\nabla_x^2/2m-\mu)\delta(x-y) - \hat{\Sigma}(x,y;\epsilon_n)\Big]\,\hat{G}(y,x';\epsilon_n) = \delta(x-x')\cdot\hat{1} \tag{1}$$

$$\sum_{n''\lambda''}\Big[(M_n\omega_n^2\,\delta_{n,n''}\,\delta_{\lambda,\lambda''} + \Pi(n\lambda,n''\lambda'';\omega_n)\Big]\,D(n''\lambda'',n'\lambda';\omega_n) = \delta_{n,n'}\delta_{\lambda,\lambda'}. \tag{2}$$

ere μ is the chemical potential, $\epsilon_n=(2n+1)\pi k_BT$ are the Matsubara requencies for fermions, and $\omega_n=2n\pi k_BT$ are the Matsubara frequencies for osons (the notation is as in Ref.[17]). The self energies Σ and Π contain ll effects of electron-electron, ion-ion, and electron-ion Coulomb nteractions. Dyson's equations are formally single-particle equations, ith Σ and Π acting as effective potentials and force constants, espectively. The self energies have diagrammatic representations in terms f the full Green's functions G and D, and of fundamental interaction ertices [17]. Their calculation to leading order in the above expansion arameters is the central task in Eliashberg's theory.

It turns out to be useful to decompose the Green's functions in requency-momentum space into "low-" and "high-"energy Green's functions, epending on whether the arguments are in the low- (superconducting) or igh- (normal-state) energy range. A systematic order of magnitude nalysis for the self energies (valid for three-dimensional metals) leads hen to one phonon and three electron self-energy diagrams, which epresent the leading contributions in the fundamental expansion parameter

Π =

$\hat{\Sigma}$ = + +

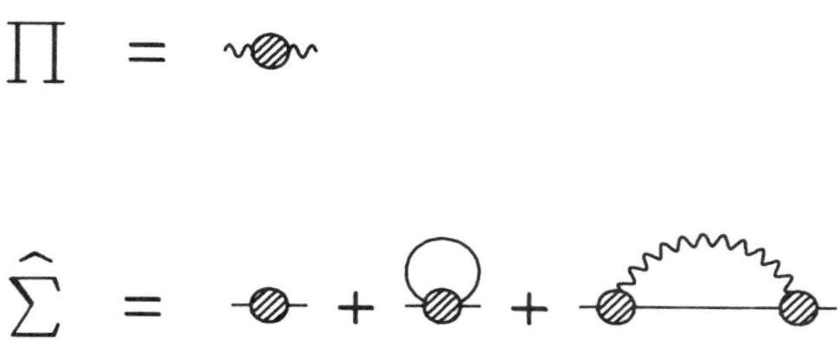

ig.1 The phonon and electron self energies in leading order in Eliashberg's fundamental expansion parameter (see text).

[17]. These diagrams are shown in Fig.1. The dashed circles in Fig.1 stand for high-energy vertices, which are constructed from fundamental interaction vertices and high-energy Green's functions. One is therefore left with a closed theory for the low-energy Green's functions, provided one knows the four high-energy vertices in Fig.1. Dyson's equations, together with the leading self energies in Fig.1 represent a coupled system of equations for the low-energy Green's functions, which determine in principle all superconducting properties in the framework of Eliashberg's theory.

In order to establish the connection to band-structure theory and to the theory of the electron-phonon coupling, Dyson's equation (1) is rewritten in the following way [17]:

$$\int d^3y\left[(i\epsilon_n\hat{\tau}_3\delta(x-y) - h_0(x,y) - \hat{\sigma}^{ee}(x,y) - \hat{\sigma}^{ep}(x,y;\epsilon_n)\right] \hat{g}(y,x';\epsilon_n) = \delta(x-x')\cdot\hat{1}. \tag{3}$$

Here $\hat{g}(y,x';\epsilon_n)$ is the "quasiparticle Green's function", which is the renormalized version of the low-energy electron Green's function in Equ.(1), and $\hat{\sigma}^{ee}(x,y)$ and $\hat{\sigma}^{ep}(x,y;\epsilon_n)$ are the renormalized electronic self energies in leading order in the fundamental expansion parameter; $\hat{\sigma}^{ee}(x,y)$ and $\hat{\sigma}^{ep}(x,y;\epsilon_n)$ are obtained by renormalizing the second and, respectively, the third contribution to Σ in Fig.1 (the first contribution is absorbed in the "quasiparticle renormalization function" and in h_0). The Hamiltonian h_0 is the "ideal band-structure Hamiltonian", which has the following representation in terms of the exact crystal potential v_0 (excitation potential):

$$h_0(x,x') = (-\nabla_x^2/2m - \mu)\ \delta(x-x') + v_0(x,x'). \tag{4}$$

Dyson's equation for quasiparticles Equ.(3) now resembles that of uncorrelated particles, and the new high-energy vertices in this equation describe *quasiparticle* interactions and the *quasiparticle*-phonon coupling.

Another important step in formulating a numerically tractable theory is to consider quantities in a narrow shell (of the order of the low-energy scale) around the Fermi surface. This procedure leads to a theory which involves only functions *at* the Fermi surface, and thus reduces the three dimensional problem to a two dimensional one. Furthermore, it turns out that in a consistant approximation the energy variable ξ in the Green's function determining the distance from the Fermi surface can be eliminated by introducing Eilenberger's (ξ-integrated) Green's function $g(s,\epsilon_n)$ (the quasiclassical propagator), where s is the Fermi-surface variable. The quasiclassical propagator $g(s,\epsilon_n)$ obeys the quasiclassical equations, which replace Dyson's equation (Eilenberger [24], Larkin and Ovchinnikov [25]) in this approximation (see section 3). The electron-phonon self energy then takes on the following form:

$$\hat{\sigma}^{ep}(s;\epsilon_n) = T\sum_{\omega_m}\int d^2s'n(s')\ \lambda(s,s';\omega_m)\ \hat{g}(s';\epsilon_n-\omega_m), \tag{5}$$

where $n(s)$ is the fractional bandstructure density of states at the Fermi-surface point s, and $\lambda(s,s';\omega_m)$ is the electron-phonon interaction function. The averaged electron-phonon interaction function $\lambda(\omega_m)$, which is one of the central quantities in the strong-coupling theory, has the following well-known spectral representation in terms of the Eliashberg function $\alpha^2F(\omega)$ [4]:

$$\overline{\lambda(s,s',\omega_m)}^{s,s'} = \lambda(\omega_m) = 2\int_0^\infty d\omega\, \alpha^2F(\omega)\, \frac{\omega}{\omega^2 + \omega_m^2} \,. \tag{6}$$

The electron-electron self energy takes on the following form (for singlet superconductors):

$$\hat{\sigma}^{ee}(s) = \int d^2s' n(s')\, \mu^*(s,s')\, T \sum_{\epsilon_n}{}' \hat{f}(s';\epsilon_n), \tag{7}$$

where $\hat{f}$ are the off-diagonal components of the propagator $\hat{g}$, and μ^* is the Coulomb pseudopotential; the frequency sum in Equ.(7) extends to some low-energy cut-off.

At this point explicit connection can be made with bandstructure theory and the theory of the electron-phonon interaction; the central quantities of Eliashberg's theory, $\alpha^2F(\omega)$ and μ^*, which enter Eliashberg's equations (5),(7), have to be calculated from first principles in an ab-initio theory of superconducting properties. This involves solving the high-energy equations of the correlated electron liquid coupled to the ionic background. It appears that this problem is not yet fully settled, even for rather "simple" metals (see [17,26-29]). However, efficient techniques have been developed in the past in order to circumvent these present deficiencies of the ab-initio theory. These techniques involve calculating $\alpha^2F(\omega)$ and μ^* from experimental tunneling data in a selfconsistent way using Eliashberg's equations [18,19].

Present problems in calculating the input for Eliashberg's theory, $\alpha^2F(\omega)$ and μ^* from first-principles considerations should be clearly separated from approximations made in Eliashberg's theory itself. As pointed out above, Eliashberg's theory is asymptotically correct to leading order in the (usually very small) fundamental expansion parameter defined earlier in this section. A satisfactory ab-initio theory of superconducting properties, of which Eliashberg's theory is only the final part, therefore, has to await substantial improvements in bandstructure theory and in the theory of the electron-phonon interaction in metals.

---

[4] $F(\omega)$ is the phonon density of states and $\alpha(\omega)$ describes the electron-phonon coupling.

3. THE GORKOV-EILENBERGER EQUATIONS FOR INHOMOGENEOUS SUPERCONDUCTORS

The strong-coupling theory discussed in the previous section yields closed expressions for the Green's functions and Eliashberg's self energies, provided certain high-energy quantities ($\alpha^2F(\omega)$ and μ^*) are known. The above equations determine properties of the homogeneous superconducting state. In the case that an external magnetic field of intermediate strength is applied to the superconductor, a lattice of magnetic flux lines is observed, depending on material parameters such as the Ginzburg-Landau parameter κ (type-II behaviour). The nature of this spatial inhomogeneous state as well as its magnetic and thermodynamic properties are governed by the Gorkov-Eilenberger equations (quasi-classical equations), which can be derived from Dyson's equation [24,25,30]. Using the notation of Serene and Rainer [31] and Alexander et al. [32], the quasiclassical equations are given by:

$$-i[\epsilon_n\hat{\tau}_3 - \hat{\sigma}(s,\vec{R};\epsilon_n) - \hat{v}(s,\vec{R}), \hat{g}(s,\vec{R};\epsilon_n)] + \vec{v}_F(s)\cdot\vec{\nabla}_R\hat{g}(s,\vec{R};\epsilon_n) = 0 \qquad (8)$$

and

$$\left[\hat{g}(s,\vec{R};\epsilon_n)\right]^2 = -\pi^2\,\hat{1}. \qquad (9)$$

Here we have introduced a spatial dependence in all functions, since we are interested in inhomogeneous properties. The self energy σ in Equ.(8) includes Eliashberg's self energies discussed in section 2, and the impurity self energy due to impurity-quasiparticle scattering processes [32]. The function v comprises external perturbations; in the case of an external magnetic field $\vec{B}(\vec{R})$, $\hat{v}$ is given by:

$$\hat{v}(s,\vec{R}) = -\frac{e}{c}\,\vec{v}_F(s)\cdot\vec{A}(\vec{R})\,\hat{\tau}_3\;, \quad \text{with} \quad \vec{B}(\vec{R}) = \text{curl}\,\vec{A}(\vec{R}), \qquad (10)$$

and $\vec{v}_F(s)$ is the (bare) Fermi velocity.

The above equations can be solved numerically, and one obtains for given $\alpha^2F(\omega)$ and μ^* essentially all physical quantities of interest such as the critical temperature T_c, the superconducting energy gap, the critical fields, the magnetic decay length, the density of states, the electromagnetic absorption spectrum, etc. In particular limiting cases have been studied extensively, since the full solution of the quasi-classical equations for arbitrary parameters including the full anisotropy (as derived from model calculations or experiments) is probably at the limit of present computer capacities. Concerning the weak-coupling approximation of the above theory, a great deal of work has been done, however (see e.g. [33,34]).

As regards the applicability of conventional strong-coupling theory for high temperature oxide superconductors, the fundamental limitations of the theory have to be considered. There is evidence that Eliashberg's expansion parameter in these materials is much greater than in conventional superconductors. Specifically, the ratio $\Delta(0)/E_F \sim k_B T_c/E_F$ for $La_{2-x}Sr_xCuO_4$ has been estimated to be of the order ~0.01-~1 [35,36] (similar results are obtained for other high temperature superconductors). Accordingly, critical behaviour in thermodynamic quantities is expected in an appreciable region near T_c, indicative of the breakdown of mean-field theory. Indeed, specific-heat measurements on a single crystal of $YBa_2Cu_3O_{7-y}$ in the field-free case near T_c have been successfully analyzed in terms of (three-dimensional) Gaussian fluctuations [37]; in this experiment the width of the critical region was infered to be just a few Kelvin, however. These data would imply that $k_B T_c/E_F$ is a good expansion parameter. It was also noted that strong-coupling effects should reduce the critical region, as compared to weak-coupling theory [36].

The above considerations as well as similar additional arguments (for a summary see e.g. [38,39]) indicate that the question whether or not Eliashberg's strong-coupling theory is applicable to high temperature superconductors cannot be definitely resolved as yet. One motivation for working with this theory, however, is that the results themselves can be used to address this question. Criteria can be established that would allow one to identify the occurence of strong-coupling effects (of the "Eliashberg type") in experimental data, or their absence. Moreover, due to the lack of better justified strong-coupling theories in the framework of BCS theory, Eliashberg's equations have been used purely as a model, aimed at calculating qualitative trends to a first approximation. Also non-phonon or combined phonon non-phonon pairing mechanisms have been studied using Eliashberg's equations (see section 5).

4. STUDIES OF PHONON PROPERTIES

Experimental studies of phonon properties are particularly important for assessing the role of the phonons in the high temperature superconductors. Extensive investigations using Raman and infrared scattering techniques have been published, including mode-assignment studies and the determination of crystal symmetries [40-45]. A common feature in these experiments is the observation of a softening of certain phonon modes (in particular the Raman-active mode at ~335cm^{-1} in $YBa_2Cu_3O_{7-y}$) when the sample is cooled below T_c. These phonon anomalies are interpreted as evidence for a coupling of these modes to the electronic system, thus supporting a phonon mechanism for superconductivity.

As pointed out recently by Zeyher and Zwicknagl [46], these observations are in accordance with Eliashberg's theory, and indicate a strong electron-phonon coupling. Using Eliashberg's theory, Zeyher and Zwicknagl calculated the phonon softening in $YBa_2Cu_3O_{7-y}$. They found that while the weak-coupling BCS theory predicts a softening of phonons below the gap (2Δ) and a hardening of phonons above the gap, in strong-coupling theory phonon softening is expected also for modes substantially above the gap, in particular if the electron-phonon interaction is strong. A T_c of ~90K corresponds to a BCS gap $2\Delta=3.5k_BT_c$ of ~220cm^{-1}. From the softening of the infrared active phonons at 271 and 305cm^{-1}, as well as the Raman active phonon at 335cm^{-1}, which are all above the gap, theauthors conclude that the weak-coupling limit is not realized in $YBa_2Cu_3O_{7-y}$. Specifically, according to Zeyher and Zwicknagl, a high-lying electronic excitation, which would lead to weak-coupling behaviour, cannot be the dominant mechanism for superconductivity.

Besides Raman and infrared-scattering techniques, another important tool for studying phonon properties are inelastic neutron-scattering experiments. A number of groups have published extensive investigations of the lattice dynamics in high temperature superconductors by means of this technique (see e.g. [47-49] for the La-Sr-Cu-O family, and [50-54] for the Y-Ba-Cu-O family). Theoretical studies of the lattice dynamics and the electron-phonon interaction have been published [55-57]. Of particular interest with regard to the nature of the pairing mechanism is whether soft-mode behaviour, as observed in Raman and infrared-scattering experiments can be detected. These phonon anomalies should show up in characteristic changes in the (generalized) phonon density of states as a function of temperature or in comparison with non-superconducting reference systems. In $La_{1.85}Sr_{0.15}CuO_4$ the temperature effect was found to be very weak, however [47]. On the other hand, a comparison of the generalized phonon density of states of this material with that of undoped (semiconducting) La_2CuO_4 reveals slight but distinct changes which exclusively affect high-frequency ($\hbar\omega \gtrsim 50$meV) optical modes of oxygen. From these data it has been concluded that, although there is evidence of increased electron-phonon coupling, on the basis of this experiment it remains doubtful whether a T_c of ~40K can be explained.

The experiment described above was carried out on polycrystalline samples, since large enough single crystals of $La_{1.85}Sr_{0.15}CuO_4$ or La_2CuO_4 are not available; the phonon dispersion curves could therefore not be determined. For La_2NiO_4, which is isostructural to La_2CuO_4 and can therefore be considered as a reference system, however, large single crystals are available. The phonon dispersion curves of La_2NiO_4 have been measured [58] and compared with Weber's theory for the electron-phonon coupling in La_2CuO_4 [55]. In good agreement with theory, the authors of Ref. [58] find pronounced anomalies in La_2NiO_4 in those phonon branches which involve in-plane oxygen-nickel stretching vibrations. These data support the notion of strong electron-phonon interactions in $La_{1.85}Sr_{0.15}CuO_4$ [58].

Similar results are obtained for $(Y,RE)Ba_2Cu_3O_{7-y}$ superconductors. In particular, no remarkable shift in peak positions in the generalized phonon density of states for $YBa_2Cu_3O_7$ could be detected on cooling, indicating that any soft-mode behaviour, if present at all, must be restricted to a very small volume in phase space [50]. Furthermore, the phonon density of states for the yttrium-based 90-K superconductor has been compared with that for yttrium-substituted compounds (Fig.2a and b) [50]. Also shown in Fig.2 (c) is the generalized phonon density of states for the oxygen-deficient 123 compound $YBa_2Cu_3O_6$ with T_c<2K. It can be seen from Fig.2a and b that despite the strong variation in T_c no drastic changes in the density of states are found. On the other hand, there are pronounced differences in the data between the y=0 and y=1 123 compounds, which cannot be explained from the loss of a single oxygen atom or the minor changes from orthorhombic to tetragonal symmetry [50].

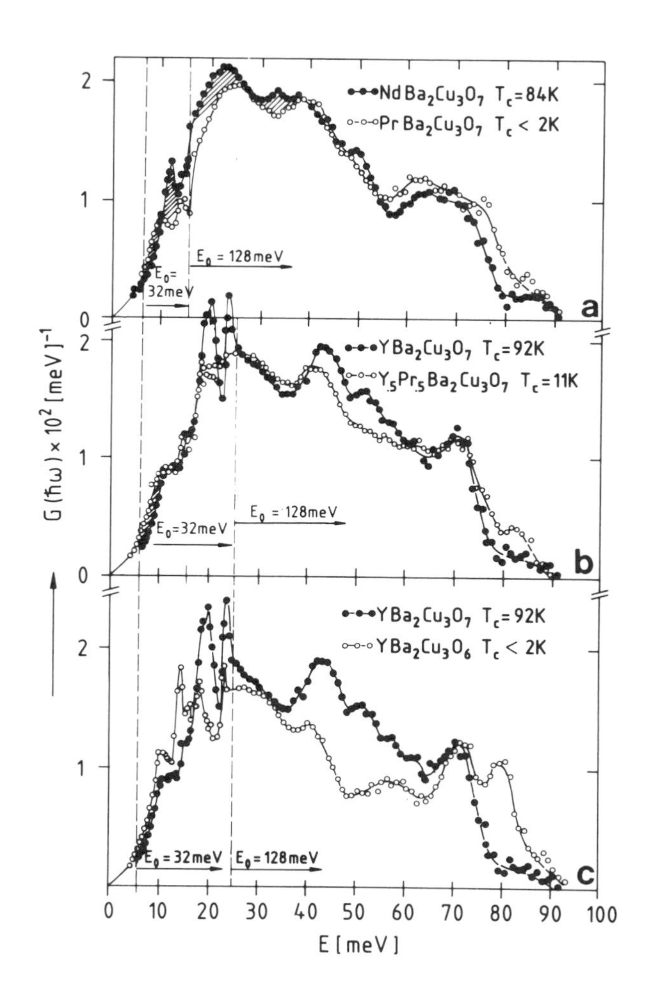

Fig.2a-c Generalized phonon density of states $G(\hbar\omega)$ for high-T_c samples and low-T_c references at 6 K from inelastic neutron scattering experiments. Vertical dashed lines separate measurements with different neutron incident energies E_0 (from [50]).

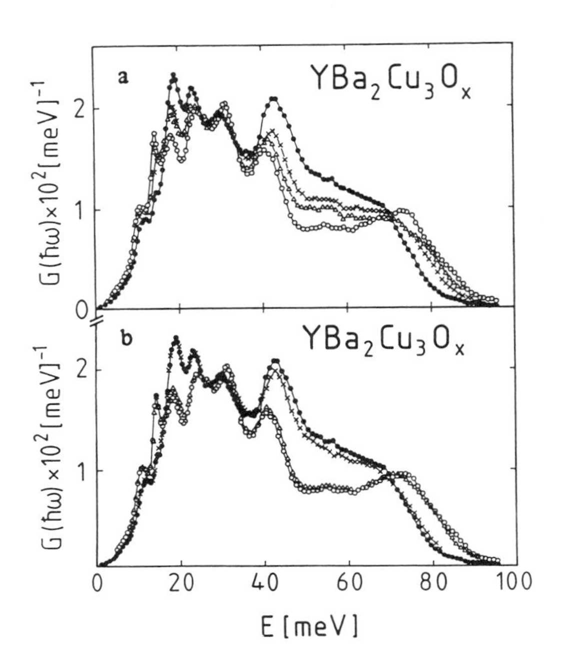

Fig.3a-b Generalized phonon density of states for $YBa_2Cu_3O_x$ at 300 K (from [52]).

dots: x=6.97 (T_c=92 K); open circles: x=6.07 (no T_c)

a) crosses: x=6.53 (T_c=~58-~68K)
   triangles: x=6.40 (T_c=~0-~35K)

b) crosses: x=6.81 (T_c=~92K)
   triangles: x=6.16 (no T_c measurable)

A more detailed study of the effect of varying the oxygen concentration between y=0 and y=1 has been carried out recently in Ref.[52]. These data are displayed in Fig.3 a and b. Model calculations indeed show that the pronounced features evident in Fig.3 cannot be explained by structural changes only, but are likely to reflect marked differences in the electron-phonon interaction in these compounds [52].

In summary, from presently available data on phonon properties, no firm conclusions regarding the phonon-pairing mechanism or phonon-assisted pairing mechanisms are as yet possible. It seems to be clear, however, that the phonons do play a role in the superconductivity of high temperature oxide superconductors. Certainly more data, especially on high-quality single crystals are desirable.

5. PHONON VERSUS NON-PHONON MECHANISMS

5.1 Criteria for the Phonon-Exchange Mechanism

For a consistent interpretation of experiments on high temperature superconductors in terms of a phonon or a non-phonon exchange mechanism, it is important to estimate the magnitude of strong-coupling effects that might occur. If the phonon-pairing mechanism is responsible for the T_c's, strong-coupling corrections to the results of weak-coupling BCS theory can be calculated routinely for many physical quantities. As input for Eliashberg's theory, results of model calculations for the electron-phonon spectral function $\alpha^2F(\omega)$ are available for $La_{2-x}(Ba,Sr)_xCuO_4$ [55] and $YBa_2Cu_3O_7$ [56]. Alternatively, phonon density of states data (e.g.[47,50]) may be employed to construct $\alpha^2F(\omega)$ model functions. Since no reliable results for the Coulomb parameter μ^* are available, standard choices may be used for this quantity.

In Ref.[59], representative model $\alpha^2F(\omega)$ functions based on the generalized phonon density of states for $La_{1.85}Sr_{0.15}CuO_4$ [47] (which served as a reference spectral function) have been designed. These model spectra all fit the observed T_c's, but differ in the relative amount of coupling to high and low frequency phonon modes, and in the high frequency (non-phonon) parameter μ^*. In Table 1, characteristic features of typical model Eliashberg functions defined in this way are listed; also included in Table 1 is Weber's theoretical spectral function [55]. For a spectrum of type B (type C), the coupling of modes above 50meV to the conduction electron system has been reduced (enhanced) by a factor of 1/3 (factor of 3), as compared to the reference spectrum. Using standard numerical routines, the thermodynamic characteristics of these model strong-coupling superconductors can be calculated. Typical results concerning $H_c(T)$, $H_{c2}(T)$, $\Delta C/\gamma T_c$, Δ_0/k_BT_c and other quantities can be found in Ref.[59]. It was shown that on the basis of these model spectra strong-coupling effects are expected to be of the order 100% or less, depending on the quantity of interest. Specifically, e.g., the deviation of the thermodynamic critical field $H_c(T)$ from a parabola form ("deviation function") was found to lie between that of lead and weak-coupling theory. The gap ratio $2\Delta_0/k_BT_c$ for these spectra was found to be not greater than 5.1 (F135 in Table 1) in 35-K superconductors and not greater than 6.5 (B291 in Table 1) in 91-K superconductors. Higher values result if the coupling function has more weight at low frequencies. Equivalent results were reported in [60,61] (see also the contribution by J.P. Carbotte and F. Marsiglio in this volume).

Theoretical studies of strong-coupling effects in the framework of Eliashberg's theory were also published in [62-72], emphasizing different aspects. These include, among many others, the energy gap in unconventional pairing states [66], the analysis of tunneling experiments [67], fluctuation effects [63], and anisotropy effects [71,72]. In Ref.[72], Yonemitsu and Wada studied anisotropy effects in the electron-phonon coupling in $La_{2-x}Sr_xCuO_4$. They found that the effect on T_c of anisotropies in the electron-phonon interaction is very small (of the order of a few percent), compared to the isotropic contribution.

TABLE 1. Model Eliashberg functions; B and C type spectra are based on the phonon density of states of Ref.[47]; the F type spectrum is the theoretical spectrum from Ref.[55].

Eliashberg function	"type"	T_c(K)	λ	μ^*
B035	"soft"	35	1.3	0.0
B235		35	2.0	0.2
C035	"hard"	35	1.0	0.0
C235		35	1.7	0.2
F135	"very soft"	35	2.3	0.13
B091	"soft"	91	4.0	0.0
B291		91	6.1	0.2
C091	"hard"	91	2.5	0.0
C291		91	4.0	0.2

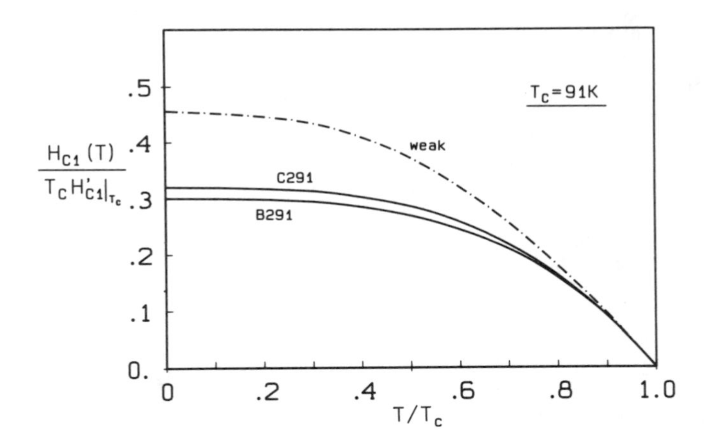

Fig.4 Lower critical field for a 91-K superconductor in weak- and strong-coupling theory in the dirty limit for κ=10; Eliashberg functions (C291, B291) are defined in Table 1.

Several groups reported measurements of the magnetic decay length of flux lines $\lambda(T)$ (penetration depth) by means of muon-spin-relaxation (μSR) techniques in polycrystalline high-T_c samples [73-77]. For general remarks on this technique see [78]. I have analyzed the experiment by Harshman et al. [76], who determined $\lambda(T)$ in $YBa_2Cu_3O_{7-y}$ ($y=0.1\pm0.05$) in terms of conventional (s-wave) weak- and strong-coupling theory, using Eliashberg functions defined in Table 1 [79]. It was shown in Ref. [79] that the experimental data are not consistent with a clean- or dirty-limit weak-coupling description. In fact, the deviations from weak-coupling theory are compatible with a strong-coupling mechanism. Other experiments show different results, however. In particular, a recent experimental study by Fiory et al. [77] by means of an alternative technique on a thin film of $YBa_2Cu_3O_7$ suggests a weak-coupling behaviour of $\lambda(T)$.

The lower critical field H_{c1} is the only critical field in high temperature superconductors which is experimentally accessible at all temperatures. It is therefore tempting to look for strong-coupling effects in the temperature dependence of H_{c1}. The first theoretical results concerning the vortex state of type-II superconductors in the framework of the Gorkov-Eilenberger theory including Eliashberg's self energies have become available only recently [80-82]. These numerical calculations represent extensions of earlier computer codes for the lower critical field in the weak-coupling limit [83,84]. In Ref. [82] it was shown that strong-coupling effects indeed cause a characteristic signature in the temperature dependence of H_{c1}; one finds a marked depression in the function $H_{c1}(T)/T_cH'_{c1}(T_c)$ at low temperatures, as compared to the weak-coupling theory (see Fig.4). These results refer to the dirty limit, where the electronic mean free path is assumed to be much smaller than the coherence length. Later it was shown that the same effect also occurs for arbitrary values of the mean free path [81]. In addition, it is independent of (uniaxial) anisotropies in the electronic density of states at the Fermi surface [81].

Unfortunately, presently available experimental results on the temperature dependence of the lower critical field in high temperature superconductors do not yet show uniform evidence. It appears to be very difficult to measure the "intrinsic" lower critical field, due to the strong pinning forces present, as well as other problems (see e.g. the contribution by H.W. Weber in this volume).

5.2 Non-Phonon and Combined Phonon Non-Phonon Mechanisms

In conventional BCS theory the pairing of electrons arises from phonon exchange. A standard scheme in the literature on non-conventional mechanisms causing superconductivity is to replace the phonon by a specified electronic excitation. Since in the BCS model (applied to arbitrary exchange mechanisms) T_c scales with a characteristic frequency of the excitation mediating the pair formation, electronic mechanisms are particularly favorable to yield high T_c's; the typical electronic energy scale in metals is often two or three orders of magnitude above the typical phonon energy scale. Accordingly, electronic pairing mechanisms have attracted much attention since the advent of the high temperature oxide superconductors.

It has been suggested that excitons [85,86], plasmons [87,89], charge-transfer excitations [90,91], spin fluctuations [92] etc mediate the pair formation in high temperature superconductors. For a recent review on non-phonon mechanisms see e.g. [93,94]. Furthermore, it has been suggested that the phonon and some non-phonon mechanism may cooperate in achieving the high T_c's [89,93]. Using Eliashberg's theory as a model, thermodynamic and other properties of non-phonon superconductors have been calculated [95-98].

Marsiglio and Carbotte [96] studied the deviation of characteristic superconducting parameters from their weak-coupling BCS values as a function of the relative coupling strength to a phonon and a non-phonon excitation (exciton). In their model, the phonon and the exciton are located at 8meV and 500meV, respectively. The authors introduce coupling constants λ_{ex} for the exciton, λ_{ph} for the phonon ($\lambda_{tot}=\lambda_{ex}+\lambda_{ph}$), and vary the ratio $\lambda_{ex}/\lambda_{tot}$ while keeping T_c=96K constant (μ^*=0.1). For $\lambda_{ex}/\lambda_{tot}=1$ (no phonon contribution), the weak-coupling BCS results are recovered because of the very high excitonic frequency. As $\lambda_{ex}/\lambda_{tot}$ is lowered, thus introducing the (low-frequency) phonon contribution, strong-coupling effects develop, but very slowly. For instance, for $\lambda_{ex}/\lambda_{tot}=0.5$ Marsiglio and Carbotte find $2\Delta_0/k_BT_c=3.8$ and $\Delta C/\gamma T_c=1.18$ (the weak-coupling values are 3.53 and 1.43, respectively). As $\lambda_{ex}/\lambda_{tot}$ is reduced further, the deviations from the weak-coupling results become progressively more pronounced, and at $\lambda_{ex}/\lambda_{tot}=0$ (no excitonic contribution), values of ~10 and ~0.6 for $2\Delta_0/k_BT_c$ and $\Delta C/\gamma T_c$, respectively, are obtained. It should be noted that the strong-coupling effect on $\Delta C/\gamma T_c$ goes in the opposite direction than would be expected for ordinary phonon superconductors; this anomaly is due to the very low phonon frequency adopted in the model, which is in fact of the order of T_c [96].

If the superconducting state in high temperature superconductors is due to a combined phonon non-phonon exchange mechanism, the question arises about how the two contributions could be separated experimentally. Kresin [99] suggested a procedure for identifying an electronic contribution (of arbitrary origin) by means of a sequence of standard experiments. His approach is based on the validity of Eliashberg's equations. For a given superconductor one needs the electron-phonon spectral function $\alpha^2F(\omega)$ and the electronic component of the heat capacity $C_e(T)$. The function $\alpha^2F(\omega)$ is obtained from the usual inversion procedure of the tunneling density of states employing Eliashberg's equations (McMillan and Rowell [18]). The electronic heat capacity $C_e(T)$ can be determined by first measuring the total heat capacity, and then subtracting the lattice contribution. The evaluation of the lattice contribution involves the phonon density of states, which can be determined in a neutron-scattering experiment.

Now the point in Kresin's method is that $C_e(T)$ can alternatively be calculated from Eliashberg's spectral function $\alpha^2F(\omega)$ [99,100]. Since the evaluation of $\alpha^2F(\omega)$ was based on Eliashberg's equations assuming no electronic contribution to the exchange mechanism, the function $C_e(T)$ determined from both methods should only be identical if there is no electronic contribution present[5]. Kresin thus concludes that an unambiguous discrepancy between the two measurements would be evidence of an additional electronic excitation supporting superconductivity.

The method described above for identifying an electronic contribution to a phonon-exchange mechanism in high temperature superconductors is probably beyond present experimental capabilities. In particular, tunneling measurements are confronted with serious difficulties [16].

5.3 Isotope Effect

The isotope effect plays a considerable role in the discussion of whether or not a phonon-mediated pairing mechanism is realized in high temperature superconductors. According to conventional BCS theory, the substitution of an atom in a superconductor by a different isotope should lead to a (small) change in T_c, reflecting the changes of the phonon frequencies associated with this atom. An observation of this so-called isotope effect on T_c would therefore suggest[6] that the phonon-exchange mechanism is relevant for superconductivity (possibly in combination with non-phonon mechanisms).

One defines the "selective isotope exponents" β_n by the relation

$$d(\log T_c) = -\sum_n \beta_n d(\log M_n) \tag{11}$$

where the index n refers to the various constituents (n=O,Cu,...) with atomic masses M_n of the superconductor. The total isotope exponent $\beta_{tot}=\sum \beta_n$ is expected to be about 0.5-0.4 for phonon-mediated pairing and a non-phonon parameter μ^* between 0 and 0.2. The selective components β_n depend on the detailed lattice dynamics and on the coupling strength of conduction electrons to vibrations of atoms of type n. A linear-response scheme for calculating the isotope exponents in the framework of Eliashberg's strong-coupling theory was derived in Ref.[101]. This scheme will be used in the following to calculate isotope exponents from model $\alpha^2F(\omega)$ functions[7]. Note that this theory yields the so-called "direct" isotope effect. For "indirect" effects see below.

From studies of the lattice dynamics in high temperature superconductors (see e.g. [47,50,53]) it has been concluded that the

---

[5] Note that such a contribution, if it is just outside the energy range covered by the tunneling experiment and thus escapes detection, is expected to distort $\alpha^2F(\omega)$. This distortion should however be small if the energy of the electronic contribution is far above k_BT_c; a strong attractive high-frequency mechanism would lead to a negative μ^* in McMillan and Rowell's inversion procedure, but would essentially not affect the shape of $\alpha^2F(\omega)$.

[6] This implication is not a rigorous one, however (see below).

[7] The computer code for these calculations was written by D. Rainer.

high-frequency range in the phonon density of states is dominated by oxygen vibrations. Therefore, assuming some coupling strength of these modes to the conduction electron system, and adjusting an overall scaling factor for the phonon density of states such that measured T_c values are obtained, the total and the oxygen isotope exponent of $\alpha^2F(\omega)$ functions defined in this way can be calculated. This method was employed in Ref.[97] using (among others) model $\alpha^2F(\omega)$ functions introduced in Table 1. For a 91-K superconductor it was found that the oxygen isotope exponent β_O exeeds the value 0.1 only if the coupling to oxygen modes is enhanced as compared to the reference spectrum (the generalized phonon density of states). The total isotope exponent β_{tot} is always close to 0.5.

In order to calculate the isotope effect in model superconductors in which a phonon and a high-frequency non-phonon mechanism cooperate, the following scheme may be adopted [97,98]: we simulate an attractive high-frequency non-phonon mechanism by a negative μ^* (this is equivalent to adding a high-frequency peak to $\alpha^2F(\omega)$); since we want to study how the isotope effect varies as a function of μ^* (>0 or <0), i.e. how it depends on the relative importance of the two mechanisms for a given superconductor, we adjust the phonon parameter λ according to the requirement that T_c stays constant; for the reference electron-phonon

TABLE 2. Oxygen and total isotope exponents for a combined phonon and non-phonon mechanism; T_c=91 K

λ	0.4	1.4	2.8	4.2	5.5	6.5
μ^*	-0.22	-0.15	-0.05	+0.1	+0.2	+0.3
β_{tot}	0.21	0.43	0.50	0.49	0.48	0.47
β_O	0.05	0.09	0.10	0.09	0.09	0.08

spectral function, we then obtain the results shown in Table 2 (T_c=91K). First, it can be seen that the total isotope exponent goes down to ~0.2 for μ^*=-0.22. It is still quite substantial, even if the non-phonon mechanism strongly dominates (a λ of 0.4 with μ^*=0 yields a T_c of only 5K). Second, surprisingly the oxygen isotope effect does not depend very much on the relative weight of the phonon and the non-phonon mechanisms. Specifically, β_O is of the same order for both λ=0.4 (μ^*=-0.22) and λ=6.5 (μ^*=0.3), which shows that a measurement of β_O can definitely not be used to estimate the relative importance of the phonon pairing mechanism. As long as phonons play any role in the 90-K superconductor, the (direct) oxygen isotope effect is expected to be of the order ~0.05-~0.10, unless there is essentially exclusive coupling to optical modes of oxygen.

The above calculations yield the direct (i.e., due to the exchange of harmonic phonons only) isotope exponents, since indirect effects cannot be calculated from Eliashberg's theory alone. Using a simple anharmonic potential for the oxygen ions, Drechsler and Plakida [102] recently obtained qualitatively different results for the isotope effect as compared to the harmonic theory. Specifically, in their model the isotope exponent will be *negative* in the weak to moderate coupling regime, and somewhat suppressed in the strong-coupling regime; in the extreme strong-coupling regime, they obtain a normal isotope effect.

It has also been pointed out that purely electronic exchange mechanisms involving atomic displacements could lead to non-zero isotope exponents as well [39,103]. In Ref. [103], Batlogg et al. suggest that the oxygen isotope effect in $Ba_{0.6}K_{0.4}BiO_3$ and $BaPb_{0.75}Bi_{0.25}O_3$ ($\beta_O \sim 0.2$ in both cases) is due to a "parasitic" phonon involvement along with electronic excitations mediating superconductivity. From their measurements, combined with results from band structure calculations, Batlogg et al. conclude that both superconductors are in the weak-coupling limit ($\lambda \lesssim 0.6$-0.8), which would exclude the phonon-exchange mechanism (according to the authors)[8].

A number of groups have measured the isotope effect in high temperature superconductors. In particular the effect on T_c of substituting ^{18}O for ^{16}O in various high-T_c superconductors is well established by now (see e.g.[105-108] for $YBa_2Cu_3O_7$, [109-110] for $La_{2-x}Sr_xCuO_4$, and [103] for $Ba_{0.6}K_{0.4}BiO_3$ and $BaPb_{0.75}Bi_{0.25}O_3$). It was found that while the oxygen isotope exponent in $YBa_2Cu_3O_7$ is close to zero, $La_{2-x}Sr_xCuO_4$, $Ba_{0.6}K_{0.4}BiO_3$ and $BaPb_{0.75}Bi_{0.25}O_3$ exhibit substantial oxygen isotope exponents (0.14-0.35 [109,110], 0.21±0.03 and 0.22±0.03 [103], respectively).

6. CONCLUSION

The scope of this review was to discuss selected aspects of a strong-coupling approach to high temperature superconductors. In general, strong-coupling theory is understood as a theory based on the BCS pairing model which comprises all effects of low-lying normal-state energies on superconducting properties, regardless of the origin of the pairing mechanism. As emphasized in section 2, Eliashberg's strong-coupling theory allows for computing strong-coupling effects in superconductors provided typical energies involved in the superconducting state are small compared to the Fermi energy or the band width. These restrictions possibly limit the applicability of Eliashberg's theory to high temperature superconductors.

---

[8] In contrast to this conclusion, on the basis of the phonon density of states for $BaPb_{0.75}Bi_{0.25}O_3$ [104] an $\alpha^2F(\omega)$ function can be designed which yields both the measured oxygen isotope exponent β_O, and has a λ of as small as 0.5 (T_c=12K, μ^*=0); for μ^*=0.13, one obtains λ=0.7 for the same values of T_c and β_O. For T_c=30K, one obtains λ=0.8 (with μ^*=0) or λ=1.1 (with μ^*=0.13). These unexpected results are due to the fact that a very strong coupling of (high frequency) oxygen modes to the electron system is necessary in order to obtain the large oxygen isotope effect; this in turn yields a low λ.

I have also discussed experimental criteria which would allow one t assess the role of the phonons for superconductivity in high temperatur superconductors. There is experimental evidence for a correlation betwee the onset of superconductivity and changes in the phonon system (se section 4), which has been interpreted in the literature in terms of strong electron-phonon interaction. Properties of superconductors whic exhibit phonon mediated or phonon-exciton meditated pairing mechanism have been considered in section 5.

Our present knowledge about the normal as well as the superconductin state in high temperature oxide superconductors is incomplete. Probabl new concepts based upon successful theories will be needed if we are t further enhance our understanding of the unusual properties observed.

ACKNOWLEDGMENTS

I have profited from discussions with D. Rainer, H.F. Braun, P. Esquinazi, A. Gupta, J. Voit and Y. Wada. Furthermore, I would like to thank D. Rainer for critical comments after reading the manuscript, and H. Rietschel for the permission to use Fig.'s 2 and 3 in this article.

This work was supported by the "Deutsche Forschungsgemeinschaft."

REFERENCES

[1] J.G. Bednorz and K.A. Müller, Z. Phys. B 64, 189 (1986)
[2] J.G. Bednorz and K.A. Müller, Rev. Mod. Phys. 60, 585 (1988)
[3] J. Bardeen, L.N. Cooper and J.R. Schrieffer, Phys. Rev. 108, 1175 (1957)
[4] B.K. Chakraverty, J. Phys. Lett. 40. L99 (1979)
[5] A. Alexandrov and J. Ranninger, Phys. Rev. B 23, 1796 (1981)
[6] A.W. Sleight, J.L. Gillson and P.E. Bierstedt, Solid State Commun. 17, 27 (1975)
[7] M.K. Wu, J.R. Ashburn, C.J. Torng, P.H. Hor, R.L. Meng, L. Gao, Z.J. Huang, Y.Q. Wang and C.W. Chu, Phys. Rev. Lett. 58, 908 (1987)
[8] M.A. Subramanian, J.C. Calabrese, C.C. Torardi, J. Gopalakrishnan, T.R. Askew, R.B. Flippen, K.J. Morrissey, U. Chowdhry and A.W. Sleight, Nature 332, 420 (1988);
Z.Z. Sheng and A.M. Hermann, Nature 332, 138 (1988)
[9] D. Esteve, J.M Martinis, C. Urbina, M.H. Devoret, G. Collin, P. Monod, M. Ribault and A. Revcolevschi, Europhys. Lett. 3, 1237 (1987)
[10] J.S. Tsai, Y. Kubo and J. Tabuchi, Phys. Rev. Lett. 58, 1979 (1987)
[11] J.T. Chen, L.E. Wengerm, C.J. McEwan and E.M. Logothetis, Phys. Rev. Lett. 58, 1972 (1987)
[12] N.D. Kataria,V.S. Tomar, A.K. Gupta and Mukesh Kumar, J. Phys. C 21, L523 (1988)
[13] P.L. Gammel, D.J. Bishop, G.J. Dolan, J.R. Kwo, C.A.Murray, L.F. Schneemeyer and J.V. Waszczak, Phys. Rev. Lett. 59, 2592 (1987)
[14] H.F.C. Hoevers, P.J.M. van Bentum, L.E.C. van de Leemput, H. van Kempen, A.J.G. Schellingerhout and D. van der Marel, Physica C 152,

105 (1988)
[15] P.J.M. van Bentum, H.F.C. Hoevers, H. van Kempen, L.E.C. van de Leemput, M.J.M.F. de Nivelle, L.W.M. Schreurs, R.T.M. Smokers and P.A.A. Teunissen, Physica C 153-155, 1718 (1988)
[16] A. Barone, Physica C 153-155, 1712 (1988)
[17] D. Rainer, Principles of Ab-initio Calculations of Superconducting Transition Temperatures, in Progress in Low Temperature Physics, Vol.10, D.F. Brewer (ed.), (North-Holland 1986), p.371
[18] W.L. McMillan and J.M. Rowell, in Superconductivity, by R.D. Parks (ed.), (Dekker, New York, 1969), Vol 1, p.561
[19] E.L. Wolf, Principles of Tunneling Spectroscopy (Oxford Univ. Press, New York, 1985)
[20] G.M. Eliashberg, Zh. Eksp. Teor. Fiz. 38, 966 (1960), [Sov. Phys. JETP 11, 696 (1960)]
[21] D. Scalapino, in Superconductivity, by R.D. Parks (ed.), (Dekker, New York, 1969), Vol 1, p.449
[22] A.L. Fetter and J.D. Walecka, Quantum Theory of Many Particle Systems (McGraw-Hill, New York, 1981)
[23] L.D. Landau and E.M. Lifshitz, Course of Theoretical Physics, Vol. 9: Statistical Physics (Pergamon, Oxford, 1980), part 2
[24] G. Eilenberger, Z. Phys. 214, 195 (1968)
[25] A.I. Larkin and Yu.N. Ovchinnikov, Zh. Eksp. Teor. Fiz. 55, 2262 (1968) [Sov. Phys.-JETP 28, 1200 (1969)]
[26] D. Rainer, Physica 109+110B, 1671 (1982)
[27] D.A. Papaconstantopoulos, L.L. Boyer, B.M. Klein, A.R. Williams, V.R. Mozurri and J.F. Janak, Phys. Rev. B 15, 4221 (1977)
[28] M. Peter, J. Ashkenazi and M. Dacorogna, Helv. Phys. Acta 50, 267 (1977)
[29] D. Glötzel, D. Rainer and H.R. Schober, Z. Phys. B 35, 317 (1979)
[30] A.L. Shelankov, J. Low Temp. Phys. 60, 29 (1985)
[31] J.W. Serene and D. Rainer, Phys. Rep. 101, 221 (1983)
[32] J.A.X. Alexander, T.P. Orlando, D. Rainer and P.M. Tedrow, Phys. Rev. B31, 5811 (1985)
[33] E.H. Brandt and A. Seeger, Advances in Physics, 35, 189 (1986)
[34] U. Klein, L. Kramer, W. Pesch, D. Rainer and J. Rammer, Internal Report, Inst. Theoret. Phys., Univ. Bayreuth (1987);
U. Klein, J. Low Temp. Phys. 69, 1 (1987)
[35] A. Kapitulnik, M.R. Beasley, C. Castellani and C. Di Castro, Phys. Rev. B 37, 537 (1988)
[36] G. Deutscher, in Novel Superconductivity, by S.A. Wolf and V.Z. Kresin (eds.), Plenum Press, New York and London (1987), p.293
[37] D.M. Ginsberg, S.E. Inderhees, M.B. Salamon, Nigel Goldenfeld, J.P. Rice and B.G. Pazol, Physica C 153-155, 1082 (1988)
[38] M.T. Béal-Monod, Phys. Rev. B 36, 8788 (1987)
[39] W. Weber: On the Theory of High T_c Superconductors, in Advances in Solid State Physics (Festkörperprobleme) 28, 141 (1988)
[40] D.M. Krol, M. Stavola, W. Weber, L.F. Schneemeyer, J.V. Waszczak, S.M. Zahurak S.G. Kosiniski, Phys. Rev. B 36, 8325 (1987)
[41] A. Yamanaka, F. Minami, K. Watanabe, K. Inoue, S. Takekawa and N. Iyi, Jpn. J. of Appl. Phys. 26, L1404 (1987)
[42] C. Thomsen, M. Cardona, B. Gegenheimer and R. Liu, Physica C 153-155, 262 (1988)
[43] C. Thomsen, M. Cardona, B. Gegenheimer, R. Liu and A. Simon, Phys. Rev. B 37, 9860 (1988)

[44] M. Krantz, H.J. Rosen, R.M. Macfarlane and V.Y. Lee, Phys. Rev. B 38, 4992 (1988)
[45] Z.V. Popovic, C. Thomsen, M. Cardona, R. Liu, G. Stanisic, R. Kremer and W. König, Solid St. Commun.66, 965 (1988)
[46] R. Zeyher and G. Zwicknagel, Solid St. Commun. 66, 617 (1988)
[47] B. Renker, F. Gompf, E. Gering, N. Nücker, B. Ewert, W. Reichhardt and H. Rietschel, Z. Phys. B 67, 15 (1987)
[48] A. P. Ramirez, B. Batlogg, G. Aeppli, R. J. Cava and E. A. Rietman, Phys. Rev. B 35, 8833 (1987)
[49] Torben Brun, M. Grimsditch, K.E. Gray, R. Bhadra, V. Maroni and C.-K- Loong, Phys. Rev. B 35, 8837 (1987)
[50] B. Renker, F. Gompf, E. Gering, G. Roth, W. Reichhardt, D. Ewert, H. Rietschel and H. Mutka, Z. Phys. B 71, 437 (1988)
[51] J.J. Rhyne, D.A. Neumann, J.A. Gotaas, F. Beech, L. Toth, S. Lawrence, S. Wolf, M. Osofsky and D.U. Gubser, Phys. Rev. B 36, 2294 (1987)
[52] B. Renker, F. Gompf, E. Gering, D. Ewert, H. Rietschel and A. Dianoux, to be published
[53] P. Brüesch and W. Bührer, Z. Phys. B 70, 1 (1988)
[54] P. Strobel, P. Monceau, J.L. Tholence, R. Currat, A.J. Dianoux, J.J. Caponi and J.G. Bednorz, Physica C 153-155, 282 (1988)
[55] W. Weber, Phys. Rev. Lett. 58, 1371 (1987), erratum: Phys. Rev. Lett. 58, 215 (1987)
[56] W. Weber and L.F. Mattheiss, Phys. Rev. B 37, 599 (1988)
[57] R.E. Cohen, W.E. Pickett, H. Krakauer and L.L. Boyer, Physica C 153-155, 202 (1988)
[58] L. Pintschovius, J.-M. Bassat, P. Odier, F. Gervais, B. Hennion and W. Reichardt, Europhys. Lett. 5, 247 (1988)
[59] B. Ashauer, W. Lee and J. Rammer, Z. Phys. B 67, 147 (1987)
[60] M. Schossmann, F. Marsiglio and J.P. Carbotte, Phys. Rev. B, 36, 3627 (1987)
[61] R. Akis, F. Marsiglio, E. Schachinger and J.P. Carbotte, Physica C 153-155, 225 (1988)
[62] V.Z. Kresin, Phys. Lett. A 122, 434 (1987)
[63] L.N. Bulaevskii and O.V. Dolgov, Solid St. Commun 67, 63 (1988)
[64] A.A. Golubov, Physica C 156, 286 (1988)
[65] J.C. Phillips, Solid St.Commun. 65, 227 (1988)
[66] O.V. Dolgov and A.A. Golubov, to be published
[67] L.N. Bulaevskii, O.V. Dolgov, A.A. Golubov, M.O. Ptitsyn and S.I. Vedeneev, to be published
[68] F. Marsiglio, R. Akis and J.P. Carbotte, Physica C 153-155, 227, (1988)
[69] F. Marsiglio, M. Schossmann, E. Schachinger and J.P. Carbotte, Phys. Rev. B 35, 3226 (1987)
[70] M.A. St. Peters, Physica C 153-155, 249, (1988)
[71] W. Pint, M. Prohammer and E. Schachinger, Physica C 153-155, 713, (1988)
[72] K. Yonemitsu and Y. Wada, to be published in the J. Phys. Soc. Jpn.
[73] F.N. Gygax, B. Hitti, E. Lippelt, A. Schenck, D. Cattani, J. Cors, M. Decroux, Ø. Fischer and S. Barth, Europhys. Lett. 4, 473 (1987)
[74] G. Aeppli, R.J. Cava, E.J. Ansaldo, J.H. Brewer, S.R. Kreitzman, G.M. Luke, D.R. Noakes and R.F. Kiefl, Phys. Rev. B 35, 7129 (1987)
[75] W.J. Kossler, J.R. Kempton, X.H. Yu, H.E. Schone, Y.J. Uemura, A.R. Moodenbaugh and M. Suenaga, Phys. Rev. B 35, 7133 (1987)

[76] D.R. Harshman, G. Aeppli, E.J. Ansaldo, B. Batlogg, J.H. Brewer, J.F. Carolan, R.J. Cava, M. Celio, A.C.D. Chaklader, W.N. Hardy, S.R. Kreitzman, G.M. Luke, D.R. Noakes and M. Senba, Phys. Rev. B 36, 2386 (1987)
[77] A.T. Fiory, A.F. Hebard, P.M. Mankiewich and R.E. Howard, Phys. Rev. Lett. 61, 1419 (1988)
[78] E.H. Brandt, Phys. Rev. B 37, 2349 (1988)
[79] J. Rammer, Europhys. Lett. 5, 77 (1988)
[80] J. Rammer, J. Low Temp. Phys. 71, 323 (1988)
[81] J. Rammer, to be published
[82] J. Rammer, Phys. Rev. B 36, 5665 (1987)
[83] W. Pesch and L. Kramer, J. Low Temp. Phys. 15, 367 (1974)
[84] L. Kramer, W. Pesch and R.J. Watts-Tobin, J. Low. Temp. Phys. 14, 29 (1974)
[85] W.A. Little, in Novel Superconductivity, by S.A. Wolf and V.Z. Kresin, Eds., Plenum Press, New York and London (1987), p.341
[86] J. Bardeen, D.M. Ginsberg and M.B. Salamon, in Novel Superconductivity, by S.A. Wolf and V.Z. Kresin, Eds., Plenum Press, New York and London (1987), p.333
[87] J. Ruvalds, Phys. Rev. B 35, 8869 (1987)
[88] J. Ashkenazi, C.G. Kuper and R Tyk, Solid St. Commun. 63, 1145 (1987)
[89] V.Z. Kresin and H. Morawitz, in Novel Superconductivity, by S.A. Wolf and V.Z. Kresin, Eds., Plenum Press, New York and London (1987), p.445
[90] C.M. Varma, S. Schmitt-Rink and Elihu Abrahams, Solid St. Commun. 62, 681 (1987)
[91] Jaejun Yu, S. Massidda, A.J. Freeman, Phys. Lett. A 122, 203 (1987)
[92] K.S. Bedell and D. Pines, Phys. Rev. B 37, 3730 (1988)
[93] V.Z. Kresin, in Novel Superconductivity, by S.A. Wolf and V.Z. Kresin, Eds., Plenum Press, New York and London (1987), p.309
[94] M.L. Cohen, in Novel Superconductivity, by S.A. Wolf and V.Z. Kresin, Eds., Plenum Press, New York and London (1987), p.1095
[95] F. Marsiglio, R. Akis and J.P. Carbotte, Solid St. Commun. 64, 905, (1987)
[96] F. Marsiglio and J.P. Carbotte, Phys. Rev. B 36, 3937 (1987)
[97] B. Ashauer, W. Lee, D. Rainer and J. Rammer, Physica 148B, 243 (1987)
[98] J. Rammer, Physica C 153-155, 1625, (1988)
[99] V.Z. Kresin, Phys. Rev. B 30, 450 (1984)
[100]K.E. Kihlstrom, P.D. Hovda, V.Z. Kresin and S.A. Wolf, Phys. Rev. B 38, 4588 (1988)
[101]D. Rainer and F.J. Culetto, Phys. Rev. B 19, 2540 (1979)
[102]S.L. Drechsler and N.M. Plakida, Physica C 153-155, 206, (1988)
[103]B. Batlogg, R.J. Cava, L.W. Rupp, Jr., A.M. Mujsce, J.J. Krajewski, J.P. Remeika, W.F. Peck, Jr., A.S. Cooper and G.P. Espinosa, Phys. Rev. Lett 61, 1670 (1988)
[104]W. Reichardt and W. Weber, Jpn. J. Appl. Phys. 26, 1121 (1987) Supplement 26-3
[105]B. Batlogg, R.J. Cava, A. Jayaraman, R.B. van Dover, G.A. Kourouklis, S. Sunshine, D.W. Murphy, L.W. Rupp, H.S. Chen, A. White, K.T. Short, A.M. Mujsce and E.A. Rietman, Phys. Rev Lett 58, 2333 (1987)
[106]L.C. Bourne, M.F. Crommie, A. Zettl, Hans-Conrad zur Loye, S. W. Keller, K.L. Leary, Angelica M. Stacy, K.J. Chang, Marvin L. Cohen

and Donald E. Morris, Phys. Rev. Lett. 58, 2337 (1987)
[107]K.L. Leary, Hans-Conrad zur Loye, S. W. Keller, Tanya A. Faltens, William K. Ham, James N. Michaels and Angelica M. Stacy, Phys. Rev. Lett. 59, 1236 (1987)
[108]E.L. Benitez, J.J. Lin, S.J. Poon, W.E. Farneth, M.K. Crawford and E.M. McCarron, Phys. Rev. B 38, 5025 (1988)
[109]T.A. Faltens, W.K. Ham, S. W. Keller, K.L. Leary, J.N. Michaels, A.M. Stacy, H.-C. zur Loye, D.E. Morris, T.W. Barbee III, L.C. Bourne, M.L. Cohen, S. Hoen. A. Zettl, Phys. Rev. Lett. 59, 915 (1987)
[110]B. Batlogg, G. Kourouklis, W. Weber, R.J. Cava, A. Jayaraman, A. White, K.T. Short, L.W. Rupp,and E.A. Rietman, Phys. Rev. Lett. 59, 912 (1987)

ION BEAM MODIFICATION AND ANALYSIS OF THIN $YBa_2Cu_3O_7$ FILMS

O. Meyer

Kernforschungszentrum Karlsruhe, Institut fur Nukleare Festkorperphysik, P.O.B. 3640, D-7500 Karlsruhe, FRG

INTRODUCTION

There has been a continuous interest in irradiation experiments of superconductors like A15 and Chevrel phases or refractory carbides and nitrides from both the scientific and applied points of view. The knowledge of the degradation of the superconducting transition temperature, T_c, by irradiation induced defects is of importance if superconductors are applied in a radiation environment. On the other hand defects which may act as pinning centers have been introduced into superconductors to improve such properties like the critical current. (For a review see ref. [1]).

It is well known that the superconducting properties of materials are strongly affected by chemical and structural disorder. Thus from a more basic point of view it is important to know the correlations between T_c depression and defect structures with other, e.g. transport properties, which can provide information on the microscopic parameters which in a given class of superconductors are important for high T_c achievement. Particle irradiation is often used to produce intrinsic damage in a controllable fashion in order to study the influence of disorder on the superconducting properties in a wide damage concentration, ranging up to levels where phase transformations, e.g. amorphization, occur. Previously, the change of the critical temperature, T_c, as a function of the radiation induced increase of the residual resistivity, $\Delta\rho_0$, revealed in general a universal behaviour depending on whether the Fermi level, E_F, was located either in a peak or a valley of the electronic density of states (DOS). Lattice defects lead to a lifetime broadening of electronic states and thus to a smearing of the peaks in the DOS. Under the assumption that the electron-phonon coupling constant, λ, was proportional to the density of states at E_F, T_c was calculated as a function of $\Delta\rho_0$ [2].

The structure of the defects causing large changes of T_c in A15 superconductors has been studied by TEM, X-ray diffraction and by Rutherford backscattering and channeling spectroscopy. The T_c decrease of high-T_c A15 superconductors is usually correlated with the decrease of the Bragg-Williams long-range order parameter S, i.e. with the formation of antisite defects by irradiation [3] or by deviations in composition from stoichiometry [4]. More recently it has been shown for Nb_3Ir that both, T_c increase and T_c decrease with increasing ion fluence are accompanied by a decrease of S[5] . Further, the changes of T_c are accompanied by a defect structure consisting of static displacements of the lattice atoms with average rms amplitudes of 0.005 to 0.009 nm, which can be interpreted as a static Debye Waller factor [6,7] . Both the depressions and enhancements in T_c have been explained by changes in the electronic density of states at the Fermi energy $N(E_F)$ due to smearing effects The elastic properties of V_3Si and Nb_3Ir are also strongly affected by particle irradiation leading to a large defect-induced stiffening of the average shear elastic constant for V_3Si [8] , while for Nb_3Ir a large softening of this long-wave length shear mode is observed [9] . In both cases the changes of the elastic constants are again correlated with the changes of $N(E_F)$ and of T_c correspondingly.

In comparison with the former superconductors, the high-T_c oxide superconductors have a rather low DOS at E_F and it is well known that by lowering the oxygen concentration $N(E_F)$ can be reduced [10] and a transition from the orthorhombic metallic phase to the tetragonal semiconducting phase can easily be enforced [e.g. 11,12] . The oxygen atoms are rather mobile and diffuse even near room temperature [13] . Thus it is concievable that oxygen atoms which are displaced during irradiation can strongly change the electronic properties of these oxides. A universal behaviour between T_c depression and resistivity increase cannot be expected due to such competing phase transitions Numerous irradiation studies have already been performed using electrons [14,15,16] ions [17-27] and neutrons [28,29,30] . In general, a rapid decrease of T_c, a considerable broadening of the transition width, δT_c, and large increase of the residual resistivity, ρ_0, with increasing particle fluence is observed. Although large differences are noted concerning the T_c dependence on the deposited energy density during irradition, there is a general agreement that the oxides range in their sensitivity scale to radiation damage between the Chevrel phases and the A15 superconductors. The origin for the observed T_c depression is still under debate. From electron irradition experiments it was concluded that the decrease of T_c is mainly due to point defects (bulk effect) while from n irradiation results it was concluded that the superconducting properties are mainly affected by intragrain radiation damage. Intragrain radiation damage was directly seen as amorphous zones on the grain boundaries by transmission electron microscopy [20] . The differences in the radiation sensitivity of various oxide materials can be attributed to the material quality ranging from loosely coupled intergrain material to single-crystalline quality with quite different initial values of T_c, δT_c and resistivity ρ. For example a dramatic loss of the superconductivity was observed by irradiating plasma-arc sprayed films with low fluences ($<10^{14}$ cm^{-2}) of H and He ions [26] . Therefore it seems to be important to separate bulk effects from intergrain effects on the changes of the superconducting properties upon irradiation.

In the following the influence of radiation damage on the electronic transport properties and the structure of polycrystalline, highly textured as well as large area single crystalline thin film is presented. Due to the strong dependences of the irradiation effects on the material quality it is necessary to characterize the thin films in great detail. Thus the preparation and the ion beam analysis concerning structure and properties of the films used for the irradiation experiments are described in the first chapter. In the second chapter the effects of ion irradiation on T_c , δT_c and ρ are described considering the initial film properties as parameter. The irradiation induced metal-semiconductor transition is presented in detail. Defect structure analyses by TEM, X-ray diffraction and high energy (2-3 MeV) He-ion channeling will be discussed. Some applications including the use of ion implantation are summarized in the last chapter.

1. Thin Film Synthesis and Analysis

Thin films have provided valuable information on the superconducting properties. Highly textured and epitaxial films exhibited critical transport currents well above 1 x 10^6 A/cm^2 at 77 K [31] , far larger than those measured for bulk samples (~10^3 A/cm^2) or single crystalline bulk samples [32] . A variety of thin film deposition techniques has been used up to now to produce thin films with T_c-onset values similar to those obtained for bulk samples. Among these techniques are r.f. sputtering [18,33,34] , a.c.-sputtering [35] and d.c. sputtering [36] from composite targets as well as d.c.-magnetron sputtering from multi-targets [37] . Electron beam evaporation from multiple sources has been used to produce thin films [38,39] as well as molecular beam epitaxy [40] . Other techniques such as pulsed laser evaporation of bulk material [41] , and screen printing [42] have also been successfully applied.

1.1 Thin Film Preparation

In order to obtain high quality films it is necessary not only to have the correct composition but to perform a proper annealing treatment in oxygen and to avoid chemical reactions with the substrate. Thus the choice of the substrate material and of the deposition process plays an important role, and the reproducible preparation of high quality high-T_c oxide superconductors in thin film form on desired substrates is still a problem of considerable challenge.

Most of the deposition techniques use a 3-step process: (a) Highly disordered or amorphous material with the stoichiometric composition is deposited at substrate temperatures up to 500°C. (b) The films are then transformed into the tetragonal phase at about 900°C in 1 atm. oxygen (O_2) and (c) are finally converted into the orthorhombic high-T_c phase by annealing the tetragonal phase in O_2 at temperatures below 600°C. The phase transformation at 900°C causes considerable interface reactions with various substrates such as Al_2O_3, MgO, ZrO_2 and $SrTiO_3$ [42,43,44] . Thus the quality of the films deposited in a three step procedure with thicknesses below 0.5 μm is rather limited. Further the nucleation of the crystalline films does not only occur at the film /substrate interface but may occur as well at the surface or at appropriate nucleation centers within the highly disordered film. Therefore the growth of a large area single crystalline high-T_c oxide film is strongly limited.

Recently a 2-step process was developed using a planar type magnetron sputter gun where the highly crystalline tetragonal phase could successfully be deposited at substrate temperatures of 800°C and below [34,45,46] . Due to these substantially lower substrate temperatures interface reactions could considerably be suppressed and the quality of the films becomes independent of the film thickness for thicknesses above 0.1 μm and independent of various substrate materials used. Using this technique, thin superconducting films with the nominal composition $YBa_2Cu_3O_7$ have been deposited by magnetron sputtering from a composite target on (100)-$SrTiO_3$ substrates at 800°C and were heated in situ at 430°C in pure O_2 atmosphere. Details of the deposition process are described in Ref. [45] . The gas pressure in the discharge was 2 x 10^{-1} Torr O_2 and 4 x 10^{-1} Torr Ar. A cathode voltage of 100 Volts and a current of 0.5 A was used. The substrates were placed 20-30 mm apart from the target on a Pt-stripe which could be resistively heated up to 1100°C. The deposition rate was typically 5 Å/s. The deposition process was further improved using a hollow cathode type magnetron gun [46] . The film composition was examined by Rutherford backscattering with 2 Mev $^4He^+$ ions.

Backscattering spectrometry is a well-known analysis technique for thin films [47] . For the development of the sputter guns this technique has successfully been applied in routine analysis of numberless thin films in order to optimize the many parameters of the sputtering process and the annealing procedures. The resonant scattering of 3.04 MeV He-ions on O has been used to enhance the detection sensitivity for oxygen. Further, the epitaxial growth of these films is analyzed using high energy ion backscattering combined with channeling. This allows the analyses of intrinsic defect structures or of defects produced by ion irradiation. The advantages of ion channeling for interface and epitaxial growth analysis are described in detail elsewhere [48] . In brief , variations of compositions at interfaces due to reactions can be analyzed nondestructively together with the crystalline quality of the deposited film as a function of depth. Lattice mismatch and the associated strain can be determined as well as the extent of mosaic spread (misorientation of the crystallites with respect to the substrate orientation) and the occurrence of special defect structures e.g. dislocations within the film.

As an example the yield of 2 MeV He-ions, backscattered from Y-Ba-Cu-O thin films which were deposited onto Al_2O_3 at different substrate temperatures is shown in Fig 1. The channel numbers can be converted in a mass scale and in a depth scale. It can be seen that the yields from Ba, Y, Cu and Al (from the underlying substrates) are well separated, especially for the film deposited at 760°C. From the corresponding step heights, the relative composition of Cu, Y and Ba can be determined. From the changes of the step heights with increasing substrate temperature it can be evaluated that the Ba/Y-ratio increases from about 1.7 to 3.5 with T_s increasing from 760°C to 870°C reaching the stoichiometric ratio at 820°C. Further it is noted that the composition profile near the interface deteriorates with increasing T_s. This could be due to either interface reactions or due to large variations of the film thickness as observed for island growth [49]. Sputter Auger experiments which also indicate interface diffusion [46] will have similar problems in profile evaluation if the films are inhomogeneous in thickness or contain pinholes.

1.2 Epitaxial Growth and Intrinsic Defects

The two-step process was well suited to grow single crystalline $YBa_2Cu_3O_7$ films on (100) and (110)-$SrTiO_3$ substrates . The growth has been studied by ion channeling spectroscopy [50] and X-ray diffraction [51] and it was shown that the substrate temperature and the substrate orientation are important parameters. On (110) substrate films grow either in the (110) or in the mixed (110)/(013) direction. On (100) substrates c-axis orientation is observed at elevated temperatures (Ts $\simeq$ 780-830°C)while a-axis orientation occurs at lower temperatures ($T_s < 720$°C) [51] .

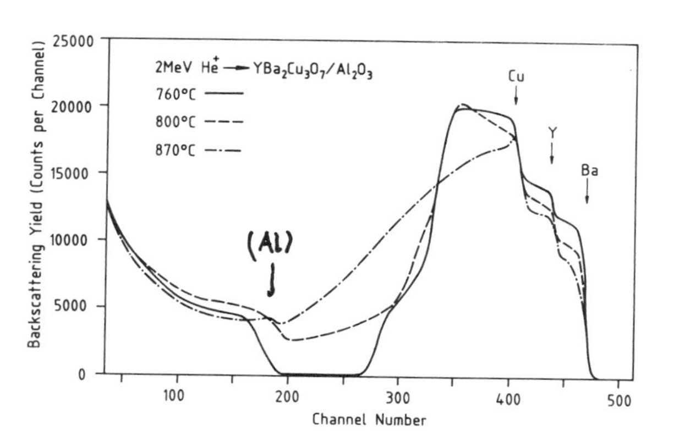

Fig.1 Backscattering spectra of $YBa_2Cu_3O_7$ thin films sputtered onto Al_2O_3 substrates at various substrate temperatures.

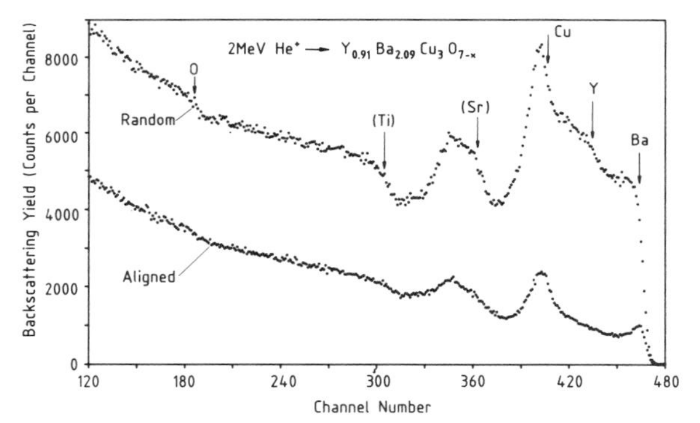

Fig. 2 Random and <100> aligned backscattering spectra from $Y_{0.91}Ba_{2.09}O_{7-x}$ thin films sputtered on (100) $SrTiO_3$ (analyzing He ion beam energy is 2.0 MeV).

The growth of the films in the single crystalline phase has also been analyzed by channeling experiments. Fig. 2 shows the random and (100)-aligned backscattering spectra of 2 MeV He ions of a single crystalline $YBa_2Cu_3O_7$ film grown on $SrTiO_3$. The front edges of the various components in the film and the substrate are indicated by arrows. The Cu-yield is overlapped with the Sr-yield which in turn together with the Ti-yield is shifted to lower energies as the He-particles have to penetrate the $YBa_2Cu_3O_7$ film in order to reach the substrate. From the step heights of the random spectrum a composition of $Y_{0.91}Ba_{2.09}Cu_3O_{7-x}$ can be determined. From the ratio of the (100)-aligned yield to the random yield behind the surface peak (≈channel 450) a relative minimum yield value, X_{min} for Ba of 16% is obtained. This is considerably better than values observed previously for films deposited by molecular beam epitaxy [40] and pulsed laser evaporation [41] where the films have been annealed at high temperatures after deposition. It clearly shows that the transformation from the disordered into the tetragonal phase at about 900°C does not exclusively start at the film/substrate interface. This represents a severe disadvantage of the 3 step preparation method. It has been shown previously that the minimum yield values decrease with decreasing analyzing beam energy which indicates that mainly dislocations contribute to the backscattered yield of the aligned spectra [50] . These dislocations could act as pinning centers as will be discussed below.

A further question is concerned with the quality of the epitaxial growth, especially the misorientation of the $YBa_2Cu_3O_7$ crystallites with respect to the $SrTiO_3$ crystal orientation. Ion channeling is a very sensitive technique to explore this problem [48] . In this case the Ba-yield is measured as a function of the angle between the incident He ion beam and the crystal orientation. The result of such a measurement is shown in Fig. 3 and is called angular yield curve. The angular yield curve is defined by the critical angle, $\Psi_{1/2}$ which is the half angle at half height between the random yield and the minimum yield,[48] . The channeling process can be evaluated in more detail by performing Monte Carlo simulation calculations through the various crystal directions in well defined tilt planes. The one-dimensional vibration amplitudes used for this calculation were 0.083 Å for Ba, 0.076 Å for Y, an averaged value of 0.067 Å for Cu(1) and Cu(2) and an averaged value of 0.088 Å for O(1), O(2), O(3) and O(4). These values are all based on high resolution neutron diffraction measurements [52] . The results of such a calculation in c-direction for the Ba- and Y-yield choosing a tilt plane 15 degrees off the (100) plane is also shown in Fig. 3 (solid line). The calculated values of 0.88° for Ba and 0.92° for Y are in rather good agreement with the measured values. In the a- or b-directions the calculated values are 0.87° and 0.75° for the Ba and Y rows, respectively. These data clearly show that the c axis of the deposited film consisting of Ba_2Y rows with similar steering force for Ba and Y is perpendicular to the substrate surface. This result is supported by the RHEED small angle diffraction pattern [50] and by X-ray diffraction [51] . As the analyzing beam of about 1 mm^2 averages over thousands of crystallites it is reasonable to assume a Gaussian distribution of the crystallite orientations with a standard deviation σ. Convoluting the calculated angular yield curve for Ba with a Gaussian curve (σ = 0.25°), the result represented by the solid line in Fig. 3 is observed. As can be seen the measured results can be simulated rather well using σ = 0.25° and random contribution to the yield of 12% due to scattering from dislocations.

The superconducting properties of thin films are not only determined by the composition and the defect structures present but especially by the amount and the lattice location of the oxygen. For bulk samples usually thermogravimetric methods are successfully used to determine the oxygen content. For thin films nuclear reactions or the resonant enhanced scattering cross section may be used to get some information about the oxygen concentration and the lattice location of oxygen. In Fig. 4 the random and <100>-aligned backscattering spectra are shown for a $YBa_2Cu_3O_7$ thin film on $SrTiO_3$ using an incident He-ion energy of 3.04 MeV. The contributions of the various components in the film and the substrate are indicated by arrows. The yield from oxygen is now strongly enhanced and can be used to determine the relative concentration of oxygen. The sensitivity enhancement factor (SEF) of 16 has been obtained from a similar analysis of $SrTiO_3$ with stoichiometric composition [50] . From the peak area of the oxygen resonance peak for the random spectrum in Fig. 4, together with the SEF-factor of 16 as deter- mined above from the analysis of $SrTiO_3$, a relative composition of $Y_{0.91}Ba_{2.09}Cu_3O_{7.0 \pm 0.1}$ was obtained for this film.

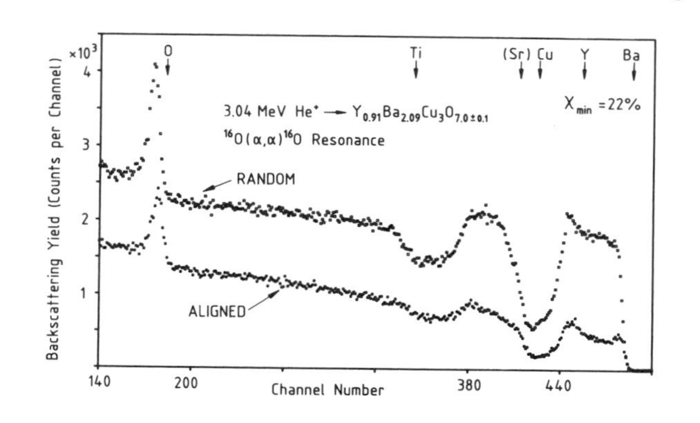

Fig. 3 Random and <100> aligned backscattering spectra from $Y_{0.91}Ba_{2.09}Cu_3O_{7.0}$ thin films sputtered on (100)$SrTiO_3$ (analyzing He ion energy is 3.04 MeV).

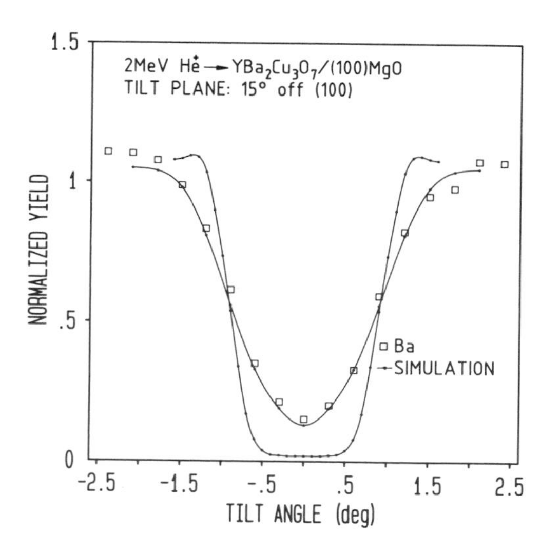

Fig. 4 Angular scan through the <100> axial direction of a $YBa_2Cu_3O_7$ thin film. The angular yield curve for Ba is shown together with the calculated angular yield curve of a perfect crystal. The latter curve is modified to fit the experimental values (solid line) assuming a mosaic spread of 0.25° and a displaced atom fraction of 12%.

Prior to deposition, the (100) $SrTiO_3$ substrates were analyzed by ion channeling and backscattering and the influence of various surface preparation techniques such as polishing, etching, sputtering and annealing in 1 atm O_2 was studied [50] . $SrTiO_3$ single crystalline substrates obtained from different manufactorers revealed quite different crystalline quality. The minimum yield values, varied between 1.5% and 30%. The dechanneling yield as a function of depth indicated that high minimum yield values were mainly due to high densities of partial dislocations throughout the bulk region. Some $SrTiO_3$ single crystals were of excellent quality as was checked by measuring and analyzing the surface peak areas of Sr and Ti in the aligned backscattering spectra of (100) $SrTiO_3$. The results compared with Monte Carlo calculations indicated that a nearly perfect (100) $SrTiO_3$ surface could be obtained by ion etching followed by annealing in O_2. Similar results have been obtained for highly polished and annealed $SrTiO_3$ surfaces. The results clearly show that a perfect substrate surface is a necessary precondition for good epitaxial growth of $YBa_2Cu_3O_7$ film on (100) $SrTiO_3$

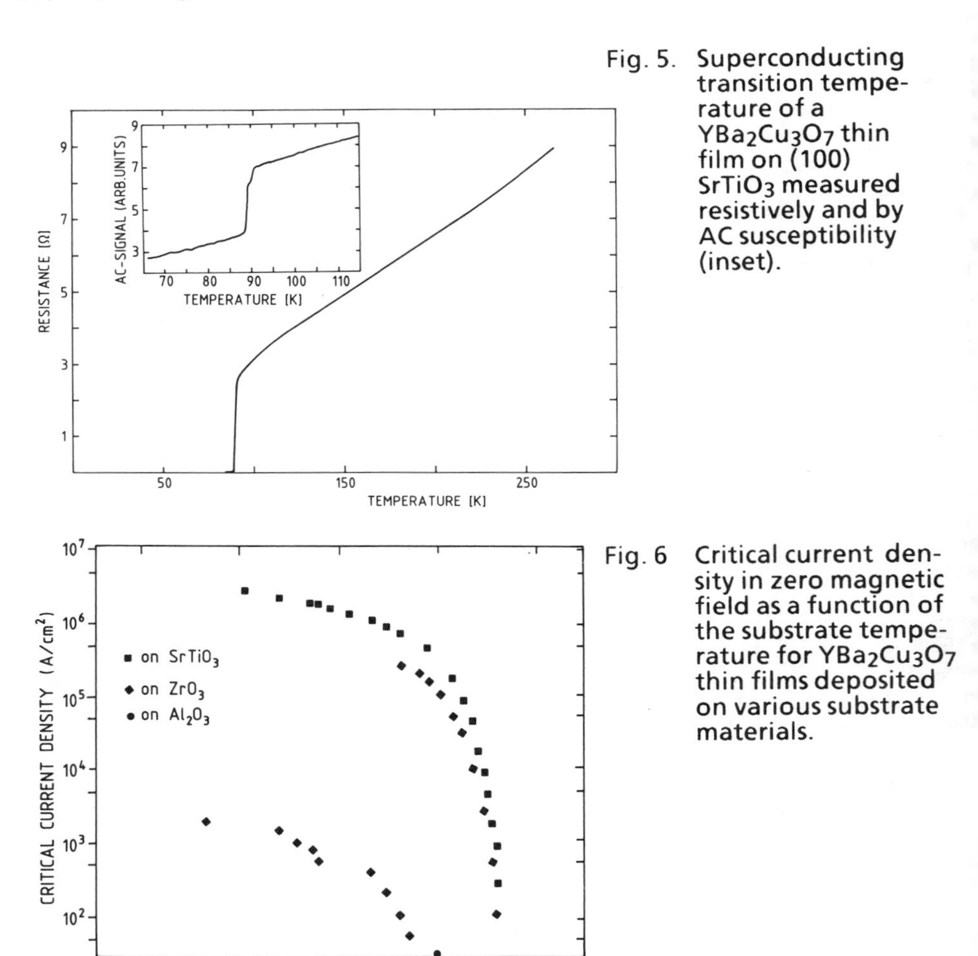

Fig. 5. Superconducting transition temperature of a $YBa_2Cu_3O_7$ thin film on (100) $SrTiO_3$ measured resistively and by AC susceptibility (inset).

Fig. 6 Critical current density in zero magnetic field as a function of the substrate temperature for $YBa_2Cu_3O_7$ thin films deposited on various substrate materials.

1.3 Thin Film Properties

The superconducting transition temperatures were determined by AC-susceptibility measurements and by the resistive method using 4 point contacts of Al-stripes connected to the films by silver paste. DC currents of 1-100 µA were applied for these measurements. As the films were found to be laterally homogeneous over typical distances of 15 mm the current and voltage contacts could be placed in most cases several millimeters apart. Films grown on (100)

TiO_3.reveal the best superconducting properties. These films show full
perconductivity at 89 K with a transition width of 1.2 K. Further a sharp drop
the Al-susceptibility signal is observed at 88 K (see e.g. Fig. 5). Resistivity
lues at 100 K are between 70 and 150 $\mu\Omega$cm and resistivity ratios $\rho_{RT}/\rho_{100\ K}$
ith values of about 3 were obtained routinely. Such films with thicknesses of
3 µm were patterned by ion beam irradiation using 600 keV Ar ions and a
uence of 1×10^{15} Ar^+/cm^2 at 293 K and by masking and etching with bridge
zes of 0.8 mm x 0.03 mm and 0.8 mm x 0.3 mm, respectively. The
perconducting transition temperature, , the width of the transition, and the
itical current in zero magnetic field, I_c, were determined by the resistive
ethod using Al strips connected with silver paste to the films. The results of
ch measurements are presented in Fig. 6 where the critical current is shown as
function of temperature for films on (100) $SrTiO_3$, ZrO_2 and on Al_2O_3. The low
itical current density observed for $YBa_2Cu_3O_7$ thin films on Al_2O_3 can be
tributed to Josephson intergrain coupling ($I_c \approx (1-T/T_c)^{1.5}$. The highest values
r the critical transport current measured up to now at 77 K in zero magnetic
eld were 5.4×10^6 A/cm^2 for $YBa_2Cu_3O_7$ films on (100)-$SrTiO_3$,5×10^5 A/cm^2
r films on (100) ZrO_2 and 1×10^6 A/cm^2 for films on (100) MgO [34]

Effects of Ion Irradiation

The slowing down processes of ions in matter are well understood and a
asonable agreement exists between the measured and calculated values for
e ion ranges, R_p and range distributions ΔR_p, as well as for the damage
stributions. Values for R_p and ΔR_p as well as for the energy density S_n
posited in nuclear collisions may be found in tables [e.g. 53,54] or can be
lculated using computer programs [e.g. 55] . The slowing down process is
scribed in terms of two uncorrelated processes [56] (i) the inelastic
teractions of the incident ions with electrons and (ii) the elastic collisions
tween the incident ions and the target nuclei . In the latter process large
ergies may be transferred to the host atoms and defects are produced by
splacing host atoms from their lattice sites which in turn are able to displace

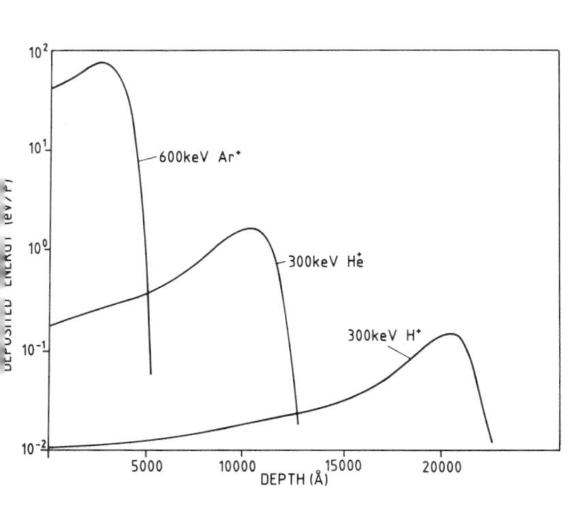

Fig. 7 Calculated deposited energy per particle for Ar-, He- and H-ions as a function of depth in $YBa_2Cu_3O_7$ using the TRIM code [55].

further lattice atoms in an avalanche-like process [57] as long as the transferred energy is greater than a minimum amount of energy, called the threshhold displacement energy, E_d. Most frequently the linear cascade model is used in the calculations where the cascade development is considered as a chain of independent two-body collisions. Recent molecular-dynamic computer simultations [58]however indicate interactions between recoils and collective excitations which supports the longstanding ideas of the existence of a thermal spike in high density cascades [59]. These basic processes in the cascade are important to estimate the amount of damage which survives the spontaneous relaxation processes and the possible formation of damage structures. Especially in a compound system having a complicated crystalline structure special defect structures or even new phases may emerge by ultra rapid melting and cooling processes in the thermal spike region. In the following most of the irradiations have been performed with 300 keV H- and He-ions and with 600 keV Ar-ions. The energy per incident ion as a function of depth has been calculated using the TRIM program [55]. For illustration the results of such calculations are shown in Fig. 7. It is seen that He- and H-ions penetrate well the film thicknesses of about 0.3 µm generally used for the irradiation experiments. For Ar-irradiations films with thicknesses of about 0.2 µm are used. Deposited energy densities have been averaged over the film thickness. This is justified as the damage profiles especially for protons are rather homogeneous. Assuming a threshold displacement energy, E_d of 20 eV, for all the atoms of the target the number of displacements per atom (dpa) is given by $(0.8 \times S_n)/(2 \times E_d)$ [60]. The ratios of the energy densities deposited in electronic excitation to those deposited in nuclear collisions are quite different for these particles and energies and are about 1, 1×10^3 and 2×10^4 for Ar, He and H in $YBa_2Cu_3O_7$, respectively. From the fact that the T_c-decrease observed after irradiation with various ions scaled with the deposited energy into nuclear collisions it was concluded that mainly displaced atoms and not electronic excitations are responsible for the observed changes of the superconducting transition temperature [17-19] .

2.1 Transition Temperature and Conductivity

In the previous chapter it was pointed out that the superconducting properties and the resistance of the $YBa_2Cu_3O_7$ thin films strongly depend on the composition and on the growth conditions. Thin films on Al_2O_3 generally are less textured and reveal higher resistivities and lower critical currents than highly epitaxial films grown on (100) $SrTiO_3$. It is conceivable, that the rate of the T_c depression and resistance increase upon irradiation would depend on the initial properties of the starting material. This has been observed for A15 compounds, for example for Nb_3Al and Nb_3Pt where the rate of Tc depression decreases as a function of deviation from stoichiometry [61]. In order to test this assumption films with different initial values of T_c, δT_c and ρ are used for the irradiation experiment.

The effect of H-ion irradiation on T_c and the transition width, δT_c, is shown in Fig. 8 for four thin films. Three films revealing rather large values of ρ and small values of r are deposited on Al_2O_3, while the high quality film was deposited on (100) $SrTiO_3$. The irradiations were performed at 77 K, however the data were taken after annealing to RT which causes considerable annealing of both T_c and ρ as will be discussed below. The initial unirradiated T_c-and δT_c-values of these films differ slightly. With increasing Ar-ion fluence T_c decreases and δT_c increases. The rate of suppression with fluence ($dTc/d\phi$) and the increase of δT_c

clearly depend on the quality of the films. The increase of both parameters is larger for films with large values of ρ(100 K) and small values of r.

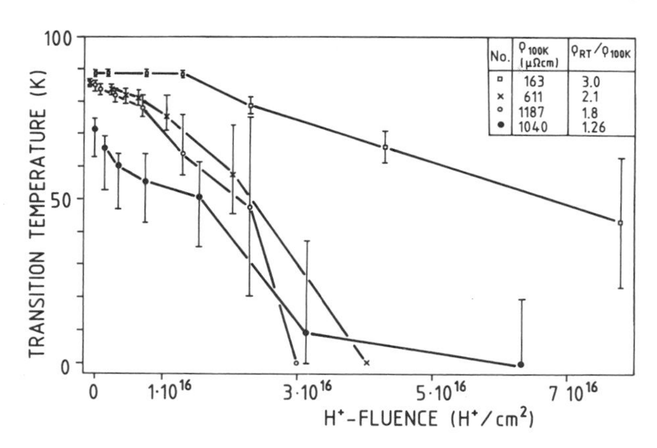

Fig. 8 Superconducting transition temperature and transition width as a function of the H ion fluence for various $YBa_2Cu_3O_7$ films of different quality.

Thus the suppression of T_c below 4.2 K is reached at smaller fluences for films of low quality. The rate of suppression for the high quality film is small at low and medium H-fluences and increases suddenly in the high fluence region. This dependence of the T_c depression rate on the film quality has been observed previously for $YBa_2Cu_3O_7$ thin films irradiated with Ar ions [27] .The large increase of δT_c with fluence as observed for the polycrystalline films, especially the fact that the T_c-onset values are less depressed than the midpoint and the downset values, may be attributed to the destruction of weak links causing the intergrain coupling. Similar arguments have been used to explain the results obtained for neutron-irradiated sintered bulk samples [29] and ion irradiated thin films [21]. The enhanced formation of defects, especially of amorphous zones on the grain boundaries has been observed after room temperature irradiation of thin polycrystalline films with oxygen ions [20] . It may be speculated that radiation enhanced outdiffusion of oxygen near grain boundaries may support this intergrain effect. The elimination of superconductivity at higher fluences of course is clearly a bulk effect. The irradiation results obtained for the single-crystalline film are also attributed to a bulk effect especially to the radiation induced transition from the metallic into the superconducting phase as will be discussed below. The increase of the T_c depression rate for films with decreasing quality is in contrast to results observed for A15 materials mentioned above. For the A15 phases Nb-Al and Nb-Pt with various compositions it was possible to construct a master curve between T_c and ϕ suggesting that the defects which are associated with the T_c depression for irradiated and for off-stoichiometric A15 structures are of similar type (anti-site defects).

It is generally agreed that the radiation sensitivity of high T_c oxides is rather high, however a quantitative comparison with that of other superconductors was hampered due to the uncertainties in the separation of bulk effects and intergrain effects of the T_c depression. In Fig. 9 we compare our results for the irradiation induced T_c depressions observed for high quality single-crystalline films of $YBa_2Cu_3O_7$ and $La_{1.8}Sr_{0.2}Cu_1O_4$ [62] with similar results obtained for other classes of superconducting materials. The normalized T_c values are shown as a function of the deposited energy in dpa for $PbMo_6S_8$ [63] , Nb_3Ge [64], V_3Si [65], NbC [66] and NbN [67]. It is seen that the radiation sensitivity of the oxide phases is higher than that of the A15 materials, however slightly lower than that of the Chevrel phases. Further it should be noted that the slope of the T_c depression curve, especially for $YBa_2Cu_3O_7$ is rather steep which may indicate that the rapid destruction of the superconducting properties ultimately is due to a phase transition and not due to an accumulation of point defects.

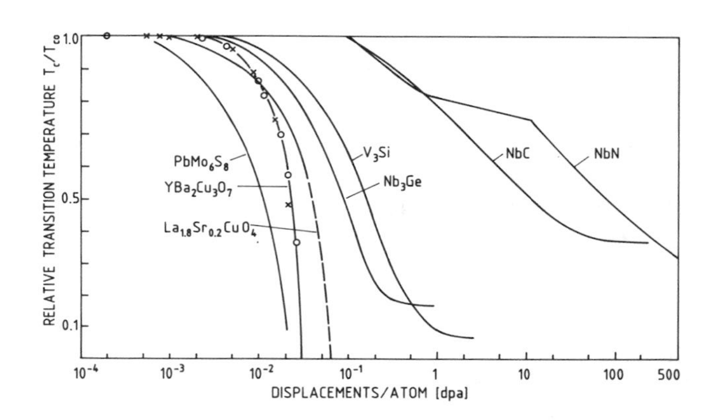

Fig. 9 Relative decrease of T_c by ion irradiation as a function of the deposited energy density in displacements per atom (dpa) for various superconducting materials.

The radiation induced increase of the resistivity also depends strongly on the initial resistivity values of the films. This is demonstrated in Fig. 10 where the increase of the resistivity at 100 K ($\Delta\rho = \rho(100\ K, \phi) - \rho(100\ K, 0)$ as a function of the proton fluence is shown. For the film with the highest initial resistivity, a jump like increase of $\Delta\rho$ is seen at low fluences which may be due to the destruction of weak links between grain boundaries. For high quality films $\Delta\rho$ is comparatively small although for all films the resistivity increase with fluence is stronger than linear and does not saturate as is usually observed for other superconducting materials under irradiation. For these materials the increase of the resistance is proportional to ϕ and thus to the number of defect centers at low fluences and reaches a saturation at large fluences. It is interesting to correlate the T_c-decrease with the resistance increase and to compare this result with that of other superconductors [22]. Previously, the electron-phonon coupling constant λ was assumed to be proportional to $N(E_F)$ and T_c was calculated as a function of $\Delta\rho$. In Fig. 11 the relative T_c decrease, T_c/T_{co}, is presented as a function of $\Delta\rho$ in comparison to other superconductors. The results obtained for single-crystalline $YBa_2Cu_3O_7$ thin films reveal a rather small slope $d(T_c/T_{co})/d(\Delta\rho)$ of 0.01 to 0.2 $K/\Delta\Omega cm$ as compared to that of other superconductors. As can be seen in Fig. 11 remarkable differences exist in the T_c vs. $\Delta\rho$ dependence even for high quality single crystalline films. The results seem to indicate the existence of quite different threshold fluences for the T_c depression. Additional results obtained up to now show that the T_c dependence on $\Delta\rho$ depends on the incident beam direction with respect to the film orientation [68]. More experiments have to be performed in order to clarify this situation. For neutron irradiated $YBa_2Cu_3O_7$- bulk samples a nearly linear T_c vs. $\Delta\rho$ correlation was observed with no indication of a threshold fluences [30]. As the slope of $dT_c/d(\Delta\rho)$ depends strongly on the material quality, no conclusion could be drawn on the existence of a universal correlation between T_c and the residual resistivity.

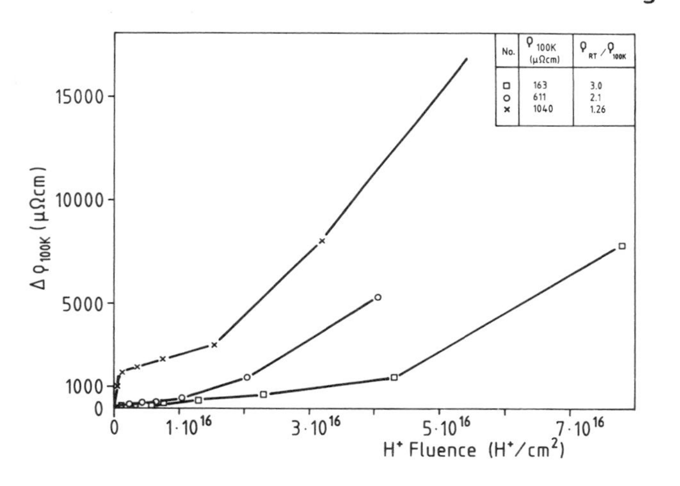

Fig. 10 Irradiation induced increase of the resistivity at 100 K as a function of the H-ion fluence for various $YBa_2Cu_3O_7$ films of different quality.

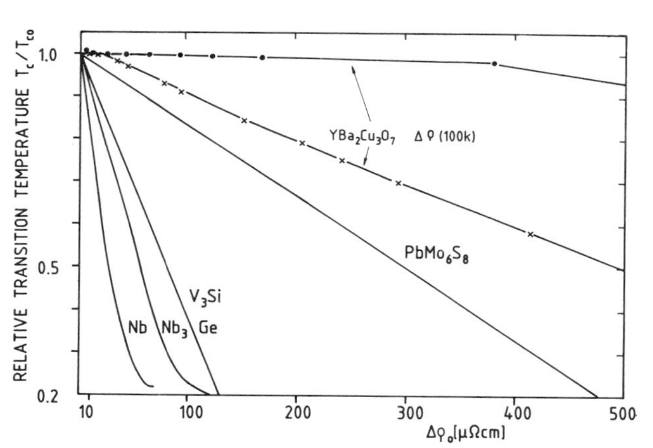

Fig. 11 Relative T_c-values as a function of the radiation induced increase of the residual resistivity for various superconducting materials.

2.2 Metal-Semiconductor Transition

In the previous chapter it was suggested that the radiation induced metal to semiconductor transition would influence the T_c depression. Detailed results on this phase transition will be presented in this and the next chapter. Before doing so we have to discuss the problem of defect mobility at and below room temperature as recovery of point defects may influence the transition process. It is known that large recovery effects, both for T_c and $\Delta\rho$ occur after irradiation at low temperature (5K, 77 K) during annealing to 293 K [22] . This recovery has also been observed for electron irradiated bulk samples even for temperatures as low as 100 K [15] and for O ion irradiation at 293 K [24]. In situ ion irradiation of thin $YBa_2Cu_3O_7$ films at low temperatures causes a number of effects which can clearly be recognized from the results presented in Fig. 12. , where the resistance of a film is shown as a function of temperature before and after irradiation with 600 keV Ar-ions at 77 K. With increasing Ar-fluence the following main features can be observed: (i) The transition temperature to superconductivity decreases and the width of the transition increases . (ii) The resistance increases and at low ion fluences the slope dR/dT stays nearly constant. A slight increase of the slope has been observed previously after irradiation with protons at 77 K [22]. At higher fluences a semiconducting phase appears, mixed with the superconducting phase. This is obvious from the negative slope dR/dT at temperatures above 70 K for curve (c). (iii) A strong annealing stage can be noted at temperatures above 150 K. This feature can be evaluated from the annealing curve (c) in comparison with the heating and cooling curve (d). Curve (c) was obtained after irradiation with a total dose of 6 x 10^{13} Ar-ions/cm^2 at 77 K then cooled to 6 K and annealed to 296 K. The sample was then remeasured and curve (d) was obtained. In comparing curves (c) and (d) it is seen that about 60% of the irradiation induced resistance increase at 100 K recovered upon warming up to 296 K. From our previous results [22] we can state that after irradiation at 6 K the heating and cooling curves between 5 and 120 K are the same and we concluded that there is no strong defect recovery stage below about 120 K.

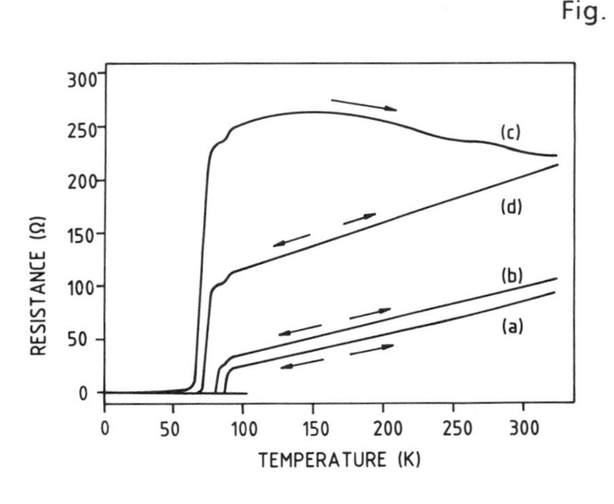

Fig. 12 Resistance vs. temperature: heating and cooling curves of a sample before (a) and after irradiation with a total fluence of 6 x 10^{12} Ar^+/cm^2 at 6 K and annealed to 325K (b). Heating curve of the same sample after irradiation with 3 x $10^{13}$$Ar^+$/cm^2 at 6 K (c) and cooling and heating curve after annealing to 325 K(d).

In order to avoid the influence of recovery on the damage production rate, irradiations with protons have been performed at 77 K and the resistance increase was determined at 100 K without any further annealing to higher temperatures. The results of this measurement are shown in Fig. 13, where $\Delta\rho$ is plotted as a function of the deposited energy density in dpa. For comparison the behaviour of the transition temperature for two irradiated single crystalline films is shown on the same dpa scale. Three different dpa regions of the $\Delta\rho$ increase can be recognized (this is more clearly seen from the slope $d(\Delta\rho/d\phi)$ and we will tentatively call the first region the metallic one (for dpa-values up to 1.5×10^{-2} dpa) followed by the transition region from the metallic to the semiconducting phase which is completed at about 4×10^{-2} dpa, followed by a transition into the amorphous phase which is not yet completed at about 0.13 dpa. More arguments for this arrangement will be presented in the next chapter on defect structures. It can be seen that T_c starts to decrease already in

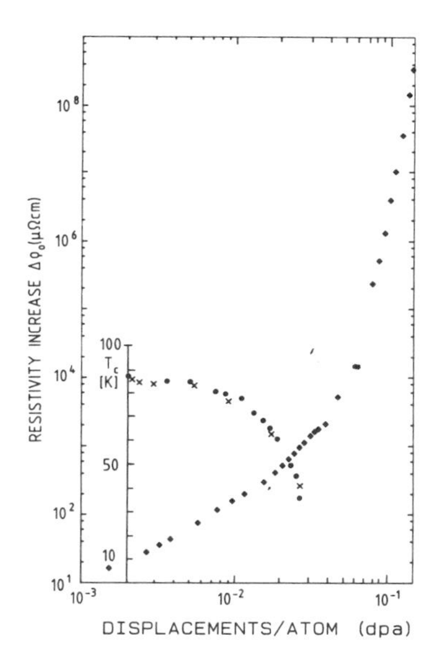

Fig. 13. Resistance change at 100 K and transition temperature as a function of the deposited energy density in dpa.

the metallic region due to point defect production. The elimination of the superconductivity occurs in the transition region where metallic and semiconducting phases coexist and is probably finished with the completion of the phase transition [22,27] . The mixture of a superconducting (metallic) phase and semiconducting phases is clearly demonstrated by the resistance vs. temperature curve shown in Fig. 14a. This sample was irradiated with 1.1×10^{17} H^+/cm^2 at 5 K (≈ 0.02 dpa). After annealing to 325 K a fraction of the material recovered and became again superconducting. It can be seen in Fig. 14a that now the cooling and heating curve consists of at least two phases. The steep increase of the resistance with decreasing temperature in the region below 10 K and also above 70 K indicate an activated conduction. The question arises if hopping conduction prevails $\sigma \approx \exp(-A/T^{1/4})$ or excitation to a gap energy $\sigma \approx \exp(E_g/kT)$. The data of Fig. 14a are replotted in Fig. 14b where the natural logarithm of the resistance, ℓnR, is given as a function of 1/T. In this plot two linear regions are noted: the linear region above about 100 K which can be fitted using a gap value of about 10 meV, while for the region between 5 K and

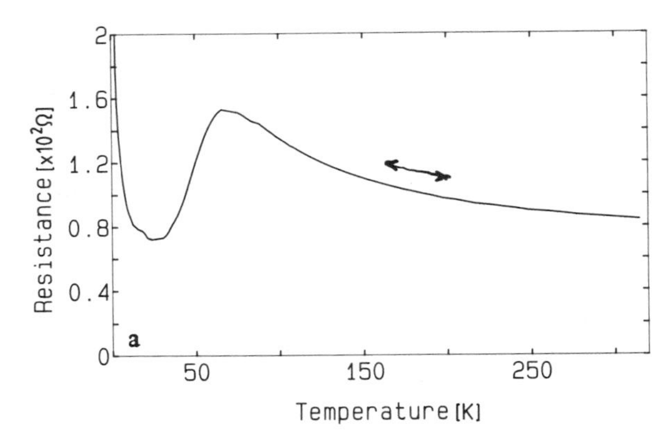

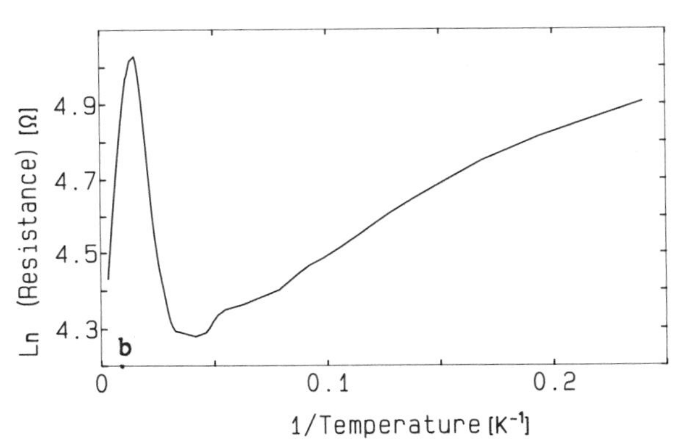

Fig. 14 (a) Resistance vs. temperature after irradiating the sample with a total dose of 1.1 x 10^{17} H^{+}/cm^{2}, 300 keV at 77 K and annealing to 293 K. (b) Resistance vs. the inverse temperature of the same sample.

10 K. gap energy E_g of 0.36 meV can be obtained. Here we tentatively assume that a semiconducting phase forms and from the activation energies immediately the gap energies can be deduced. This may be questioned and is discussed in more detail below. For the low temperature region deviations from the straight line are noted at temperatures below 5.5 K, which indicates that additional processes contribute to the thermally activated conduction. Thus the temperature dependence of the resistivity curves shown in Fig. 14a and b seem to indicate the coexistence of three phases: (a) the metallic (superconducting) phase, (b) a phase having a rather small activation energy and (c) a semiconducting phase. The latter phase dominates with increasing fluence. The energy gap values change by annealing as well as by irradiation. Before annealing to 325 K the sample shown in Fig. 14a reveals a gap value of 0.9 meV in the low temperature region. The gap value in the high temperature region which prevails at larger fluences increases with increasing deposited energy density as is shown in Fig. 15. From the results obtained up to now an energy gap is observed which increases with a slope of about 1.5 eV/dpa. No saturation is seen as the completely amorphous state is not yet reached. Large gap values of 0.2 and 0.42 eV have been published recently for deposited energy densities 0.26 and 0.37 dpa, respectively [24]. These gap values are

smaller than those calculated using the slope given above which indicates that a saturation is approached.

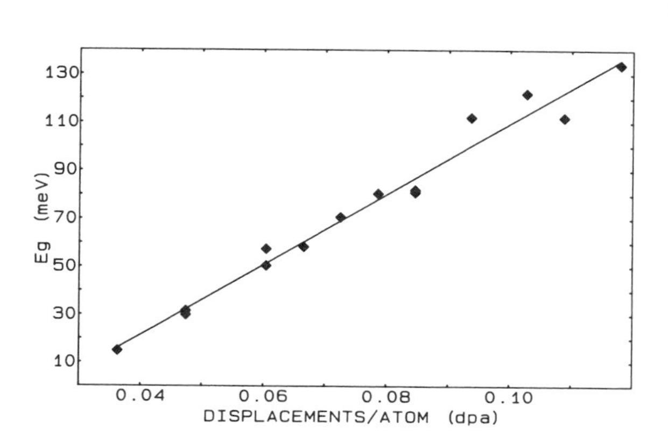

Fig. 15 Energy gap as a function of the deposited energy density in dpa.

Several processes may contribute to the ion induced metal-insulator transition. It is well known that by reducing the oxygen content, N from 7 to 6.5 an orthorhombic to tetragonal phase transition occurs where the resistivity of the compound changes from metallic to semiconducting. The activation energy for the semiconducting phase changes from ≈0.07 eV to 0.37 eV for N from 6.5 to 6.25 [69]. The band gaps E_g deduced from the activation energies would range between 0.14 and 0.72 eV, the latter value is far larger than that obtained by high dose ion irradiation. Nevertheless, it is suggested that the radiation induced phases are due to oxygen displaced from their lattice sites. Oxygen loss by outdiffusion during irradiation at 293 K and below can be excluded by experiment [70] . Previously it was suggested that knocked out O-atoms would not go into stable interstitial positions because there are many O-vacancies between the chains which would act as deep traps [15]. Indications for a random occupation of lattice sites in the Cu(1) plane has been observed by X-ray diffraction [27] and will be discussed below.

Another possible mechanism causing the metal to semiconductor transition is the shift of the Fermi energy in regions with negligible density of electronic states . This may be caused by an increase of the unit cell as observed by X-ray diffraction [24,27]. The metal to semiconductor transition at N = 6.5 as observed by varying the oxygen content has been interpreted as supporting the Hubbard model for the bandstructure near the Fermi level [69].

For the ion induced semiconducting phases the experimental situation is not yet settled. The observed dependences of ℓnR on T vary from T^{-1}, to $T^{-1/2}$ [71a] to $T^{-1/4}$ [71b]. The latter dependence was obtained for low fluences and temperatures . A $T^{-1/4}$ dependence would be in accordance with charge carrier localization [72] where the conduction at low temperatures is due to variable range hopping. Our data at low doses and low temperatures are better described by a 1/T dependence of ℓnR, although deviations from this dependence occur at temperatures below 5 K. Hall effect and/or thermoelectric power measurements are necessary to study the dependence of the carrier density and the mobility on the fluence and on the temperature in order to decide if the thermally activated conduction is by exciting electrons across a gap into the conducting band or correspondingly creating holes in the valence band or is due to a strong temperature dependence of the carrier mobility reflecting a diffusion process through a potential energy barrier. First Hall measurements on irradiated samples [69a,b] indicate only a slight dependence of the Hall number on the ion fluence as well as on the temperature which could not account for the large changes observed for $\rho(\phi)$ and for $\rho(T)$.These preliminary results suggest that the mobility and its temperature dependence probably affect the conduction process to a larger extent.

2.3 Defect Structures

Radiation induced defect structures in $YBa_2Cu_3O_7$ have been studied by TEM [20,73,74] by X-ray diffraction [19,27,75] and by ion channeling [21,50]. Extended defect formation is observed after prolonged electron irradiation of the single crystalline orthorhombic phase below 50°C [73] . Threshold energies determined for oxygen atom displacement were near 20 eV. The observed motion of twin boundaries and the shrinkage of the twinned volume implies a movement of oxygen in the basal plane. Irradiation of $YBa_2Cu_3O_y$ ($5.8 < y < 6.5$) with 1 MeV electrons in a high electron energy transmission electron microscopy indicated the disordering of the BaO- and Y-layers, forming a cubic superlattice structure [74]. No extended defects ,e.g., dislocation loops were observed by TEM studies of oxygen irradiated polycrystalline thin films [20,75] . Low doses of oxygen ions produced thin amorphous layers on the grain boundaries.

The crystalline to amorphous transition was also studied with X-ray diffraction. The intensity of the (006) reflex decreased suddenly after a critical dose of oxygen ions of about 0.1 dpa, which is low compared to other oxides with perovskite structure [75] . The decrease of X-ray line intensities is demonstrated in Fig. 16, where some spectra are shown before (a) and after irradiation of polycrystalline $YBa_2Cu_3O_7$ thin films with 300 keV He ions at 77 K at fluences of 8×10^{16}(b), 1.5×10^{17}(c) and 3.6×10^{17} H^+/cm^2(d).

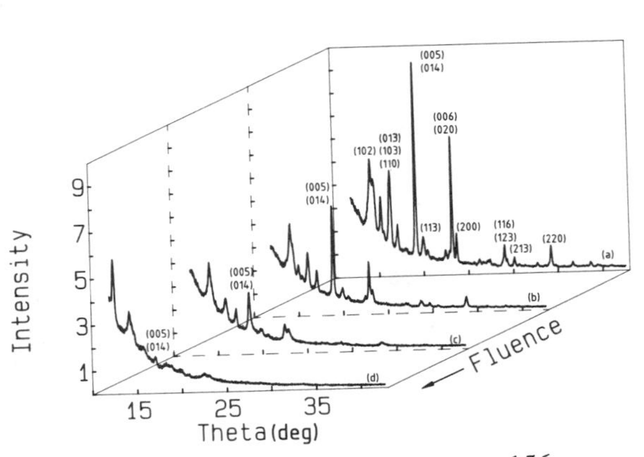

Fig. 16 Relative X-ray intensities as a function of the scattering angle Θ. (a) for the as-grown $YBa_2Cu_3O_7$ film and (b,c,d) after irradiation with 8×10^{16}, 1.5×10^{17} and 3.6×10^{17} H^+/cm^2.

After irradiation the films were warmed up to room temperature and transferred to a Seemann-Bohlin thin film diffractometer. Fig. 16 shows that the line intensities decrease strongly with increasing proton fluence which indicates that a certain fraction of atoms is displaced from their lattice sites and do no longer contribute to the coherent scattering. Complete amorphization is not yet obtained although the total deposited energy was about 0.1 dpa. This may be due to the light ion mass used, where the average transferred energy and therefore the average cascade density is smaller than for the oxygen ion irradiations mentioned above. The line intensity weakening in irradiated samples, especially at high diffraction angles suggested the use of the temperature factor concept applying the modified Wilson plot. This technique was previously successfully used to determine local atomic displacements around vacancies in refractory materials [76] as well as static displacements and amorphous fractions in V_3Si [77] . First attempts [27] to obtain the amorphous fraction in irradiated $YBa_2Cu_3O_7$ thin films using modified Wilson plots reveal a rather good agreement with the results obtained by ion channeling [27] although the accuracy was limited. The intensity ratio of a group of X-ray lines I(006,(020))/I(200) which is about 0.63 in the tetragonal phase and 4.4 in the orthorhombic phase was studied as a function of the fluence after proton irradiation at 77 K and annealing to 296 K. The intensity ratio of the non-separated (020) and (006) lines to the (200) line decrease from 4.4 to 1.4 (Fig. 16).

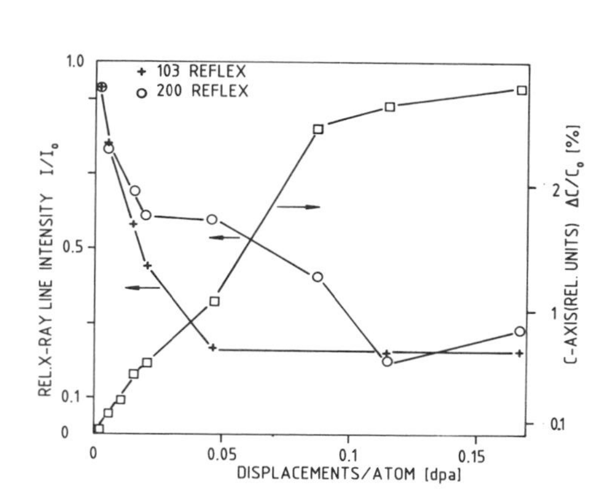

Fig. 17 Relative X-ray intensities for the (103) and (200) reflex and relative change of the c-axis as a function of the deposited energy density in dpa.

This indicates that there is a tendency for a random occupation of oxygen vacancies with oxygen atoms during irradiation, equivalent to an orthorhombic to tetragonal phase transformation during irradiation [27] . The intensity decrease with increasing fluence varies strongly for different reflexes. In Fig. 17 the relative intensities for the (103) and (200) reflexes are shown as a function of the deposited energy density in dpa. Although a strong decrease of the line intensities is noted at small dpa-values, the lines do not disappear completely indicating that some polycrystalline material is still present even at 0.15 dpa. Further it is shown in Fig. 17 that the lattice continuously expands under irradiation. A saturation value of about 2.8% for the expansion is reached with increasing fluence. This value is in good agreement with results obtained after O and As ion irradiation [75,24] .

Defect structures in the past have been determined by X-ray diffraction and by ion-channeling. For the analysis of defects in A15 superconductors these techniques yielded complementary results in different defect density regions [77] . Channeling analysis of irradiated single crystalline $YBa_2Cu_3O_7$ thin films at 2 MeV has been applied to determine the fraction of displaced atoms in the metal sublattice. It was clearly shown that the same number of atoms in the Ba, Y and Cu sublattice were displaced during the irradiation with 2 MeV He ions at room temperature [50] . In Fig. 18 the random and <100> aligned backscattering and channeling spectra before and after irradiation with 2.6 x 10^{16} He^+/cm^2 at 3.04 MeV are shown. As discussed in the first chapter He ion channeling together with the resonance scattering of He on oxygen at 3.04 MeV was used to determine the oxygen content and to study the damage production and annealing in the oxygen sublattice. From the increase of the peak area of the oxygen peak and from the increase of χ_{min} in the Ba, Y and Cu-region it is concluded that about 13% of the metal atoms and about 16% of the oxygen atoms have been displaced by depositing an energy density of about 0.01 dpa. Irradiation with $3.4 x 10^{16}$ He^+/cm^2, 1 MeV at 77 K (0.04 dpa) leads to an increase of 30% for the displaced atom fraction in the metal sublattices and of about 50% in the oxygen sublattice. Considering the higher sensitivity for the scattering of He-ions on slightly displaced O-atoms (due to the flux distribution in the channel) it is concluded that nearly the same amount of atoms is displaced in each sublattice after irradiation at 296 K. However, if we assume that oxygen atoms are randomly displaced, then the metal atoms should have an average threshold energy for displacement slightly above that of oxygen. Warming up to 293 K and remeasuring at 77 K did not reveal any recovery effect. This is in contrast to the large recovery stage of the resistance observed below 296 K. Therefore it is assumed that the recovery of the radiation enhanced resistance is due to the recombination of O-atoms which had been displaced to positions not visible by channeling analysis along the c-axis. Therefore the visible displaced oxygen atoms would belong to disordered regions which will not anneal by warming up to 293 K, while the annealing stage at 293 K and below is due to oxygen displaced to oxygen vacancies (not visible by channeling) and performing some ordering jumps upon annealing in order to achieve local charge neutrality.

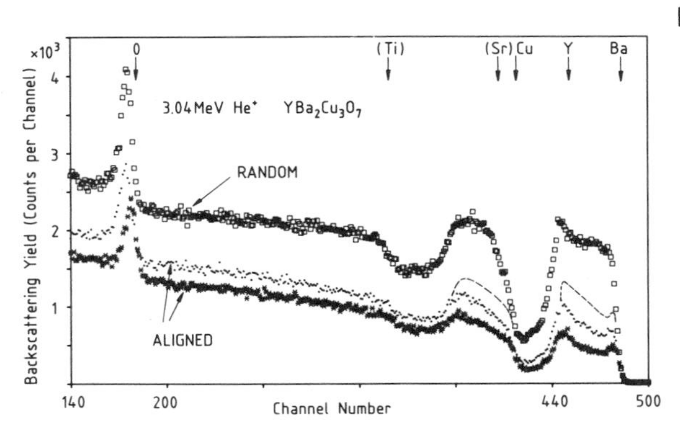

Fig. 18 Random and <100> aligned backscattering spectra from an $YBa_2Cu_3O_7$ thin film. The aligned spectra are shown after irradiation with 1 MeV He ions.

Most of the irradiation experiments have been conducted using the superconducting othorhombic phase. The question arose if the intrinsic damage introduced during thin film growth as well as the radiation induced damage is different in the tetragonal phase. First channeling results indicate that the standard deviation of the crystallite orientation distribution from the (100) $SrTiO_3$ direction also for the tetragonal phase is 0.2°, similar as that observed for the orthorhombic phase. The main defects incorporated during film growth are low angle dislocations [23]. Concerning the radiation induced defect structure as determined by ion channeling, no difference is noted concerning the defect production rate. The angular yield curves for the Ba-sublattice of the tetragonal phase are shown in Fig. 19.

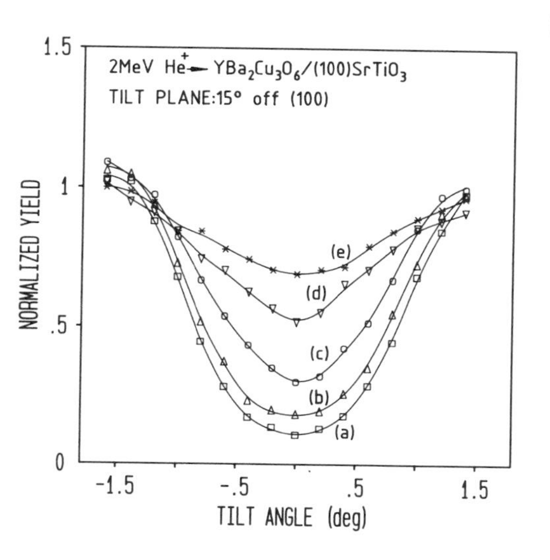

Fig. 19 Angular scan curves through the <100> axial direction of an $YBa_2Cu_3O_6$ thin film. The yield for Ba is shown after irradiation with various fluences of 2 MeV He ions ($1.5x10^{16}$(b), $3.4x10^{16}$(c), $5.2x10^{16}$(d) and $7.5x10^{16}$ He^+/cm^2.

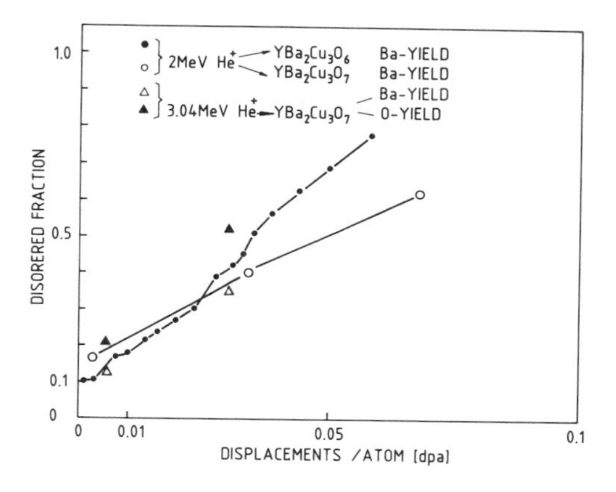

Fig. 20 Disordered fraction of Ba and O atoms as a function of the deposited energy density in dpa after irradiation with 2 and 3.05 MeV He-ions.

It can be seen that the minimum yield increases continuously with 2 MeV He ion irradiation at room temperature. Further it is noted that the critical angles decrease indicating an increase of a static Debye Waller factor. In order to determine if the increase of the minimum yield, X_{min} is due to the fraction of displaced atoms and not due to the production of extended defects, energy dependent measurements have to be performed which are in progress [78]. The increase of the disordered fraction which is equal to X_{min} is plotted as a function of the deposited ion energy in dpa (Fig. 20). Included in this figure are some results for the orthorhombic phase (open circles) and for oxygen. At about 0.1 dpa (extrapolated) amorphization would occur in good agreement with X-ray diffraction data. It should be noted that at about 0.03 dpa where the superconductivity is eliminated, already about 30 to 40 vol.% of the material is disordered corresponding to about 2.3×10^{21} displaced atoms/cm^3. However, the dpa- value of 0.03 corresponds to about 2.3×10^{22} displacements/cm^3 which is about a factor of 10 smaller than the value obtained from X_{min}. Currently the source of this discrepancy is not clear. One hypotheses could be the formation of disordered region by rapid quenching from non-linear cascade regions [58] . Further studies are obviously required to solve this question. In general, it is remarkable that the damage structures observedfor $YBa_2Cu_3O_7$, such as point defects, amorphous zones and static displacements where all host atoms are displaced from their lattice sites by about 0.07 Å , resemble very closely similar damage structures observed for other cluster compounds. This provides an explanation for the fact that the superconducting properties of these compounds reveal similar sensitivity to radiation induced disorder.

3. Applications

Ion implantation is a versatile non-equilibrium technique to modify the physical properties of materials. The changes of the material properties are due to chemical or alloying effects caused by the implanted ion species and due to disorder, produced by the interaction of the ions with the host atoms during the slowing down process. The influence of both effects on the property changes have to be separated. In the previous chapter mainly the influence of disorder has been discussed as the incident ions penetrated the thin films and came to rest in the substrate. In the application of ion beams for materials modification both processes, implantation and irradiation have to be considered. The results obtained by ion implantation into conventional superconductors have been reviewed [79,80]. Ion implantation has been used to produce new and metastable alloys, e.g. supersaturated solid solutions and to improve compounds by compensating deviations in composition from stoichiometry. Three main areas for the application of ion implantation have been explored: (i) increasing the pinning forces to enhance the surface current carrying capacity (ii) reduction of rf losses at superconducting surfaces and (iii) production of weak links. The results are summarized in ref. [79] . With the availability of single crystalline thin oxide films the use of ion beams has been explored for the modification of the high T_c superconductors.

First implantation experiments have been reported for the fabrication of $La_{1.8}Sr_{0.2}CuO_4$ films (nominal composition) and of $YBa_2Cu_3O_7$ thin films [25]. $YBa_2Cu_3O_7$ thin films have also been synthesized by ion beam mixing [81]. After implanting Sr ions in La/Cu multilayer films and annealing in flowing oxygen at 500°C the films became superconducting revealing a broad transition. The transition curves were very sensitive to the annealing conditions. The superconducting layer was buried beneath the surface [25] . Since the oxygen incorporation from the environment may be limited by surface barrier diffusion oxygen was implanted in amorphous $La_{1.8}Sr_{0.2}CuO_4$ layers at 77 K and a two step annealing in air was applied. An improvement of the T_c onset value was observed in contrast to similarly treated non-implanted thin films [34] . By ion implantation $YBa_2Cu_3O_7$ becomes amorphous and an important question is if solid state epitaxy would occur starting preferentially at the interface between amorphous film and oriented substrate surface. First attempts to test this were not successful. Although some regrowth did occur, channeling and backscattering indicated that the substrate layer was not a preferred nucleation site for monolayer by monolayer growth [25] . Ion beam mixing is a well known technique to intermix atomic species at the interface of thin film couples within the collision cascade and simultaneously avoid the material loss due to sputtering at the bombarded surface [82] . This techique has been used to intermix BaO/Y_2O_3/Cu layered structures by bombarding these couples with Xe- or O-ions (200-300 keV) at temperatures below 500°C. After annealing the mixed films in O_2 at temperatures below 600°C, the films were orthorhombic and superconducting [81] .

Particle implantations and irradiation of superconductors often leads to an enhancement of the critical current capacity due to the production of damage structures which act as pinning centers. By neutron irradiation of $YBa_2Cu_3O_{7-x}$ bulk single crystals the critical magnetization current has been enhanced [28,84] This rises the question how critical transport currents in single crystalline $YBa_2Cu_3O_7$ thin films are affected by intrinsic defects and by defects produced during irradiation. In order to study the influence of radiation induced defects on I_c, narrow bridges of the $YBa_2Cu_3O_7$ thin films on (100)$SrTiO_3$ and (100)ZrO_2 have been irradiated at 293 K with 300 keV protons. In Fig. 21 the critical current at 77 K in zero field is shown as a function of the deposition energy

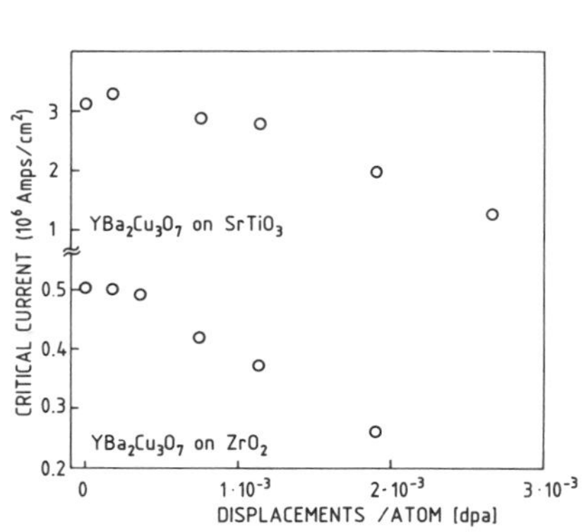

Fig. 21 Critical current as a function of the deposited energy density in dpa for $YBa_2Cu_3O_7$ thin films on $SrTiO_3$ and ZrO_2.

density in dpa. It is seen that the critical current immediately decreases even in a dpa-region where only a small decrease of T_c is observed (see Fig.13 for comparison). Similar results have also been observed after irradiation of epitaxial $YBa_2Cu_3O_7$ films with 1 MeV Ne ions at 300 K [83] . The average spacing between fluxoids is of the order of 300 Å [31]. For the lowest irradiation fluence used in Fig. 21 we have about 1.5×10^{19} displacements/cm3, corresponding to an average spacing of about 46 Å between displacements. Assuming the displacements to act as pinning centers, then with increasing density the fluxoids would overlap at a scale of the coherence length ($\approx$30 Å) and move easily between pinning centers. Thus the critical current would greatly be reduced. It seems that the intrinsic defect density in single-crystalline thin films is optimal for flux pinning. Besides twins, low angle dislocations may act as pinning centers.

Ion beam patterning has been used to form fine structures like narrow bridges for critical current measurements and for the production of superconducting quantum interference devices (SQUID) [84,24]. Here the radiation induced metal to insulator transition is used for patterning circuits onto thin films of $YBa_2Cu_3O_7$.. The patterning process is rather simple and has been described already in chapter 1. A more elaborate processing sequence was used for SQUID production such as an evaporation of a gold film, (b) negative organic resist spun onto the gold and photolithography (c) argon milling of the exposed gold and then finally (d) ion implantation [84]. Using the controlled reduction of the critical current by ion irradiation without affecting the transition temperature to a large extent, the sensitivity of SQUID's could be enhanced [83].

4. Conclusions

Ion beam modification and analyses has provided a wealth of information within a short period of time. The analysis of numerous thin films by ion backscattering spectrometry made it possible to optimize the various sputtering parameters in such a way that films with stoichiometric composition were grown. Film compositions and interface reactions were monitored to optimize different annealing and growth conditions. The preparation of nearly defect free single crystalline surfaces of the various substrates was analyzed and optimized using ion channeling combined with backscattering. The growth of $YBa_2Cu_3O_7$ thin films on these substrate surfaces was monitored and intrinsic damage structures were analyzed using the ion beam technique. The films optimized in this way revealed a high single crystalline quality and good superconducting properties which often exceeded those of bulk single crystals.

The high T_c oxides are sensitive to damage produced by ion irradiation. Defect structures as analyzed by ion channeling are mainly point defects, small static displacements of all lattice atoms from their lattice sites by about 0.07 Å and amorphous zones. Nuclear collisions and probably dense cascade effects are the main source for the observed property modifications. The T_c depression and the resistance increase with increasing ion fluence depends on the microcrystalline structure indicating that grain boundaries play a large role in polycrystalline films. For single crystalline films the elimination of superconductivity occurred in a fluence region where both metallic and semiconducting phases are mixed, which indicated that a phase transition is reponsible for the disappearance of the superconductivity. With further increasing the ion fluence the activation energy of the insulating phase increased and a transformation in the amorphous phase occured. This metal to insulator transition had been successfully used for ion beam patterning to fabricate narrow bridges, to measure critical currents and for the production of SQUID's.

The critical current immediately decreased upon low dose irradiation which indicated that the intrinsic damage in the as-grown films provided optimal pinning of fluxoids. The sensitivity of the SQUID's was improved by lowering the critical current without affecting T_c to a large extent. Ion implantation and ion beam mixing was used to produce high T_c oxide materials. Much remains to be learned about both the electronic and the structural properties of ion irradiated and implanted oxide materials. An increase in understanding in this respect will certainly be followed by an improvement of superconducting properties.

ACKNOWLEDGEMENT

The author would like to thank his colleagues L. Abu-Hassan, B. Egner, J. Geerk, H.C. Li, Qi Li, G. Linker, F. Ratzel, J. Remmel, R. Smithey, B. Strehlau, F. Weschenfelder, G.C. Xiong and X.X. Xi for valuable discussions and for providing some of their results prior to publication.

REFERENCES

[1] Proc. of the International Discussion Meeting on Radiation Effects on Superconductors, J. Nucl. Mat. 72 (1978) 1-300

[2] L.F. Mattheiss and R.L. Testardi, Phys. Rev. B20 (1979) 2196, Phys. Rev. Lett. 41 (1978) 1612

[3] A.R. Sweedler and D.E. Cox, Phys. Rev. B12 (1975) 147

[4] R.D. Blaugher, R.E. Heim, J.E. Cox and R.M. Waterstrat, J. Low Temp. Phys. 1 (1969) 539

[5] R. Schneider, G. Linker and O. Meyer, Phys. Rev. B35 (1987) 55

[6] O. Meyer, J. Nucl. Mat. 72 (1978) 182

[7] O. Meyer, R. Kaufmann, B.R. Appleton and Y.K. Chang, Solid State Comm. 39 (1981) 825

[8] A. Guha, M.P. Sarachik, F.W. Smith and L. R. Testardi, Phys. Rev. B18 (1978) 9

[9] M. Grimsditch, K.E. Gray, R. Bhadra, R.T. Kampwirth and L.E. Rehn, Phys. Rev. B35 (1987) 883

[10] N. Nücker, J. Fink, J.C. Fuggle, P.J. Durham, W.M. Temmermann, Phys. Rev.
B37 (1988) 5158

[11] J.M. Tarascon, W.R. McKinnon, L.H. Greene, G.W. Hull, and E.M. Vogel, Phys. Rev. B36 (1987) 883

[12] J. van den Berg, G.J. van der Beek, P.H. Kes, G.J. Nienwenhuys, J.A. Mydosh, H.W. Zandenbergen, F.P.F. van Berkel, R. Steems and D.J.W. Ijdo, Europhys. Lett. 4 (1987) 737

[13] B.G. Bagley, L.H. Greene, J.M. Tarascon and G.W. Hull, Appl. Phys. Lett. 51 (1987) 622

[14] Y. Quere, Nucl. Instr. and Meth. B33 (1988) 906 and references therein

[15] N. Moser, A. Hofmann, P. Schüle, R. Henes, and H. Kronmüller, Z. Phys. B 71 (1988) 37; Physica C153-155 (1988) 341

[16] V.F. Zelenskij, I.M. Neklyndov, Yu. T. Petrusenko, A.N. Sleptsov and V.A. Finkel, Physica C153-155 (1988) 850

[17] G.J. Clark, A.D.Marwick, R.H. Koch and R.B. Laibowitz, Appl. Phys. Lett. 51 (1987) 139

[20] G.J. Clark, F.K. LeGoues, A.D. Marwick, R.B. Laibowitz, and R. Koch, Appl. Phys. Lett. 51 (1987) 1462
[21] A.E. White, K.T. Short, D.C. Jacobson, J.M. Poate, R.C. Dynes, R.M. Mankiewich, W.J. Skocpol, R.E. Howard, M. Anzlowar, K.W. Baldwin, A.F.J. Levi, J.R. Kow, T. Hsieh, M. Hong, Phys. Rev. B (1988) 3755
[22] G.C. Xiong, H.C. Li, G. Linker, O. Meyer, Phys. Rev. B38 (1) (1988); Physica C153-155 (1988) 1447
[23] O. Meyer, B. Egner, J. Geerk, R. Gerber, G. Linker, F. Weschenfelder, X.X. Xi, and G.C. Xiong, 7th Int. Conf. on Ion Implantation Technology, IIT'88; June 1988 Kyoto Japan, to be published in Nucl. Instr. and Meth. B; F. Weschenfelder et al. to be published.
[24] A.D. Marwick and G.J. Clark, ibid. to be published in Nucl. Instr. and Meth. B
[25] A.E. White, K.T. Short, J.P. Gamo, R.C. Dynes, L.F. Schneemeyer, J. Waszczak, A.F.J. Levi, M. Anzlowar, and K.W. Baldwin, ibid. to be published in Nucl. Instr. and Meth. B
[26] D.B. Chrisey, G.P. Summers, W.G. Maisch, E.A. Burke, W.T. Elam, H. Herman, J.P.Kirkland and R.A. Neiser, Appl. Phys. Lett. 53 (1988) 1001
[27] O. Meyer, B. Egner, G.C. Xiong, X.X. Xi, G. Linker and J. Geerk, 6th Int. Conf. on Ion Beam Modification of Materials, June 1988, Tokyo, Japan, to be published in Nucl. Instr. and Meth. B
[28] A. Umezawa, G.W. Crabtree, J.Z. Lin, H.W. Weber, W.K. Kwok, L.H. Nunez, T.J. Moran, C.H. Sowers, H. Claus, Phys. Rev. B 36 (1987) 167
[30] P. Müller, H. Gerstenberg, M. Fischer, W. Schindler, J. Stöbel, G. Saemann-Ischenko, H. Kammermeier, Solid State Commun. 65 (1988) 223, Physica C153-155 (1988) 343
[31] P. Chaudhari, R.H. Koch, R.B. Laibowitz, T.R. McGuire, and R.J. Gambino, Phys. Rev. Lett. 58 (1987) 2684; Jap. J. Appl. Phys. Z6 (1987) Suppl. 26-3, p. 2023
[32] T.R. Dinger, T.K. Worthington, W.j. Gallagher and R.L. Sandstrom, Phys. Rev. Lett. 58 (1987) 2687
[33] H. Adachi, K. Setsune, T. Mitsuyu, K. Hirochi, Y. Ichikawa, T. Kamada and K. Wasa, Japanese J. of Appl. Phys. 26 (1987) L709
[34] JX.X. Xi, G. Linker,O. Meyer, E. Nold. B. Obst, F. Ratzel, R. Smithey, B. Strehlau, to be published in Z. f. Phys. - Condensed Matter
[35] M. Kawasaki, S. Nagota, Y. Sato, M. Funabashi, T. Hasegawa, K. Kishio, K. Kitazawa, K. Fueki and H. Koinuma, Japanese J. of Appl. Phys. 26 (1987) L738
[36] L. Li, B. Zhao, Y. Lu, H. Wang, Y. Zhao, and Y. Shi, Chinese Phys. Lett., Vol. 4, No. 5 (1987)
[37] R.E. Somekh, M.G. Blamire, Z.H. Barber, K. Butler, J.H. James, G.W. Morris, E.J. Tomlinson, A.P. Schwarzenberger, W.M. Stobbs, and J.E. Evetts, Nature, Vol. 326 (1987) 857
[38] M. Naito, D.P.E. Smith, M.D. Kirk, B. Oh, M.R. Hahn, K. Char, D.B. Mitzi, J.Z. Sun, D.J. Webb, M.R. Beasley, O. Fischer, T.H.Geballe, R.H. Hammond, A. Kapitulnik, C.F. Quate, Phys. Rev. B35 (1987) 7228
[39] O. Fischer, T.H. Geballe, R.H. Hammond, A. Kapitulnik, and C.F. Quate, Phys. Rev. B35 (1987) 7228
[40] J. Kwo, T.C. Hsieh, R.M. Fleming, M. Hong, S.H. Liou, B.A. Davidson, and L.C. Feldman, Phys. Rev. B36 (1987) 4039
[41] D. Dijkkamp, T. Venkatesan, X.D. Wu, S.A. Shaheen, N. Jisrawi, Y.H. Min-Lee, W.L. McLean and M. Croft, Appl. Phys. Lett. 51 (1987) 619
[42] K. Koinuma, T. Hashimoto, T. Nakamura, K. Kishio, K. Kitazawa, and K. Fueki, Japanese J. of Appl. Phys. 26 (1987) 2761

[43] J. Geerk, H.C. Li, G. Linker, O. Meyer, C. Politis, F. Ratzel, R. Smithey, B. Strehlau, X.X. Xi and G.C. Xiong, Proc. 8th Int. Symp. on Plasma Chemicals, Tokyo 1987, eds. K. Akashi and A. Kinbara, p. 2349.
[44] B. Oh, M. Naito, S. Arnason, P. Rosenthal, R. Barton, M.R. Beasley, T.H. Geballe, R.H. Hammond and A. Kapitulnik, Appl. Phys. Lett. 51 (1987) 852
[45] H.C. Li, G. Linker, F. Ratzel, R. Smithey, and J. Geerk, Appl. Phys. Lett. 52 (1988) 1098
[46] X.X. Xi, H.C. Li, J. Geerk, G. Linker, O. Meyer, B. Obst, F. Ratzel, R. Smithey, and F. Weschenfelder, Physica C153-155 (1988) 794;
[47] W.K. Chu, J.W. Mayer, and M.A. Nicolet, Backscattering Spectroscopy, Academic Press, New York, 1978
[48] L.C. Feldman, J.W. Mayer and S.T. Picraux, Materials Analysis by Ion Channeling (Academic Press, New York, 1982).
[49] O. Meyer, H. Mann and G. Linker, Appl. Phys. Lett. 20 (1972) 259
[50] O. Meyer, F. Weschenfelder, J. Geerk, H.C. Li, and G.C. Xiong, Phys. Rev. B37 (1988) 97, Nucl. Instr. and Meth. B (1988)
[51] G. Linker, X.X. Xi, O. Meyer, Q. Li, J. Geerk, Solid State Comm. (1988)
[52] M.A. Beno, L. Soderholm, D.W. Capone II, D.G. Hinks, J.D. Jorgensen, J.D. Grace, I.K. Schuller, C.U. Serge and K. Zang, Appl. Phys. Lett. 51 (1987) 57
[53] K.B. Winterbon, Ion Implantation Range and Energy Deposition Distribution, Vol. 2, Plenum, N.Y. (1975)
[54] J.F. Ziegler (ed.), The Stopping and Ranges of Ions in Matter, Vol. 1-6, Pergamon Press (1980)
[55] J.P. Biersack, L.G. Haggmark, Nucl. Instr. and Meth. 174 (1980) 257
[56] J. Lindhard, M. Scharff, H.E. Schiott, Mat. Fys. Medd. 33 (1963) no. 14
[57] G.H. Kinchin, R.S. Pease, Rep. Progr. Phys. 18 (1955) 1
[58] T. Diaz de la Rubia, R.S. Averback, R. Benedek and W.E. King, Phys. Rev. Lett. 59 (1987) 1930
[59] F. Seitz and J.S. Koehler in Solid State Physics, F. Seitz and D. Turnbull (eds.), Academic N.Y. (1956) Vol. 2
[60] M.T. Robinson and I.M. Torrens, Phys.Rev. B9 (1974) 5008
[61] A.R. Sweedler, D.E. Cox and S. Moehlecke , J. Nucl. Mat. 72 (1978) 50
[62] J. Remmel et al., to be published
[63] G. Hertel, A. Adrian, J. Bieger, C. Nölscher, L. Söldner, G. Saemann-Ischenko, Phys. Rev. B27 (1983) 212
[64] J. Pflüger, O. Meyer, Solid State Commun. 32 (1979) 1143
[65] O. Meyer, G. Linker, J. Low Temp. Phys. 38 (1980) 747
[66] N. Kobayashi, R. Kaufmann, G. Linker, J. Nucl. Mat. 133/134 (1985) 732
[67] V. Jung, LT17, U. Eckern, A. Schmid, W. Weber, H. Wühl (eds.) North Holland, Amsterdam (1984) p. 109
[68] B. Egner et al., to be published
[69] P.P. Freitas, T.S. Plaskett, Phys. Rev. B37 (1987) 3657
[70] A.D. Marwick et al., to be published
[71] (a) A. White (b) A.D. Marwick and G. Clark, private communication
[72] P.W. Anderson, Phys. Rev. 109 (1958) 1492
[73] M.A. Kirk, M.C. Baker, J.Z. Liu, D.J. Lam, H.W. Weber, Materials Research Soc. Meeting, Boston (1987)
[74] Y. Matsui, E. Takayama,-Muromachi, K. Kato, Jap. J. of Appl. Phys. 26 (1987) L1183
[75] G.J. Clark, A.D. Marwick , F. Legoues, R.B. Laibowitz, R. Koch, P. Madakson, Nucl. Instr. and Meth. B32 (1988) 405
[76] R. Kaufmann, G. Linker and O. Meyer, Nucl. Instr. and Meth. 191 (1981) 532

SUPERCONDUCTING PROPERTIES ASSOCIATED WITH SHORT COHERENCE LENGTH- FLUCTUATION EFFECT AND FLUX CREEP PHENOMENON IN HTSC

Yasuhiro Iye

The Institute for Solid State Physics
The University of Tokyo
Roppongi, Minato-ku, Tokyo 106, Japan

1. INTRODUCTION

Salient features of the high temperature superconductors (HTSCs), aside from their extraordinary high transition temperatures, are their quasi-two dimensional character and short coherence lengths. The former reflects the layered crystal structure of the HTSCs. That the latter is a direct consequence of the high T_c and low carrier density of the HTSCs can be readily seen by recalling the BCS expression for the coherence length, $\xi_0 \sim hv_F/\pi\Delta$ [1,2]. Here, v_F is the Fermi velocity and Δ is the superconducting energy gap. For ordinary superconductors, $v_F \sim 10^8$ cm/sec and $\Delta \sim 1$ to 10 K yield $\xi_0 \sim 10^3$ to 10^4 A, whereas for the HTSCs $v_F \sim 10^7$ cm/sec and $\Delta \sim 100$ K lead to $\xi_0 \sim 10$ A. Experimental values of the coherence length extracted from the upper critical field measurements [3,4] are consistent with this crude estimate. Such a short coherence length comparable to the atomic distances gives rise to certain behavior unobservable in conventional bulk superconductors. The phenomenology of the high temperature superconductivity, therefore, can be quite different in certain aspects from that of the conventional superconductivity. This chapter deals with such unique superconducting properties of the HTSCs associated with the short coherence length[5-7], i.e. fluctuation effect and flux creep phenomenon.

2. FLUCTUATION EFFECT

2.1 Superconducting Fluctuation

Superconductivity is known as a textbook example of a successful mean field theory. The Ginzburg-Landau (GL) theory[1,2], which basically neglects fluctuations, gives an extremely good phenomenological description of superconductivity. The basic reason for this lies in the large BCS coherence length ($\sim 10^3$ to 10^4 A) mentioned above. The number of electrons forming

Cooper pairs is roughly a fraction 10^{-4} of the itinerant electron density, 10^{22} to 10^{23} cm^{-3}. In a pictorial language, the BCS coherence length can be viewed as the average distance of two electrons forming a Cooper pair. The above values of the radius and the density of Cooper pairs give a picture that the volume occupied by one Cooper pair contains the centers of 10^4 to 10^6 other pairs. Superconductivity caused by such highly overlapping Cooper pairs is described extremely well by a mean field theory except in a very narrow critical region near T_c. Superconducting fluctuations which occur over the volume of ξ^3 costs energy given by the condensation energy per unit volume times this volume, that is $[H_c^2(T)/8\pi]\cdot\xi^3(T)$. The critical region around T_c in which the mean field description breaks down can be estimated from the condition that this energy becomes comparable to the thermal energy,

$$[H_c^2(T)/8\pi]\cdot\xi^3(T)\sim k_BT. \quad (1)$$

Using the mean field temperature dependences,

$$H_{c2}(T)\sim H_{c2}(0)\,|T-T_c|/T_c,$$
$$\xi(T)\sim\xi(0)\{|T-T_c|/T_c\}^{-1/2}, \quad (2)$$

the condition (the Ginzburg criterion [8]) for the breakdown of the mean field treatment can be written as

$$|T-T_c|/T_c\sim\{8\pi k_BT_c/H_c^2(0)\xi^3(0)\}^2. \quad (3)$$

For conventional clean superconductors, this criterion for the breakdown of mean field theory yields $|T-T_c|/T_c \sim 10^{-8}$. In the case of dirty superconductors, this condition is somewhat relaxed to $|T-T_c|/T_c \sim 10^{-6}$, which still does not permit an easy experimental access.

To observe the fluctuation effect in conventional superconductors, it is convenient to study systems with reduced dimensionality. When one or more dimensions of the specimen are smaller than the coherence length, the fluctuating volume becomes size-limited for those directions. In such cases, the fluctuation effect is more easily observable. The reduction of dimensionality can be either geometrical as in the cases of thin films, thin wires and fine particles, or structural as in the cases of layered materials and materials with chain-like structures. The most well-investigated fluctuation effects include those manifested above T_c as excess conductivity, precursor diamagnetic susceptibility. Fluctuations are also important below T_c in such phenomena as appearance of finite resistance associated with phase slip.

2.2 Ginzburg-Landau Theory

Fluctuations can be treated within the framework of the GL theory, as long as they are small[2,9]. The GL free energy is given by

$$F=\alpha|\Psi|^2+(\beta/2)|\Psi|^4+(1/2m^*)|\{(h/i)\nabla-(e^*/c)A\}\Psi|^2+(\mathrm{rot}A)^2/8\pi \quad (4)$$

where $\alpha=\alpha_0(T-T_c)/T_c$ $(\alpha_0>0)$. The GL parameters α and β are related with ξ and H_c by

$$\alpha = h^2/2m^* \xi^2 \quad \text{and} \quad \alpha^2/2\beta = H_c^2/8\pi. \tag{5}$$

with the temperature dependent GL coherence length ξ and the thermodynamic critical field H_c.

Since we are interested in the temperature range $T \sim T_c$ where $|\Psi|$ is small, we can neglect the $|\Psi|^4$ term in the GL free energy. In the absence of magenetic fields, the GL free energy becomes

$$F = \alpha|\Psi|^2 + (h^2/2m^*)|\nabla\Psi|^2. \tag{6}$$

We expand Ψ in Fourier series

$$\Psi(r) = \Sigma_k \Psi_k \exp(ikr), \tag{7}$$

and insert this into Eq. (6). Integrating over unit volume we obtain

$$\begin{aligned} F &= \{\alpha + h^2k^2/2m^*\}|\Psi_k|^2 \\ &= (h^2/2m^*)\{k^2 + 1/\xi^2\}|\Psi_k|^2. \end{aligned} \tag{8}$$

We can calculate the thermodynamic average of $|\Psi_k|^2$ using the standard method of statistical mechanics using the Boltzmann factor $\exp(-F/k_BT)$. The result is

$$\langle|\Psi_k|^2\rangle = 2m^*k_BT/h^2\{k^2 + 1/\xi^2\}, \tag{9}$$

which can be also obtained by an argument based on the equipartition law.

Equation (9) states that the fluctuation modes with wavelength shorter than ξ have smaller statistical weight. Yet the integrated contribution of short wavelength fluctuations yields a divergent result. Such an unphysical divergence is circumvented by introducing a cutoff to the upper limit of the k-integration. A proper choice of the cuttoff may be $k \sim 1/\xi(0)$, since the fluctuations of length scale shorter than $\xi(0)$ cannnot be properly treated by the GL theory. The physical meaning of Eq. (9) can be understood if we evaluate the spatial correlation of fluctuations by taking the Fourier transform of $|\Psi_k|^2$,

$$\begin{aligned} \langle\Psi(0)\Psi(r)\rangle &= \Sigma_k |\Psi_k|^2 \exp(ikr) \\ &= (m^*k_BT/2\pi h^2)\exp(-r/\xi)/r. \end{aligned} \tag{10}$$

Equation (10) shows that the spatial correlation of fluctuation extends roughly over a distance ξ. In other words, the system above T_c can be viewed as a collection of independently fluctuating superconducting droplets of size $\sim \xi$.

2.3 Fluctuation Diamagnetism

We first discuss the fluctuation diamagnetism because it is rather simple in the sense that it does not require knowledge of time dependence of fluctuations. As a simplest model let us consider superconducting particles whose radius R is smaller than the coherence length ξ, so that spatial variation of fluctuations can be neglected. Based on the London equation, the diamagnetic susceptibility of spherical particles of radius R ($\ll \lambda$, λ: penetration depth) is given by

$$\chi = -(1/40\pi)R^2/\lambda^2. \tag{11}$$

Since the penetration depth is given by

$$1/\lambda^2 = (4\pi e^*/m^*c^2)\langle\Psi^2\rangle, \tag{12}$$

the diamagnetic susceptibility of zero-dimensional samples furnishes a direct measure of $\langle\Psi^2\rangle$. Although $\langle\Psi\rangle$ vanishes above T_c, $\langle\Psi^2\rangle$ remains finite due to fluctuations. The mean square amplitude of the order parameter, $\langle\Psi^2\rangle$, above T_c is evaluated by the relation,

$$(1/2)|d^2F/d\Psi^2|\langle\Psi^2\rangle \sim k_BT, \tag{13}$$

yielding

$$\langle\Psi^2\rangle = k_BT/\alpha V = (k_BT/\alpha_0 V)\{T_c/(T-T_c)\}. \tag{14}$$

Thus, the precursor diamagnetic susceptibility is expected to show a temperature dependence $\sim(T-T_c)^{-1}$.

As seen in the previous subsection, a three-dimensional superconductor above T_c can be viewed as a collection of droplets of size $\xi(T)$, which fluctuate independently. It is reasonable to take $R^2 \sim \xi^2$, $V \sim \xi^3$ in Eqs.(11)-(14). Together with $\alpha = h^2/2m^*\xi^2$, the fluctuation susceptibility of a three dimensional system is given by

$$\chi_{3D} = -(k_BT/\Phi_0^2)\xi(T) \sim (T-T_c)^{-1/2}. \tag{15}$$

The above simple argument essentially reproduces the more rigorous calculation by Schmid[10] apart from numerical factors. The prefactor of this $\{(T-T_c)/T_c\}^{-1/2}$ dependence of the fluctuation susceptibility is of the same order of magnitude as the Landau diamagnetism of normal metals.

In the case of two-dimensional systems like a thin film with thickness $d < \xi$, the fluctuation volume becomes $V \sim d\xi^2$. This results in a fluctuation susceptibility with temperature dependence different from the above 3D case. For H perpendicular to the 2D plane, the fluctuation diamagnetism is found to be

$$\chi_{2D} = -(k_BT/\Phi_0^2 d)\xi^2(T) \sim (T-T_c)^{-1}. \tag{16}$$

The fluctuation susceptibility of a 2D system for H $\parallel$ plane can be obtained by a similar argument. It turns out to be temperature independent, and hence is difficult to separate out experimentally.

Layered materials and artificial superlattices can be regarded as two-dimensional systems, if the superconducting layers are sufficiently well-decoupled. The key parameter is the ratio of the layer separation s and the coherence length ξ_c in the c-axis direction which is perpendicular to the layer planes. Calculation based on the Lawrence-Doniach model[11] of layer superconductor yields the following form of fluctuation diamagnetism

$$\chi_{LD} = -(\pi k_B T/6\Phi_0^2)(M/m)^{1/2}\xi\{1+(s/2\xi_c)^2\}^{-1/2}, \quad (17$$

which reduces to the 2D or the 3D form according to ξ_c 《 or 》 s. Studies fluctuation diamagnetism in layered superconductors are found in [12,13] a references therein.

Studies of fluctuation diamagnetism in $YBa_2Cu_3O_{7-\delta}$ by Freitas et al. [1 and by Kanoda et al. [15] Figure 1 shows the fluctuation diamagnetism of t several samples studied by the latter authors. The fluctuation-induc diamagnetic susceptibility χ' is extracted by subtracting out the Pauli-li paramagnetic background from the measured total susceptibility χ. The authors note that the log-log plot of $\chi'(T)$ of the fully oxydized sample (43) exhibits a slope of -1/2 characteristic of a 3D system, while in the ca of the oxygen deficient sample (Y-41), a crossover from -1/2 slope to slope, i.e. from 3D to 2D behavior, is observed as $T-T_c$ is increased. A te tative explanation for the change of dimensionality of the superconductivi with the oxygen stoichiometry is as follows: In $YBa_2Cu_3O_{7-\delta}$ the first oxyg defects are known to be produced predominantly in CuO chain layers.

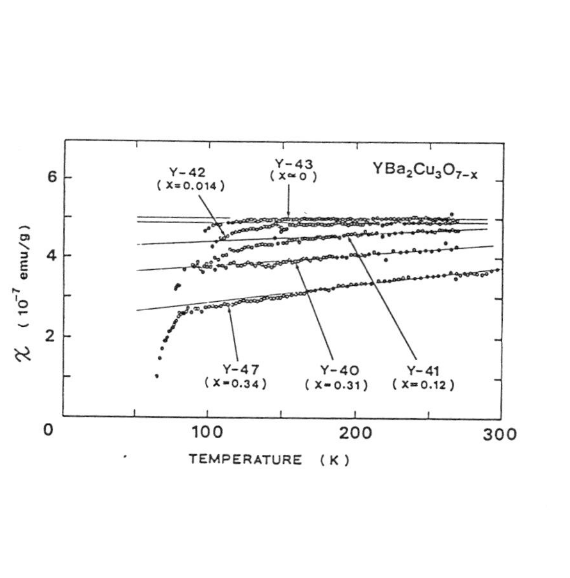

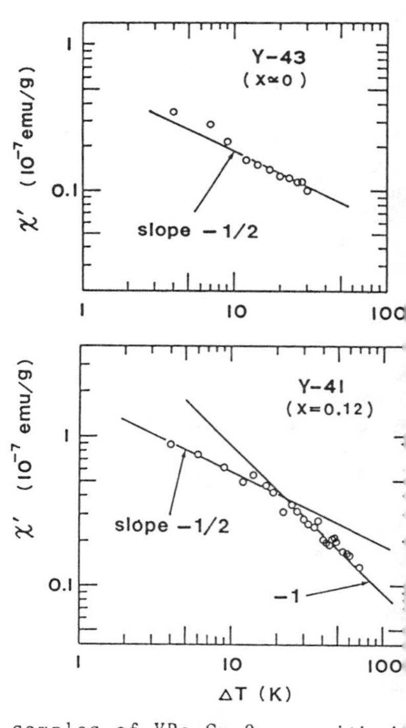

Fig. 1 Magnetic susceptibility of several samples of $YBa_2Cu_3O_{7-\delta}$ with di ferent oxygen contents. The two figures on the right shows the fluctu tion induced excess diamagnetic susceptibility extracted by subtracti the background drawn by the straight lines in the left figure. The full oxidized sample (Y-43) shows a slope of -1/2, indicative of 3D fluctua tions, while the oxygen deficient sample shows the apparent dimensiona crossover behavior around 25 K. (ref. [15])

increase of oxygen defects in the CuO chain layers may weaken the interlayer coupling between the superconducting CuO_2 layers and makes the system more two-dimensional.

It should be reminded here that the nature of the "Pauli-like" normal state susceptibility is a currently debated issue. It is not clear whether it can be understood as a usual Pauli susceptibility of a Fermi liquid, or it rather reflects a Bonner-Fisher type susceptibility[16] of a quantum spin liquid as suggested by the resonating valence bond (RVB) theory[17-19]. In fact, as seen in Fig. 1 and noted by many other groups, the normal state susceptibility very often has a term linearly decreasing with T, whose origin is not clarified yet. Thus, in an attempt to analyse the fluctuation diamagnetism, we should keep in mind that we do not yet have a clear picture of the "normal" magnetism of HTSCs.

2.4 Fluctuation-induced Excess Conductivity

We now turn to the fluctuation effect on conductivity. Because vanishing resistance is the most striking attribute of superconductivity, experimental studies of superconducting phenomona most frequently begin with transport measurements. The initial development of the study of fluctuation was no exception and it emerged from measurements of excess conductivity above T_c, called paraconductivity.

To discuss non-equilibrium properties such as conductivity, we need to consider the time dependence of fluctuation. Usually this is done by use of the so-called time dependent Ginzburg-Landau (TDGL) equation

$$h\gamma\, d\Psi/dt = -[(1/2m^*)\{(h/i)\nabla-(e^*/c)A\}^2+|\alpha|]\Psi \tag{18}$$

here

$$h\gamma/|\alpha| = \pi h/8k_B(T-T_c) \equiv \tau_0. \tag{19}$$

is a measure of relaxation time of superconducting fluctuation. Here, the equation has been linearized by dropping the term of order Ψ^3. In the case of $A=0$, Eq. (18) becomes

$$\tau_0 d\Psi/dt = -(1-\xi^2\nabla^2)\Psi \tag{20}$$

Equation (20) states that a fluctuation mode with finite k, i.e. $\Psi_k\exp(ikr)$, decays more rapidly than the long wavelength fluctuation mode. The relaxation time is given by

$$\tau_k = \tau_0/(1+k^2\xi^2). \tag{21}$$

The conductivity of normal electrons is given by

$$\sigma_n = (ne^2/m)\tau_{tr}, \tag{22}$$

where τ_{tr} is the transport relaxation time. By analogy, contribution of the fluctuation-induced Cooper pairs to conductivity may be written as

$$\sigma' = (e^{*2}/m^*)\Sigma_k \langle |\Psi_k|^2\rangle \tau_k/2. \quad (23)$$

Inserting Eqs.(9) and (21) into this equation, we obtain

$$\sigma' = (e^{*2}\xi^2\tau_0/h^2)\Sigma_k(1+k^2\xi^2)^{-2}. \quad (24)$$

The sum over k can be converted to an appropriate integration depending on th dimensionality of the system. The results are

$$\sigma'^{AL}_{3D} = \{e^2/32h\xi(0)\}\ \varepsilon^{-1/2}, \quad (25)$$

$$\sigma'^{AL}_{2D} = \{e^2/16hd\}\ \varepsilon^{-1}, \quad (26)$$

where $\varepsilon = (T-T_c)/T_c$ is the reduced temperature. The fluctuation contributic to the conductivity expressed by Eqs.(25) and (26) were first derived by As lamazov and Larkin[20]. It is noteworthy that the paraconductivity in tw dimensions does not depend on material parameters.

The Aslamazov-Larkin (AL) result (Eq.(26)) was very successful in ex plaining the universal excess conductivity above T_c in dirty 2D systems suc as amorphous or alloy superconducting films. However, in the case of cleane films, an additional contribution elucidated by Maki[21] and Thompson[22 gains importance. This second contribution, called Maki-Thompson (MT) tern turns out to be divergent in one and two dimensions. To obtain a finit result, a long wavelength cutoff $k_c = \varepsilon_c/\xi(0)$ is introduced, where ε_c $(T_{c0}-T_c)/T_c$ is the reduced shift of T_c due to pair-breaking. In tw dimensions, the MT term is given by

$$\sigma'^{MT}_{2D} = \{e^2/8hd\}\ (\varepsilon - \varepsilon_c)^{-1} \log(\varepsilon/\varepsilon_c). \quad (27)$$

The MT term is associated with the increase in the normal electron conduc tivity induced by the superconducting fluctuations, while the AL term repre sents the direct contribution of the fluctuation-induced Cooper pairs to th conductivity. The total paraconductivity is given by the sum of AL term ar the MT term.

The fluctuation-induced enhancement of conductivity in HTSCs has been ir vestigated by many authors using ceramic samples[14,23-25], thin films[26] and single crystals[27,28]. Figure 2 shows the data of Freitas et al.[14] c ceramic samples. These authors conclude that the fluctuation conductivit fits the 3D AL formula (Eq.(25)) with $\xi(0) = 22$ A. Goldenfeld et al. reache a similar conclusion except they obtained $\xi(0) = 13.4\pm4.8$ A. Ausloos ar Laurent[24] made a similar measurement, but interpreted their data as a evidence of 2D fluctuation near T_c. Figure 3 shows the data on thin fil samples reported by Oh et al.[26] These authors compare the result with th following expression of the AL term based on the Lawrence-Doniach model.

$$\sigma'^{AL}_{LD} = \{e^2/16hs\}\ [\varepsilon\{\varepsilon + (2\xi_c/s)^2\}]^{-1/2}. \quad (28)$$

This formula interpolates between the 2D (Eq.(26)) result for $\xi_c \ll s$ and th 3D (Eq.(25)) result for $\xi_c \gg s$, and predicts a 2D/3D crossover at $\xi_c \sim s/$ The data in Fig.3 show apparent dimensional crossover behavior in qualitati agreement with theory. Based on fitting of Eq.(28) to their data, Oh et al.

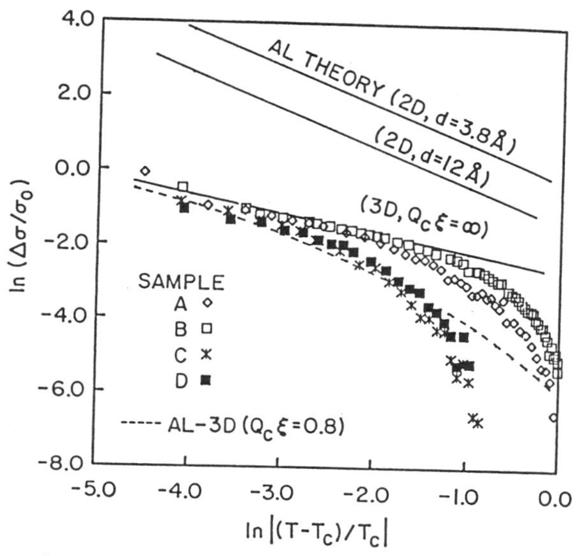

Fig. 2 Fluctuation induced excess conductivity of polycrystalline $YBa_2Cu_3O_{7-\delta}$ samples. Predictions of the Aslamazov-Larkin theory for 3D and 2D cases are shown by the solid lines. For the 3D case, $\xi(0)$=22 A is used. The dashed curve refers to a modified theory incorporating short-wavelength cutoff. (ref. [14])

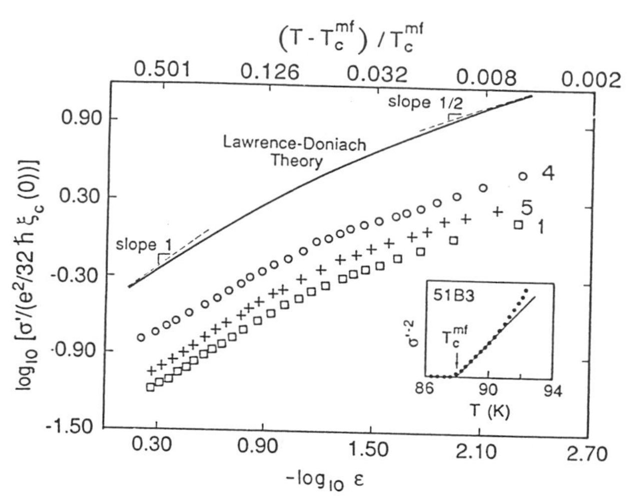

Fig. 3 Fluctuation conductivity of thin film samples of $YBa_2Cu_3O_{7-\delta}$. The solid line represents the AL term calculated in the Lawrence-Doniach model, showing the dimensional crossover behavior. In plotting this figure, $\xi_c(0)$ is taken as 1.85 A. (ref. [26])

estimated the c-axis coherence length as $\xi_c(0) \sim 1.5$-2 Å. A quantitativ agreement was not reached, however, for the absolute scale of the fluctuatio conductivity. Matsuda et al.[25] obtained a similar result using cerami samples. Ong et al.[27] studied the paraconductivity in single crysta samples, but failed to observe simple temperature dependences comparable with the expected power-law dependence on ε. Instead, they noted logarithmi dependence, $\sigma' \sim \log \varepsilon$ over a wide range of ε. A comment from a purely tech nical point may be appropriate here. Transport measurements on single crysta HTSCs can be distorted by the combined effect of anisotropy and sampl inhomogeneity. One should keep this in mind, especially when one attempts resistivity measurement by the Montgomery method.

Further insight can be gained by studying magnetoresistance in the fluc tuation regime. The effect of magnetic field on the fluctuation-induce excess conductivity has been well studied for the conventional supercon ductors[29-31]. Theoretical calculation for the case of HTSCs was done b Hikami and Larkin[32] and by Tsuneto[33]. The former authors gave an expres sion for the magnetic field dependent AL term appropriate to a syste described by the Lawrence-Doniach model. Hikita and Suzuki[28] found reasonable agreement between the expression given by Hikami and Larkin[32] an their data on single crystal sample, from which they extracted $\xi_{ab}(0) = 13$ and $\xi_c(0) = 2$ Å.

2.5 Fluctuation Specific Heat and Critical Fluctuation Regime

When one approaches T_c more closely, fluctuation can no longer be treate as a small correction to the mean field behavior. As discussed earlier, th width of the critical region is given by the Ginzburg criterion, Eq.(3). A estimate by Lobb[5] for the width of the critical region for $YBa_2Cu_3O_{7-\delta}$ i $\Delta T \sim 0.1$ K. As one enters the critical fluctuation regime, the exponent which characterize the temperature dependence of various physical quantitie deviate from their mean field values. For example, the temperature dependenc of the coherence length transfers from the GL behavior $\xi(T) \sim |T-T_c|^{-1/2}$ t the critical behavior $\xi(T) \sim |T-T_c|^{-0.67}$.

Fluctuation effect can be also manifested in specific heat. The exces specific heat due to fluctuations outside the critical region is given by

$$\Delta C = C^{\pm} |\varepsilon|^{-(2-d/2)}, \quad C^{+} = k_B/8\pi\xi^3(0), \qquad (29)$$

where d denotes the dimensionality of the system[34]. Inderhees et al.[35 measured the specific heat of single crystal samples of $YBa_2Cu_3O_{7-\delta}$, an found an excess specific heat components on both sides of T_c as shown i Fig. 4. They fitted this excess specific heat component to Eq.(29) to find a exponent 0.5, consistent with 3D Gaussian fluctuations. They found the rati of the prefactors, C^+/C^-, to be 2.8 ± 0.8 and claimed it to be significantl larger than $1/\sqrt{2}$ expected for a conventional superconductor. They propos that the order parameter has more than two components indicating a very uncon ventional pairing. Salamon et al.[36] studied the effect of magnetic field o the fluctuation specific heat of the same single crystal as ref.[35]. In con trast to the Gaussian fluctuation in the zero magnetic field case, the fluc tuation specific heat in magnetic fields was found to satisfy a scaling rela tion given by Lee and Shenoy[37] appropriate to the critical behavior. This

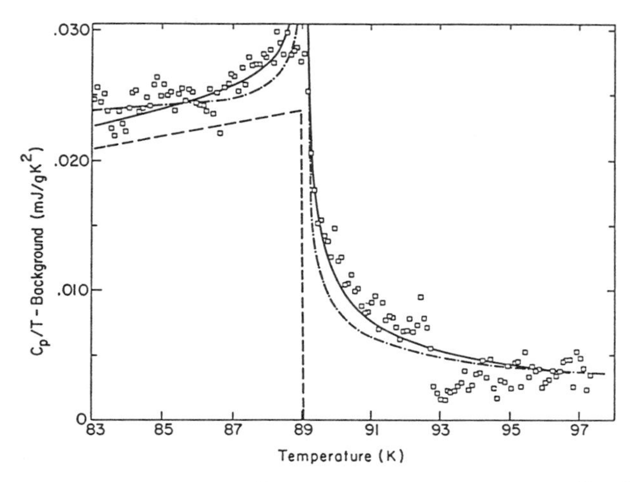

ig. 4 Specific heat data near T_c, of a single crystal $YBa_2Cu_3O_{7-\delta}$. The broken curve is the fitted BCS-type contribution. The solid curve corresponds to the 3D Gaussian fluctuation, while the dash-dotted curve represents the 2D case. (ref. [35])

as interpreted by Salamon et al. [36] in terms of the broadening of the criti al fluctuation regime by the magnetic field.

While these results open up a very interesting possiblity for and ex- erimnental observation of the Gaussian and even the critical fluctuation ef- ect in the specific heat of a bulk superconductor manifested, one should take hese conclusions with some words of caution. First of all, the analysis of he data involves a rather delicate subtraction of a background plus a fitted CS-type contribution from the raw data. Secondly the determination of the ritical exponent is very sensitive to the choice of the mean field T_c. And hirdly, the possible influence of sample inhomogeneity must be carefully ssessed. Recent data of the specific heat jump at T_c of the best polycrys- alline samples[38] appear to be sharper (i.e. closer to a BCS-type step) than he one shown in Fig. 4. Evidently more experimental studies are needed to lucidate the fluctuation contribution to the specific heat.

Another interesting possibility of critical fluctuation has been put orth by Oh et al. [26] in relation to the temperature dependence of the upper ritical field. In the critical fluctuation regime, the temperature depen- ence of the upper critical field is expected to be $H_{c2}(T) \sim (T_c-T)^{1.33}$ as con- rasted to the GL behavior $H_{c2}(T) \sim (T_c-T)$. The upward curvature of the $H_{c2}(T)$ urve of HTSCs have been noticed by many experimentalal groups[3, 4, 26]. Oh et l. [26] fitted their own $H_{c2}(T)$ data and those of Iye et al. [3] to a func- ional form $(T_c-T)^{2\nu}$ and obtained $2\nu \sim 1.3 \pm 0.04$. Although very tempting, nis line of interpretation has some difficulties. For example, if this in- erpretation is correct, the fluctuation region below T_c is much wider than nat above T_c, while the specific heat experiment suggests critical region to e no wider than a tenth of a degree both above and below T_c. As for the up- ard curvature of $H_{c2}(T)$, a different interpretation along the line of the lux creep physics is presented, which will be discussed in the next section.

A possibility of two-dimensional Kosterlitz-Thouless transition is discusse by Stamp et al. [39] based on their resistive measurements on single crysta $YBa_2Cu_3O_{7-\delta}$.

3. FLUX CREEP PHENOMENON

Many practical applications of superconductors rely upon their ability t carry large transport current without dissipation in the presence of magneti field. The dissipation-free current in magnetic field, however, does not com automatically with high H_{c2}. In the mixed state of a type II superconducto with transport current, the Lorentz force is exerted on magnetic flux lines It causes flux motion and leads to energy dissipation. It is only in th presence of strong pinning of the flux lines that a type II superconductor i a magnetic field can carry large transport current up to $J_c(H,T)$ withou dissipation. The flux pinning arises from the spatial inhomogeneity of th condensation energy. It is customary to distinguish the two categories o flux motion, i.e. "flux flow" and "flux creep". In the former case, th Lorentz force dominates. The flux lines are driven by the Lorentz force unde the influence of "frictions" exerted by the pinning centers, and move more o less steadily. In the latter case, the flux pinning is strong and the flu lines move only by thermally activated jump from one pinning site to another Extensive reviews of the flux motion in the conventional type II superconduc tors are given in refs. [40] and [41].

A theoretical model for the flux creep process has been formulated by An derson and Kim[42]. The model assumes thermally activated jumps of flux line with the rate given by $\nu = \nu_0 \exp(-U_0/k_BT)$, where U_0 is the activation energ corresponding to the barrier height for the flux jump process and ν_0 repre sents some sort of attempt frequency. In the absense of the Lorentz force flux jump to the right is as likely to occur as one to the left, and hence th net flux creep velocity is zero. When there is a gradient of the magneti flux density or when a transport current is present, the Lorentz force is ex erted on the flux lines. This creates an imbalance of the barrier height fo the flux jump to the right and that to the left, which leads to a net flu jump to one direction with a rate given by

$$\nu = \nu_+ - \nu_- = \nu_0 \exp(-U_0/k_BT)\,\{\exp(\Delta U/k_BT) - \exp(-\Delta U/k_BT)\}\,. \quad (30)$$

The change in the activation energy is given by $\Delta U \sim FL$. Here, F is th force acting on the flux lines and L is the hopping length. The drivin force is given by

$$\begin{aligned} F &= -B \times (\nabla \times B)/4\pi \\ &= J \times B/c\,, \end{aligned} \quad (31)$$

where J is the transport current. Inserting this into Eq. (30), we obtain th hopping frequency as

$$\begin{aligned} \nu &= 2\nu_0 \exp(-U_0/k_BT)\sinh(JBVL/k_BT) \\ &\sim (2\nu_0 VL^2 JB/k_BT)\exp(-U_0/k_BT), \end{aligned} \quad (32)$$

where the hyperbolic sine is linearized in the second line. The motion of the flux lines accompanies the phase slip giving rise to a finite time-averaged dc voltage (finite resistance).

The key parameter of the thermally activated flux creep is U_0/k_BT. A rough estimate of U_0 can be made by the following argument. If each flux line moves independently, it is reasonable to take $U_0 \sim (H_c^2/8\pi)\xi^3$, which is the condensation energy over the volume ξ^3 around a pinning center. When the flux lattice spacing, $a_0 = 1.075(\Phi_0/B)^{1/2}$ becomes significantly smaller than λ, the flux lines are expected to move in a correlated way as a bundle. In this case, an order of magnitude estimate of U_0 may be given (apart from numerical factors) by

$$U_0 \sim (H_c^2/8\pi)a_0^2\xi$$
$$\sim \{H_c^2(0)/8\pi\}(\Phi_0/B)\xi(0)\{(T_c-T)/T_c\}^{3/2}. \qquad (33)$$

This formula gives some idea of the magnetic field- and temperature-dependence of the pinning strength.

The above model was first proposed by Yeshurun and Malozemoff[43] in order to account for the characteristic "irreversibility boundary", $(T_c-T)/T_c \sim H^{2/3}$, at which magnetic irreversibility sets in. Such magnetic irreversibility boundary was observed earlier in ceramic samples[44]. It has been attributed to a glassy nature of the superconductivity in granular materials, and called a "quasi de Almeida-Thouless line" in an analogy to the spin glass physics. Yeshurun and Malozemoff[43] later observed a similar phenomenon in single crystals, and attributed it to "giant flux creep". The key ingredient in the model is the low value of U_0 in HTSCs. Taking $U_0 = \{H_c^2(0)/8\pi\}\xi^3(0)$ for the purpose of comparison, we obtain typically several electron volts for conventional type II supercondcutors. On the other hand, U_0 for HTSCs is about an order of magnitude lower because of the short coherence lengths. The low value of U_0 and high operating temperature leads to the "giant flux creep" phenomenon in HTSCs.

On the basis of this model, Thinkham[45] calculated the flux creep resistance, using a formula worked out by Ambegaokar and Halperin[46] for the case of thermally activated phase-slip events in a single heavily damped current-driven Josephson junction. The result is expressed as

$$\rho/\rho_n = \{I_0(U_0/2k_BT)\}^{-2}$$
$$= \{I_0[\alpha\{(T_c-T)/T_c\}^{3/2}/H]\}^{-2}, \qquad (34)$$

where Eq.(33) was used to clarify the magnetic field- and temperature-dependence. Here, ρ_n denotes the normal resistance, $I_0(x)$ is the modified Bessel function. For large values of U_0/k_BT, ρ/ρ_n falls exponentially as $\sim\exp(-U_0/k_BT)$. The following points are immediately observed. (1)Since the temperature and field dependences of ρ enters through $U_0(H,T)/k_BT$, the resistively measured upper critical field near T_c will vary as $(T_c-T)^{3/2}$. This is in good agreement with the experimentaly observed upward curvature of $H_{c2}(T)$[3,4,26], which was interpreted in ref.[26] in terms of fluctuation effect. (2)The "width" of the resistive transition will increase with field

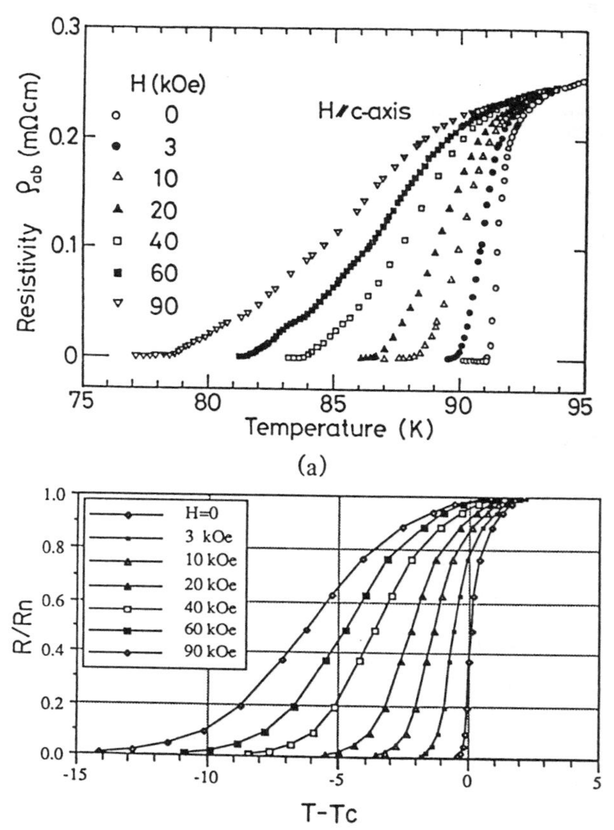

Fig. 5 Temperature dependence of the basal plane resistivity of a single crystal $YBa_2Cu_3O_{7-\delta}$ in magnetic fields along the c-axis. (ref. [3]) The bottom part of the figure shows the theoretical curves calculated by Eq. (34), based on the flux creep model. (ref. [45])

as $\Delta T_c(H) \sim H^{2/3}$. This is seen in Fig. 5 where the traces of the resistive transition in a single crystal $YBa_2Cu_3O_{7-\delta}$ measured by Iye et al. [3] are shown together with theoretical curves calculated from Eq. (34) with single fitting parameter α.

A detailed study of the thermally activated flux creep was carried out by Palstra et al. [47] for the $Bi_2Sr_2CaCu_2O_{8+y}$ system, which shows the effect even more dramatically than $YBa_2Cu_3O_{7-\delta}$. Figure 6 shows the traces of the resistive transition of a single crystal sample of $Bi_{2.2}Sr_2Ca_{0.8}Cu_2O_{8+y}$ in magnetic fields parallel and perpendicular to the c-axis. As seen in the lower part of the figure, the low resistivity end of the resistive transition follows the exponential law. These data are found to be represented by a universal func

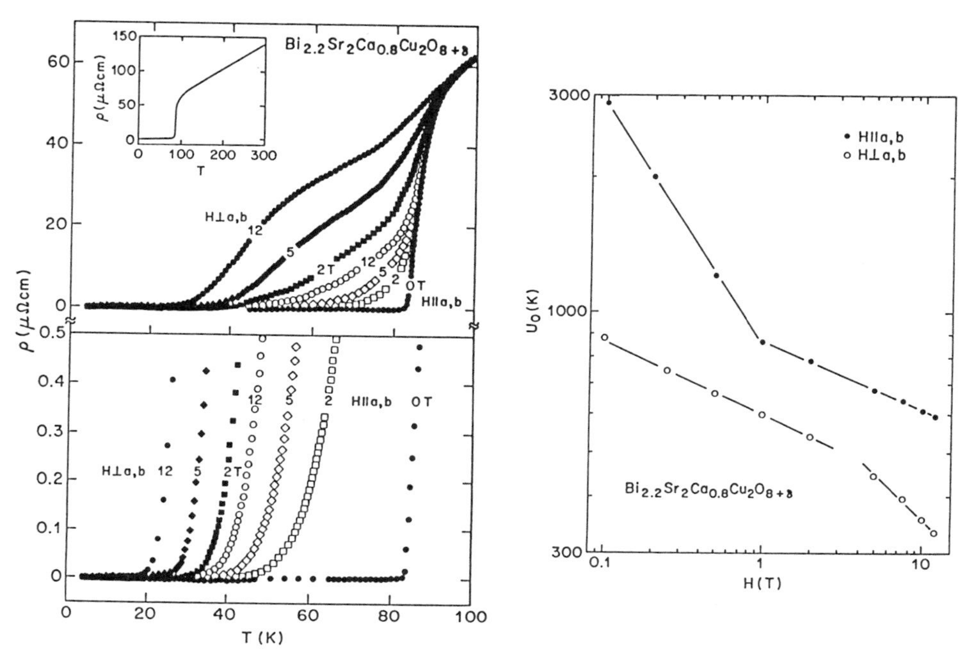

ig. 6 The left part of the figure shows the temperature dependence of the resistivity of a single crystal sample of $Bi_{2.2}Sr_2Ca_{0.8}Cu_2O_{8+y}$ in magnetic fields oriented parallel (open symbols) and perpendicular (solid symbols) to the basal plane. . The lower part of the figure is a magnification of the low resistance part to emphasize the exponential behavior. The insert is the zero-field resistivity up to room temperature. The right part of the figure shows the magnetic field dependence of the activation energy determined from the Arrhenius plot of the data shown in the left part. The value of U_0 is independent of temperature and only weakly dependent on the strength and the orientation of magnetic field. (ref. [47])

ion of the form,

$$\rho = \rho_0 \exp[-U_0(H, \theta)/k_B T]. \tag{35}$$

e activation energy U_0 determined from the Arrhenius plot is given in the ght part of Fig. 6 as a function of the magnetic field for the two principal eld orientaions. The first thing to note is that the values of U_0 are deed very low ($\sim$1000 K), which of course is the reason why the phenomenon experimentally observable. It is also seen that U_0 is only weakly depennt on magnetic field, i.e. the field dependence is $\sim H^{-0.15}$ to $H^{-0.5}$ rather an the 1/H dependence assumed in Eq. (34). The value of U_0 is considerabley wer for $Bi_2Sr_2CaCu_2O_{8+y}$ than for $YBa_2Cu_3O_{7-\delta}$, so that the temperature range which the flux creep phenomenon is experimentally observed corresponds to rger values of $(T_c-T)/T_c$ for the former material than the latter. Thus, the mperature dependence of U_0 in the case of $Bi_2Sr_2CaCu_2O_{8+y}$ is expected to be ss important. A surprising result is that the prefactor $\rho_0 \sim 0.1$ Ω cm is dependent of magnetic field strength and orientation. The value of ρ_0 is roughly three orders of magnitude larger than the normal resistivity of the material. There is no clear picture for the physical meaning of this quantity. It should be recognized that the weak flux pinning as seen above is intimately tied to the short coherence length and is an inevitable consequence of high T_c. This may impose a serious limitation to the prospective large current application of the HTSCs at high temperature and high magnetic fields.

4. CONCLUDING REMARKS

In this chapter, we picked up a few topics related with the short coherence lengths of the HTSCs. As emphasized earlier, the short coherence length stems from the high transition temperature and is one of the inherent properties of HTSCs. The small coherence volume and the high operating temperature lend a particular importance to the thermodynamic fluctuations in the various aspects of superconductivity. This opens up an intigueing oppotunity to study the fluctuation effects, particularly the critical fluctuations, in a bulk superconductor. In order to investigate the fluctuation effects in HTSCs, we should better understand the normal state properties, since the fluctuation effect above T_c appears as deviations from the normal behavior. Another important manifestation of the short coherence lengths is the unusually weak flux pinning. With this new ingredient, the phenomenology for the transport in HTSCs will be very different from the case of the conventional hard superconductors. Given the importance of the flux creep process for the high-field, high-current applications of HTSCs, the transport theory has to be renovated.

REFERENCES

[1] R. D. Parks (ed.), "Superconductivity", (Marcel Dekker, New York, 1968).
[2] M. Tinkham, "Introduction to Superconductivity", (McGrow-Hill, New York, 1975).
[3] Y. Iye, T. Tamegai, H. Takeya and H. Takei, Jpn. J. Appl. Phys. 26 (1987) L1057; Y. Iye, T. Tamegai, T. Sakakibara, T. Goto, N. Miura, H. Takeya and H. Takei, Physica C153-155 (1988) 26.
[4] W. J. Gallagher, T. K. Worthington, T. R. Dinger, F. Holtzberg, D. L. Kaiser and R. L. Sandstrom, Physica 148B (1987) 228; T. K. Worthington, W. J. Gallagher, D. L. Kaiser, F. H. Holtzberg and T. R. Dinger, Physica C153-155 (1988) 32.
[5] C. J. Lobb, Phys. Rev. B36 (1987) 3930.
[6] A. Kapitulnik, M. R. Beasley, C. Castellani and C. Di Castro, Phys. Rev. B3 (1988) 537.
[7] G. Deutscher, Physica C153-155 (1988) 15.
[8] V. L. Ginzburg, Fiz. Tverd. Tela 2 (1960) 2031 [Sov. Phys. Solid State (1961) 1824.
[9] W. J. Skocpol and M. Tinkham, Rep. Prog. Phys. 38 (1975) 1049.
[10] A. Schmid, Phys. Rev. 180 (1969) 527.
[11] J. Lawrence and S. Doniach, in Proc. 12th Int. Conf. Low Temp. Phys. Kyoto, 1970 (Keigaku, Tokyo, 1971) p. 361.
[12] R. R. Gerhardts, Phys. Rev. B9 (1974) 2945.
[13] D. E. Prober, M. R. Beasley and R. E. Schwall, Phys. Rev. B15 (1977) 5245.
[14] P. P. Freitas, C. C. Tsuei and T. S. Plaskett, Phys. Rev. B36 (1987) 833.
[15] K. Kanoda, T. Takahashi, T. Mizoguchi, M. Hasumi and S. Kagoshima, Physica C153-155 (1988) 749.; K. Kanoda, T. Kawagoe, M. Hasumi, T. Takahashi,

S. Kagoshima, T. Mizoguchi, J. Phys. Soc. Jpn. 57 (1988) 1554.
[16] J. C. Bonner and M. E. Fisher, Phys. Rev. 135A (1964) 640.
[17] P. W. Anderson, Science 235 (1987) 1196.
[18] P. W. Anderson, G. Baskaran, Z. Zou, J. Wheatley, T. Hsu, B. S. Shastry, B. Doucot and S. Liang, Physica C153-155 (1988) 527.
[19] H. Fukuyama, Y. Hasegawa and Y. Suzumura, Physica C153-155 (1988) 1630.
[20] L. G. Aslamazov and A. I. Larkin, Phys. Lett. 26A (1968) 238.
[21] K. Maki, Prog. Theor. Phys. 39 (1968) 897; 40 (1968) 193.
[22] R. S. Thompson, Phys. Rev. B1 (1970) 327.
[23] N. Goldenfeld, P. D. Olmsted, T. A. Friedman and D. M. Ginsberg, Solid State Commun. 65 (1988) 465.
[24] M. Ausloos and Ch. Laurent, Phys. Rev. B37 (1988) 611.
[25] Y. Matsuda, T. Hirai and S. Komiyama, Solid State Commun. 68 (1988) 103.
[26] B. Oh, K. Char, A. D. Kent, M. Naito, M. Beasley, T. H. Geballe, R. H. Hammond A. Kapitulnik and J. M. Graybeal, Phys. Rev. B37 (1988) 7861.
[27] N. P. Ong, Z. Z. Wang, S. Hagen, T. W. Jing, J. Clayhold and J. Horvath, Physica C153-155 (1988) 1072.
[28] M. Hikita and M. Suzuki, submitted to Phys. Rev. B.
[29] E. Abrahams, R. E. Prange and M. J. Stephen, Physica 55 (1971) 230.
[30] J. E. Crow, R. S. Thompson, M. A. Klenin and A. K. Bhatnagar, Phys. Rev. Lett. 24 (1970) 371.
[31] R. A. Craven, G. A. Thomas and R. D. Parks, Phys. Rev. B7 (1973) 157.
[32] S. Hikami and A. I. Larkin, Mod. Phys. Lett. B2 (1988) 693.
[33] T. Tsuneto, J. Phys. Soc. Jpn., to be published.
[34] L. G. Aslamazov and A. I. Larkin, Fiz. Tverd. Tela. 10 (1968) 1104 [Sov. Phys. Solid State 10 (1968) 875].
[35] S. E. Inderhees, M. B. Salamon, N. Goldenfeld, J. P. Rice, B. G. Pazol, D. M. Ginsberg, J. Z. Liu and G. W. Crabtree, Phys. Rev. Lett. 60 (1988) 1178.
[36] M. B. Salamon, S. E. Inderhees, J. P. Rice, B. G. Pazol, D. M. Ginsberg and N. Goldenfeld, Phys. Rev. B38 (1988) 885.
[37] P. A. Lee and S. R. Shenoy, Phys. Rev. Lett. 28 (1972) 1025.
[38] Y. Nakazawa and M. Ishikawa, unpublished data.
[39] P. C. E. Stamp, L. Forro and C. Ayache, Phys. Rev. B38 (1988) 2847.
[40] Y. B. Kim and M. J. Stephen, in ref. [1], p. 1107.
[41] A. M. Campbell and J. E. Evetts, Adv. Phys. 21 (1972) 199.
[42] P. W. Anderson and Y. B. Kim, Rev. Mod. Phys. 36 (1964) 39.
[43] Y. Yeshurun and A. P. Malozemoff, Phys. Rev. Lett. 60 (1988) 2202.
[44] K. A. Mueller, M. Takashige and J. G. Bednorz, Phys. Rev. Lett. 58 (1987) 1143.
[45] M. Tinkham, Phys. Rev. Lett. 61 (1988) 1658.
[46] V. Ambegaokar and B. I. Halperin, Phys. Rev. Lett. 22 (1969) 1364.
[47] T. T. M. Palstra, B. Batlogg, L. F. Schneemeyer and J. V. Waszczak, Phys. Rev. Lett. 61 (1988) 1662.

BASIC THIN FILM PROCESSING FOR HIGH-Tc SUPERCONDUCTORS

K.Wasa, H. Adachi, Y. Ichikawa, K.Setsune, and K. Hirochi

Matsushita Electric Industrial Co., Ltd.
Moriguchi, Osaka

INTRODUCTION

Much attention has been paid for the thin films of perovskite-type oxides especially for the thin films of the high-Tc superconducting ceramics. Historically the thin films of the perovskite-type oxides have been studied as a basic research for ferroelectric materials. Thin films of $BaTiO_3$ and $PbTiO_3$ were tried to deposited and there ferroelectricity was evaluated[1,2]. Recently this kind of perovskite thin films, including PZT ($PbTiO_3$-$PbZrO_3$) and PLZT [$(Pb,La)(Zr,Ti)O_3$] have been studied in relation to the synthesis of thin film dielectrics, pyroelectrics, piezo-electrics, electro-optic materials, and acousto-optic materials[3,4,5]. Thin films of BPB ($BaPbO_3$-$BaBiO_3$) were studied as oxide superconductors[6].

At present the thin films of the rare-earth high-Tc superconductors of LSC ($La_{1-x}Sr_xCuO_4$) and YBC ($YBa_2Cu_3O_{7-\delta}$) have been successfully synthesized owing to the previous studies on the ferroelectric thin films of the perovskite-type oxides[7,8]. Similar to the rare-earth high-Tc superconductors thin films of the rare-earth-free high-Tc superconductors of BSCC (Bi-Sr-Ca-Cu-O)[9] and TBCC (Tl-Ba-Ca-Cu-O)[10] system have been synthesized[11,12].

In this section the basic processes for the fabrication of the high-Tc perovskite superconducting thin films are described.

THIN FILM PROCESSES

Basic processes

The perovskite crystal comprises ABO_3 compounds. Typical basic perovskite crystal is $BaTiO_3$. The structure is cubic, with Ba^{2+} ions at the cubic corners, O^{2-} ions at the face centers, and a Ti^{4+} ion at the body center. While the typical perovskite type high-Tc superconductor YBC exhibites orthorhombic structure.

Basic processes for the deposition of the perovskite thin films are shown in Fig.1. Thin films of amorphous phase are deposited at the substrate temperature Ts below the crystallizing temperature Tcr. The Tcr for the perovskite-type oxides is around 500 - 600°C. In some case different crystal structure appears at the substrate temperature below the Tcr for the perovskite structure. For the thin films of $PbTiO_3$ the pyrochrore phase appears at lower substrate temperature around 400°C[13].

Thin films of polycrystalline phase are deposited at Ts > Tcr. The polycrystalline phase is also achieved by the deposition of the amorphous phase followed by postannealing at the annealing temperature above Tcr.

Thin films of single crystalline phase are epitaxially deposited on a single crystal substrate at the substrate temperature above an epitaxial temperature Te (Te $\geq$ Tcr). The amorphous thin films deposited on the single crystal substrate will be converted into the single crystalline thin films after the postannealing at the annealing temperature above Te owing to a solid phase epitaxy.

Several kinds of the deposition processes are proposed for the perovskite-type oxides, including an electron beam deposition[14], a laser beam deposition[15], a cathodic sputtering[16], and a chemical vapour deposition[17]. It is considered that the oxygen density control is necessary for the deposition of the high-Tc superconductors, YBC. For this purpose oxygen and/or oxygen ions are supplied onto the growing surface of the thin films during the deposition in the electron beam deposition system.

In the cathodic sputtering the thin films of perovskites are deposited directly from the compound ceramic target of the perovskites in an rf-system. The rf-magnetron sputtering is commonly used for the deposition from the compound ceramics target. The sintered ceramics plate or sintered ceramics powder is used for the sputtering target. In the dc-magnetron sputtering the metal targets of A-site and B-site elements are sputtered in an oxydizing atmosphere.

In the chemical vapour deposition the metal-organic compounds such as Y $(C_{11}H_{19}O_2)_3$, Ba $(C_{11}H_{19}O_2)_2$, and Cu $(C_{11}H_{19}O_2)_2$ are tentatively used as the source for the deposition of the YBC thin films[18]. The halides such as $BiCl_3$, CuI, CaI_2, and SrI_2 are used as the source for the deposition of the BSCC thin films[19]. Figure 2 shows a construction of typical deposition systems for the high-Tc superconductors.

Among these processes the sputtering deposition is most commonly used, since the sputtering process easily achieves the stoichiometric composition even for the complex compounds of the perovskites.

For the deposition of the single crystal films the selection of the substrate crystal will affect the crystal properties of the resultant films. The crystallographic properties of the crystal substrates used for the epitaxial growth of the perovskite-type oxides are shown in Tab. 1[20].

1.	AMORPHOUS PHASE	Ts < Tcr
2.	POLYCRYSTALLINE	Ts > Tcr Ts < Tcr, postannealing
3.	SINGLE CRYSTALS (single crystal sub.)	Ts > Te Ts < Tcr, postannealing (solid-phase epitaxy)

Fig.1. Basic processes for the deposition of the perovskite thin films.

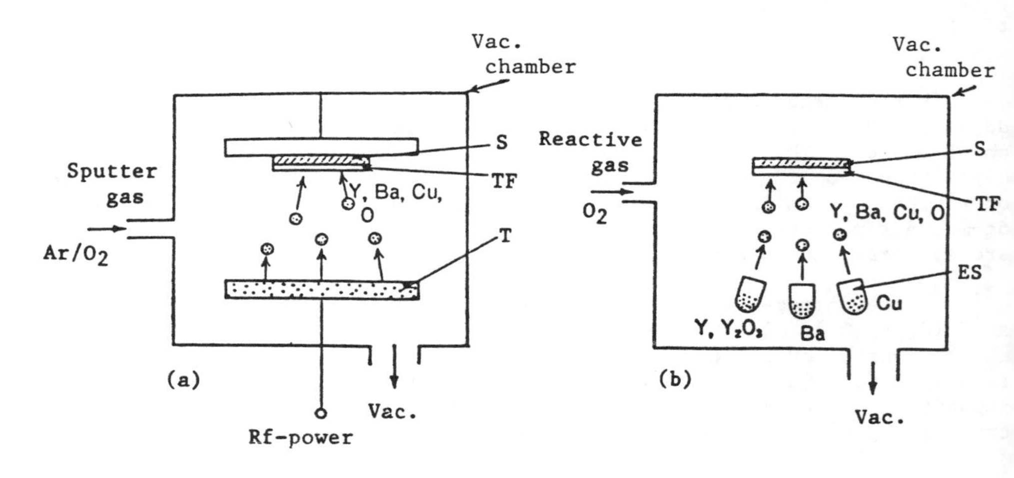

Fig.2 Typical deposition systems for the high-Tc superconductors; (a) sputtering system , (b) reactive evaporation system . S: substrate, TF: thin film, T: target. ES: evaporation source.

Table 1 Substrates for the deposition of the high-Tc superconducting thn films.

	Crystal System	Structure	Lattice Constants	Thermal Expansion Coefficient
$La_{1.8}Sr_{0.2}CuO_4$	tetragonal #	K_2NiF_4	a=3.78A c=13.23A	$10\text{-}15\times10^{-6}$/K
$YBa_2Cu_3O_x$	orthorhombic	oxygen deficient perovskite	a=3.82A b=3.89A c=11.68A	$10\text{-}15\times10^{-6}$/K
Bi-Sr-Ca-Cu-O	pseudo tetra.	Bi-layered structure	a≃5.4A c≃30A, 36A	12×10^{-6}/K
Tl-Ba-Ca-Cu-O	pseudo tetra.	Bi-layered structure	a≃5.4A c≃36A	
sapphire (α-Al_2O_3)	trigonal	corundum	hex. axes a'=4.763A c'=13.003A	//c 8×10^{-6}/K ⊥c 7.5×10^{-6}/K
MgO	cubic	NaCl	a=4.203A	13.8×10^{-6}/K
$MgAl_2O_4$	cubic	spinel	a=8.059A	7.6×10^{-6}/K
YSZ	cubic	fluorite	a=5.16A	10×10^{-6}/K
$SrTiO_3$	cubic	perovskite	a=3.905A	10.8×10^{-6}/K

Superconductivity at orthorhombic phase.

Figure 3 shows a typical epitaxial relations between the high-Tc superconductors and the cubic substrates. The c-axis of the epitaxial films will be perpendicular to the (100) plain of the substrate crystals. The isotropic superconducting currents will flow in the (001) plain of th deposited films. On the (110) plain of the cubic crystal substrates the c-axis of the epitaxial films will be in the films. The large anisotropy will be expected for the current flow in the (110) plain of the epitaxial films.

It should be noted that most of the crystal substrates exhibite cubi structure. The twin will be reasonably formed in the epitaxial films of the orthorhombic superconductors. The orthorhombic substrates are important for the reduction of the twin density in the epitaxial films. Besides the crystallographic properties the possibility of the mutual diffusion at the film and substrate interface should be considered for the selection of the substrates[21].

Numbers of experiments were done for the deposition of the high-Tc thin films. These deposition processes are classified into three as indicated in Tab. 2. The process for the ceramics is composed of three stages; (i) mixing, (ii) annealing for crystallization and sintering, (iii) annealing for the control of oxygen vacancies and/or crystal structure. For the thin films three processes are considered; the proces (1), deposition at a low substrate temperature followed by the post-annealing. The process (1) is commonly used for the deposition of the high-Tc superconducting thin films, since the stoichiometric composition of the thin films is easily achieved. It is expected that the single crystal thin films will be obtained under a condition of a solid phase epitaxy during the postannealing process. At present, however, the resultant films show the polycrystalline phase. The single crystal films are possibly obtained by a vapour phase epitaxy achieved in the process (2) and (3).

Synthesis temperature

As indicated in Tab. 2 the synthesis temperature of the high-Tc superconducting ceramics is around 850 - 950°C. Lowering the synthesis temperature is much important not only for the scientific interests but for the fabrication of the thin film superconducting devices. It is seen in Tab. 2 that the maximum temperature in the thin film process may be governed by the postannealing process.

In the thin film process (2) for the rare-earth $YBa_2Cu_3O_x$ superconductors, the maximum temperature is governed by the postannealing process for the control of oxygen vacancy, if the as-deposited thin films are crystallized. The structural analyses for the $YBa_2Cu_3O_x$ ceramics suggest that the structural transition from the non-superconducting tetragonal phase ($x \lesssim 6.3$) into the superconducting orthorhombic phase ($7 > x > 6.3$) occurs around 700°C and the orthorhombic phase is predominant at the annealing temperature below 600°C[22.23]. Similar results are obtained in the Gd-Ba-Cu-O (GBC) thin films[24]. In the GBC thin films the high-Tc orthorhombic phase is obtained at the annealing temperature of 350 - 550°C in O_2.

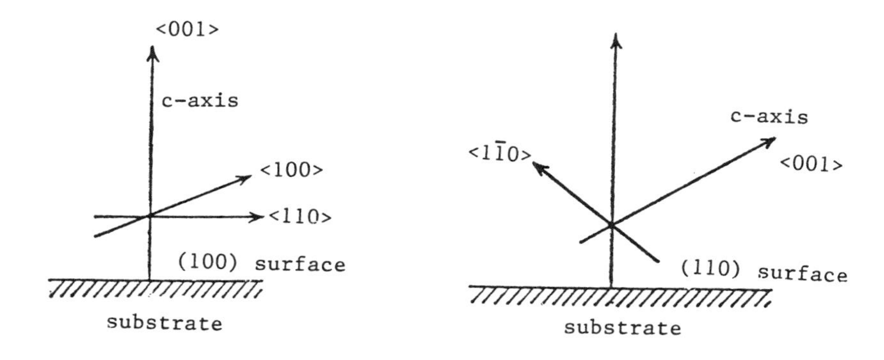

Fig. 3. Epitaxial relations of the high-Tc superconducting thin films on crystal substrates.

Table 2. Fabrication processes for the high-Tc superconducting ceramics and thin films.

		Chemical Composition	Crystallization	Oxygen Vacancy and/or structural Control
Ceramics		mixing	sintering (850 - 950°C)	annealing * (850 - 950°C)
Thin films	1	deposition (Ts > Tcr)	annealing (850 - 950°C)	annealing * (400 - 950°C)
	2	deposition (Ts > Tcr)		annealing * (400 - 950°C)
	3	deposition ** (Ts > Tcr)		

Ts: substrate temperature during deposition
Tcr: crystallizing temperature (500 - 600°C)
* slow cooling
** quenching

These structural analyses suggest that the synthesis temperature is not governed by the postannealing temperature for the YBC films but by the crystallizing temperature.

For the rare-earth $La_{2-x}Sr_xCuO_4$ type thin films the as-deposited films happen to show the oxygen deficiency. The oxidation is taken place in the postannealing process. The synthesis temperature for the LSC type thin films may be also governed by the crystallizing temperature since the oxidation will be conducted below the crystallizing temperature. The crystallizing temperature for these rare-earth high-Tc oxides is around 500 - 600°C.

For the rare-earth-free high-Tc superconductors the synthesis temperature may also correspond the crystallizing temperature of perovskites, 500 - 600°C. It is noted that in the YBC thin films the partial substitution of O-sites by S shows a tendency of decreasing the crystallizing temperature[25].

Low temperature processes

It is reasonably considered that lowering the synthesis temperature is achieved by the deposition at the crystallizing temperature of 500 - 700°C followed by the postannealing in O_2 at the temperature of 350 - 550°C. The maximum temperature for the rare-earth high-Tc film processes are governed by the crystallizing temperature of 500 - 700°C.

Several works have been done on the low temperature synthesis of rar earth high-Tc superconductors. These processes are classified into two cases:

(i) deposition at the substrate temperature above the crystallizing temperature Tcr (500 - 700°C), followed by the postannealing at a low temperature (400 - 600°C) in O_2.

(ii) deposition at the substrate temperature above the crystallizing temperature without any additional postannealing process.

The case (i) is corresponding to the process (2). In one case the postannealing is conducted successively in the deposition equipment without breaking the vacuum. In the other case the postannealing is proceeded in the oxygen furnace after the deposition. The former is called "in situ annealing" and the latter is called "ex situ annealing". The case (ii) is corresponding to the process (3) which is called "in situ deposition".

The low temperature process with the in situ annealing or the ex sit annealing was studied by the several deposition processes including the sputtering[26,27], the pulsed laser deposition[28], and the reactively deposition[29]. In these processes the temperature of the substrates durin the deposition is 500 - 700°C. The annealing is done at around 400 - 500°C in O_2. However, the low temperature process without annealing, in situ deposition, is much available for making thin film electronic device since the in situ deposition achieves a formation of multi-layered structure of the high-Tc superconductors.

It is considered that if the substrate temperature during the deposition Ts satisfies the following relation

$$Te \lesssim Ts \lesssim Tt , \qquad (1)$$

where Te denotes the epitaxial temperature and Tt, the transition temperature from the tetragonal to the orthorhombic phase, and the enough oxygen is supplied onto the film surface during the deposition so as to oxidize the deposited films, the as-deposited films show the single crystal phase and reasonably exhibit the superconductivity without the postannealing process which corresponds to the in situ deposition.

The in situ deposition has been tried by the magnetron sputtering. It is confirmed that the irradiation of oxygen ions and/or plasma onto the deposited film during the deposition is important for the achievement of the in situ deposition. Under a suitable irradiation of the oxygen plasma onto the film surface, an excellent superconductive transition was observed in the as deposited Er-Ba-Cu-O thin films: The onset temperature was 95K with zero resistance temperature of 86K, for the films deposited at 650°C[30].

Several good points have been found in the in situ deposition such as a smooth surface of the deposited films and a small interdiffusion between deposited superconductive films and substrates[31]. The in situ deposition, however, has exhibited a low critical current density due to a presence of the crystal boundaries in the deposited films[32]. The crystal boundaries will form weak-links. The magnitude of the critical currents is storongly affected by an application of an external magnetic field when the direction of the magnetic field is parallel to the c-axis of the oriented superconducting films[33]. These weak points observed in the in situ deposition may result from the imperfect crystallinity of the deposited films. The crystallinity is essentially improved by the refinement of the deposition system.

In the rare-earth-free high-Tc superconductors several different superconducting phases are simultaneously formed during the postannealing process. Although the basic thin film processes for the rare-earth-free high-Tc superconductors are essentially same to those of the rare-earth high-Tc superconductors, the simultaneous growth of the different superconducting phase causes the difficulty in the controlled deposition of the single phase high-Tc superconducting thin films.

DEPOSITIONS AND SUPERCONDUCTING PROPERTIES

Rare-earth high-Tc superconductors

The most simple method for making the rare-earth high-Tc films was the deposition of amorphous films by the sputtering at low substrate temperature followed by the postannealing [process (1)]. Typical sputtering conditions for the YBC films are shown in Tab. 3. The as-sputtered films are insulative with brown color. After the postannealing at 900°C for 1 hour in O_2 the films show superconductivity as shown in Fig.4. The onset temperature is 94K with zero-resistivity at 70K[34].

Table 3. Sputtering conditions, process (1)

Target	$(Y_{0.4}Ba_{0.6})_3Cu_3O_x$ (100 mm in dia.)
Substrate	$(1\bar{1}02)$ plane of sapphire
Substrate temperature	200°C
Sputtering gas	Ar
Gas pressure	0.4 Pa
Rf input power	150 W
Growth rate	150 A/min

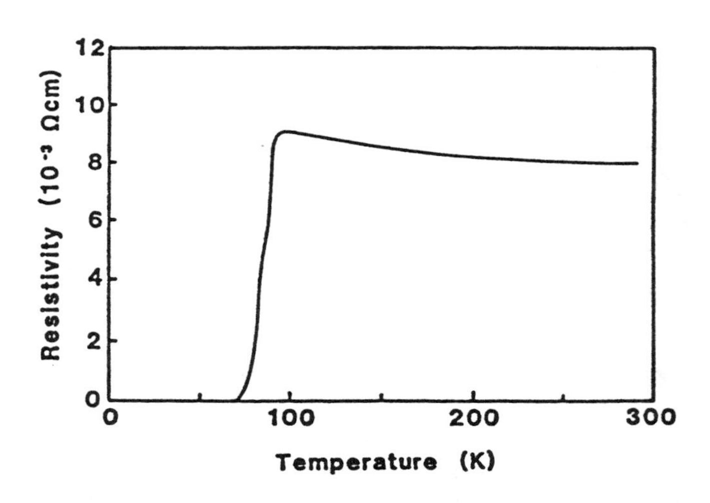

Fig.4. Temperature dependence of resistivity for sputtered Y-Ba-Cu-O thin films on $(1\bar{1}02)$ sapphire, process (1).

These annealed YBC films show polycrystalline phase with the preferred orientation of <103> crystal axis perpendicular to the substrate as shown in Fig.5. This is reasonably considered the (103) surface of the YBC corresponds to the closest packed plane. The temperature dependence of the resistivity in Fig.4 suggests that the YBC films are composed of the superconducting orthorhombic phase and the semiconducting tetragonal phase since the temperature dependence above the transition temperature is semiconductive. The relatively high resistivity, 2 mΩcm at the transition temperature may also result from the presence of grainboundaries in the YBC films. The sharp superconducting transition with single superconducting ortho-II phase is obtained for the stoichiometric composition in the deposited films. The stoichiometric composition is successfully achieved by the multi-layer deposition[35].

The improvement of the crystallinity of the high-Tc films was achieved by the deposition at higher substrate temperature [process (2)]. Table 4 shows the typical sputtering conditions for the improvement of the crystalline properties of the GBC films. Since the concentration of Ba and Cu in the GBC films are reduced at the higher substrate temperature, the composition of the target is modified so as to achieve the stoichiometric composition for the sputtered GBC films[36]. Typical superconducting properties of these GBC films are shown in Fig.6. The low resistivity less than 0.5 mΩcm at the transition temperature which corresponds to the bulk resistivity is observed for the YBC films. The temperature dependence of the resistivity is metallic at the temperature above the transition temperature. Similar properties are also obtained by the electron beam deposition[37,38].

However, the epitaxial YBC thin films with single phase YBC are generally difficult to be deposited, since the composition of the films often differ from the stoichiometric value of the YBC at the substrate temperature above Tcr (Tcr ≃ 500 - 600°C). In contrast the epitaxial LSC thin films are easily deposited, since the LSC is composed of a solid solution. Typical sputtering conditions are shown in Tab. 5[39]. The as-sputtered films are conductive with a black color similar to the target. Electron-probe X-ray microanalyses show the concentration of La, Sr, and Cu are close to the target composition. The electron diffraction pattern suggests that an excellent single crystal is epitaxially grown on the (100) $SrTiO_3$. The epitaxial relations are as follows:

$$(001)(LaSr)_2CuO_4 /\!/ (100)SrTiO_3$$

$$\langle 010\rangle (LaSr)_2CuO_4 /\!/ \langle 010\rangle SrTiO_3 \; . \qquad (2)$$

Typical electron diffraction pattern is shown in Fig.7.

These as sputtered films show semiconductive behaviour. The superconductivity is observed after postannealing in air at 900°C for 3 days. Typical results are shown in Fig.8. The single crystal LSC films grown on (100) $SrTiO_3$ exhibite excellent superconducting properties. The onset temperature is 34K with Tc=25K. The narrow transition width less than 3K suggests that these sputtered films are composed of the single phase of the layered perovskites K_2NiF_4.

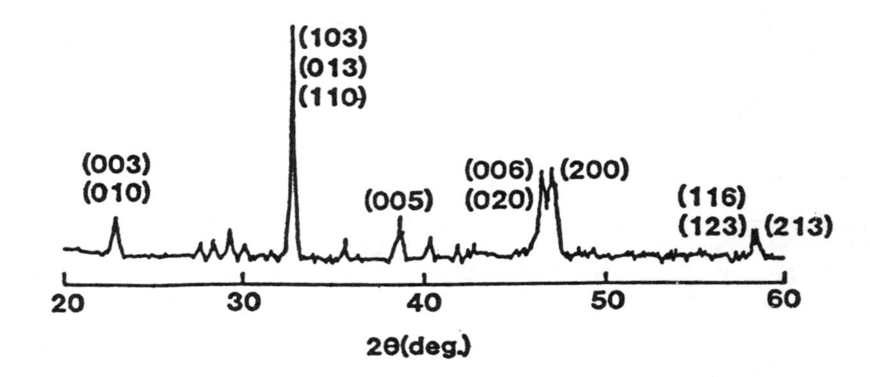

Fig.5. Typical X-ray diffraction pattern of sputtered Y-Ba-Cu-O thin film on (100) $SrTiO_3$, process (1).

Table 4. Sputtering conditions, process (2)

Target	$(Gd_{0.33}Ba_{0.66})_{0.66}CuO_x$ (100 mm in dia.)
Substrate	(100) MgO
Sputtering gas	Ar + O_2 3 : 2
Gas pressure	0.4 Pa
Rf input power	130 W
Substrate temperature	600°C
Growth rate	80 Å/min

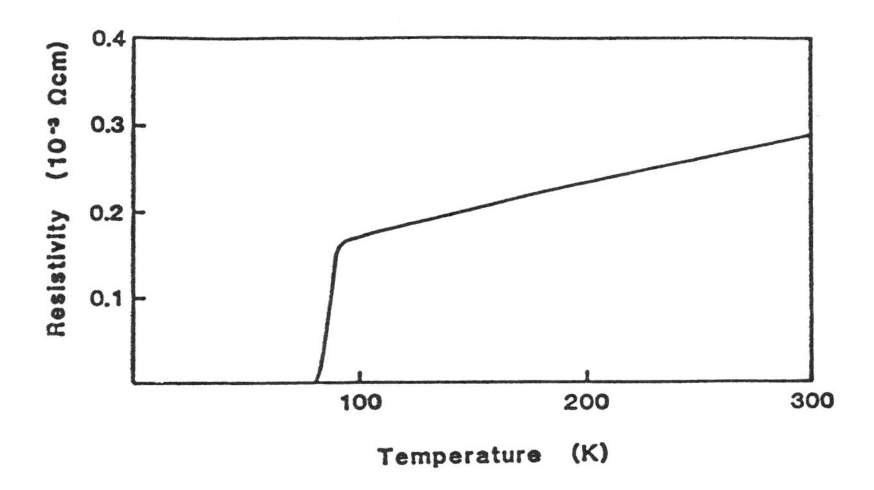

Fig.6. Temperature dependence of resistivity for sputtered Gd-Ba-Cu-O thin films on (100) MgO, process (2).

Table 5. Sputtering conditions, process (2)

Target	$(La_{0.9}Sr_{0.1})_2CuO_4$ (100 mm in dia.)
Substrate	(100) plane of $SrTiO_3$
Substrate temperature	600°C
Sputtering gas	Ar
Gas pressure	0.4 Pa
Rf input power	150 W
Growth rate	100 Å/min

However, these processes need the troublesome postannealing process. The postannealing induced the diffusion at film and substrate interface when the postannealing is conducted at the annealing temperature above 800 - 900°C. This causes the board transition due to a mutual diffusion between substrate and the deposited films[40].

It is considered that the discharge plasma of oxygen in sputtering is suitable for the oxidation of the thin films during the deposition. So, if the thin films are immersed in the oxygen plasma during the deposition the in situ deposition [process (3)] will be reasonably achieved[41].

Sputtering deposition with two kinds of target-substrate spacing, which are 35 mm and 40 mm, is selected in the preparation of Er-Ba-Cu-O films on MgO[42]. Typical sputtering conditions are shown in Tab. 6. The surface of the substrate is exposed to discharge plasma for the spacing of 35 mm. For the spacing of 40 mm the substrate is positioned at the outside of the plasma.

Table 6. Sputtering conditions, process (3)

Target	$Er_1Ba_2Cu_{4.5}O$ (100 mm in dia.)
Substrate	(100)MgO and (110)$SrTiO_3$
Sputtering gas	Ar + O_2 4 : 1
Gas pressure	0.4 Pa
Rf input power	175 W
Substrate temperature	650 °C
Growth rate	70 A/min

The temperature dependence of resistivity for the as-sputtered films is shown in Fig.9. The film made with the spacing of 35 mm shows a sharp superconductive transition with onset at 92K and Tc=86K. On the other hand, the film with the spacing of 40 mm exhibited a much broader superconductive transition and zero-R is realized as low as 57K. It is considered that the effect of the spacing on the superconducting properties resulted from the difference of oxidation in the films. We can roughly presume oxidation of the films from crystalline information. Sufficient oxidation leads surely to the superconducting orthorhombic structure, while oxygen defects cause the semiconducting tetragonal stracture.

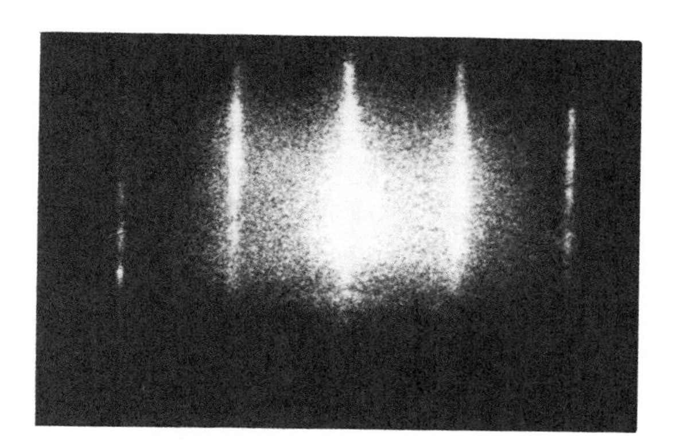

Fig.7. Typical electron diffraction pattern of La-Sr-Cu-O thin film sputtered on (100) $SrTiO_3$, process (2).

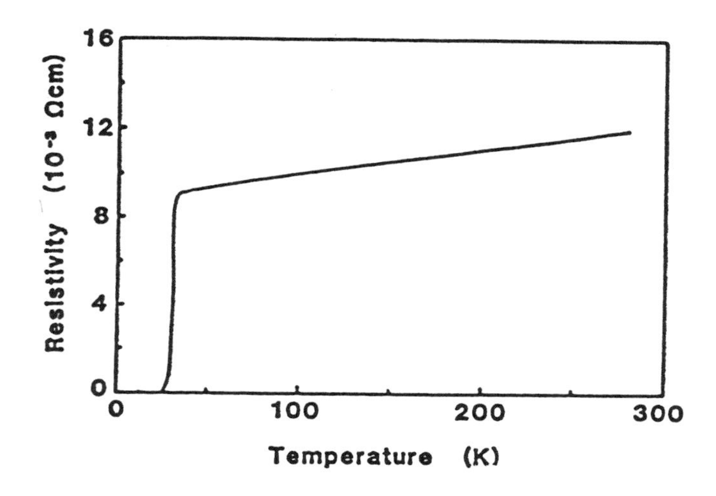

Fig.8. Temperature dependence of resistivity for sputtered La-Sr-Cu-O thin film on (100) $SrTiO_3$, process (2).

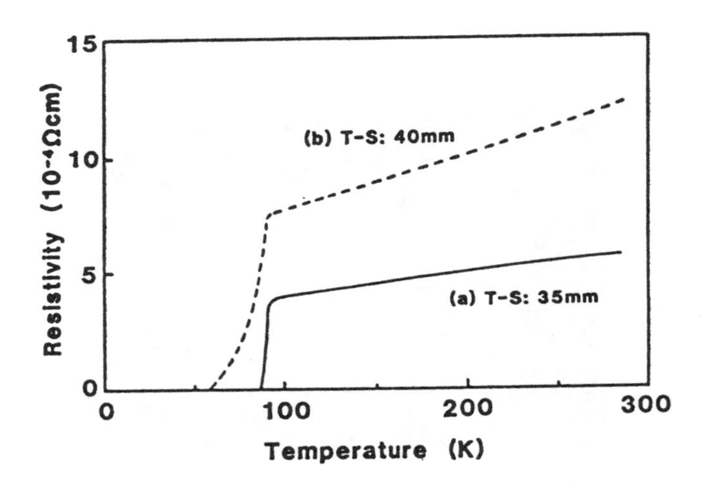

Fig.9. Temperature dependence of the resistivity for Er-Ba-Cu-O thin film on (100) MgO, process (3).

Figures 10 (a) and 10 (b) show the x-ray diffraction patterns of the films made with the spacing of 35 mm and 40 mm, respectively. The c-axis is primarily oriented perpendicular to the film plane. The crystal system can be discriminated by the lattice constant c, i.e., c=11.68A for the orthorhombic structure (o) and c=11.8-11.9A for the tetragonal structure (t). The film made with the spacing of 35 mm shows mixed structure with dominant orthorhombic and minor tetragonal phases. On the other hand, the film with the spacing of 40 mm shows the tetragonal structure. From this result it is evident that oxidation has progressed more for the spacing of 35 mm than for 40 mm. For comparison, the x-ray diffraction pattern of the film (Tc, zero=55K) made with the T-S of 35 mm and at a higher substrate temperature of 700°C is shown in Fig.10 (c). The film shows the tetragonal structure. Since the deposition is carried out at higher temperature than the T-O transition, the oxidation in passing through the T-O point is not sufficient for quick cooling.

On $SrTiO_3$ (110) substrates, epitaxial films were prepared by the same process. Figure 11 shows the RHEED pattern of the epitaxial Er-Ba-Cu-O film. The temperature dependence of resistivity for the as-deposited films shows similar characteristics and zero-R is realized below 80K. Y-Ba-Cu-O films are also prepared by this process.

These facts suggest the possibility of the in situ deposition owing to the process (3) in Tab. 2, although the in situ deposited films are not composed of the single phase of the orthorhombic structure. Recent

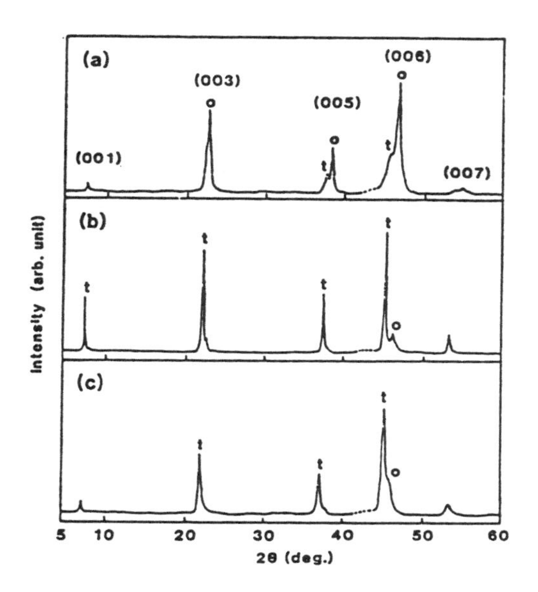

Fig.10. X-ray diffraction patterns for Er-Ba-Cu-O thin films on (100) MgO, process (3).

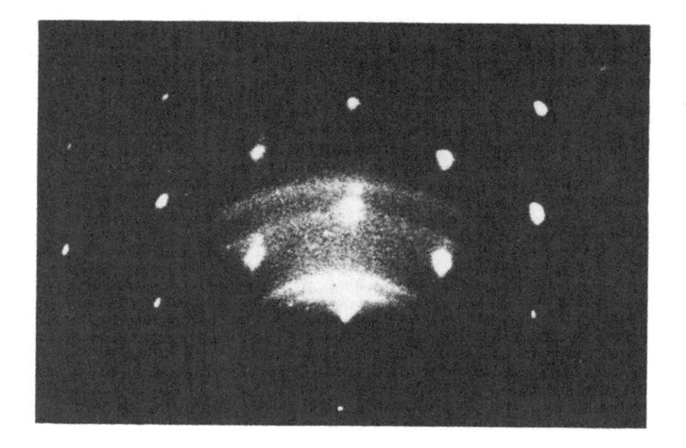

Fig.11. Electron diffraction pattern of the epitaxial Er-Ba-Cu-O thin film on (110) $SrTiO_3$, process (3).

experiments on Gd-Ba-Cu-O thin films suggest that the in situ postannealing in O_2 at relatively low temperature of 400 - 600°C increases the orthorhombic phase and improves the superconducting properties[43]. The effects of the low temperature postannealing are also verified in the pulsed laser deposition[44].

Rare-earth-free high-Tc superconductors

In the rare-earth-free high-Tc superconductors several superconducting phases are present for the different chemical compositions. Typical chemical compositions for the Bi-system and the Tl-system are listed in Tab. 7 . Their superconducting properties have not been wholly understood yet.

Table 7. Rare-earth-free high-Tc superconductors

		Tc (k)	Institute	Date
Bi-system				
: $Bi_2O_2 \cdot 2SrO \cdot (n-1)Ca \cdot nCuO_2$				
$Bi_2Sr_2CuO_6$	(2 2 0 1)	7∿ 22	Caen Univ. (France) Aoyamagakuin Univ. (Japan)	1987.5
$Bi_2Sr_2CaCu_2O_8$	(2 2 1 2)	80	National Res. Institute for Metals (Japan)	1988.1
$Bi_2Sr_2Ca_2Cu_3O_{10}$	(2 2 2 3)	110	National Res. Institute for Metals (Japan)	1988.3
$Bi_2Sr_2Ca_3Cu_4O_{12}$	(2 2 3 4)	∿ 90	Matsushita Elec. (Japan)	1988.9
Tℓ-system				
: $T\ell_2O_2 \cdot 2BaO \cdot (n-1)Ca \cdot nCuO_2$				
$T\ell_2Ba_2CuO_6$	(2 2 0 1)	20∿ 90	Institute for Molecular Sci. (Japan) Arkansas Univ. (U.S.A.)	1987.12
$T\ell_2Ba_2CaCu_2O_8$	(2 2 1 2)	105	Arkansas Univ. (U.S.A.)	1988.2
$T\ell_2Ba_2Ca_2Cu_3O_{10}$	(2 2 2 3)	125	Arkansas Univ. IBM (U.S.A.)	1988.3
: $T\ell O \cdot 2BaO \cdot (n-1)Ca \cdot nCuO_2$				
$T\ell Ba_2CaCu_2O_7$	(1 2 1 2)	70∿ 80	IBM (U.S.A.)	1988.5
$T\ell Ba_2Ca_2Cu_3O_9$	(1 2 2 3)	110∿116	IBM (U.S.A.)	1988.3
$T\ell Ba_2Ca_3Cu_4O_{11}$	(1 2 3 4)	120	ETL (Japan)	1988.5
$T\ell Ba_2Ca_4Cu_5O_{13}$	(1 2 4 5)	<120	ETL (Japan)	1988.5

Thin films of Bi-Sr-Ca-Cu-O system are prepared by an rf-planar magnetron sputtering similar to the YBC films. The target is complex oxides of Bi-Sr-Ca-Cu-O which is made by sintering a mixture of Bi_2O_3 (99.999%), $SrCO_3$ (99.9%), $CaCO_3$ (99%) and CuO (99.9%) at 880°C for 8 h in air.

Typical sputtering conditions are shown in Tab. 8. The target is complex oxides of Bi-Sr-Ca-Cu-O. The composition is around 1-1-1-2 ratio of Bi-Sr-Ca-Cu. The process (1) and/or (2) are used for the deposition. Single crystals of (100) MgO are used as the substrates. The superconducting properties are improved by the postannealing at 850 - 900°C in 5 hr in O_2[45].

It is known that the superconducting properties are strongly affected by the substrate temperature during the deposition. Figure 12 shows typical X-ray diffraction patterns with resistivity-temperature characteristics for the Bi-Sr-Ca-Cu-O thin films of around 0.4 μm thick deposited at various substrate temperature. It is seen that the films deposited at 200°C exhibit $Bi_2Sr_2CaCu_2O_x$ structure with the lattice constant C ≃ 30.64 Å which corresponds to the low Tc phase[46]. The films show the zero resistance temperature of ≃ 70K [Fig.12 (a)]. When the substrate temperature is raised up during the deposition the high Tc phase with Tc ≃ 110K, the $Bi_2Sr_2Ca_2Cu_3O_x$ structure with the lattice constant C ≃ 36 Å, is superposed on the X-ray diffraction pattern [Fig.12 (b)][47]. At the substrate temperature of around 800°C a single high-Tc phase is observed. The films show the zero resistance temperature of ≃ 104K [Fig.12 (c)].

Table 8. Sputtering conditions

Target	Bi:Sr:Ca:Cu:=1-1.7:1:1-1.7:2 (100 mm in dia.)
Sputtering gas	Ar/O_2=1-1.5
Gas pressure	0.5 Pa
Rf input power	150 W
Substrate temperature	200 - 800°C
Growth rate	80 Å/min

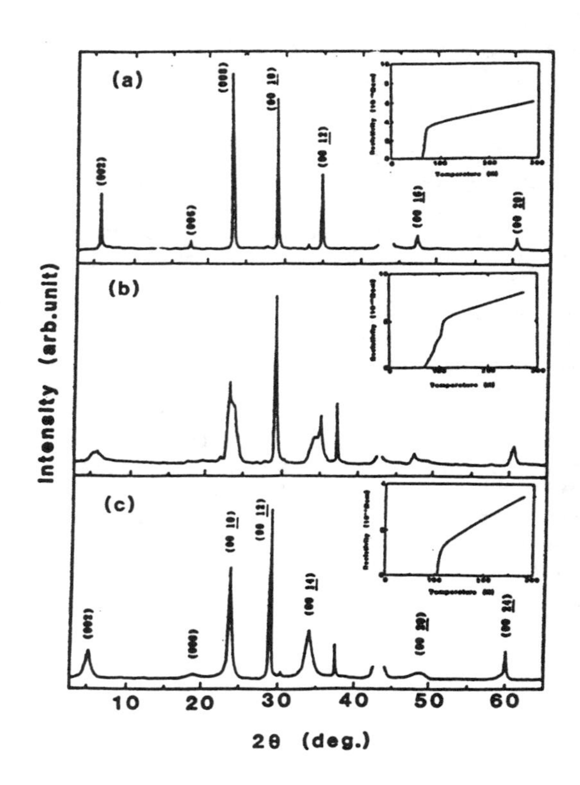

Fig.12. X-ray diffraction patterns with resistivity versus temperature for the annealed Bi-Sr-Ca-Cu-O films.

Similar to the Bi-Sr-Ca-Cu-O system, thin films of Tl-Ba-Ca-Cu-O system are prepared by the rf-magnetron sputtering on the MgO substrate. Typical sputtering conditions are shown in Tab. 9. However, their chemical composition is quite unstable during the deposition and the postannealing process due to the high vapour pressure of Tl. Thin films of the Tl system are deposited without intentional heating of substrates (∿ 200°C) and annealed at 890 - 900°C in Tl vapour[48]. It is seen that the superconducting phase of the resultant films strongly depends on the postannealing conditions. Figure 13 shows typical X-ray diffraction patterns with resistivity-temperature characteristics for the Tl-Ba-Ca-Cu-O thin films annealed at different conditions. The 0.4 μm thick film exhibits the low temperature phase. $Tl_2Ba_2CaCu_2O_x$ structure, with the lattice constant C ≃ 29 Å after slight annealing at 900°C 1 min [Fig.

Table 9. Sputtering conditions

Target	Tl:Ba:Ca:Cu:=2:1-2:2:3 (100 mm in dia.)
Sputtering gas	Ar/O_2=1
Gas pressure	0.5 Pa
Rf input power	100 W
Substrate temperature	200°C
Growth rate	70 Å/min

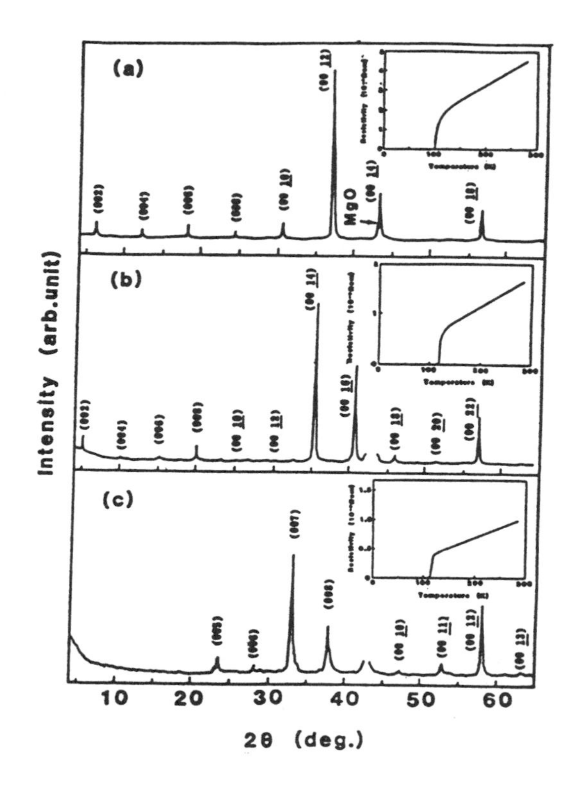

Fig.13. X-ray diffraction patterns with resistivity versus temperature for the annealed Tl-Ba-Ca-Cu-O films.

13 (a)]. The 2 μm thick films heavily annealed at 900°C 13 min show the high temperature phase, $Tl_2Ba_2Ca_2Cu_3O_x$ structure, with the lattice constant C ≃ 36 Å [Fig.13 (b)]. In the specific annealing condition the other superconducting phase $TlBa_2Ca_3Cu_4O_x$ structure with the lattice constant C ≃ 19 Å is also obtained [Fig.13 (c)][49].

SUMMARIES AND DISCUSSIONS

The thin film processing of the high-Tc superconductors is classified into three processes: deposition at low substrate temperature with post-annealing [process (1)], deposition at high temperature with postannealing [process (2)], deposition at high temperature without postannealing [process (3)]. One of the most important problems to be solved for the thin film processing is lowering the synthesis temperature. At present lowering the synthesis temperature could be achieved by both process (2) and process (3) for the rare-earth high-Tc superconductors.

In the process (2) lowering the synthesis temperature could be achieved with the low temperature postannealing at around 400 - 600°C. The minimum synthesis temperature is determined by the crystallizing temperature of the high-Tc superconductors in these processes. The crystallizing temperature of YBC, for instance, is around 500 - 600°C. It is noted that in the process as-deposited films show the superconducting properties without any postannealing, i.e. in situ deposition.

As seen in the X-ray diffraction pattern, the in situ deposited films comprise the orthorhombic phase and the tetragonal phase. The SEM image suggests that these films comprise small crystallites. The TEM image of the sputtered films also denotes the presence of the crystal boundaries. This may reduce the critical current Jc. The Jc for the in situ deposited Er-Ba-Cu-O thin films is proportional to $(1 - T/Tc)^{1.8}$, which is close to $(1 - T/Tc)^{1.5}$. This indicates that the current transport is partially governed by weak-link of superconductive regions[50]. At present high critical current is obtained in the process (2) using the postannealing[51].

Although the low temperature synthesized films are not the perfect single crystal, the low temperature process gives several favorable properties such as the suppression of the diffusion at film and substrate interface as shown in Fig.14.

As described in previous section for the rare-earth high-Tc superconductors the superconducting orthorhombic phase is reasonably stabilized during the postannealing process. For the rare-earth-free high-Tc superconductors of Bi-system the superconducting phases of the sputtered films are controlled by the substrate temperature during the deposition; the low-Tc phase of $Bi_2Sr_2Ca_1Cu_2O_x$ system is obtained at the substrate temperature below 600°C, and the high-Tc phase of $Bi_2Sr_2Ca_2Cu_3O_x$ system is obtained at the substrate temperature above 750°C.

The SEM image suggests that the thin films of the Bi-system comprise mica-like crystallites as shown in Fig.15. The c-axis of the crystallites is perpendicular to the crystal plane. The large crystallites allow the

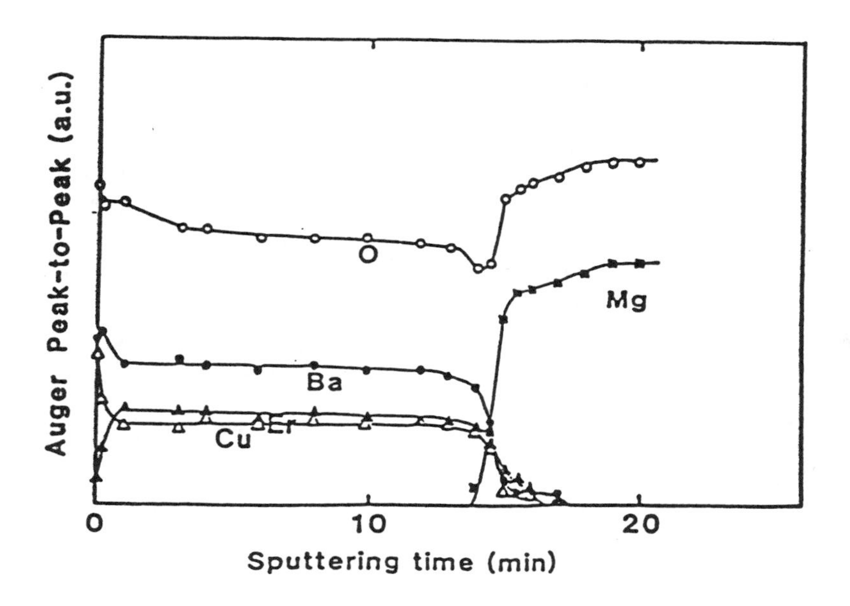

Fig. 14. Auger depth profile for as grown Er-Ba-Cu-O thin films sputtered on (100)MgO substrate at 650°C (2000 A thick).

Bi-Sr-Ca-Cu-O film

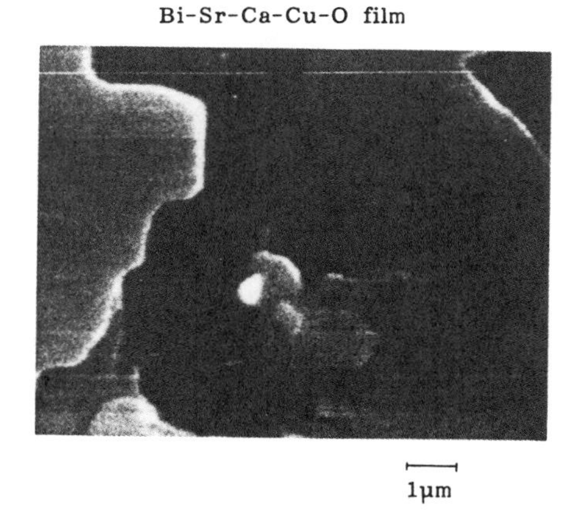

1μm

Fig. 15. SEM images of the sputtered Bi-Sr-Ca-Cu-O thin films.

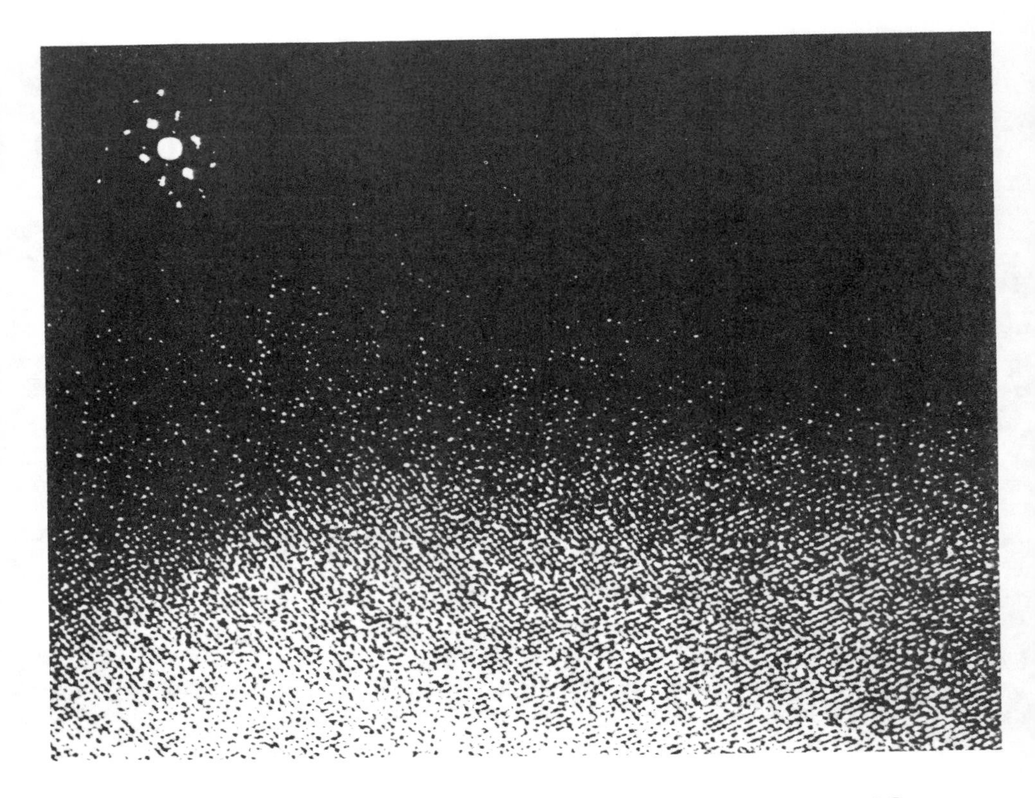

Fig. 16. TEM images of the sputtered Bi-Sr-Ca-Cu-O thin films of 2-2-1-2 structure.

large critical current.

The critical current density measured for the Bi-Sr-Ca-Cu-O films is as high as 2×10^5 A/cm^2 at 77K and 6×10^6 A/cm^2 at 4.2K. The critical current density at 77K will be governed by the high Tc phase. The current will flow through the current channel presented in the sputtered Bi-Sr-Ca-Cu-O films since the films are composed of mixture of the high-Tc phase and the low Tc phase. More critical current density will be possible in the case of films with single high-Tc phase. The diamagnetic measurements suggest that the contents of the high-Tc phase are 5 - 10%[52]. This denotes that the net critical current of the high-Tc phase will be $2\times10^6 \sim 4\times10^6$ A/cm^2 at 77K and $6\times10^7 \sim 1.2\times10^8$ A/cm^2 at 4.2K.

It is noted that the temperature variations of the Jc are governed by $(1 - T/Tc)^2$ [53]. The square power dependence is different from the 3/2 power dependence predicted by well-studied proximity junction tunneling model based on the BCS theory. The presence of the layered structure will cause the square power dependence. Similar properties are observed in the YBC films[54].

In the crystallites of the low-Tc Bi-Sr-Ca-Cu 2-2-1-2 phase the atomic arrangements are found to be uniform as indicated in the TEM image shown in Fig.16. However, in the crystallites of the high-Tc Bi-Sr-Ca-Cu 2-2-2-3 phase the crystallites comprise the different superconducting phases including Bi-Sr-Ca-Cu 2-2-1-2, and 2-2-3-4 phases as seen in the TEM image shown in Fig.17, although the resistivity-temperature characteristics correspond the single superconducting phase of the 2-2-2-3 structure. The presence of the mixed phase is also confirmed by the spreading skirt observed in the X-ray diffraction pattern at the low angle peak around $2\theta \fallingdotseq 4°$. It is reasonably considered that the presence of the mixed phases results from the specific growth process of the present rare-earth-free superconducting thin films: The rare-earth-free superconducting thin films may be molten during the annealing process. The superconducting phase will be formed during the cooling cycle[55].

For the Tl-Ba-Ca-Cu-O system the sputtered films exhibit rough surface morphology. The critical current is lower than that of the Bi-system[56]. XPS measurements for the crystallized Bi-Sr-Ca-Cu-O films suggest that the annealing process modifies the crystal structure near the $Cu{-}O_2$ layer and increases the density of Cu^{3+}. The Bi-O layered structure is stable during the annealing[57]. This implies that the single superconducting phase will be synthesized when the Bi-O basic structure is crystallized and the stoichiometric composition is kept for the unit cell of the Bi-Sr-Ca-Cu-O.

These considerations have been confirmed by the layer-by-layer deposition conducted by the multi-target sputtering system. Typical sputtering conditions are shown in Tab.10. The deposition rate is selected so as to pile up the B-O, Sr-O, Cu-O, and Ca layer in an atomic scale range. The substrate temperature is kept around crystallizing temperature of 650°C. Figure 18 shows the typical results for the layer by layer deposition. It is noted that the phase control is achieved simply by the amounts of Cu-Ca-O during the layer-by-layer deposition. The experiments show that the Tc does not increase monotonously with the

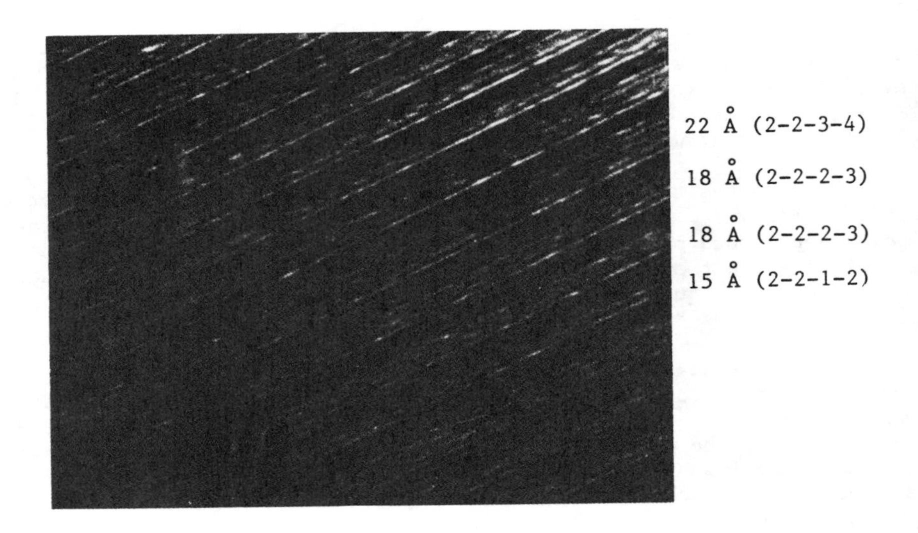

Fig. 17. TEM images of the sputtered Bi-Sr-Ca-Cu-O thin films indicating the presence of the mixed phases.

Table 10. Sputtering conditions

Target	Bi, SrCu, and CaCu (60 mm in dia.)
Substrate	(100) plane of MgO 10 mm × 10 mm
Target-substrate spacing	100 mm
Sputtering gas	Ar + O_2 (5 : 1)
Gas pressure	3 Pa
Input power	Bi: 5 - 6 W SrCu: 9 - 10 W CaCu: 30 - 60 W

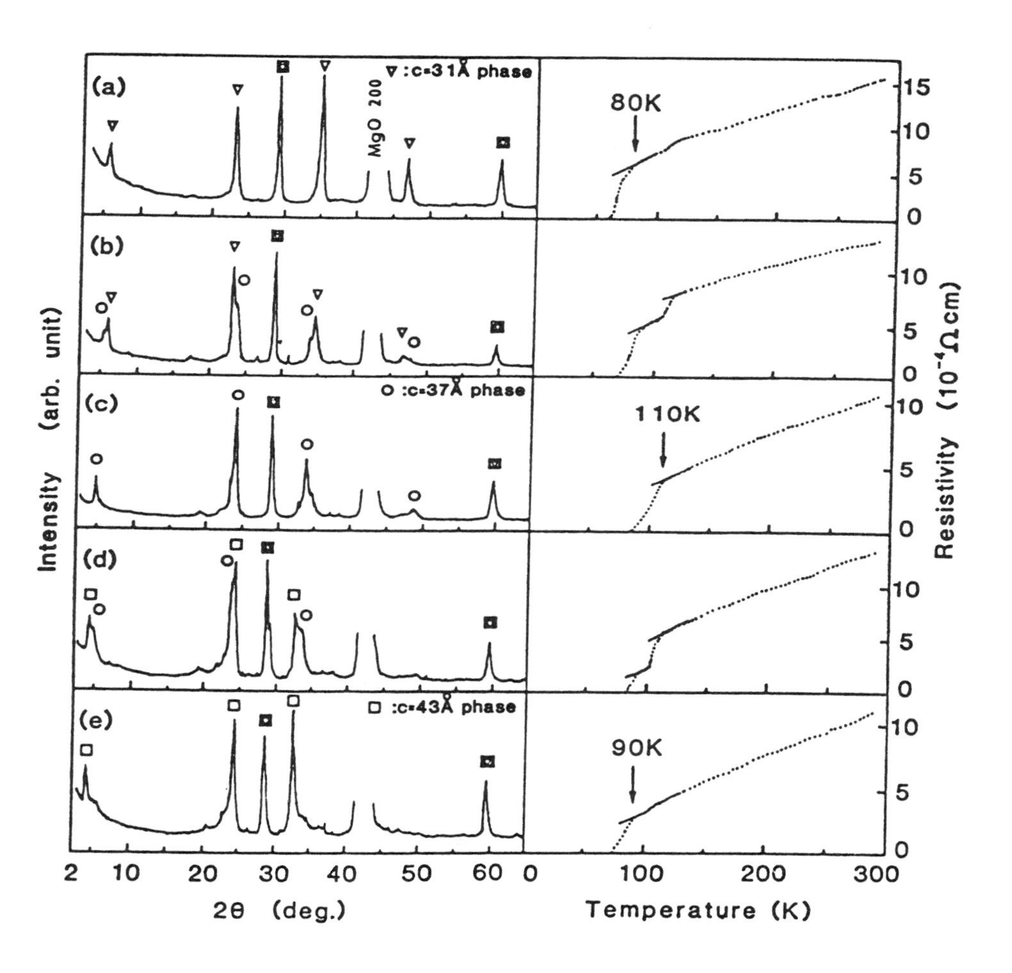

Fig. 18. The X-ray diffraction patterns and temperature dependence of the resistivity for the Bi-Sr-Ca-Cu-O thin films with various c-axis lattice spacing prepared by the layer-by-layer deposition. The deposited films are postannealed in O_2 at 855°C for 5 hours (film thickness, 200 - 300 A).

numbers of the Cu-O layer. In the Bi bi-layer system the Tc shows maximum, 110K, at three layers of Cu-O, $Bi_2Sr_2Ca_2Cu_3O_x$. At the four layers of Cu-O, $Bi_2Sr_2Ca_3Cu_4O_x$, Tc becomes 90K[58].

CONCLUSIONS

The layer by layer deposition is one of the most promising process for a fine control of the superconducting phase of the rare-earth-free superconductors. Man made high-Tc superconductors will be synthesized by the layer by layer deposition. Further studies on the layer by layer deposition of the high-Tc superconductors are necessary not only for a fabrication of thin film devices, but also for the scientific interests in the superconductivity of the high-Tc superconductors.

ACKNOWLEDGEMENTS

The authors thank S. Hayakawa and T. Nitta for their continuous encouragements. They also thank S. Hatta, T. Mitsuyu, M. Kitabatake, H. Higashino and T. Hirao for useful discussions and the members of Matsushita Technoresearch, Inc. for the analyses of the specimens.

REFERENCES

1. K. Kusao, K. Wasa, and S. Hayakawa, Jpn. J. Appl. Phys. 7, (1968) 437.
2. K. Iijima, R. Takayama, Y. Tomita, and I. Ueda, J. Appl. Phys. 60, (1986) 2914.
3. K. Wasa, O. Yamazaki, H. Adachi, T. Kawaguchi, and K. Setsune, Ieee J. Lightwave Tech. LT-2, (1984) 710.
4. H. Adachi, T. Mitsuyu, O. Yamazaki, and K. Wasa, Jpn. J. Appl. Phys. Suppl. 24-2, (1985) 287.
5. H. Adachi, T. Mitsuyu, O. Yamazaki, and K. Wasa, J. Appl. Phys. 60, (1986) 736.
6. M. Suzuki, Y. Enomoto, and T. Murakami, J. Appl. Phys. 56, (1984) 2083.
7. H. Adachi, K. Setsune, and K. Wasa, Phys. Rev. B 35, (1987) 8824.
8. M. Hong, S.H. Liou, J. Kwo, and B.A. Davidson, Appl. Phys. Lett. 51, (1987) 694.
   G.K. Wehner, Y.H. Kim, D.H. Kim, and A.M. Goldman, Appl. Phys. Lett. 54, (1988) 1187.
9. H. Maeda, Y. Tanaka, M. Fukutomi, and T. Asano, Jpn. J. Appl. Phys. 27, (1988) L209.
10. Z.Z. Sheng and A.M. Hermann, Nature 332, (1988) 138.
11. Y. Ichikawa, H. Adachi, K. Hirochi, K. Setsune, S. Hatta, and K. Wasa Phys. Rev. B., (in press).
12. H. Adachi, K. Wasa, Y. Ichikawa, K. Hirochi, and K. Setsune, J. Cryst Growth, (in press).
13. M. Kitabatake, T. Mitsuyu, and K. Wasa, J. Non-Cryst. Solids 53, (1982) 1.
14. M. Oikawa, and K. Toda, Appl. Phys. Lett. 29, (1976) 491.

15. D. Dijkkamp, T. Venkatesan, X.D. Wu, S.A. Shaheen, N. Jisrawi, Y.H. Min-Lee, W.L. McLean, and M. Croft, Appl. Phys. Lett. 51, (1987) 619.
16. T. Fukami, and T. Sakuma, Jpn. J. Appl. Phys. 20, (1981) 1599.
17. T. Nakagawa, J. Yamaguchi, M. Okuyama, and Y. Hamakawa, Jpn. J. Appl. Phys. 21, (1982) L655.
18. H. Abe, T. Tsuruoka, and T. Nakamori, Jpn. J. Appl. Phys. 27, L1473 (1988).
19. M. Ihara and T. Kimura, FED HiTcSC-ED Workshop, Miyagi-ZaO (1988).
20. K. Wasa, H. Adachi, Y. Ichikawa, K. Setsune, and K. Hirochi, Proc. International Workshop on High-Tc Superconductors Srinagar, 1988 May (World Scientific, A.K. Gupta, S.K. Joshi, C.N. Rao Ed., Singapore, 1988, p.353).
21. Y. Ichikawa, H. Adachi, T. Mitsuyu, and K. Wasa, Jpn. J. Appl. Phys., 27, (1988) L381.
22. E. Takayama-Muromachi, Y. Uchida, K. Yukino, T. Tanaka, and K. Kato, Jpn. J. Appl. Phys. 26, (1987) L665.
23. J.D. Jorbensen, M.A. Beno, D.G. Hinks, L. Soderholm, K.J. Volin, R.L. Hitterman, J.C. Grace, I.K. Shhuller, C.V. Segre, K. Zhang, and M.S. Kleefish, Phys. Rev. B, 36 3608 (1987).
24. S. Hayashi, T. Kamada, K. Setsune, T. Hirao, K. Wasa, and A. Matsuda, Jpn. J. Appl. Phys., 27, (1988) L1257.
25. Y. Ichikawa, M. Kitabatake, H. Adachi, S. Hatta, and K. Wasa, Extended Abstracts (The 35th Spring Meeting, 1988) Jpn. Soc. Appl. Phys. (1988).
26. T. Kamada, K. Setsune, T. Hirao, and K. Wasa, Appl. Phys. Lett. (in press).
27. H.C. Li, G. Linker, F. Ratzel, R. Smithey, and J. Geek, Appl. Phys. Lett. 52, (1988) 1098.
28. X.D. Wu, A. Inam, T. Venkatesan, C.C. Chang, E.W. Chase, P. Barboux, J.M. Tarascon, and B. Wilkens, Appl. Phys. Lett. 52, (1988) 754.
29. T. Terashima, K. Iijima, K. Yamamoto, Y. Bando and H. Mazaki, Jpn. J. Appl. Phys. 27, (1988) L91.
30. H. Adachi, K. Hirochi, K. Setsune, M. Kitabatake, and K. Wasa, Appl. Phys. Lett. 51, (1987) 2263.
31. K. Hirochi, H. Adachi, K. Setsune, O. Yamazaki, and K. Wasa, Jpn. J. Appl. Phys. 26, (1987) L1837.
32. H. Adachi, K. Setsune, K. Hirochi, T. Kamada, and K. Wasa, Proc. of International Conf. on High-Temp. Superconductors and Materials and Mechanisms of Superconductivity, E12, Interlaken, Switzerland, 1988.
33. S. Hatta and K. Wasa, Proc. 55th Magnetics Symposium, Tokyo, (1988) p.7.
34. H. Adachi, K. Setsune, T. Mitsuyu, K. Hirochi, Y. Ichikawa, T. Kamada, and K. Wasa, Jpn. J. Appl. Phys. 26, (1987) L709.
35. B-Y Tsaur, M.S. Dilorio, and A.J. Strauss, Appl. Phys. Lett., 51, (1987) 858.
36. K. Setsune, H. Adachi, T. Kamada, K. Hirochi, Y. Ichikawa, and K. Wasa, in Extended Abstracts, 8th International Symposium on Plasma Chemistry (Tokyo, 1987), p.2335.
37. P. Chaudhari, R.H. Koch, R.B. Laibowitz, T.R. McGuire, and R.J. Gambino, Phys. Rev. Lett. 58, (1987) 2684.
38. B. Oh, M. Naito, S. Arnason, P. Rosenthal, R. Barton, M.R. Beasley, T.H. Geballe, R.H. Hammond, and A. Kapitulnik, Appl. Phys. Lett. 51, (1987) 852.
39. H. Adachi, K. Setsune, and K. Wasa, Jpn. J. Appl. Phys., Suppl.

26-3, (1987) 1139.
40. K. Hirochi, H. Adachi, K. Setsune, O. Yamazaki, and K. Wasa, Jpn. J. Appl. Phys. 26, (1987) L1837.
41. K. Wasa, M. Kitabatake, H. Adachi, K. Setsune, and K. Hirochi, Proc. American Vac. Soc. 34th National Symposium, Anaheim, 1987, HTS-FrM3.
42. H. Adachi, K. Hirochi, K. Setsune, M. Kitabatake, and K. Wasa, Appl. Phys. Lett. 51, (1987) 2263.
43. T. Kamada, K. Setsune, T. Hirao, and K. Wasa, Appl. Phys. Lett., 52, (1988) 1726.
44. C. C. Chang, X. D. Wu, A. Inam, D. M. Hwang, T. Venkatesan, P. Barboux, and J. M. Tarascon, Appl. Phys. Lett. 53, (1988) 517.
45. Y. Ichikawa, H. Adachi, K. Hirochi, K. Setsune, and K. Wasa: Proc. MRS Int. Meeting on Advanced Materials, June 1988, Tokyo (in press).
46. E. Takayama-Muromachi, Y. Uchida, A. Ono, F. Izumi, M. Onoda, Y. Matsui, K. Kosuda, S. Takekawa, and K. Kato, Jpn. J. Appl. Phys. 27 (1988) L365.
47. H. Adachi, Y. Ichikawa, K. Setsune, S. Hatta, K. Hirochi, and K. Wasa, Jpn. J. Appl. Phys., 27, (1988) L643.
48. Y. Ichikawa, H. Adachi, K. Setsune, S. Hatta, K. Hirochi, and K. Wasa, Appl. Phys. Lett., 53, (1988) 919.
49. J. Zhou, Y. Ichikawa, H. Adachi, T. Mitsuyu, and K. Wasa, Jpn. J. Appl. Phys., (in press)
50. H. Adachi, K. Setsune, K. Hirochi, T. Kamada, and K. Wasa, Proc. of International Conf. on High-Temp. Superconductors and Materials and Mechanisms of Superconductivity, E12, Interlaken, Switzerland, 1988.
51. Y. Enomoto, T. Murakami, M. Suzuki, and K. Moriwaki, Jpn. J. Appl. Phys. 26, (1987) L1248.
52. S. Hatta, Y. Ichikawa, K. Hirochi, H. Adachi, and K. Wasa, Jpn. J. Appl. Phys., 27, (1988) L855.
53. K. Setsune, K. Hirochi, H. Adachi, Y. Ichikawa, and K. Wasa, Appl. Phys. Lett. 15, (1988) 600.
54. S.B. Ogale, D. Dijkkamp, T. Venkatesan, X.D. Wu, and A. Inam, Phys. Rev. B36, (1987) 7210.
55. K. Wasa, H. Adachi, Y. Ichikawa, K. Hirochi, and K. Setsune: Proc. ISS'88, Aug. 28-31, Nagoya (1988).
56. Y. Ichikawa, H. Adachi, K. Setsune, S. Hatta, K. Hirochi, and K. Wasa, Appl. Phys. Lett., 53, 919 (1988).
57. S. Kohiki, K. Hirochi, H. Adachi, K. Setsune, and K. Wasa, Phys. Rev. B, 38 (1988) (in press).
58. H. Adachi, S. Kohiki, K. Setsune, T. Mitsuyu, and K. Wasa, Jpn. J. Appl. Phys., 27 (1988) L1883.

THREE TERMINAL HIGH-Tc SUPERCONDUCTING DEVICES WITH LARGE CURRENT GAIN

Takeshi Kobayashi and Uki Kabasawa

Faculty of Engineering Science, Osaka Univerrsity
1-1 Machikaneyama-cho, Toyonaka, Osaka 560, Japan

INTRODUCTION

Since a discovery of a new oxide superconductor by Bednorz and Müller[1], there has been a rush of new compound with the higher critical temperature than the triggering material. The first one whose critical temperature T_{co} (zero resistance temperature) surpassed the liquid nitrogen temperature (77 K) was an $Y_1Ba_2Cu_3O_y$ discovered by the groupe of Chu [2]. Most recently, synthesis of $Bi_2Sr_2Ca_2Cu_3O_y$ and $Tl_2Ba_2Ca_2Cu_3O_y$ systems recorded the zero resistance temperature higher than 100 K and really drew the superconductivity electronics up for the practical use [3,4].

One of the advantages inherent in the high T_c superconducting materials is a simple formation of the Josephson junction by only making a waist portion in the poly-crystalline superconducting film or carved ceramic bulk. The junction made thus is named a grain-boundary Josephson junction (GBJJ for short). Till now, many papers have been published dealing with the fabrication of the GBJJ and application to the millimeter wave detection [5] and the magnetic sensor SQUID operative at 77 K [6]. However, the Josephson junction is, in principle, a two terminal diode, and therefore, it can not work well as an active device unless the junction is installed in a special circuit. This is the reason why the limited applications are alloted to the GBJJ.

To solve these problems, new superconducting devices other than the Josephson diode have been explored. A three terminal device is a promising candidate. We had so far five kinds of the three terminal superconducting devices, all of which comprised the metal superconductors: the Josephson field effect transistor (JOFET)[7], the quasi-particle transistor (QUITERON)[8], the superconductor-base semiconductor insulating transistor (SUBSIT)[9], the superconductor-base hot-electron transistor (Super-HET)[10,11], and the microbridge with the current injection gate [12]. The latter three devices are characterized by the monolithic combination of the superconductor and semiconductor. The basic technology required for the fabrication of any of five is the multi-layered film growth, which has been, to certain extent, established for the metal superconductors, but not for the high T_c superconductors. The exotic high T_c superconducting transistor can not be reallized until the basic technologies are built-up [13].

In the present work, we did the first attempt in fabricating the high T_c superconducting transistor by using the state-of-the-art technologies. As is now well known, the high T_c superconducting materials commonly have the characteristic features of the layered conduction, low carrier concentration, very short coherence length, high critical current J_c and critical field H_{c2}, and ease of the substitution of the constituent atoms. Among them, the second one "low carrier concentration" gives positive aid to the present device. This peculiarity helps the development of the non-equilibrium superconductivity through the hot quasi-particle injection, which, in turn, modulates the superconducting characteristics such as the critical current density. One type of the device comprises the GBJJ with the hot quasi-particle injector, and the current modulation was brought about through the non-equilibrium superconductivity developed by the accumulation of the excess quasi-particles around the junction area [14]. The gate current density as low as 3 A/cm² was sufficient to get 90 % modulation. The second type, being addressed here as a main subject, is made with the epitaxial YBaCuO thin films, and is designed to directly modulate the superconducting critical current flowing through the narrow channel. Although the experiment is still in the preliminary stage, the current modulation gain of 5~7 has been obtained.

It is worth noting here that the three terminal device proposed in the present work is promising so much from view point not only of the pure electronic application but also of exploring the high T_c superconducting mechanism. This is because the non-equilibrium superconductivity is a kind of minority-carrier physics, and therefore, the new device based on the non-equilibrium superconductivity can provide us a lot of information concerning the superconductivity from the minority-carrier point of view. In the present work, a simple analysis is given to explain the observed current modulation characteristics of the transistor device. The results based on the BCS non-equilibrium superconductivity were qualitatively in good agreement with the experimental results at least in appearance. However, it does not definitely mean that the high T_c superconductivity originates from the phonon-coupling effect. We need data much more to do it.

DEVICE STRUCTURE

Figure 1 shows the schematic drawing of the fabricated superconducting transistor and the photo micrograph of the completed device. In the figure, a narrow portion 10 μm wide is an active region of the device. The device has silver ohmic contacts for the source and drain electrodes, and alminum gate as a quasi-particle injector. We prepared poly-crystalline high T_c films for use to the three terminal GBJJ. The GBJJ was formed in the 10 μm wide active region to obtain the higher modulation effect. Otherwise, the epitaxially grown films were used. Five transistors were integrated in a chip with a common drain electrode.

This new type of device is, in principle, capable of working as a current switching transistor with high gain. So, we hereafter, tentatively name this device "SCST (Superconducting Current Switching

Transistor) " for short.

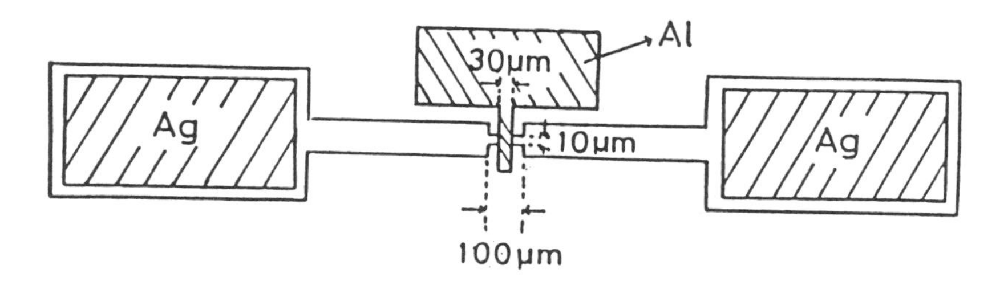

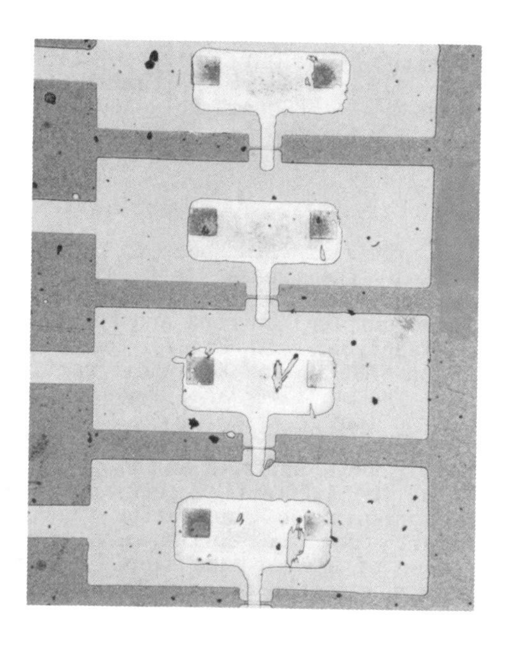

Fig.1 Schematic drawing of the superconducting transistor (SCST) and photo micrograph. Five device elements are integrated in a chip.

SUPERCONDUCTOR THIN FILM PREPARATION

As high T_c superconductor films, we used the most popular oxide superconductors, YBaCuO and ErBaCuO. These 1-2-3 phase superconductors are particularly benificial to form their own crystal in a single crystal phase by virtue of the advanced epitaxial growth technology. The films were grown on (100) oriented MgO substrates in argon and oxygen mixed atmosphere by the conventional rf magnetron sputtering under the following conditions summarized in Table I.

Table I. Growth condition of high T_c superconducting thin film .

Target	$Y_1Ba_2Cu_{4.5}O_y$, $Er_1Ba_2Cu_{4.5}O_y$
Substrate	(100) MgO
Discharge Gas	Ar + O_2(37%)
Gas Pressure	4 Pa
Sub. Temperature	< 670 ℃
rf Power	50 W
Growth Rate	~ 6 nm/min
Film Orientation	(103) Poly-crystalline Film
	(001) Epitaxial Film
Film Thickness	800 nm Poly-crystalline Film
	160 nm Epitaxial Film

The (001) oriented epitaxial film seems to have great advantages over other kind of films in the sense that the film can carry the superconducting critical current which is several orders of magnitude higher than the Josephson critical current and there is no restriction for the current flowing direction. In Fig.2, the reflection high energy electron beam diffraction (RHEED) and X-ray diffraction (XRD) patterns of the as-grown epitaxial YBaCuO film are shown. The electron beam is parallel to [100] direction. From this RHEED observation, it is understandable that the film is epitaxially grown with the orientation to (001) and it has pretty smooth surface morphology. In addition, in the XRD pattern, we can find the diffraction peaks of (00n) in series and no sign of other crystal orientation, indicating that the film orients to (001) direction. Apart from the epitaxially grown films, the poly-crystalline film was also of use to make simply a GBJJ. In our persent work, a reduced substrate temperature down to 600 ℃ resulted in growing the poly-crystalline film. On the (100) MgO substrate, the film preferentially oriented to (103) axis was obtained. Although the film itself was poly-crystalline, its surface was very flat as far as we observed it by the photo-microscope.

The critical temperature T_{c0} of the as-grown film was not so high, being at most 50 K. Particularly, this problem was more serious for the as-grown epitaxial film. The reason of this might be a poor crystallinity of the as-grown film. This is clearly seen in the XRD pattern given in Fig.2 which exposes a broad diffraction pattern, indicating the poor

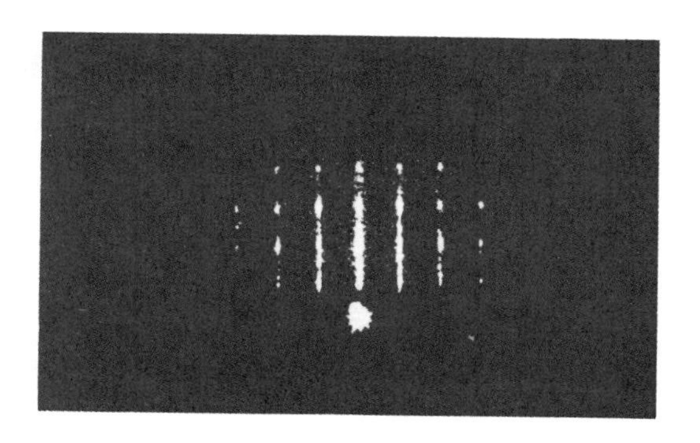

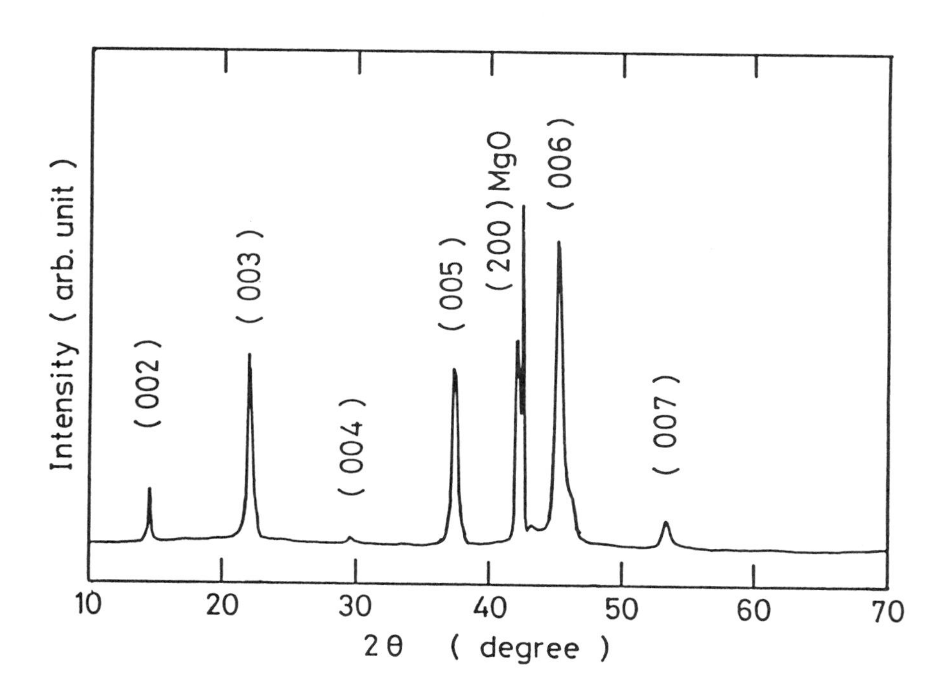

Fig.2 Reflection electron beam diffraction (upper) and X-ray diffraction (lower) patterns taken from the epitaxially grown film. The electron beam was irradiated to [100] azimuth. The epitaxial film is (001) oriented.

crystalline quality of the as-grown film. It might be due to an original imperfection, plasma irradiation damage, oxygen missing, and so on. Whatever the reason might be, the post annealing process was necessary to obtain a high T_{c0} value for any film. So, the as-grown films were routinely subjected to the annealing in the flowing oxygen at 800 ℃ and slowly cooled down to room temperature in the furnace. Despite of introduction of the post anneal, the surface morphology and RHEED pattern of the film very fortunately remained unchanged before and after the annealing process.

DEVICE FABRICATION

We fabricated the new three terminal devices by procedures mentioned in the followings and in Fig.3. The grown film was first patterned by the wet chemical etching employing the diluted phospholic acid (H_3PO_4). This wet etching is a simple technique we proposed suitable for the oxide superconductors [15]. Indeed, there is no sign of degradation in the superconductivity after the etching process. As the etching mask, we can use the positive-type photoresist AZ1350. It was not only quite resistive to the H_3PO_4 solution, but also very safe to the superconductivity of the high T_c film. Then, the silver was evaporated on the film and by using the lift-off technique the drain and source electrodes were formed, followed by the annealing at 500 ℃ for 4 hrs in the oxygen atmosphere. This resulted in a very low specific contact resistance lower than 10^{-6} $\Omega \cdot cm^2$. In the actual devices, the contact resistance was less than a few tens of $\mu\Omega$, being negligibly small. This helped us to employ the two probe measurement way for the source-drain I-V characteristics instead of the four probe technique commonly used in this field.

In the present devices, the gate electrode serves as a tunneling quasi-particle injector. To form this kind of injector, the usual device has been equipped with an ultra-thin insulating film between the oxide superconducting film and the gate electrode. However, at least at present, the fabrication of the gate electrode was extremely simplified. Namely, as shown in Fig.3, the aluminum was directly deposited onto the superconductor film without intentionally inserting the insulating thin film between them. Nevertheless, the gate electrode worked well as a tunneling injector. It is understood as follows: During the device process, the film surface is likely to lose not only its superconductivity but also the normal conductivity due to intensive dissociation of surface layer or surface oxygen fugacity in the vacuum. However, it rather facilitates the tunneling injector formation by a simple aluminum deposition onto the film. Another view is that the aluminum deposition may extract a lot of oxygen atoms from the oxide superconductor and form automatically a thin aluminum oxide (Al_2O_3) tunneling barrier at the interface. At present, it is not known which is more reasonable explanation for the native barrier formation. Though the present process technique is very simple, it has a disadvantage. A drawback of this kind of barrier formation is an uncontrolability of the thickness, under which the quasi-particle injection is entirely put. Replacement of this native layer by the artificial insulating film must be a key to produce the well-controlled gate electrode.

Hereafter, we name the device "Type-I SCST" the channel of which is made up with the poly-crystalline high T_c superconducting film. Otherwise (use of the epitaxial film), it is "Type-II SCST".

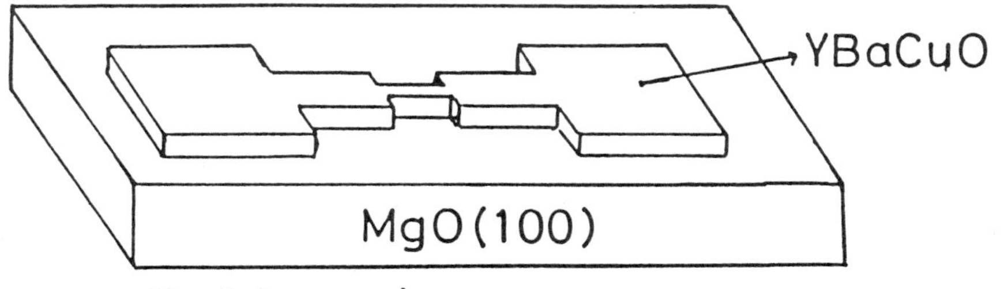

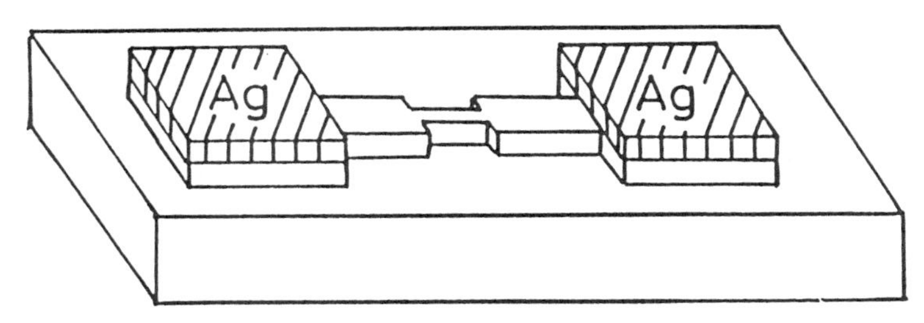

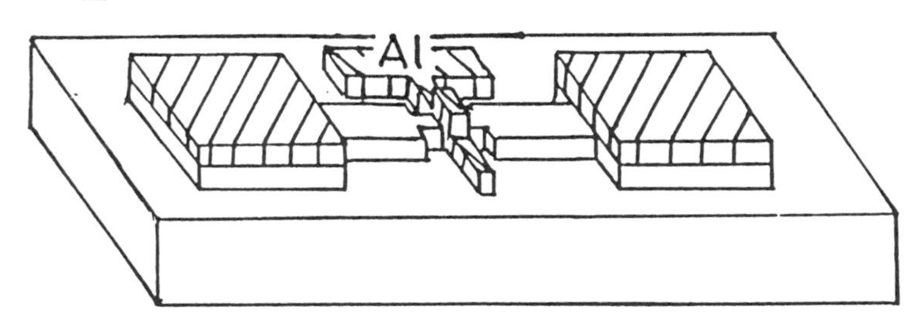

Fig.3 Transistor fabrication process. In the actual process, five elements are integrated in a chip. The silver and aluminum electrodes were formed by using the lift-off technique.

THREE TERMINAL DEVICE CHARACTERISTICS

The completed device was diced from the wafer and mounted on DIL IC heade and wire-bonded with the gold wire (25 µm). Most of measurements were mad at liquid helium temperature where the non-equilibrium superconductivit is more likely to advances under the quasi-particle injection. Very fe measurements were done at the elevated temperatures.

Figure 4 shows an I-V characteristic of an injector of the Type- device, where the gate current flows from the gate to drain electrode through the narrow (10 µm wide) channel. This characteristic proves tha the hot quasi-particles can be injected, and it might arise from th Metal-Insulator-Superconductor (MIS) structure whose insulator is forme at the aluminum gate-YBaCuO film interface as mentioned above. Althoug the I_g-V_g characteristic is plotted from -600 µA to 600 µA of I_g in thi figure, the real injection was mostly done in the range below 50 µA, unde

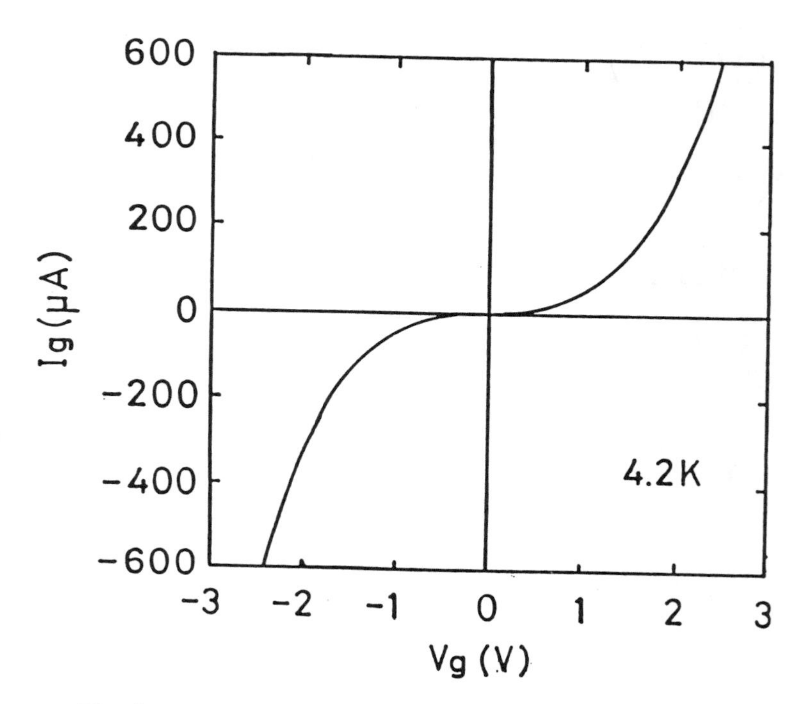

Fig.4 Current- voltage characteristics of the gate electrode. The circuit is connected to the gate and drain electrodes. The hot quasi-particles are injected into the oxide superconductor adjacent to the gate electrode.

which the net input power density was as low as 10 W/cm^2 (less than 30 μW / gate in the actual device). The input power per gate was roughly comparable to that used in the famous QUITERON experiment [8]. Whereas, the input power density was much lower, nearly one tenth of that. This comparison guarantees, to certain extent, that the following observation of the transistor action is coming from a modulation of superconductivity due to some kind of physical mechanism caused by the quasi-particle injection other than a simple heating effect. Of cource, as mentioned before, the electrical characteristics of the gate circuit change from device to device, and therefore, the input power has the variation in its value with the multiplication factor from 0.5 to 2 at the same injection current.

The quasi-particle injection effects on the drain current - drain voltage characteristics are depicted in Figs.5 to 7, wherein the results of Type-I device are given in Fig.5 and in remaining two the Type-II devices are shown. These two Type-II SCST's are named for convenience the Type-IIA (Fig.6) and Type-IIB (Fig.7). The Type-IIB curve was obtained at the elevated temperature ($T/T_c \sim 0.8$). In any cases, the stress of the gate current injection resulted in the efficient modulation of the critical current flowing through the channel. As a whole, the current modulation curve of the SCST looks like that of the semiconducting Bipolar Transistor or the Field Effect Transistor. However, one point we must notice is that an essential difference of the SCST is the zero voltage drop across the channel up to the current of the critical value. On the other hand, the flux-flow resistance (in the non-zero voltage range) remains almost unchanged irrespective of the amount of the quasi-particle injection except for the Type-I SCST.

As to the Type-I SCST, one can see somewhat large softening of the I_d-V_d curve at the critical current. This hinders us to strictly estimate the value of the critical current. It should be kept in mind that the residual resistance of the ohmic contact at both the source and drain electrodes and/or the gold lead wire takes part in this. Anyway, with increasing the gate current, the non-linearity in the I-V curve is getting less pronounced. The non-linear feature can no longer be found in the figure when I_g increases to 50 μA. Contrasting with Type-I SCST, two Type-II SCST's exhibit a clear breaking point in the I-V curve at around the critical current. Moreover, owing to the excellent crystallinity of the epitaxial film, much higher critical current can flow.

To get more knowledge on the injection effect of the SCST's, let us reproduce the experimental data concerning the relationship between the critical current and the gate current in Figs. 5(b), 6(b) and 7(b). To do this for Type-I SCST, we should first introduce a parameter I_d as a measure of the critical current of the device. The value of I_d is reasonably defined by [I(+50 μV) - I(-50 μV)]/2, but we must bear in mind that I_d never vanishes due to a presence of the normal current component even when the true value of the critical current goes to zero. In Fig.5(b), one can see clearly a current modulation effect. A slope of $\partial I_d/\partial I_g$ in the gate current range of -8 ~ -4 μA reaches 1.4, which is larger than unity. More efficient current modulation was obtained in Type-IIA SCST. It is not difficult to know a large current gain (5 ~ 7) from the curve in Fig.6(b). It is expected from this result that much higher current gain will be possible to obtain by simply making the gate width narrower and narrower just as has done in the sub-micron gate Si MOSFET's. Our present result was obtained for the gate 30 μm wide. This

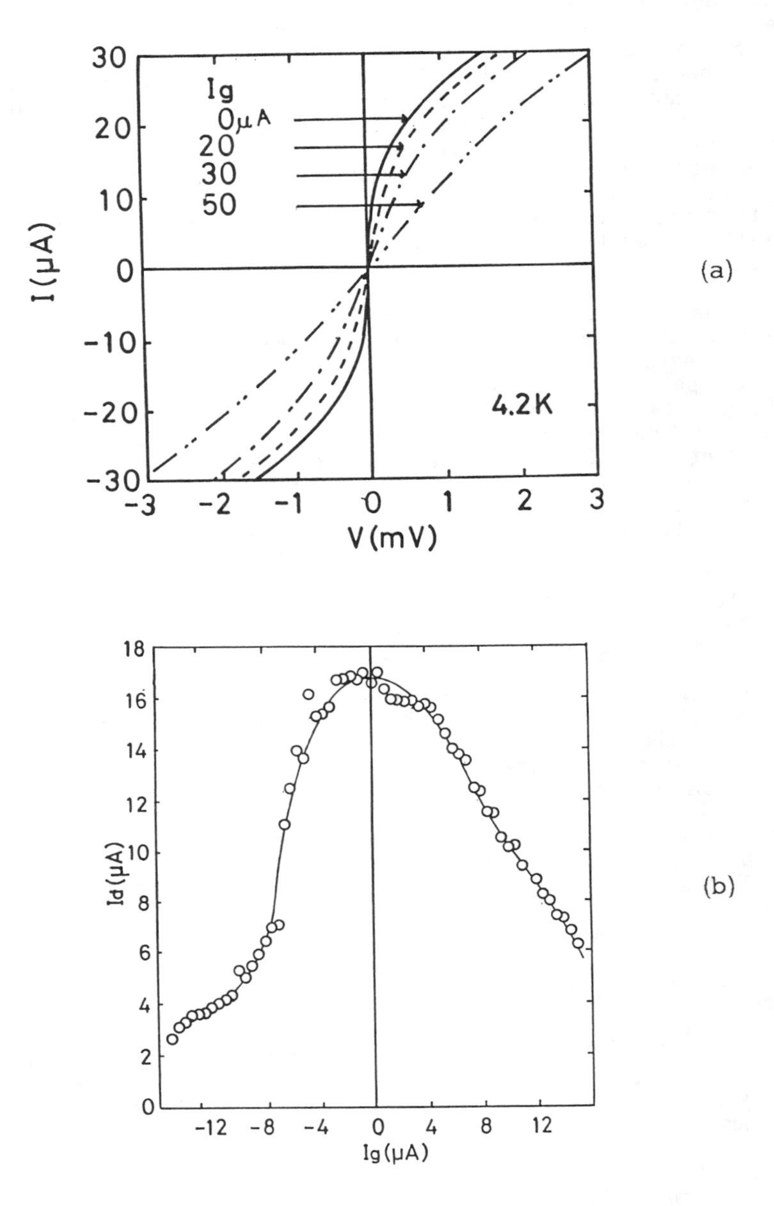

Fig.5 Current modulation characteristics of the Type-I device. Poly-crystalline film was used. (a) drain current-drain voltage characteristics, and (b) critical current modulation by the gate current.

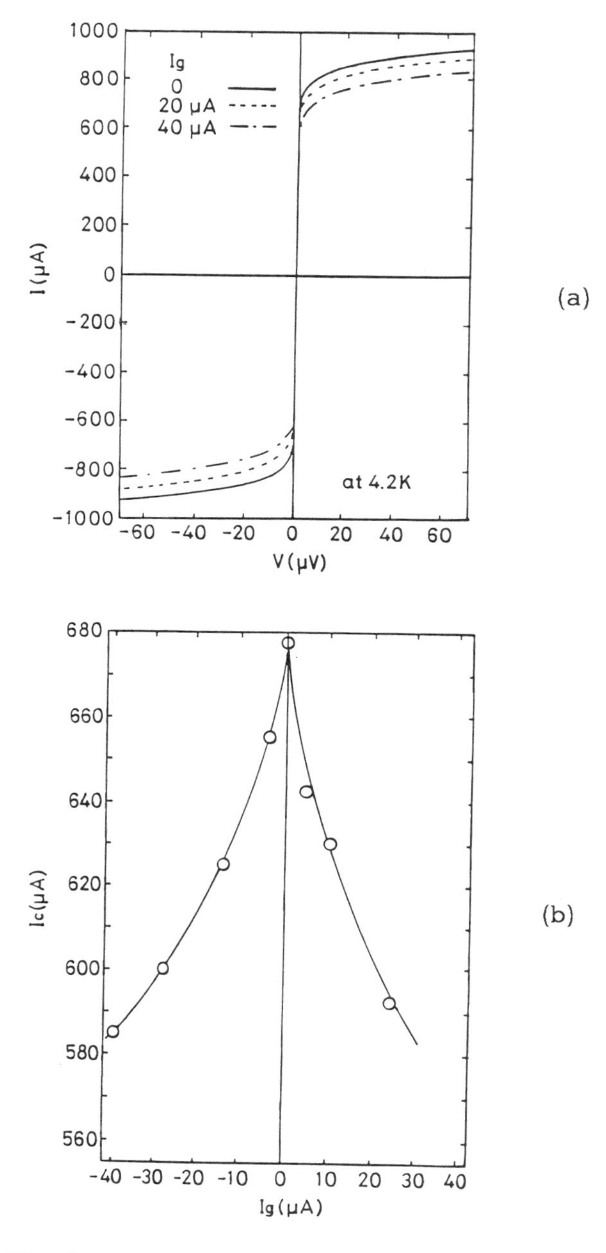

Fig.6 Current modulation characteristics of the Type-IIA device. Epitaxial film was used. (a) drain current-drain voltage characteristics, and (b) critical current modulation by the gate current.

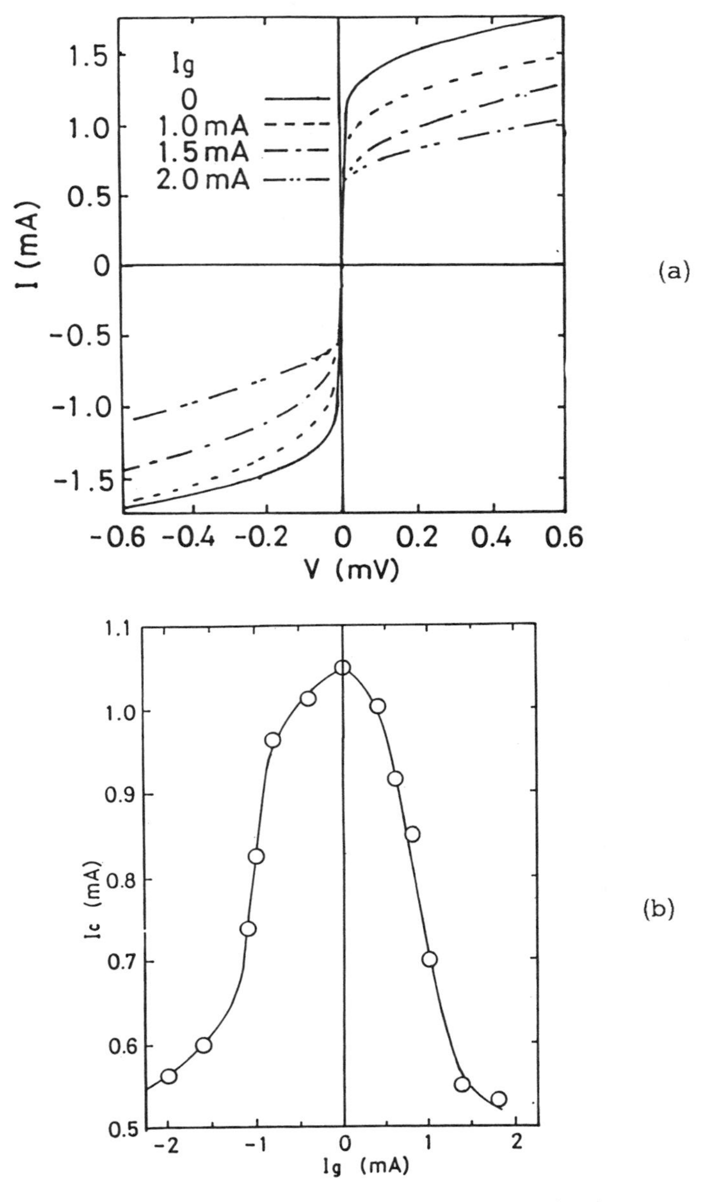

Fig.7 Current modulation characteristics of the Type-IIB device. Epitaxial film was used. (a) drain current-drain voltage characteristics, and (b) critical current modulation by the gate current.

is the very hopeful and realistic expectation. Thinning the channel may be an alternative of improving the current gain.

A glance of Fig.6(b) and 7(b) tells that the current modulation characteristics, particularly the shape of I_c vs I_g curve, depends to large extent on the operating temperatures. Both SCST's used for the measurement were prepared from the same kind of epitaxial wafers. Moreover, the specimens were completed by using the same process manner. However, the curve of Type-IIA is a steep hump like a cross-section of the sharp sword edge and, contrary to this, the Type-IIB exhibites a shape like a bell structure. This new question arising here may have a close corelation with the mechanism itself of the observed current modulation under the quasi-particle injection. What is the major role of the quasi-particle injection in the SCST? We would here propose three posibilities which can take part in the observed current modulation: the non-equilibrium superconductivity, a simple thermal effect, and the current induced magnetic field effect. The following supplemental experiments will give us some clue to understand the modulation mechanism.

A plenty of experimental data involving what are not shown here imply that the observed current modulation did not significantly depend on the injection power but rather on the current density at the gate electrode. From this fact, it is hard to say that the current modulation in the SCST comes from the simple heating effect. Second, no modulation was observed at all under the application of the magnetic field self-induced by the gate current flow. As to this verification, a sham examination has been done where the gate electrode was directly connected to the drain by the gold wire just on the device surface. The gate current, in this case, just served as the induction of the magnetic field, which, in turn, might affect the I-V characteristics of the channel. Contrary to our expectation, however, no meaningful change was brought about into the SCST via the self-induced magnetic field effect. The remainder, the non-equilibrium superconductivity, is now considered to be the most plausible mechanism of the current modulation in SCST's. In the following section, a detailed discussion is given and some aspects of the experiment are partly explained by the simple analysis.

POSSIBLE INTERPRETATION OF CURRENT MODULATION MECHANISM

As discussed above, a possible mechanism of the current modulation in the SCST device is the non-equilibrium superconductivity developed or retreted with respect to the gate injection current. If so, it is of great interest to know to what extent the tentative model based on the non-equilibrium superconductivity can explain the experimental results. This is the central question addressed in this section.

In Figs.6(b) and 7(b), we already showed the two types of modulation characteristics of the SCST. One is Type-IIA, having the I_c-I_g curve whose slope is becoming larger as I_g goes to zero. This modulation curve looks like a cross-section of the sharp wedge placed upside-down. The other, Type-IIB, has a shape of the bell: The slope of the modulation curve becomes null near the corner of $I_g = 0$. Although devices were prepared in a common way and with the same kind of epitaxial films, they offered the modulation characteristics different from each other. One thing which two experiments did not share in common was a temperature T/T_{c0}, where T and T_{c0} are the measurement temperature and the zero resistance temperature of the superconducting channel of the completed device, respectively. The values of T/T_{c0} were ~ 0.1 and ~ 0.8 for the Type-IIA and Type-IIB, respectively

At least at present, the knowledge on the mechanism of the high T_c superconductivity itself is still insufficient, much less the non-equilibrium superconductivity in the oxide material. This really implies the rigorous discussion is far distant from this text. However, as a first step, it is reasonably acceptable to employ the familiar BCS theory for the description of the relevant materials. This is because there exist many reports to date intimating that the superconducting behavior of the oxides has much similarity with the BCS superconductors from the phenomenological point of view. In this section, we show a trial calculation of the property of the three terminal device on the basis of the non-equilibrium BCS superconductivity and give some explanation to the observed current modulation characteristics.

Let us start with the μ^* model proposed by Owen and Scalapino for the description of the non-equilibrium superconductivity [16]. This model, based on the BCS gap equation, supposes an exsistence of the quasi-Fermi level μ^* according to an accumulation of the excess quasi-particle. Hence, the gap equation is given by

$$1 = N(0)V \int_0^{\hbar\omega_c} \frac{d\xi}{\sqrt{\xi^2+\Delta^2}}\left(1 - \frac{2}{1+\exp\{(\sqrt{\xi^2+\Delta^2}-\mu^*)/kT\}}\right), \quad (1)$$

where
- $N(0)$: density of states of electron at the Fermi level,
- V : interaction constant,
- Δ : gap energy,
- μ^* : quasi-Fermi level for quasi-particle
- ω_c : cut-off frequency of the atractive interaction between electrons

ω_c is regarded as Debye frequency in case of the usual metal superconductor. Whatever the interaction source might be, there should be a cut-off frequency as long as the strength of the interaction is finite. Equation (1) suggests that the gap parameter Δ is a function of the temperature T and qusi-Fermi level μ^*.

As to the number densities of the quasi-particle N_q and phonon N_ω under the injection of the quasi-particle I_g(cm^{-3}) are given by the following simultaneous equations [17]:

$$\frac{dN_q}{dt} = -RN_q^2 + \frac{2N_\omega}{\tau_B} + I_g, \quad (2a)$$

$$\frac{dN_\omega}{dt} = \frac{RN_q^2}{2} - \frac{N_\omega}{\tau_B} - \frac{N_\omega - N_{\omega t}}{\tau_{es}}, \quad (2b)$$

where N_q : number density of quasi-particle,
N_ω : number density of phonon with energy higher than 2Δ,
$N_{\omega t}$: number density of equilibrium phonon,
R : recombination constant of quasi-particles,
τ_B : phonon lifetime through pair breaking,
τ_{es}: phonon escape time.

In each expression in eqs.(2), the first term stands for the process where two quasi-particles form a Cooper pair and create a phonon, and the second term represents the reverse process. The third term in eq.(2b) denotes the phonon relaxation. In the steady state under the thermal equilibrium, by putting d/dt = I_g = 0 in eqs.2, we obtain

$$N\omega t = \frac{R\tau_B}{2} Nqt^2, \tag{3}$$

where N_{qt} is the number density of the quasi-particle at an equilibrium state. Using the above expression, we also obtain from eq.2(a)

$$Ig = R\frac{\tau_B}{\tau_B + \tau_{es}}(Nq^2 - Nqt^2), \tag{4}$$

N_q and N_{qt} are related to the gap equation as follows:

$$Nq = 2\sum_k \frac{1}{1 + \exp\{(\sqrt{\xi_k^2 + \Delta_k^2(\mu^*, T)} - \mu^*)/kT\}}$$

$$Nqt = 2\sum_k \frac{1}{1 + \exp\{\sqrt{\xi_k^2 + \Delta_k^2(\mu^* = 0, T)}/kT\}}$$

As for the description of the critical current I_c, we can roughly write down in the form of

$$Ic = LHc, \tag{5}$$

where L is a constant with a dimension of length. As well known, this formula is correct for the type-I superconductor with a cylindrical shape. Differed from this, the oxide superconductor $Y_1Ba_2Cu_3O_y$ belongs to the type-II groupe, and therefore, eq.(5) is not adequate in the strict sense. If the flux pinning is weak enough or strong, however, the lower and upper critical field H_{c1} and H_{c2} can reasonably play the role in place of H_c in eq.(5), respectively, even for the type-II superconductors. A relation between the H_c and the free energy is written as

$$\frac{Hc^2}{8\pi} = Fn - Fs, \tag{6}$$

where F_n and F_s are the free energy of the normal and superconducting states, respectively. Using eqs.(5) and (6), the critical current is rewritten by

$$Ic = L\sqrt{8\pi(Fn-Fs)} \tag{7}$$

The free energy F_s is generally given by

$$\begin{aligned} Fs &= U-TS \\ &= \sum_k \left(|\xi_k| - \frac{\xi_k^2}{E_k} + \frac{2\xi_k^2}{E_k} f_k\right) - \sum_k \frac{\Delta_k^2}{2E_k}(1-2f_k) \\ &\quad +2kT\sum_k \{f_k \ln(f_k) + (1-f_k)\ln(1-f_k)\} \end{aligned}$$

where

$$f_k = \frac{1}{1+\exp\{(E_k-\mu^*)/kT\}}$$

$$E_k = \sqrt{\xi_k^2+\Delta_k^2}$$

and F_n can be easily obtained by putting $\mu^*= 0$ and $\Delta = 0$ in the above equations.

From eqs.(1), (4) and (7), the modulation characteristics of I_c against the quasi-particle injection I_g is estimated. The numerically calculated results of I_c vs I_g characteristics are depicted in Fig.8 as a parameter of the reduced temperature T/T_c. In this figure, I_c is normalized by the value of I_{c0}, the critical current under no injection current at each temperature, and I_g is normalized by I_{g0}, the minimum value of I_g which gives rise to complete suppression of $I_c (= 0)$. It is quite obvious in this figure that the steeper slope of the curve exsists at the region of lower I_g when the device works at lower reduced temperature. When the temperature becomes closer to the critical temperature, the figure shows the disappearance of the steep slop portion in the modulation curve at $I_g \sim 0$. Though it can not still surpass the qualitative discussion, these features coincides well with the observed two kinds of modulation characteristics of the SCST devices. In order to meet the quantitative aspects, further research in both the experimental and theoretical sides is required. It is a future subject. Anyway, the qualitative agreement between the experimental and theoretical results strongly intimates that the mechanism of the current modulation in the SCST is the enhanced non-equilibrium superconductivity. The BCS theory surprisingly well explained the experiments at least in the appearance. It must be noted, however, that the present work does not imply the importance of phonon coupling as a main mechanism of the high T_c superconductivity. All we can say at this stage is that the non-equilibrium superconductivity may take place in the oxide superconductor as well as in the metal system and our SCST devices work via manipulation of the non-equilibrium superconductivity.

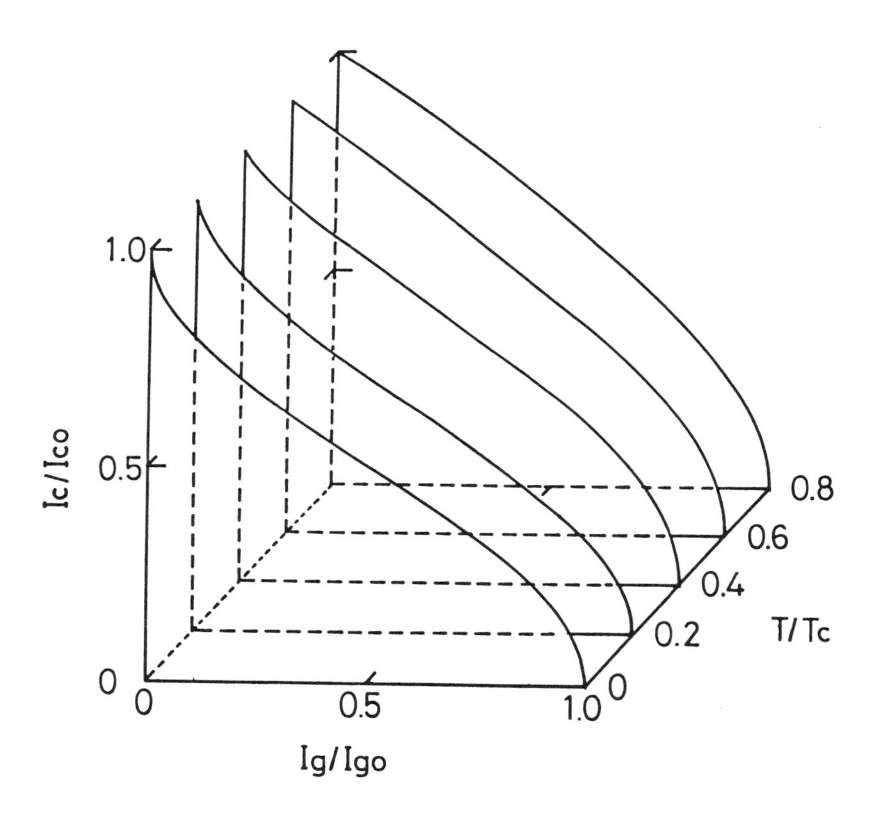

Fig.8 Calculated modulation characteristics of the three terminal device. The non-equilibrium superconductivity based on the BCS model was employed.

ACKNOWLEDGMENTS

The authors would like to thank Mr. Hashimoto and Dr. Tonouchi for the experimental work and discussions. This work was partly supported by Science Research Grant-in-Aid from Ministry of Education, Science and Calture of Japan.

REFERENCES

[1] J. B. Bednorz and K. A. Müller, Z. Phys. B64 (1986) 189.
[2] M. K. Wu, J. R. Ashburn, C. J. Torng, P. H. Hor, R. L. Meng, L. Gao, Z. J. Huang, Y. Q. Wang, and C. W. Chu, Phys. Rev. Lett., 58 (1987) 908.
[3] H. Maeda, Y. Tanaka, M. Fukutomi and T. Asano, Jpn. J. Appl. Phys.,27 (1988) L209.
[4] Z. Z. Cheng, A. M. Hermann, A. EI Ali, C. Almasan, J. Estrada, T. Datta and R. J. Matson, Phys. Rev. Lett., 60 (1988) 937.
[5] S. Kita, H. Tanabe and T. Kobayashi, Appl. Supercon. Conf. Abstract ED-14 (1988, San Fransisco).
[6] G. B. Donaldson, Extended Abstract of Int. Supercon. Electron. Conf. 10 (1987, Tokyo).
[7] T. D. Clark, R. J. Prace and A. D. C. Grassie, J. Appl. Phys.,51 (1980) 2736.
[8] S. M. Faris, S. I. Raider, W. J. Gallagher and R. E. Drake, IEEE Trans. Magn., MAG-18 (1983) 1293.
[9] D. J. Frank, M. J. Bradt and A. D. Davidson, IEEE Trans. Magn., MAG-21 (1985) 721.
[10] T. Kobayashi, H. Sakai and M. Tonouchi, Electron. Lett., 22 (1986) 659.
[11] M. Tonouchi, H. Sakai and T. Kobayashi, Jpn. J. Appl. Phys., 25(1986) 705.
[12] S. Sakai and H. Tateno, Jpn. J. Appl. phys., 21-1(Supplement) (1981) 331.
[13] T. Kobayashi, This Issue (Chapter: MULTIPLE HETEROEPITAXIAL GROWTH OF HIGH T_c SUPERCONDUCTORS) *
[14] T. Kobayashi, K. Hashimoto, U. Kabasawa and M. Tonouchi, Appl. Supercon. Conf. (1988, San Fransisco)
[15] Y. Yoshizako, M. Tonouchi and T. Kobayashi, Jpn. J. Appl. Phys., 26 (1987) L1533.
[16] C. S. Owen and D. J. Scalapino, Phys. Rev. Lett., 28 (1972) 1559.
[17] A. Rothwarf and B. N. Taylor, Phys. Rev. Lett, 19 (1967) 27.

* see Volume-2

TWINS IN HIGH-Tc YBa2Cu3O7- SUPERCONDUCTORS

C.J. Jou and J. Washburn

Department of Materials Science and Mineral Engineering
University of California
and
Center for Advanced Materials
Materials and Chemical Sciences Division
Lawrence Berkeley Laboratory
1 Cyclotron Road
Berkeley, CA 94720

I. INTRODUCTION

Since discovery of the La-Ba-Cu-O (LBCO) system of high-T_c oxide superconductors with $T_c \sim 28°$ K in 1986 [1], a world-wide search for other oxide superconductors has resulted in the discovery of the Y-Ba-Cu-O (YBCO) system with $T_c \sim 93°$ K in 1987 [2], two new classes in 1988, i.e., the Bi-Ca-Sr-Cu-O (BCS) system [3] and the Tl-Ca-Ba-Cu-O (TCB) systems [4], with the highest reported $T_c \approx 110°$ K and 125° K respectively. Also, a perovskite-related Ba-K-Bi-O (BKBO) system [5] with $T_c \sim 30°$ K became the first known high-T_c oxide superconductor in the bismuth oxide family. The discovery of each class provided a new challenge to the fundamental understanding of superconductivity. For example, the discovery of the YBCO system initiated a race to find a superconducting mechanism related to the oxygen-ordered Cu-O chain structure. When the BCS and TCB systems were discovered, which did not contain the linear chain structure of the YBCO system but had even higher T_c, emphasis shifted to the CuO_2 layers that were common to all the high T_c copper oxide superconductors. All except BKBO share a common pseudo-two-dimensional structure with CuO_2 planes sandwiched between metallic layers. This structure appears to play a critical role for the observed superconductivity. In all of these materials the structure can be described by stacked perovskite-related unit cells. They all show bad metallic properties above T_c and are all oxygen-deficient because of doping with randomly distributed dopants having different charges, e.g., A^{II} in $(La^{III}_{1-x} A^{II}_x)_2 CuO_{4-\delta}$ with A = Ba, Sr or Ca and K^I in $(Ba^{II}_{0.6} K^{I}_{0.4})BiO_{3-\delta}$. $La_2^{III} CuO_4$ and $Ba^{II}BiO_3$ are insulators. For the YBCO, BCS and TCB systems, Cu^{III}, Bi^{III} and Tl^{III} can be considered as substitutional dopants in the three-dimensional Cu^{II} lattice which result in randomly distributed oxygen vacancies in the capping layers, i.e., $Cu^{III}O$ in $YBa_2Cu_3O_{7-\delta}$, $Bi_2^{III}O_2$ in BCS and $Tl_2^{III}O_2$ in TCB. The result of this doping is that the uniformity of the oxidation state (III) of cations in the capping layer(s), which is also a common feature in copper oxide superconductors, is destroyed. A defected macro-resonance-cell model [6] has been proposed to qualitatively discuss superconductivity in these oxides. The cations in the capping layers all have in common various oxidation states with charge difference two, e.g., Tl with I and III, Bi with III and V, Pb with II and IV and Cu with I, II and III.

Among all the classes of high-T_c oxide superconductors, the YBCO system has the most unusual features in its structure. The cation in its capping layer, CuO, is the same as that of the CuO_2 layers. Hence, the Cu cations in this structure form a three-dimensional frame similar to the Bi frame in the BKBO system. The stacking sequence of the YBCO system has only a single period and single boundary layer, unlike the other systems, all of which have a double period with phase shift as the stacking sequence is followed through the center of the double boundary layers (see Table I.1). Also, oxygen atoms in the capping layer of the YBCO system are ordered forming a linear chain structure on the basal plane. This oxygen (vacancy) ordering mechanism results in an orthorhombic structure which in turn results in a finely twinned microstructure which is formed when the structure changes from tetragonal to orthorhombic on cooling. However, this oxygen ordering can only take place when the oxygen content, x, in $YBa_2Cu_3O_x$, is larger than about 6.5. For x below 6.5, oxygen atoms on the basal plane are randomly distributed resulting in a tetragonal structure which is nonsuperconducting. For the superconducting orthorhombic structure, the critical temperature, T_c, is strongly dependent on the oxygen content [7], and therefore on the amount and the distribution of the oxygen vacancies. In this chapter, a twin formation mechanism is discussed and an oxygen-depleted twin boundary model is proposed that may explain the observed difficulty of reaching the stoichiometric composition, x = 7, in orthorhombic YBCO.

Table I.1. Stacking sequence of layers in unit cell.

$YBa_2Cu_3O_7$ (n=2)

Ba		Y		Ba		← center
	O_2		O_2		O	← edge
O	Cu		Cu	O	Cu	← corner

$Bi_2CaSr_2Cu_2O_8$ (n=2)

							‖							
O	Sr		Ca		Sr	O	‖	Bi	O	Cu		Cu	O	Bi
		O_2		O_2			‖			O_2		O_2		
Bi	O	Cu		Cu	O	Bi	‖	O	Sr		Ca		Sr	O

$Tl_2CaBa_2Cu_2O_8$ (n=2)

							‖							
O	Ba		Ca		Ba	O	‖	Tl	O	Cu		Cu	O	Tl
		O_2		O_2			‖			O_2		O_2		
Tl	O	Cu		Cu	O	Tl	‖	O	Ba		Ca		Ba	O

La_2CuO_4 (n=1)

			‖			
O	Cu	O	‖	La		La
	O_2		‖		O_2	
La		La	‖	O	Cu	O

boundary ← → layer(s)

II. FORMATION OF COHERENT TWINS IN $YBa_2Cu_3O_{7-\delta}$ SUPERCONDUCTORS

The 1-2-3 structure of the $YBa_2Cu_3O_x$ system, $6 \leq x \leq 7$, has a unique property—the so called "oxygen sponge," i.e., when cooling from the sintering temperature, T_s, usually around 950°C, oxygen content increases monotonically (Fig. 1a) and when heating up, oxygen content decreases. The oxygen disorder-order induced phase transition from the semiconducting tetragonal structure to the superconducting orthorhombic structure is accompanied by the formation of coherent twins at

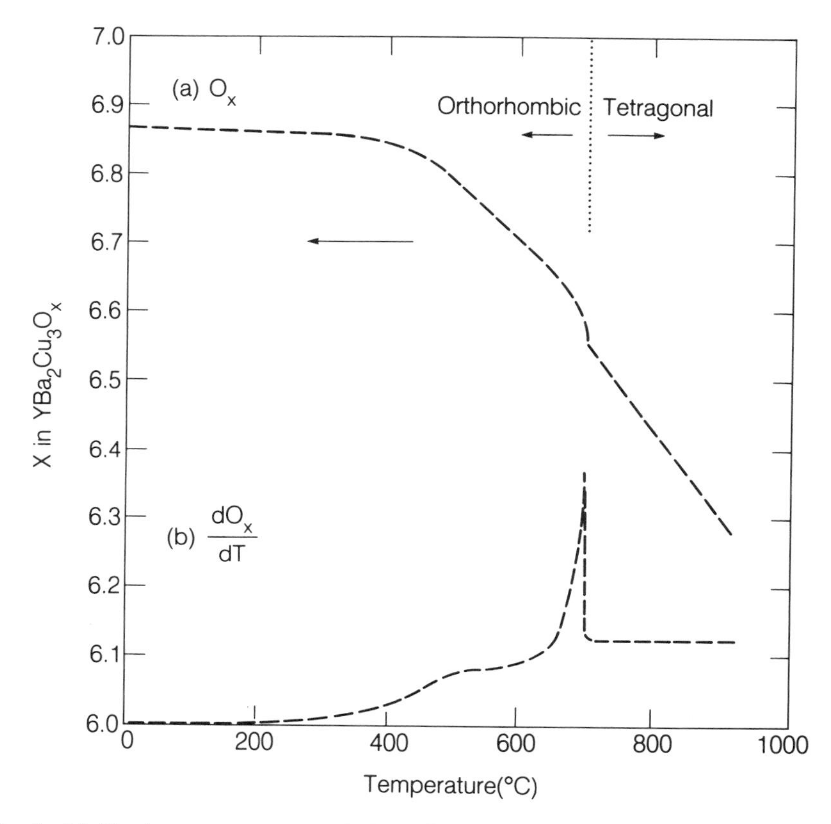

Fig. 1. (a) Total oxygen content, x, in $YBa_2Cu_3O_x$ increases monotonically when cooled slowly from sintering temperature in 1 atm O_2 ambient. Note there is a kink at the structure transition temperature T_p. Data deduced from Ref. 1. Scale of O_x is for (a) only. (b) The differentiation of the curve in (a) reveals a peak at T_p forming the characteristic "λ" curve of a second order phase transformation. Oxygen ordering on the basal plane is in progress starting at T_p.

temperature T_p and oxygen content $(O_x)_p$. Below T_p, the structure is orthorhombic with oxygen content $O_x > (O_x)_p$ when cooled slowly in an oxygen-rich ambient from the sintering temperature, T_s. Above T_p, the structure is tetragonal with $O_x < (O_x)_p$. The rate of oxygen uptake increases sharply at T_p as shown in Fig. 1b which has the characteristic "λ" shape of a second order phase transition. Below 650°C it drops back to about the same rate as just before T_p, and with further cooling decreases gradually until by about 300°C no further oxygen uptake occurs. Both T_p and $(O_x)_p$ are a function of the oxygen partial pressure, P_{O_2}, in the ambient under equilibrium conditions, e.g., for P_{O_2} = 1 atm (in 100% pure oxygen ambient), $T_p \approx 700°C$, for P_{O_2} = 0.2 atm (in air or 20% O_2–80% Ar ambient), $T_p \approx 670°C$, and for $P_{O_2} \approx 0.02$ atm (in vacuum or 2% O_2–98% Ar), $T_p \approx 620°C$, and $(O_x)_p$ in this partial pressure range is about 6.5 ~ 6.6 [8]; when P_{O_2} decreases both T_p and $(O_x)_p$ decrease.

Based on this information and in-situ transmission electron microscope (TEM) observations [9-12], a model describing the mechanism of formation of coherent twins is presented which can help to explain some of the observed experimental results. In the slow cooling step, which is essential for good quality high T_c material, the cooling rate, dT/dt, is small and usually constant. The changing oxygen content dO_x/dt can thus also be considered as dO_x/dT. Although the diffusivity of oxygen in the material also changes as temperature decreases, for slow cooling rates and in not too low a temperature range it can be assumed that equilibrium is approached at least in porous or thin specimens. As seen in Fig. 1b, $dO_x/dT \approx dO_x/dt$ is the order parameter for this phase transition.

1. Nucleation

As revealed by in-situ neutron powder diffraction [8], at temperature just above T_p, i.e., the tetragonal structure, O(4) and O(5) sites are randomly occupied by equal amounts of oxygen with an average occupancy of each site about 1/4. The O(1) sites on the BaO layer, O(2) and O(3) sites on the CuO_2 layer are all fully occupied. At T_p the occupancy of the (0, 1/2, 0), O(4), b-chain sites on the CuO basal plane layer, starts increasing at a faster rate than the decrease of occupancy on the (1/2, 0, 0), O(5), sites on the same layer. This suggests that the increased uptake of oxygen from the ambient goes primarily to the b-chain sites and an oxygen ordering process is in progress. The overall oxygen content increases at a rate much larger than just above T_p (Fig. 1b) indicating incorporation of oxygen from the ambient into the material is facilitated by the structural transformation. It is suggested here that this sudden increase of oxygen absorption rate is associated with a nucleation process, the formation of embryos of the orthorhombic phase.

Since the oxygen-ordered orthorhombic phase contains the linear chain structure on the basal plane, these embryos are assumed to consist of clusters of short parallel b-chains, i.e., Cu(1)-O(4)-Cu(1). They should first appear at the heterogeneous sites located at grain surfaces, i.e., grain boundaries and pore surfaces, since the internal stress due to the localized distortion associated with b-chain growth in the tetragonal matrix can be at least partially relaxed (Fig. 2). The formation of b-chain embryos at grain boundaries probably first develops by consuming oxygen in the O(4)-O(5) basal plane which would soon become depleted near the grain boundaries. Further growth of the orthorhombic ordered phase then would require a supply of oxygen from the ambient through the pore-grain boundary network (Fig. 2). Clusters of b-chains with b directions orthogonal to each other are equally probable because of the symmetry of the parent tetragonal structure. Furthermore, for stress compensation nucleation of an embryo with one b-chain direction tends to induce the nucleation of another with an orthogonal b-chain direction. Below T_p the oxygen-ordered configuration on O(4)-O(5) basal plane has a lower free energy than that of the oxygen-disordered configuration [13]. Once initiated at a grain surface the extension of b-chains should be rapid. An elongated embryo with an oxygen-depleted zone around it is expected.

In powder-sintered materials, depending on the pressing process and the sizes and kinds of powders, there are many voids and open channels among grains. For high porosity material, the total internal pore surface area may be larger than the total area of the grain boundaries. This free surface area may play a significant role in determining the physical properties of such a material. Grain boundaries also permit faster diffusion than that possible through a grain because of the looser structure. Significant diffusion of oxygen within a grain along the c-axis direction is unlikely to happen at temperatures below T_p. Oxygen from the environment and from pore surfaces is likely to be distributed to the grains deep inside the sample primarily through this pore-grain boundary network. To diffuse into the grain, oxygen probably has to find a basal plane edge. Hence, grain boundary diffusion would often be required for supplying oxygen from the ambient to parts of the grains.

As b-chain clusters at grain boundaries grow, embryos probably also develop homogeneously or heterogeneously inside the grain by local oxygen ordering. They are probably formed somewhat later than those at grain surfaces, since the internal stress developed by the b-chain ordering can not be partially relaxed as at grain surfaces and the incorporation of the limited available oxygen is

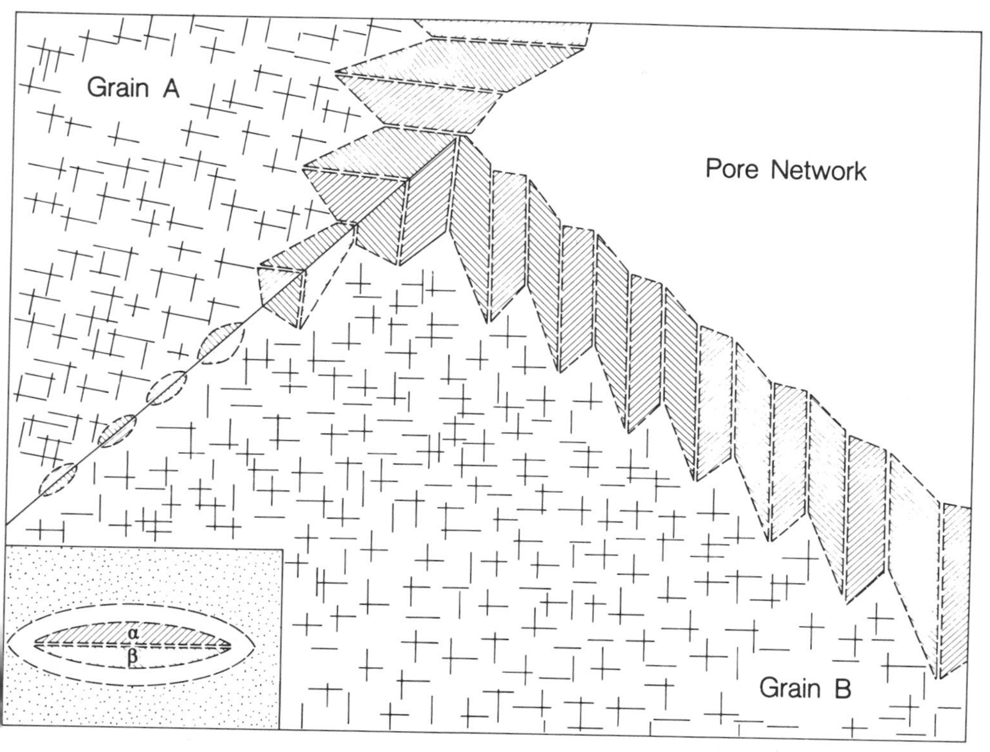

Fig. 2. Proposed nucleation of ordered oxygen clusters starts at grain surfaces, i.e., grain boundaries or pore surfaces, where the stress due to structural transformation can be partially relaxed. When two neighboring clusters with orthogonal b-chain directions meet, coherent twin boundaries with an oxygen-depleted zone are formed. Twin nuclei (insert at lower left corner) are formed later in the interior of the grain. They consume most of the available oxygen content in the grain at T_p forming a heavily interpenetrated tweed structure (Fig. 3a). An oxygen concentration gradient between the grain surface and the grain interior is set up. Further growth of the twin laths in the grain is limited by diffusion of oxygen into the grain center. The twin nuclei at the grain surfaces can grow faster with the oxygen supplied directly from ambient via the pore-grain boundary network and the oxygen-depleted twin boundaries. During the growth, twin laths in the grain interior are incorporated at the growth front.

ompetitive among embryos in the central region of the grain. The available oxygen would be uickly consumed by the formation of such nuclei. Oxygen-depleted zones would be formed round these nuclei setting up a concentration gradient between center and surface of the grains, ausing more oxygen to diffuse into the grain. These b-chain clusters have locally the orthorhombic ttice structure with the lattice constants, a and b on the basal plane, smaller and larger, respec-vely, than a_t of the parent tetragonal lattice. The tetragonal matrix surrounding the elongated ucleus experiences contraction at the ends of the b-chains and expansion at the sides parallel to the -chains. When two clusters with orthogonal b-chain directions approach each other, the oxygen epleted zones and the strain fields overlap causing a reduction of the stress along one <110> direction and enhancement along the orthogonal <110> direction; impingement of the two domains along the latter <110> direction eventually results in formation of a coherent twin boundary. The equilibrium width of the twin boundary zone is assumed in this model to be determined by a balance between the coulomb repulsion force of oxygen ions at opposite sides of the boundary and the chemical potential favoring extension of the b-chains. A "twin nucleus" within the grain with a narrow oxygen-depleted zone remaining at the twin boundary is expected (Fig. 2, left lower corner).

Either <110> direction on the basal plane can lie parallel to a twin boundary, i.e., the twin nuclei have two equally probable twin boundary orientations perpendicular to one another. These nuclei have been observed by in-situ TEM as randomly distributed and inter-penetrated orthogonal sets, a tweed-like structure (Fig. 3) [14-16]. Other combinations of b-chain nuclei, quartet, etc., are also possible, but with smaller probability than doublet. When two b-chain clusters with the same b-chain direction approach each other, coalescence takes place resulting in a larger b-chain cluster.

Since oxygen content (O_{x_p}) is about 6.5 at T_p, only half of the available b-chain sites of the orthorhombic phase in the grain are occupied by oxygen, i.e., the total volume of the orthorhombic twin nuclei is only about half the total volume of the grain. This leaves a relatively open structure for oxygen diffusion. Further transformation requires oxygen diffusion into the grain, the driving force being the tendency for further extension of b-chains, resulting in the observed sudden rise in rate of oxygen uptake at T_p as shown in Fig. 1b.

2. Growth

The b-chain clusters nucleated at grain surfaces grow most rapidly by incorporating oxygen readily supplied from the ambient. When the transformation is complete at grain surfaces, the diffusion of oxygen into the grain becomes more restricted, i.e., oxygen-depleted twin boundary zones.

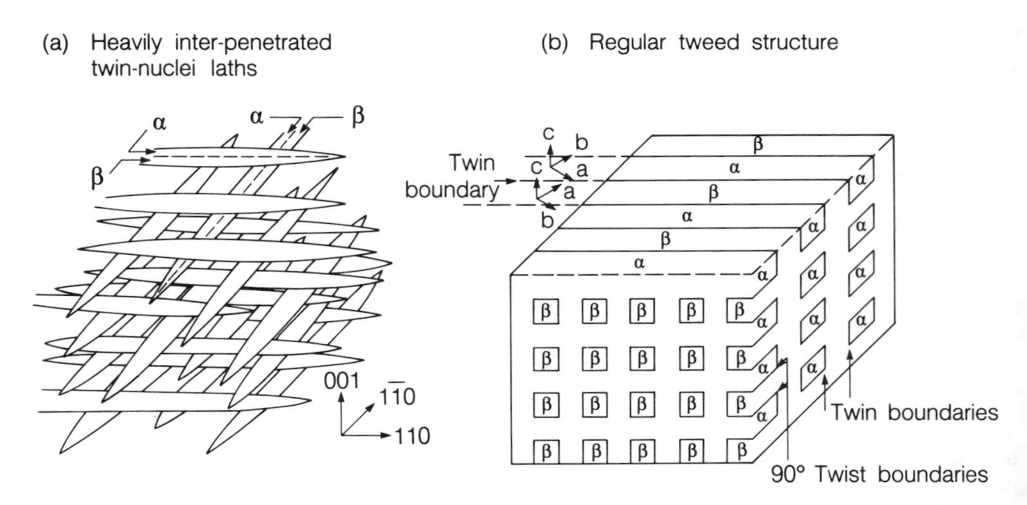

Fig. 3. (a) Interpenetrated twin laths leave the untransformed matrix oxygen-depleted inside the grain. (b) A fully grown tweed structure would form a two-dimensional mutually-modulated regular pattern, possibly observed in some in-situ transmission electron microscopy observations during cooling from high temperature. The electron diffraction pattern of this structure may reveal an effective unit cell with $\sqrt{2}a \times \sqrt{2}a \times \sim 3a$. However, the structure contains both twin boundaries and high energy 90°-rotation boundaries, therefore it would be a metastable intermediate product.

This can explain the sharp decrease of the rate of oxygen uptake below T_p (see Fig. 1b). The growth front of the twinned orthorhombic structure nucleated at grain surfaces gradually extends deeper into the grain by adding new b chains at the growth front with oxygen being supplied from ambient via the pore-grain boundary twin boundray network and by incorporating twin nuclei that have already formed inside the grain.

3. Coarsening

During the last stage of twin formation, coarsening takes place eliminating some of the twin boundaries, and thus decreasing total free energy. Small domains included within larger domains tend to disappear by localized oxygen diffusion within the boundary. As long as the temperature is still high enough to permit oxygen diffusion, this climbing process continues to coarsen the twin domains until the widths of the twins are several hundred Å. Rapid cooling severely inhibits this coarsening process; the resulting structure contains more numerous fine twins. When cooling rate is very rapid, e.g., quenching from temperature above T_p, the tetragonal structure is partially preserved with oxygen content less than 6.5. There is insufficient time for oxygen absorption and thus insufficient oxygen to complete the growth of the b-chain orthorhombic structure. Quenching from temperature between ~350°C and T_p yields intermediate T_c with oxygen contents between 6.8 and 6.6. Sintered polycrystalline $YBa_2Cu_3O_{7-\delta}$ is usually given a low temperature annealing step to enhance its superconducting properties. The annealing temperature is typically chosen below T_p but above 400°C in an oxygen-rich environment to permit further coarsening, stress relaxation and oxygen uptake.

III. MODELS AND SIMULATIONS OF OXYGEN-DEPLETED TWIN BOUNDARIES

1. Models

A model of twin boundaries in orthorhombic $YBa_2Cu_3O_x$ superconductors with a number of layers (j) containing oxygen vacancies at twin boundaries [17] (Fig. 4) avoids bringing oxygen neighbors too close together while preserving coherency at twin boundaries. The oxygen-depleted

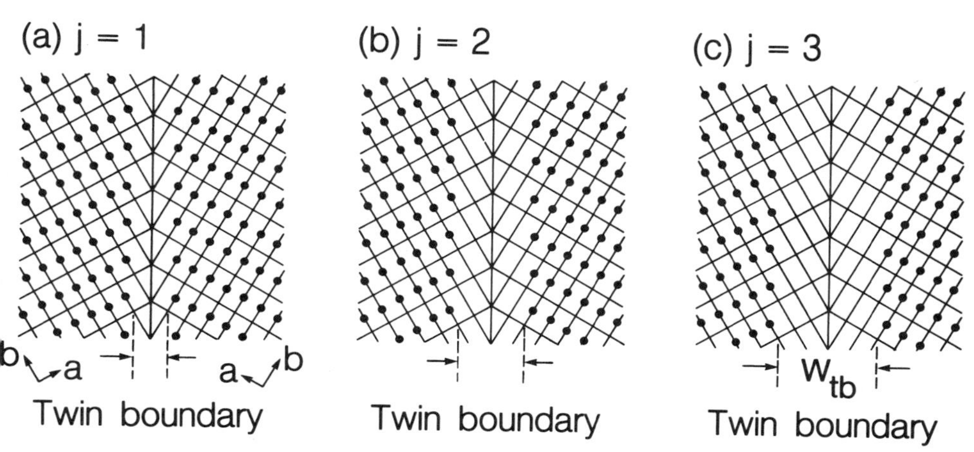

Fig. 4. Oxygen vacant twin boundary models for (a) 1 vacant layer (j=1), (b) 3 vacant layers (j=2), and (c) 5 vacant layers (j=3). Note: The angle at twin boundaries and the length difference between the lattice constants (a) and (b) are exaggerated. For simplicity, only oxygen atoms (black dots) are shown.

twin boundary width can be expressed as $W_{tb}(j) = 2j\, ab/(a^2 + b^2)^{1/2}$. Taking the estimated averag twin width [18] as: $W_t \approx ab/(b - a)$ and assuming a perfect stoichiometric $YBa_2Cu_3O_7$ within th twin domains and $YBa_2Cu_3O_6$ at twin boundaries, then the average oxygen content, x, i $YBa_2Cu_3O_x$, which would be experimentally measured, is:

$$x(j) = 6 + \frac{W_t - W_{tb}(j)}{W_t} = 7 - \frac{W_{tb}(j)}{W_t} = 7 - \frac{2j\,(b - a)}{\sqrt{a^2 + b^2}}$$

Applying a = 3.822 Å, b = 3.891 Å from the work of Cava et al. [7], the calculated values o x(j) match the experimentally measured oxygen content very well (Table III.1) suggesting that i their samples, near perfect stoichiometric $YBa_2Cu_3O_7$ may form within the twin domains but a fe layers of oxygen depletion exist at twin boundaries. The oxygen deficiency study [22] seems t indicate that under normal pressure processing, i.e., 1 atm O_2 or air, the ultimate oxyge stoichiometry is about 6.93 or equivalent to the case j = 3. However, there must always be som vacancies dispersed within the twin domains at random or in an ordered structure [29]. Thus th number of oxygen vacant layers at twin boundaries may be smaller than that calculated.

Table III.1

j	1	2	3	4	5	6	7	8	9
x(j)	6.975	6.949	6.924	6.899	6.873	6.848	6.823	6.798	6.772
x_{exp}	6.975[7] 6.97[19]	6.94	6.92 ± 0.01[21] 6.92[19,22]	(~6.9)	6.87	6.85	6.82	6.80	6.78
Ref.	7,19	20	19,21,22	23	24	25–27	28	19,26	7
T_c(°K)	90[7]	89	95[22]	87	90	91[25]	91	NA	83

2. Simulations

Simulated high resolution electron microscope (HREM) images of twin boundaries with vari ous j values, different defocussing conditions and different sample thicknesses are shown i Fig. 5a-c respectively. The simulated images of the twin boundaries assuming no oxygen vacancie [(Fig. 5a) top] show no contrast across the twin boundaries regardless of the objective lens defocu and sample thickness. However, when a few layers of oxygen at twin boundaries are taken away contrast across twin boundaries appears. The width of the contrast band increases in proportion to as shown in Fig. 5a. Contrast at the twin boundaries also varies markedly with defocussing condi tions as shown in Fig. 5b. Note that the apparent boundary widths also vary slightly. Contrast a twin boundaries due to sample thickness variation, as shown in Fig. 5c, is equally dramatic. It ca vary from almost no contrast to strong contrast. Various band widths and different contrast at twi boundaries in real HREM images are commonly observed [14,30-32] in agreement with the mode proposed here (Fig. 6). Also, twin boundaries exposed to a focussed beam for a short time, e.g., minute, show a decrease in contrast [10]. Oxygen atoms in twin domains close to the twin boun daries may gain energy through electron bombardment to jump into the twin boundary vacancies Thus, the contrast might be expected to smear out as illustrated by the simulated images in Fig. 5d.

3. Discussion

A study by Chen et al. [16] shows that the ordered vacancies within the body of the twi domains accounts for only 3% of the total volume in a $YBa_2Cu_3O_{6.72}$ sample. This discrepanc

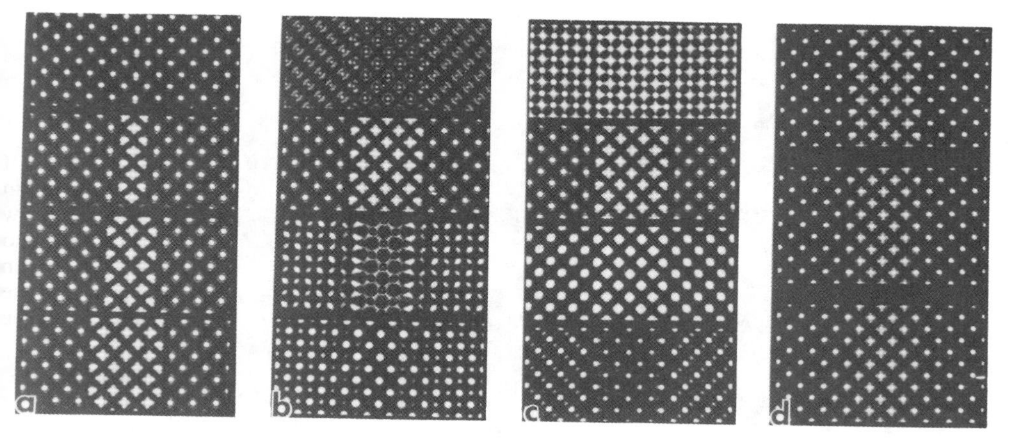

Fig. 5. Simulated HREM [001] images of twin boundaries. Simulation parameters are chosen to correspond to the operating conditions in the JEOL 4000 FX electron microscope. Sample thickness is 100Å and objective lens defocus is –800Å unless specified otherwise. (a) Changes with respect to j from j = 0 (no oxygen depleted layer, top) to j = 3 (bottom). (b) Variation with respect to objective lens defocus; –400Å (top) to –1600Å (bottom) in increments of –400Å, j = 3. (c) Variation with respect to sample thickness; 50Å (top) to 200Å (bottom) in increments of 50Å, j = 3. (d) Total oxygen atom number fixed, but abrupt concentration changes at twin boundaries are gradually relaxed (j = 3).

could be explained by the oxygen-depleted twin boundary model with $j \approx 3$. The high resolution image shown in Fig. 6 also matches the simulated images using the model proposed here with $j \approx 3$.

Since the equilibrium width of the oxygen-depleted zone at a twin boundary is determined by the balance of the Coulomb repulsion force and the chemical potential, the resulting optimized configuration may have at least one layer of oxygen vacancies condensed at the twin boundary (j = 1 case) where the repulsion energy is the highest. For $j \geq 2$, the repulsion energy decreases inversely proportional to the oxygen ion distance resulting in increasingly diffuse oxygen-depleted layers, i.e., oxygen-vacancy segregation at these layers is probably incomplete. For simplicity, however, this diffuse atmosphere of oxygen vacancies has been described as a number of layers of oxygen vacancies, j.

IV. EFFECTS OF OXYGEN-DEPLETED TWIN BOUNDARIES ON SUPERCONDUCTING PROPERTIES

A defected macro-resonance-cell (MRC) model for high-T_c oxide superconductors [6] based on their common features and minimum energy principles suggests that copper in its high oxidation state may dissociate into the fundamental, i.e., the lowest, oxidation state and contribute hole(s) as charge carriers. In the normal conducting (NC) state of $YBa_2Cu_3O_7$, each Cu on the CuO basal plane contributes two holes ($Cu^{+3} \rightarrow Cu^{+1} + 2h$) for conduction and Cu on the CuO_2 planes contributes one hole ($Cu^{+2} \rightarrow Cu^{+1} + h$). In the MRC model, these holes form charge carrier pairs and are delocalized. However, the coupling between CuO and CuO_2 layers is weak in the NC state and the propagation directions of the holes are random except in the presence of an applied field. In the superconducting (SC) state, the oxidation states of copper may oscillate between I and III forming a

Fig. 6. High resolution electron micrograph of a twin boundary in $YBa_2Cu_3O_x$ material reveals significant contrast across the twin boundary. (Photo courtesy of Dr. K. Hiraga.) Simulated image of an oxygen-depleted twin boundary using model proposed here with j = 3.

three-dimensional resonating network and the coupling between CuO and CuO_2 layers is strong. Free pairs of holes could condense into coherent superconducting pairs. In the static SC mode, i.e. no field applied, there is no net current. In the dynamic SC mode, hole pairs would be driven by the applied field resulting in a net current. In order to permit hole pairs propagation, absence of some holes on the CuO plane, yielding imperfect resonance locally, is essential. This may be realized in several ways, e.g., a small number of O4 jumping to O5 sites, residual oxygen vacancies located on O4 site, doping with elements with different charges, etc. All these defects can cause unsynchronized resonance locally delocalizing hole pairs and shifting the nonconducting (insulating) state into the conducting state. One type of site where oxygen is expected to be deficient is at the twin boundaries if oxygen is depleted on the CuO layer. The CuO_2 layers, on the other hand, already have empty overlapped-orbitals (Cu : $3d_{x^2-y^2}$ and O : 2p) since $Cu^{+2} \rightarrow Cu^{+1} + h$ contributes only one hole. CuO_2 layers must be coupled to the CuO layer to maintain a macro resonance state of the whole lattice.

1. SIS Josephson Junction and Glassy States

Oxygen-depleted twin boundaries with stoichiometric $YBa_2Cu_3O_6$ have CuO_2 planes but are missing the Cu-O chains. Cu at the twin boundaries is in the fundamental oxidation state, Cu^I. Thus the resonance of the Cu frame would not propagate across the twin boundaries, i.e., these twin boundaries are in a nonsuperconducting state. Insulating twin boundaries together with the twin domains on both sides form a superconductor-insulator-superconductor (SIS) Josephson junction. (A similar situation exists at the grain boundaries.) Inside a grain or crystal, the (110) and/or (1 –1 0) twin boundaries together with the (001) 90°-rotation boundaries or grain boundaries would form a three dimensional network dividing the grain into many superconducting clusters weakly coupled by SIS Josephson junctions. Such a superconducting glassy state including intra- and inter-grain Josephson junctions, would be expected to result in a significant reduction of average current density in polycrystalline oxide superconductors especially when temperature is near T_c [33].

For tunneling to occur, the thickness (d) of the insulating layer, i.e., the width (W_{tb}) of the oxygen-depleted zone at twin boundary or grain boundary, should be smaller than or at most near the coherence length of the oxide superconductor. Since $\xi_{\perp c} \approx 34$ Å [34] ($\xi_{\perp c}$ (77 K) ≈ 60.7 Å [35] or about 42.9 Å across twin boundary), the largest distance the superconducting current could tunnel through at liquid nitrogen temperature would be about eight layers (~ 43.6 Å) of insulating $YBa_2Cu_3O_6$. Experimentally, this expectation is supported by comparing various numbers (j) of oxygen-depleted layers at twin boundary to the J_c (O_x) data [36] (Fig. 7). J_c drops almost linearly with increasing thickness of the insulating layers and approaches zero when $O_x < 6.80$. This is consistent with Cava's [7] measurement of T_c as a function of O_x. In their study, T_c was almost constant for $O_x > 6.80$ and drops quickly when $O_x < 6.78$. For $O_x \approx 6.72$, patches of double cell structure [16,29] with stoichiometric $O_x \approx 6.50$ (homologous phase: 1/2 case) and other homologous phases, e.g., stoichiometric $O_x \approx 6.60$ (homologous phase: 3/5 case) [16], have been found with transmission electron microscopy in the oxygen-heavily-depleted material. In these materials, oxygen vacancies at (0, ½, 0) sites in twin domains and at twin boundaries are reorganized into more stable homologous phases [37], especially the double cell structure [13], with smaller numbers (~ 3) of oxygen-depleted layers at twin boundaries [6,38] which is believed to be the range over which Coulomb repulsion dominates. Presumably, this is also the cause of the peculiar behavior of resistivity in normal conduction [7].

2. Flux Pinning and Vortex State

When a magnetic field is applied to a YBCO superconductor, the nonsuperconducting areas, i.e., twin boundaries as well as grain boundaries with stoichiometric $YBa_2Cu_3O_6$, are expected to

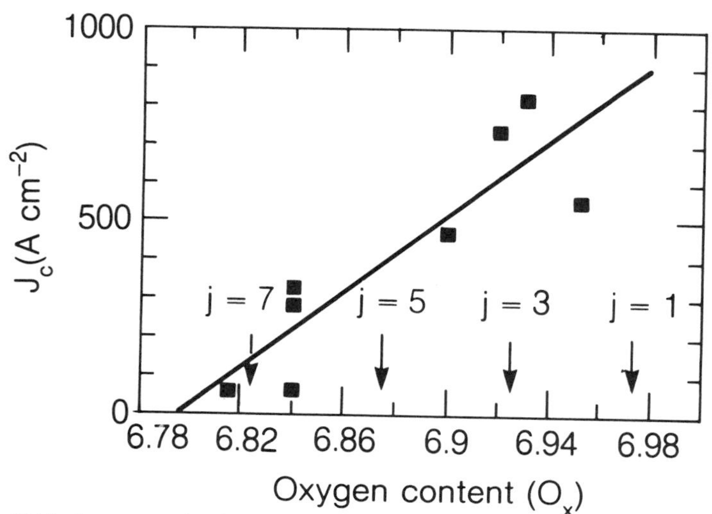

Fig. 7. Critical current density as a function of oxygen content, x, in $YBa_2Cu_3O_x$. Arrows are added to mark the oxygen content of the samples assuming x = 7 within twin domains and various numbers (j) of oxygen depleted layers at twin boundaries. Within these layers x is assumed to be 6. (Reprinted by permission from *Nature* Vol. 332 pp. 58. Copyright (c) 1988 Macmillan Magazines Ltd.)

provide the permeable local paths for magnetic flux lines. That these boundaries act as the pinning sites for flux lines has been successfully demonstrated by the decoration technique [39,40]. Because of the very fine twinning structure, numerous grain boundaries and voids, small grain size and high porosity, as well as the existence of other nonsuperconducting phases, e.g., CuO, $BaCuO_2$ and Y_2BaCuO_5, a regular flux lattice in the vortex state is difficult to obtain. Near liquid nitrogen temperature, only partial flux lattices have been observed [39]. Also, in YBCO material, because the average width of the twin domains, $w_{td} \approx 200 \sim 300$ Å, is smaller than the penetration distance, $\lambda_{\|c} \approx 1250$ Å [34], a regular flux lattice with all lattice sites pinned at the twin boundaries is unlikely since the vortex diameter is about 2λ [41]. However, the numerous twin boundaries acting as flux pinning sites should raise the lower critical field, H_{c_1}, above the intrinsic value of pure $YBa_2Cu_3O_7$ without twin boundaries.

It is well known that type-II superconductors have a negative surface energy between the normal conducting phase (NCP) and the superconducting phase (SCP). The "glassy" state due to the high density of twin boundaries thus provides a large favorable surface area for SCP nucleation when NCP is cooled from above T_c in zero field or when the field is reduced from above H_{c3} at $T < T_c$. The resulting H_{c2} is therefore expected to be smaller than the intrinsic value for a perfect $YBa_2Cu_3O_7$ domain.

The remanent magnetic moment in the hysteresis loop from the magnetization measurement can be attributed to the trapping of flux lines at these insulating boundaries as well. In addition, the flux pinned at twin boundaries can also explain the less than 100% Meissner effect or imperfect diamagnetism measured in field cooled (FC) YBCO crystals (without grain boundary) in which twin boundaries are the only strong pinning sites when flux exclusion progresses as temperature drops below T_c. In zero field cooled (ZFC) specimens, time-dependent magnetization-relaxation phenomena have been reported by several groups [42-44]. This could be due to the weak linking, i.e., the insulating boundaries, in this material. A stable magnetization configuration (state), which minimizes the energy of all the weak couplings, may not initially be achieved. As time passes, the system may transit from one metastable state to another metastable state having lower energy.

3. Superlattice, Band-folding and Minigaps

When calculating the energy band structure of $YBa_2Cu_3O_{7-\delta}$, a homogeneous matrix extending infinitely in all directions is presumed to exist. In the real material, however, dense twin structure with estimated spacing ~216 Å between coherent twin boundaries is almost uniformly present everywhere. Their presence breaks the inversion symmetry of the presumed infinite matrix, introduces boundary mirror and glide symmetry operations into the crystal symmetry, and forms polarized boundaries [45]. Since CuO_2 layers are not strongly affected by the presence of these oxygen-depleted twin boundaries, only energy bands related to the CuO b-chain plane are affected by the new symmetry operation.

Along the direction perpendicular to the twin boundary plane, the whole lattice is similar to a structure with a widely spaced superlattice, i.e., b-chains run zig-zag across twin boundaries with a repeating unit of about 156 b-chain units ($W_t \approx 216$ Å). This superlattice structure with glide symmetry thus requires the b-chain related bands folded into a $\pi/156$ b Brillouin zone in $\vec{k}_y$ direction yielding band degeneracy at zone boundary (Y) and subtle changes toward Γ-point. However, the CuO plane is coupled to the neighboring CuO_2 layers via Peierl's deformation of oxygen atoms creating a Jahn-Teller gap. A calculation employing a tight binding model with 75 b-chain unit [45] reveals minigaps at Y-point with scale as small as 200 kelvin as expected, and unknown gaps at Γ-point. These minigaps give rise to the semiconducting or semimetallic character of the CuO b-chain plane. Due to the splitting of degenerated points at zone boundaries, these densely packed, b-chain-plane-related bands are further flattened. However, T_c should be only slightly enhanced by the twinning geometry.

For an oxygen-depleted $YBa_2Cu_3O_x$ sample with $O_x \geq 6.80$, T_c is almost unchanged since the extrapolation length of the pair potential is about 37 Å [33]. The width of the oxygen depleted region at a twin boundary would never be expected to be this large. For $O_x < 6.78$, the oxygen vacancies dispersed in the twin domains start rearranging into a series of homologous phases with the double cell structure as the most stable end phase in the series and T_c drops rapidly to that of the secondary stable structure (~60° K).

V. CONCLUSION

In conclusion, we have proposed a qualitative description for the formation of the twinned structure in $YBa_2Cu_3O_{7-\delta}$. This model assumes oxygen-depleted twin boundary zones which serve as the primary oxygen diffusion paths from grain surfaces toward the interior of the grains and thus control the growth rate of the twins in the grain centers. These twin boundaries are formed during the growth of the b-chain embryos which are first nucleated heterogeneously at grain surfaces. The width of the oxygen-depleted twin boundary zones should be determined by the balance of the chemical potential energy at the ends of the b-chains at the edges of the twin domains and the coulomb repulsion energy between oxygen ions on the opposite sides of the twin boundaries. Under the ordinary processing condition, this is equivalent to the case $j = 3$ in this model. The rapid rate increase of oxygen uptake at T_p is attributed to a high oxygen vacancy concentration in the grain during early stages of the tetragonal to orthorhombic transformation caused by the formation of embryos which quickly consume a large fraction of the oxygen in the grain at T_p. Completion of the transformation by growth of twins occurs most rapidly at the grain surfaces by absorbing oxygen from the ambient and more slowly in grain interiors. Coarsening of the twin domains then occurs by elimination of small domains included within larger domains by localized transfer of oxygen atoms across the twin boundaries. This coarsening process takes place during annealing, resulting in residual stress relaxation and reduction of residual oxygen vacancies within the twin domains. The observed oxygen deficiency remaining after long time annealing in oxygen can be explained if it is assumed that the equilibrium structure of coherent twin boundaries contains one or more layers of oxygen vacancies as proposed in this model.

These twin boundaries together with the 90°-rotation boundaries and grain boundaries divide the material into many superconducting clusters weakly coupled by SIS Josephson junctions, i.e., in a superconducting glassy state. Oxygen-depleted twin boundaries may provide hole-absent sites necessary for yielding local imperfect resonance and delocalization of hole pairs. The critical current as expected from the model would decrease as the number of the oxygen-depleted layers increases causing a wider insulating thickness in the SIS Josephson junction structure. The same effect also would be reflected in the room-temperature resistivity which increases almost linearly with increasing numbers ($1 \leq j \leq 3$) of oxygen-depleted layers at twin boundaries. Imperfect Meissner effect, remanent magnetization as well as the time-dependent magnetization-relaxation phenomena all can be qualitatively understood based on this model. Oxygen-depleted twin boundaries can provide not only the pinning sites for the magnetic flux but also the nucleation sites for the superconducting phase, yielding the measured H_{c1} and H_{c2} higher and lower than the intrinsic value respectively. Finally, the model is consistent with the fact that T_c is nearly constant for oxygen contents above 6.8 but drops quickly when oxygen content is less than 6.78 as oxygen vacancies become rearranged into a more stable secondary structure. Even though the model is only qualitative, it is consistent with a large number of experimental observations and may be helpful in further development of a fundamental understanding of superconductivity in this interesting material.

ACKNOWLEDGMENTS

The authors would like to thank Dr. K. Hiraga for sending the high resolution transmission electron micrograph of a twin boundary.

This work was supported by the Director, Office of Energy Research, Office of Basic Energy Sciences, Materials Sciences Division of the U.S. Department of Energy under Contract No. DE-AC03-76SF00098.

REFERENCES

[1] J.G. Bednorz and K.A. Müller, Z. Phys. B 64, (1986) 189.

[2] M.K. Wu, J.R. Ashburn, C.J. Torng, P.H. Hor, R.L. Meng, L. Gao, Z.L. Huang, X.Q. Wang, and C.W. Chu, Phys. Rev. Lett. 58, (1987) 908.

[3] H. Maeda, Y. Tanaka, M. Fukutomi, and T. Asano, Jpn. J. Appl. Phys. 27, (1988) L209.

[4] Z.Z. Sheng and A.M. Hermann, Nature 332, (1988) 138.

[5] L.F. Mattheiss, E.M. Gyorgy, and D.W. Johnson, Jr., Phys. Rev. B 37, (1988) 3745. R.J. Cava, et al., Nature 332, (1988) 814.

[6] C.J. Jou and T. Washburn, Lawrence Berkeley Laboratory report, LBL-26078.

[7] R.J. Cava, B. Batlogg, C.H. Chen, E.A. Rietman, S.M. Zahurak, and D.J. Werder, Phys. Rev. B 36, (1987) 5719.

[8] J.D. Jorgensen, M.A. Geno, D.G. Hinks, L. Soderholm, K.J. Volin, R.L. Hitterman, J.D. Grace, I.K. Schuller, C.U. Segre, K. Zhang, and M.S. Kleefisch, Phys. Rev. B, 36, (1987) 3608.

[9] G. Van Tendeloo and S. Amelinckx, J. Electron Microsc. Tech., 8, (1988) 285.

[10] J.C. Barry, J. Electron Microsc. Tech., 8, (1988) 325.

[11] M. Sugiyama, R. Suyama, T. Inuzuka, and H. Kubo, Jpn. J. Appl. Phys., 26, (1987) L1202.

[12] T.E. Mitchell, T. Roy, R.B. Schwarz, J.F. Smith, and D. Wohlleben, J. Electron Microsc. Tech., 8, (1988) 317.

[13] D. de Fontaine, L.T. Wille and S.C. Moss, Phys. Rev. B, 36, (1987) 5709.

[14] G. Van Tendeloo, H.W. Zandbergen and S. Amelinckx, Solid State Comm., 63, (1987) 603.

[15] S. Iijima, T. Ichihashi, Y. Kubo, and J. Tabuchi, Jpn. J. Appl. Phys., 26, (1987) L1478.

[16] C.H. Chen, D.J. Werder, L.F. Schneemeyer, P.K. Gallagher, and J.V. Waszczak, Phys. Rev. B, 38, (1988) 2888.

[17] C.J. Jou, R. Kilaas, and J. Washburn, Lawrence Berkeley Laboratory Report, LBL-25149.

[18] C.S. Pande, A.K. Singh, L. Toth, D.U. Gubser, and S. Wolf, Phys. Rev. B, 36, (1987) 5669.

[19] H. Verweij, Solid State Comm., 64, (1987) 1213.

[20] D.C. Johnson, A.J. Jacobson, J.M. Newsam, J.T. Lewandowski, D.P. Goshorn, D. Xie, and W.B. Yelon, in "Chemistry of High-Temperature Superconductors," ACS Sym. Ser. 351, American Chemical Society, Washington, D.C., (1987) p. 148.

[21] J.Y. Henry, P. Burlet, A. Bourret, G. Roult, P. Bacher, M.J.G.M. Jurgens, and J. Rossat-Mignod, Solid State Comm., 64, (1987) 1037.

[22] K. Kishio, J.-I. Shimoyama, T. Hasegawa, K. Kitazawa, and K. Fueki, Jpn. J. Appl. Phys., 26, (1987) L1228.

[23] B.G. Hyde, J.G. Thompson, R.L. Withers, J.G. Fitzgerald, A.M. Stewat, D.J.M. Bevan, J.S. Anderson, J. Bitmead, and M.S. Paterson, Nature, 327, (1987) 402.

[24] H. Oyanagi, H. Ihara, T. Matsubara, T. Matsushita, M. Hirabayashi, M. Tokumoto, K. Murata, N. Terada, K. Senzaki, T. Yao, H. Iwasaki, and Y. Kimura, Jpn. J. Appl. Phys., 26, (1987) L1233.

[25] B. Raveau, C. Michel, and M. Hervieu, in "Chemistry of High Temperature Superconductors," ACS Sym. Ser. 351, American Chemical Society, Washington, D.C., (1987) p. 128.

[26] M. Hervieu, B. Domenges, C. Michel, and B. Raveau, Europhysics Lett., 4, (1987) 205.

[27] W.I.F. David, W.T.A. Harrison, J.M.F. Gunn, O. Moze, A.K. Soper, P. Day, J.D. Jorgensen, D.G. Hinks, M.A. Beno, L. Soderholm, D.W. Capone, I.K. Schuller, C.U. Segre, K. Zhang, and J.D. Grace, Nature, 327, (1987) 310.

[28] J.M. Tarascon, P. Barboux, B.G. Bagley, L.H. Greene, W.R. McKinnon, and G.W. Hull, in "Chemistry of High-Temperature Superconductors," ACS Sym. Ser. 351, American Chemical Society, Washington, D.C., (1987) p. 198.

[29] D.J. Werder, C.H. Chen, R.J. Cava, and B. Batlogg, Phys. Rev. B, 37, (1988) 2317.

[30] K. Hiraga, D. Shindo, M. Hirabayashi, M. Kikuchi, and Y. Syono, J. Electron Microsc. 36, (1987) 261.

[31] G. Van Tendeloo, H.W. Zandbergen, and S. Amelinckx, Solid State Comm. 63, (1987) 389.

[32] Y. Hirotsu, Y. Nakamura, Y. Murata, S. Nagakura, T. Nishihara, and M. Takata, Jpn. J. Appl. Phys. 26, (1987) L1168.

[33] G. Deutscher and K.A. Müller, Phys. Rev. Lett. 59, (1987) 1745.

[34] T.K. Worthington, W.J. Gallagher, and T.R. Dinger, Phys. Rev. Lett. 59, (1987) 1160.

[35] P.G. de Gennes, "Superconductivity of Metals and Alloys," Benjamin, New York, (1966) p. 229.

[36] N. McN. Alford et al., Nature 332, (1988) 58.

[37] A.G. Khachaturyan and J.W. Morris, Jr., Phys. Rev. Lett. 59, (1987) 2776.

[38] C.J. Jou and J. Washburn, Lawrence Berkeley Laboratory report, LBL-26065.

[39] C.J. Jou, E.R. Weber, J. Washburn, and W.A. Soffa, Appl. Phys. Lett. 52, (1988) 326.

[40] A. Ourmazd, J.A. Rentschler, W.J. Skocpol, and D.W. Johnson, Jr., Phys. Rev. B 36, (1987) 8914.

[41] P.G. de Gennes, "Superconductivity of Metals and Alloys," Benjamin, New York, (1966) p. 56.

[42] P. Norling, P. Svedindh, P. Nordblad, L. Lundgren, and P. Przyslupsky, Physica C, 153, (1988) 314.

[43] J.R. Fraser, T.R. Finlayson, T.F. Smith, G.N. Heintze, R. McPherson, and H.J. Whitfield, MRS Spring Meeting, Reno, Nevada, April 5–9, 1988 (in press).

[44] T.R. Finlayson, J.R. Fraser, T.F. Smith, R.H. Mair, G.N. Heintze, and R. McPherson, Australian Bicentenary Congress of Physicists, Sydney, January 25–29, 1988 (in press).

[45] F.M. Mueller, S.P. Chen, M.L. Prueitt, J.R. Smith, J.L. Smith, and D. Wohlleben, Phys. Rev. B 37, (1988) 5837.

THE ROLE OF MOTT-INSULATION, NON-STOICHIOMETRY AND ALTERED VALENCE IN HIGH-T_c SUPERCONDUCTIVITY

G.J. Hyland

Department of Physics
University of Warwick, Coventry CV4 7AL, UK

I. INTRODUCTION

Traditionally, in the quest for materials exhibiting superconductivity at elevated temperatures - *i.e.* ones with a high critical temperature, T_c, ***intuitive*** approaches - such as, for example, that developed by Matthias and coworkers [1], from consideration of empirical correlations between different elements in the Periodic Table - have been singularly more successful than approaches based on microscopic theory; indeed some of the highest T_c materials so discovered turned out to be alloys composed entirely of non-superconductive elements. In 1986, however, 75 years after the discovery of Superconductivity by Kamerlingh Onnes, this situation changed. Bednorz and Müller, convinced that no further progress could be made with intermetallics, yet aware of the fundamental importance of the electron-phonon interaction in the *BCS* theory, deliberately sought materials wherein this interaction might be expected to be strong - such as Transition Metal oxides, particularly those containing ***copper***, which is well-known to be Jahn-Teller active. In this way, they discovered superconductivity [2] near 35 *K* in a quarternary ceramic, $La_{2-x}Ba_xCuO_{4-\delta}$, containing only one superconducting element, *La* - a material which had actually already been synthesized and extensively investigated at somewhat higher temperatures by Michel and Raveau [3]. As is now well documented, subsequent replacement [4] of *La* by *Y* to form the compound $YBa_2Cu_3O_{7-\delta}$, composed ***entirely*** of ***non*** -superconductive elements, raised T_c ***above*** the boiling point of liquid nitrogen - to near 90 *K*. The deadlock in surpassing the 23 *K* critical temperature of Nb_3Ge, discovered [5] in 1973, had been broken: a new era in the history of superconductivity had begun, which was soon to be highlighted by the discovery [6] of an even higher critical temperature (≈125 *K*) in a compound of nominal composition $Ca_2Ba_2Tl_2Cu_3O_{10+y}$.

Somewhat ironically, however, despite the fact that this revolutionary advance had actually stemmed from consideration of certain fundamental aspects of the *BCS* theory - specifically, Fröhlich's phonon-mediated electron-electron attraction, originating from the electron-phonon interaction - doubts were soon raised [7] over the ability of this particular pairing mechanism to account for the (relatively) very high T_C's so discovered - especially those above the boiling point of liquid nitrogen. These doubts arose not only from consideration of T_C within the context both of the weak and strong-coupling limits of the *BCS* theory, but also from the apparently null (or at best very small) Isotope Effect exhibited by the new materials [8].

Accordingly, attention turned to ***alternate*** (pre-existing) pairing mechanisms - such as those involving ***resonating valence bonds*** [9] and ***bipolarons*** [10]. That the conventional *BCS* theory most probably does not enjoy universal applicability - even in the case of elemental superconductors - had, however, already been suspected for some time [1], but to establish its status in the case of the new materials is far from trivial, in consequence of their rather complex quarternary structure which makes the implication of a null Isotope Effect, for example, much less unambiguous than it is in the case of simpler materials, such as Transition Metals. On the other hand, to establish the relevance of alternate mechanisms is equally problematic, in consequence of their formulation within the context of highly simplified and idealised models which are far removed from the conditions obtaining in the real materials in question.

Basic to any amelioration of this situation must be the (correct) identification of the carriers which actually form the superconducting pairs; for whatever the pairing mechanism ultimately turns out to be, it most surely will not be independent of the identity of the carriers involved - a point which appears to have been studiously avoided in many attempts at microscopic theory. The aim of this Chapter is to redress this situation. The task in hand is a challenging one, not only because of the rather complex crystal structures of the new materials - which can often accommodate a rather wide range of stoichiometries - but also, in the majority of cases, because of a ***subtlety in electronic structure,*** inherited from their "parent" compounds whose ground-states differ from that predicted by Band Theory, in that they are ***insulating,*** owing to a ***localization*** of the would-be carriers by ***strong Coulomb correlations***. Accordingly, it is with respect to these so-called ***Mott-insulating*** ground-states that ***deviations*** from ***stoichiometry*** - upon which the occurrence of superconductivity is known to be crucially dependent - must be considered; for these deviations can involve a change from the formal insulating valence *i.e.* entail a certain degree of ***altered valence***. It will thus be seen that considerations of "**Mott-insulation, Non-Stoichiometry and Altered Valence**", coupled with information yielded by studies of transport and magnetic properties of the normal state, are ***indispensible*** to any elucidation the electronic structure of the new high T_C superconductors. These matters are discussed in Section III, following a more general exposition (Section II) of the properties and peculiarities of ***Mott-insulators,*** which provides an opportunity to draw attention to certain fundamental points which would appear not always to have been adequately appreciated in the recent literature on the new high T_C materials - such as, for example, the nature of the so-called "upper and lower Hubbard bands" and their connection with the altered-valent cationic states associated with deviations from stoichiometry. Additionally, the interaction with the longitudinal optic modes of the material is considered in connection with the possible ***re*** localization of certain would-be carriers into small-polaron eigenstates, escape from which

(*via* "hopping") requires an activation energy which manifests itself in the mobility.

The Chapter concludes with a Resume (Section IV) of the principal conclusions reached, stressing not only the ***oxygen*** character of the holes involved in the high T_c superconductivity exhibited by the new cuprate ceramics, but also the existence of certain features associated with the ***copper*** subsystem which are common to all three classes of materials.

II. BASIC PROPERTIES OF MOTT-INSULATORS: ALTERED VALENCE AND NON-STOICHIOMETRY

Mott-insulation represents a breakdown of the Band Model in which a ***non*** -metallic state is realized ***despite*** the existence of ***in*** completely filled atomic shells. That this is relevant, under certain conditions, in the case of some of the new ceramic superconducting materials is now becoming generally accepted, following the early remarks by Anderson [9] concerning La_2CuO_4. In this case, for example, simple valence counting arguments indicate - assuming the bonding in the material is predominantly ionic - that the electronic configuration of the *Cu* ions is $Cu^{2+}(3\ d\ )^9$. Accordingly, the Band Model - refined by the inclusion of crystal field effects - predicts that the single hole on each *Cu* ion contributes to a ½-filled conduction band - specifically, an antibonding σ -band, based on hybridized $Cu{:}3d_{x^2-y^2}$ and $O{:}2p_{x,y}$ orbitals; the anticipated metallic behaviour is ***not*** however, realized - the nominally stoichiometric material being, empirically, essentially ***insulating*** at low temperatures (T>100K), with a low electrical conductivity ($\simeq 10(\Omega\ cm)^{-1}$) and a large Seebeck coefficient ($\simeq 300\mu\ V/K$) [11].

That the metallic behaviour predicted by the Band Model is not realized is due to its inadequate treatment of the ***Coulomb interaction*** between "would-be" carriers (the *Cu* holes in the above example), - the interaction of a given carrier with the remainder being treated only approximately in terms of the latter's time-average electric field, which simply renormalizes the periodic lattice potential and reduces the many-electron problem to many ***one*** -electron problems. Thus neglected are ***dynamical*** * ***correlations*** between particulate (individual) carriers, in consequence of which two carriers with opposite spin are allowed to approach arbitrarily closely and thereby acquire a large Coulomb ***potential*** energy. The Band Model can thus be expected to provide a valid description of the system only as long as this energy is ***small*** in comparison to the itinerant ***kinetic*** energy of the delocalized carriers; this kinetic energy is essentially the familiar one-electron band-width, *W*. The major contribution, *U*, to the Coulomb potential energy, on the other hand, is associated with configurations in which two carriers with antiparallel spin occupy, in the course of their itinerant (band) motion, the same cation ***thereby changing its valence.***

---

* ***Statistical*** correlations between Bloch electrons of ***parallel*** spin ***are***, of course, included in the Band Model through the use of ***determinantal*** many-electron wavefunctions (Slater Determinants) based on one-electron Bloch functions.

Thus the validity of the Band Model qualitatively requires $W>>U$; establishment of a more quantitative criterion is the aim of many-body approaches based on the so-called "Hubbard Hamiltonian" in which both itinerant kinetic ***and*** Coulomb potential energies are built in *ab initio* [12]. Such approaches indicate that for W/U less than a certain value, the itinerant (metallic) state predicted by the simple one-electron Band Model cannot be maintained without an overall increase (of order $U-W$) in the total energy of the system - the increase in Coulomb potential energy associated with the altered-valent cations (double occupancies) permitted by the Band Model wave-functions ***outweighing*** the decrease in kinetic energy which would be achieved in the first instance by the carriers delocalizing themselves from their "parent" atoms; accordingly, under such conditions, it is energetically preferable for the carriers to remain localized, and instead of a metallic state an ***insulating*** state results. This phenomenon is now known as ***Mott- insulation*** , after N.F. Mott [13] who, already in 1949, pointed out the possibility of this type of breakdown of the Band Model in which a negligible number of (localized) non-conducting states split off to lower energies, thus opening up in the electronic density of states a ***gap*** - now called the ***"Mott-Hubbard gap"*** - which separates a continuum of many-body conducting states from one containing only non-conducting states; these two continua are known, respectively, as the ***upper*** and ***lower Hubbard "bands"*** - an appellation which is somewhat unfortunate since they are certainly ***not*** bands in the sense understood in the Band Model - as will be explained below. Mott's conclusions were based on a simple model of N s -centres, each with a single electron, arranged on a simple cubic lattice with spacing a. Using a as a parameter, Mott predicted that a first-order transition (a "Mott-transition") into a metallic state (Band Model valid) should occur as a is reduced through a critical value, a_c. For as a is reduced the overlap of the wave-functions of the localized electrons increases, screening the interaction of the resident electrons with their centres, until at $a=a_c$ bound states (electrons localized at their respective centres) can no longer be maintained, whence each and every electron is simultaneously delocalized and a metallic state established*.

In real materials the overlap integral which determines W depends not only on the interatomic separation but also on the spatial extension of the orbitals associated with the incomplete shells, which in the case of the ***d*** and ***f*** shells of Transition and Rare Earth elements can be quite small. This, coupled with relatively large inter-cation distances (due to large intervening anions) would result in ***very*** narrow bands, were it not for hybridization between metal and oxygen orbitals. In such highly polarizable materials, on the other hand, the magnitude of the "on-site"

---

* It is of interest to note that this possibility was actually pointed out long before (in 1927) by Herzfeld [14] in connection with dielectric instabilities; more recent work [15] in this area shows that the onset of the metallic phase may be preceded by an (electronic) ferro-or antiferroelectric state - a result which could well prove to be of relevance to the new high T_c materials, whose ***perovskite*** -based structure is typical of many ferroelectrics.

Coulomb potential energy, *U*, is itself reduced to well below its free ion value (given by the difference in ionization energies of the altered-valent cations) by ***relaxation*** effects [16]; essentially, one of the two electrons of a doubly occupied site distorts its surroundings, creating a displacive polarization field with which the other electron interacts ***attractively***, thereby effectively reducing its Coulomb repulsion with the former. Notwithstanding this, there are apparently many materials - *e.g. NiO*, in which the reduction in *U* is insufficient to ensure $U \ll W$ - as evidenced, empirically, by their non-metallic behaviour, despite the existence of incomplete shells [17]; ***to this class of materials must now be added those intimately related to the new high T_c materials.***

The most striking evidence that the non-metallic states in such materials is due to Coulomb correlations localizing the would-be conduction electrons is provided by their ***magnetic*** properties; for whilst electrically, both Band Model insulators and Mott insulators are identical, their magnetic properties are - even qualitatively - quite different - namely, ***diamagnetism*** in the case of a completely filled band (Band Model insulator), compared with the existence of local magnetic moments in the Mott-insulating case - the ground-state of which is known [9] to be ***antiferromagnetically*** ordered, since such order permits localized wavefunctions to extend over neighbouring atoms (thereby reducing the associated kinetic energy by $\approx W^2/U$) without encountering an electron with identical quantum numbers. Antiferromagnetically ordered states have now been observed, under certain conditions, in La_2CuO_4 and $YBa_2Cu_3O_{7-\delta}$, although the magnitude of the local moments is somewhat below $1\mu_B$ [18], consistent with a finite degree of ***covalency*** between the $Cu^{2+}(3d)^9$ and oxygen *(sp)* orbitals. Quite independent support for a long-lived Cu^{2+} configuration associated with a single ***localized*** hole on each *Cu* ion comes from spectroscopic data of various kinds, and in particular from the observation in valence band *XPS* spectra of ***satellite*** structure [19] at energies where according to the band structure calculations (based on the independent particle approximation) there is ***no*** density of states; additionally, these experiments yield estimates of $U \approx 5\text{–}8$ *eV* .

The creation of current carriers out of the highly correlated, "self-locked", Mott-insulating ground-state - which are clearly necessary for conductivity ***and*** superconductivity - can be achieved either "intrinsically" by supplying enough energy to bridge the Mott-Hubbard gap thereby creating amongst the Mott-insulating background a certain number of ***doubly*** occupied ions (electrons) and an equal number of ***vacant ones*** (holes) - or "extrinsically" by the introduction of altered-valent dopants or by variation of stoichiometry which creates only ***one*** type of carrier (***n or p*** type).

As mentioned above these Hubbard "bands" are ***not*** bands in the sense of one-electron Band Theory but are, rather, continua of ***many*** -particle states, in which ***either*** of the two electrons (holes) on a doubly occupied (vacant) ion can be considered to be in the upper (lower) Hubbard "band"; thus the number of states in the upper Hubbard "band", for example, is equal to ***twice*** the number of doubly occupied ions. It can be shown [20], however, that the thermodynamic properties of a Mott-insulator depends only on the number of double occupancies and that it is, in fact, possible to introduce an effective "single-particle" representation of these two Hubbard "bands" in which the number of states actually equals the number of double occupancies. The price paid is thus that the number of states in the upper band now depends on their occupation; the band remains a meaningful concept only provided occupied states continue to exist in the lower Hubbard "band". Accordingly, the statistical mechanics of Mott-insulators differ in a fundamental way from ordinary Band Model insulators, as evidenced by the well-known "factor of two" difference in the value of various thermodynamic quantities, according to which of the two insulating ground-states is assumed; thus - for example [21], the low temperature *($kT<<U$)* specific heat of a Mott-insulator is only one half of that of a Bloch-Wilson (Band Model) insulator!

The essential distinction between the nature of Hubbard "bands" and those envisaged in the Band Model is pointedly illustrated by comparing [22], in both cases, the total number of states of the whole system, comprising the simple cubic lattice of N s -centres (each 2-fold spin degenerate) and N electrons considered above. In the Band Model case there are ${}^{2N}C_N$ ($\simeq 2^{2N}$) states of the whole system, corresponding to the number of ways of distributing the N electrons over the 2N Bloch states; in the Mott-insulating case, on the other hand, there is a continuum (***the lower Hubbard "band"***) containing only 2^N many-electron states, deriving solely from the 2 spin degrees of freedom of each s -centre. For large N, $2^N << 2^{2N}$, and when $U \geq W$, these 2^N non-conducting, localized states split off to ***lower*** energies from the 2^{2N} extended (Band) states; to retrieve all 2^{2N} states, starting with the 2^N -fold degenerate ground-state, it is necessary to admit, in addition, all possible numbers of doubly occupied centres up to the maximum of $N/2$ *i.e.* for r double occupancies the degeneracy of the system, D_r, is given by:

$$D_r = {}^{N}C_r \; {}^{N-r}C_r \; 2^{N-2r}, \tag{1}$$

corresponding to the number of ways of distributing the r double occupancies and r vacant centres amongst the remaining $(N-2r)$ singly occupied sites; for then:

$$\sum_{r=0}^{N/2} D_r = {}^{2N}C_N \tag{2}$$

The greatest term in the summation on the *LHS* of Eq. 2 occurs for $r=N/4$ *i.e.* when 50% of the centres are singly occupied, 25% doubly occupied and 25% empty; these proportions correspond ***exactly*** to the ***average*** valence in a ½-filled s -band. For it can be shown [12] in the case of dynamically ***un*** correlated electrons belonging to a band containing ν states/atom that the probability, P_n of finding n electrons on any given atom is given by the binomial distribution - *i.e.*

$$P_n = {}^{\nu}C_n \left(\frac{s}{\nu}\right)^n \left(1-\frac{s}{\nu}\right)^{\nu-n}, \tag{3}$$

where s is the mean number of electrons/atom - *i.e.* the total number of electrons/total number of atoms. Thus for the case of the N s -centres considered above, $\nu=2$, $s=1$, whence:

$$P_1 = 0.5, \quad P_2 = P_o = 0.25 \tag{4}$$

i.e. ***in a ½-filled s-band, the atomic valence fluctuates about its mean value of unity with a frequency of order W/h, where W is the one electron***

bandwidth and h is Planck's constant. Thus whilst ***on average*** there will always be 50% singly occupied (neutral) atoms, A^o, together with 25% doubly occupied (A^-) and 25% vacant atoms, A^+, the occupancy of ***any particular*** atom fluctuates between 0 and 2 - *i.e.* underlying the ***k*** -space motion in the ½-filled band under consideration are ***real*** -space transitions of the kind:

$$2A^o \rightleftharpoons A^+ + A^- \tag{5}$$

In terms of this equilibrium, the validity of the Band Model requires that the associated "reaction" free energy falls within the range of energies available to the electrons within the band of width W.

Such a real-space picture is particularly relevant to any understanding of carrier motion in the new high T_c materials, in which there has been much discussion of the role and existence of altered valent *Cu* ions. Here the altered valence must be realized extrinsically, since U is much too large $(U >>> kT)$ to permit intrinsic production by thermal excitation across the Mott-Hubbard gap. Thus, for example, upon partial substitution of *La* in La_2CuO_4 by (divalent) *Ba*, charge neutrality can be maintained (see Section III.2) by the formation of Cu^{3+} according to:

$$\overset{3+}{La_2}\,\overset{2+}{Cu}\,\overset{2-}{O_4} \rightarrow \overset{3+}{La_{2-x}}\overset{2+}{Ba_x}\,\{\overset{2+}{Cu_{1-x}}.\overset{3+}{Cu_x}\}\,\overset{2-}{O_4} \tag{6}$$

i.e. instead of the transitions represented in Eq.5, conduction here proceeds *via* "hole" exchange of the following kind:

$$Cu^{2+}(3d)^9 + Cu^{3+}(3d)^8 \rightleftharpoons Cu^{3+}(3d)^8 + Cu^{2+}(3d)^9 \tag{7}$$

In this connection it is essential to appreciate that the chemical introduction of Cu^{3+} does not "unlock" the highly correlated Mott-insulating ground-state - which is here based on the existence of a single hole in each *Cu 3d* -shell (Cu^{2+}) - as might at first be anticipated from the ***non-integral*** number of holes/ *Cu* ion so produced, but instead ***introduces holes in the lower Hubbard "band"*** . For since a metallic ½-filled band is not *ab initio* realized it ***cannot*** be rendered ***less*** than ½-full by the introduction of Cu^{3+}!; in this way *n* -type behaviour is avoided, in accordance with experiment.

As the doping level changes so does the concentration of Cu^{3+} ions and hence so does the number of states in the effective single-particle representation of the lower Hubbard "band". It should be appreciated that this change in the number of states occurs for the reasons set out above in connection with the general properties of Mott-insulators, and is to be distinguished from the situation in the case of doped ***elemental*** semiconductors, wherein the reduction in the number of states in the conduction (valence) band, for example, is simply a consequence of the donors (acceptors) entering the lattice ***substitutionally*** !

In the absence of any further interactions there is no activation energy for the process represented by Eq. 7, since the Cu^{2+} ions neighbouring the Cu^{3+} present ***energetically equivalent*** sites for the additional hole associated with Cu^{3+}, which can thus move ***freely*** by quantum mechanical ***tunnelling,*** and contribute to transport. In view of the highly polar nature of the new high T_c materials, however, interaction of these (otherwise itinerant) Cu^{3+} holes with the longitudinal optic lattice modes cannot be neglected - especially at low carrier densities (small x) - and the possibility that these Cu^{3+} holes might be ***re*** localized in ***small polaron*** eigenstates must be investigated; empirically, such self-trapping would reveal itself as an ***activated*** hole mobility, $\mu_h \propto e^{-\varepsilon/kT}$. At higher Cu^{3+} concentrations, on the other hand, the polaron-forming hole-phonon interaction becomes effectively screened out (each hole competing for the same lattice distortion) and the carriers again move freely - a situation which might not be unconnected with the empirical existence in the new materials of a ***minimum*** dopant level below which there is effectively no transport and no superconductivity!

Finally, there has been much discussion concerning the relevance of crystallographic transformations in the high T_c materials - it originally being thought [23] that the role of the dopants was, in the case of $La_{2-x}X_xCuO_4$, for example, to stabilize the tetragonal phase. It is now known that this is ***not*** so, and given the central role played by Coulomb correlations the occurrence of such structural transformations assumes only a secondary significance [24].

III APPLICATION TO THE CUPRATE SUPERCONDUCTORS

III.1 La_2CuO_4 - Based Materials

As already noted above, the parent compound, La_2CuO_4, of the first group of high T_c materials to be discovered is, in its ground-state, a ***Mott-insulator***. The small but finite conductivity actually observed can be accounted for if the material is only ***nominally*** stoichiometric - for in the presence either of *La* deficiency or *O* excess, overall charge neutrality can be maintained by the formation of Cu^{3+}, according to:

$$La_2CuO_4 \longrightarrow \begin{cases} \overset{3+}{La}_{2-y}\{\overset{2+}{Cu}_{1-3y}.\overset{3+}{Cu}_{3y}\}\overset{2-}{O}_4 & (8a) \\ \overset{3+}{La}\{\overset{2+}{Cu}_{1-2z}.\overset{3+}{Cu}_{2z}\}\overset{2-}{O}_{4+z} & (8b) \end{cases}$$

As outlined above for the case of the divalently doped material, the Cu^{3+} so formed* introduce ***holes*** in the lower Hubbard "band" (or rather its effective single-particle equivalent), consistent with which is the observed p -type behaviour of the thermoelectric [26] and Hall [27] effects, which is ***opposite*** to that expected were the Band Model valid, when the presence of Cu^{3+} would, in effect, render the conduction band ***less*** than half full, thus entailing [27] n -type behaviour!

In passing it may be noted that the universally observed p -type behaviour is ***not*** consistent with existence of itinerant *electrons* (Cu^{1+}) moving in the upper Hubbard "band" - such as might be associated with oxygen *deficiency*, for example, according to:

---

* It would appear - in view of the apparent **in**sensitivity to *La* -content [25] - that **hyper**stoichiometry (Eq. 8b) is the more likely cause.

$$La_2CuO_{4-\delta} \longleftrightarrow \overset{3-}{La_2}\ \{\overset{1+}{Cu_{2\delta}}.\overset{2+}{Cu_{1-2\delta}}\}\overset{2-}{O_{4-\delta}} \tag{9}$$

Partial substitution of *La* with a *divalent* element *X* - such as *Ba, Ca or Sr* - similarly introduces holes (Cu^{3+}) in the lower Hubbard "band", assuming, again, that charge neutrality is **not** maintained by the concommitant formation of oxygen vacancies; at least for small $x \leq 0.15$ this indeed seems to be the case experimentally, [28] whence:

$$La_{2-x}\ X_x\ CuO_4 \longleftrightarrow \overset{3+}{La_{2-x}}\ \overset{2+}{X_x}\ \{\overset{2+}{Cu_{1-x}}.\overset{3+}{Cu_x}\}\ \overset{2-}{O_4} \tag{10}$$

In the simultaneous presence of oxygen deficiency, however, the concentration of Cu^{3+} is reduced in accordance with:

$$La_{2-x}X_xCuO_{4-\delta} \longleftrightarrow \overset{3+}{La_{2-x}}\overset{2+}{X_x}\ \{\overset{2+}{Cu_{1-(x-2\delta)}}.\overset{3+}{Cu_{x-2\delta}}\}\overset{2-}{O_{4-\delta}} \tag{11}$$

For in consequence of the stability of the Mott-insulating ground-state of the host material, any Cu^{1+} ("initially") formed upon loss of oxygen will ***spontaneously*** recombine with Cu^{3+} to reproduce Cu^{2+}, *via*

$$Cu^{1+} + Cu^{3+} \longrightarrow 2Cu^{2+} \tag{12}$$

It can be seen from Eq. 11 that the existence of a non-zero concentration of Cu^{3+} (potential carriers) requires [29] the following inequality to hold:

$$x > 2\delta \tag{13}$$

Determination of the hole concentration by wet chemical analysis [28] indicates that

$$[Cu^{3+}] \simeq \begin{cases} x, & 0 \leq x \leq 0.15 \quad (14a) \\ 0, & x \geq 0.15\ , \quad (14b) \end{cases}$$

implying that:

$$\delta \simeq \begin{cases} 0\ , & 0 \leq x \leq 0.15 \quad (15a) \\ x/2\ , & x > 0.15 \quad (15b) \end{cases}$$

Furthermore, and most significantly perhaps, it is found that ***T_c tracks [Cu^{3+}]***.

The *x* -dependence of [Cu^{3+}] for $0 \leq x \leq 0.15$ is consistent with the observed ***normal*** state behaviour of both the Hall constant [27] and the thermoelectric Seebeck coefficient, *S* [26], both of which are ***positive***, as is the case with the ***nominally*** stoichiometry undoped host material, *La_2CuO_4;* the decrease [30] in the magnitude of *S* as X_x is introduced clearly indicates a significant increase in the value of [Cu^{3+}] ***above*** that (0.03) appropriate when $x = 0$.

During the past year there has been much experimental effort devoted to ***spectroscopic*** detection of the presence of Cu^{2+} and Cu^{3+} ions in $La_{2-x}X_xCuO_4$, often with conflicting results, however - two experimental groups [31,32] using the ***same*** technique *(XANES)* **disagreeing** on whether Cu^{3+} exists in the material! In this connection the following points should be noted.

a) different samples with the ***same*** x value could have ***different*** oxygen deficiencies (δ) and hence, in accordance with Eq. 11, different $[Cu^{3+}]$ values; it is thus essential to complement any spectroscopic investigation with transport measurements on the same sample, in order to establish whether an absence of Cu^{3+}'s could indeed be due simply to (excessive) oxygen deficiency.

b) Techniques utilizing ***surfaces*** must be anticipated to be particularly δ -sensitive.

c) Detection of altered-valent cations using *XANES* requires [32] that the particular valence states in question have a lifetime of at least 10^{-14} *s*. Now the observed T -dependence of the normal state electrical conductivity of $La_{2-x}X_xCuO_4$ reveals [33] no trace in oxygen annealed samples of an activated mobility, suggesting that the holes introduced by X_x move ***freely*** in the single-particle equivalent of the lower Hubbard "band" - in which their residence time, τ_R, on any particular *Cu* ion is given approximately by:

$$\tau_R \simeq h/W, \tag{16}$$

where h is Planck's constant and W is the width of the hole band; thus $\tau_R > 10^{-14}$ *s*, requires:

$$W < 0.4 \text{ eV}, \tag{17}$$

which could well ***not*** be satisfied, thus precluding the detection of Cu^{3+} using *XANES*.

It is quite possible, however, that there is, in fact, ***no*** Cu^{3+} in $La_{2-x}X_xCuO_4$, in consequence of electron transfer [34,35] from the oxygen* sub-system which restores would-be Cu^{3+}'s to a ***divalent*** configuration, according to:

$$Cu^{3+} + O^{2-} \rightarrow Cu^{2+} + O^{-}, \tag{18}$$

whence the holes (which are, of course, still essential for the preservation of charge neutrality) are to be regarded as existing in the oxygen bands of the material which, unlike the *Cu* "bands" considered above, are ordinary, genuinely one-electron bands in the true sense of Band Theory**. In these much ***wider*** bands a higher hole mobility is anticipated, and a correspondingly greater stability against self-trapping, facilitating

---

* It is of interest in this connection to recall that an analogous situation is believed [17] to obtain in the archetypal Mott-insulator, *NiO*, when doped with *Li*.

** The existence of a non-negligible degree of hydridization between Cu^{2+} and *O* is, remember, already evidenced by the reduced value of the local *Cu* magnetic moment ($\simeq 0.43\ \mu_B$) [18].

the establishment of a superconducting state; indeed it may be that it is ***only*** within the oxygen sub-system that the holes can remain itinerant.

In this connection the question arises as to whether the increase observed in the normal state resistivity of ***undoped*** La_2CuO_4 just prior to T_c is in any way connected with the existence of an ***activated*** (hopping) mobility $\mu_h \propto e^{-\varepsilon/kT}$, especially since this increase is ***not*** found in ***doped*** materials the (polaron-forming) carrier-optic mode interaction might be anticipated to be screened out, in consequence of the higher concentration of Cu^{3+}'s. Given the intuitive difficulty in conceiving of a superconducting transition out of a normal state in which transport occurs by thermally assisted ***hopping,*** a more suggestive explanation of the observed increase is that it is due to a suppression of the itinerancy of a ***limited*** number of holes in consequence of their interaction with (excess) interstitial oxygen atoms. Provided a finite number of ***free*** holes remain (and as indicated by Eq. 8b, the concentration of Cu^{3+} is ***twice*** the oxygen excess) then a transition to a superconducting state is still, in principle, possible, although the associated supercurrent may be quite small. It should be noted, however, that the value of T_c of the undoped material is ***close*** to that of the doped implying,(if T_c is in any way dependent on the hole concentration) the existence of approximately ***equal*** concentrations in ***both*** materials. To realise this, assuming each excess oxygen traps ***one*** Cu^{3+} hole, then requires $z=x$, as can be seen immediately from Eqns 8b and 10.

Experimental evidence is now accumulating which supports the existence of O^- holes - first suggested by Fujimora et al. [36] on the basis of their photoemission results; specifically one can cite the recent work of:

a) Nücker et al. [37] who investigated the electronic structure near the Fermi surface of $La_{2-x}Sr_xCuO_4$ as a function of x, using *EELS* to study the core level spectrum of the *O 1s* level; they find a significant *O 2p* character in the density of states near E_F.

b) Yu et al. [38] who showed that the thermoelectric Seebeck coefficient, *S*, is ***independent*** of magnetic field (up to 30 *T*)- ***quite contrary*** to the behaviour expected if the holes moved in a (lower) Hubbard "band", wherein the spin contribution to *S* is ***quenched*** by strong magnetic fields.

III.2 $YBa_2Cu_3O_{7-\delta}$, $0 \leq \delta \leq 1$.

Discussion of this range of materials is somewhat more complicated than the case of $La_{2-x}X_xCuO_4$ on account of the existence of ***two*** crystallographically distinct and chemically dissimilar *Cu* sites. One site [*Cu(I)*] is surrounded by a squashed square planar oxygen configuration in the ***b-c*** plane and is linked to other similar sites in a ***one-dimensional*** fashion along the ***b*** -axis; the other site [*Cu(II)*] is 5-fold coordinated by a square pyramidial arrangement of oxygens, the apically elongated member of which is one of the two ***c -axis*** near-neighbours of *Cu(1)*. There are **twice** as many *Cu(II)* sites as *Cu(I)'s* and the conventional view [39] is that ***Cu(II) remains divalent*** **for** $0 \leq \delta \leq 1$, *whilst the Cu(I)'s change from Cu^{3+} to Cu^{1+}* as δ increases from 0 to 1, according to:

$$YBa_2Cu_3O_{7-\delta} \longrightarrow Y^{3+}Ba_2^{2+} \left\{ \begin{matrix} \{Cu^{2+}_{2\delta}(I).Cu^{3+}_{1-2\delta}(I)\} \\ \\ \{Cu^{1+}_{2\delta-1}(I).Cu^{2+}_{2-2\delta}(I)\} \end{matrix} \right\} Cu^{2+}_2(II)O^{2-}_{7-\delta} \quad \begin{matrix} ,\ 0\leq\delta\leq 0.5 & (19a) \\ \\ ,\ 0.5\leq\delta\leq 1 & (19b) \end{matrix}$$

The oxygen loss occurs from the ***b*** -axis near-neighbours *O* -sites of *Cu(I)*, which can be emphasized by rewriting the *RHS* of Eq. (19a) as follows:

$$(Y^{3+}\ O_v(II))(Ba^{2+}\ O^{2-})_2\ \{Cu^{2+}_{2\delta}(I).\ Cu^{3+}_{1-2\delta}(I)\ \ O^{2-}_{1-\delta}\ O_v(I)\ \}(Cu^{2+}(II)O^{2-}_2)_2 \qquad (20)$$

At δ =0, each Cu^{3+}(I) has ***two b*** -axis nearest neighbour oxygens and the associated one-dimensional *O-Cu(I)-O* chains are perfectly intact; at δ =0.5, however, 50% of the in-chain *O* 's are absent - this figure increasing to 100% at δ =1, where the chains are ***completely*** devoid of oxygen and each associated *Cu(I)* assumes the closed shell configuration $Cu^{1+}(3d)^{10}$. It should be noted that at $\delta\geq 0.5$, there is the possibility of realizing an ***equal*** occupancy of chain and $O_v(I)$ sites in the basal plane (vacant at δ=0), under which conditions the crystal symmetry increases from orthorhombic ($a\neq b$)to tetragonal ($a=b$) [40].

The δ=1 limit material $YBa_2Cu_3O_6$ is well-established to be **insulating**, despite the existence of incomplete Cu^{2+}(II) ions; there is, of course, no problem with the *Cu(I)* in this limit since they assume the ***complete*** shell configuration, $Cu^{1+}(I)(3d)^{10}$. Thus the existence of insulating $YBa_2Cu_3O_6$ empirically indicates the occurrence of ***Mott-insulation*** within the *Cu(II)* system [41]. The *Cu(I)*, on the other hand, themselves become ***divalent*** at δ=0.5, and if the corresponding ½-filled band is similarly unstable against localization (due to strong Coulomb correlations) *then* ***$YBa_2Cu_3O_{6.5}$ will also be Mott-insulating*** . As noted in Section II, Mott-insulation requires $U>>W$, and, in general, ***different**** values of *U* and *W* must be assumed for the two *Cu* systems; thus Mott-insulation within *Cu(II)* does ***not necessarily*** entail its occurrence in *Cu(I)*. Indeed, the actual situation realized appears to be dependent on the conditions under which the material is prepared.

Thus samples from which oxygen is removed in a controlled way at relatively low temperatures using a *Zr* -gettering technique [42] exhibit superconductivity near δ=0.5 (admittedly with a T_c which is ***lower*** (≃60 *K*) than that (≃90 *K*) associated with δ=0) - in contrast to the insulating

---

* For as discussed in Section II, the magnitude of *U* is significantly influenced by lattice polarization effects - the strength of which depends on the number of *O* near-neighbours of a given *Cu* , and on the *Cu-O* separation - both of which are different for the two *Cu* sites; at the same time, *W* depends on the *Cu-Cu* distance which again differs for the two species of *Cu*, and, like U, is weakly δ-dependent.

behaviour traditionally found near $\delta=0.5$ in samples which have been prepared at higher temperatures using flowing gas or quenching techniques; a further difference is that in the gettered samples the vanishing of T_c does ***not*** correlate with the orthorhombic → tetragonal transition as it does in the case of alternatively prepared samples (at $\delta=0.5$).

Clearly, the superconductivity found in the gettered samples near $\delta=0.5$, implies the existence here of itinerant carriers - *i.e.* that the metallic state predicted by the Band Model for the $Cu^{2+}(I)(3d)^9$ ions ***is,*** in fact, realized after all, and that Mott-insulation is confined to the *Cu(II)* system; thus:

$$U(I) \ll W(I) \tag{21a}$$
$$U(II) \gg W(II) \tag{21b}$$

It is to be reemphasized that Eq. 21b holds for ***all*** samples, irrespective of their preparation, since there is ***no*** dispute as to the ***insulating*** nature of $YBa_2Cu_3O_6$.

In connection with the realization of a metallic *Cu(I)* system in the case of the gettered samples it is perhaps significant that this particular technique produces samples in which the oxygen distribution is ***considerably more homogeneous*** than it would otherwise be, which could well favour an enhancement of the polarization reduction in the magnitude of *U(I)*, so facilitating the inequality $U(I) \ll W(I)$. It should be noted, however, that ... *"we expect samples prepared in this manner do not represent equilibrium oxygen configurations, but rather a subset of metastable states"*, [42].

Accordingly, for the moment we shall confine our attention to the more frequently encountered situation in which the material is insulating near $\delta=0.5$, and thus indicative here of ***Mott-insulation*** in ***both*** the *Cu(I)* and *Cu(II)* systems; ***$YBa_2Cu_3O_{6.5}$ will thus be regarded as the counterpart within the $YBa_2Cu_3O_{7-\delta}$ phase field of La_2CuO_4 in the case of $La_{2-x}X_xCuO_4$.***

Oxidation above $O_{6.5}$ creates Cu^{3+}(I) ions, and it is tempting, in parallel with our initial discussion of the $La_{2-x}X_xCuO_4$ materials, to attribute the metallic-like normal state electrical transport properties exhibited by $YBa_2Cu_3O_{7-\delta}$ for $\delta<0.5$ to the existence* of itinerant (Cu^{3+}) ***holes*** in the single-particle equivalent of the lower Hubbard "band" of the *Cu(I)* system. With this ***Cu(I)-based hole*** interpretation of the observed *p* -type transport [43,44], however, not only is it difficult to reconcile the insulating behaviour at $\delta>0.5$, with the existence of ***electrons*** (Cu^{1+}(I)) in the ***upper*** Hubbard "band" - which is expected on "symmetry" grounds (see Eq. 19b) - but also serious difficulties develop near $\delta=0$ (O_7), **even** for ***lower*** Hubbard "band" itself!. For in accordance with Eq. 19a, the equivalent single-particle band, which relies on the existence of **a** (majority) of ***localized*** $Cu^{2+}(3d)^9$ states, there ***ceases to exist,*** since in this limit ***all the Cu(I) are $Cu^{3+}(I)(3d)^8$,*** i.e. the localized orbitals are all now ***empty***. Thus the present interpretation

---

* $YBa_2Cu_3O_{7-\delta}$ can thus be said to be ***"self-doped "***, in the sense that the counterpart of the divalent dopant X^{2+} which creates holes in La_2CuO_4 is here ***oxygen*** (in excess of $O_{6.5}$) which is ***already indigeneous*** to the material.

entails insulating behaviour near O_7 - precisely where the highest T_c ($\simeq$90 K) are found, and where the normal state $(T>T_c)$ exhibits well-established metallic-like properties [45].

A resolution [41] of this paradox follows, however, if the holes are transferred from the *Cu(I)* system to the ***oxygen*** sublattice, according to:

$$\overset{3+}{Cu(I)} + \overset{2-}{O} \longrightarrow \overset{2+}{Cu(I)} + O^-, \qquad (23)$$

whereby ***all*** the *Cu(I)* are maintained in the ***divalent*** (Mott-insulating) configuration, $Cu^{2+}(I)(3d)^9$; thus Eq. 19a ***must*** be replaced according to:

$$\overset{3+}{Y}\,\overset{2+}{Ba_2}\{\overset{2+}{Cu(I)}_{1-2\delta}\overset{3+}{Cu(I)}_{2\delta}\}\overset{2+}{Cu(II)}_2\,\overset{2-}{O}_{7-\delta} \rightarrow \overset{3+}{Y}\,\overset{2+}{Ba_2}[\overset{2+}{Cu(I)}\overset{2+}{Cu(II)}_2]\overset{2-}{O}_{6+\delta}\overset{-}{O}_{1-2\delta} \qquad (24)$$

The concentration of oxygen holes, $[O^-]$, is thus given by:

$$[O^-] = 1-2\delta, \qquad (25)$$

which is maximum at $\delta=0$(O_7) and vanishes at $\delta=0.5$ ($O_{6.5}$).

Although - as noted above - there is growing evidence of O^- holes also in $La_{2-x}X_xCuO_4$, it is essential to appreciate that in that case the reason for their existence is ***fundamentally different*** from that in $YBa_2Cu_3O_{7-\delta}$; for in $La_{2-x}X_xCuO_4$ the lower Hubbard "band" of the Cu^{2+} system does not disappear until $x=1$, which is far in excess of the empirical upper limit for transport of $x\simeq0.15$.

Experimental support for the ***oxygen*** character of the holes in $YBa_2Cu_3O_{7-\delta}$ can be found both in transport data and in spectroscopic studies. Firstly, the observed T -dependence [41] of the thermoelectric Seebeck coefficient, for example, is much more characteristic of (relatively free) motion in a wide band (i.e. one which is validly described by the one-electron Band Model) than of highly correlated motion in a narrow Hubbard "band" (based on Cu^{2+}(I)). Secondly, Fujimori et al. [36] were probably the first to suggest, on the basis of photoemission data, that the holes in $YBa_2Cu_3O_{7-\delta}$ are ***oxygen*** based - a similar conclusion being reached by Yarmoff et al. [46] who supplemented photoemission spectroscopy with inverse photoemission and near-edge X -ray absorption *(XANES)* studies. More recently, O^- holes have been observed ***directly*** [47] by oxygen *O 1s* excitations into ***empty*** p -states *via* electron energy loss spectroscopy *(EELS)*. Whilst numerous spectroscopic investigations using *XPS* and *XANES* generally support the existence of O^- holes, there is some controversy concerning the weight of $Cu^{3+}(3d)^8$ in the ground-state; thus whilst Yarmoff et al. [46] find ***no*** Cu^{3+} at all (in agreement also with Baudelet et al. [48] and Bianconi et al.[49]), Steiner et al. [50] find about 1/3 of the *Cu* ions per formula unit in the configuration $Cu^{3+}(3d)^8{:}O(2p)^6$ and 2/3 as $Cu^{2+}(3d)^9{:}O(2p)^5$. These latter proportions were later corroborated by Lengeler et al. [51] who point out the importance of using reference materials in which the copper coordination matches that in $YBa_2Cu_3O_{7-\delta}$. Thus *e.g.* *CuO* is not a good model for Cu^{2+}(II), since its square-planar oxygen coordination is not realized in the case of *Cu(II)* which is, instead, pyramidially coordinated by 5 oxygens; in place of *CuO* they recommend $YBa_2Cu_2O_5$. This refinement does not, of course, resolve the controversy concerning the existence of trivalent Cu^{3+}(I) configurations since, in accordance with the considerations at the beginning of Section III.2, these trivalent ions originate (if at all) within the *Cu(I)* system. In terms of Eq. 23, however, coexisting *Cu* **and** *O* holes, simply means that the electron transfer from the oxygen ligands is **not** 100%, which could well account for the differing results - particularly if the actual degree of transfer is δ (and hence ***sample***) dependent.

In connection with the apparent asymmetry noted above concerning the insulating behaviour of $YBa_2Cu_3O_{7-\delta}$ at $\delta \geq 0.5$ - where transport by electrons in the upper Hubbard "band" (of Cu(I)) might be anticipated, in parallel with the ***hole*** motion at $\delta<0.5$ - it may be noted that both *XPS* [52] and *XANES* [53] indicate the presence of a significant concentration of Cu^{1+} in highly reduced samples *(e.g.* $\delta \simeq 0.8$) in accordance with Eq. 19b. The very detectability of these charge states implies that they exist for a ***sufficiently long time***, which would, of course, be the case if the "extra" electron on a Cu^{1+}(I) - instead of occupying extended itinerant states in the upper Hubbard "band" - is actually ***re*** localized (self-trapped) in a small polaron eigenstate; this would immediately account for the non-metallic behaviour found empirically at $\delta >0.5$, where - unlike the situation at $\delta <0.5$ - the formation of oxygen O^- holes is much less likely in consequence of the closure of the *d* -shell of the Cu^{1+}(I). If a similar self-trapping obtains in the case of any Cu^{3+}(I) which might still persist at $\delta <0.5$ (in consequence of incomplete electron transfer) than it follows that ***the existence of O^- holes is necessary, not only to guarantee a nonzero concentration of carriers at O_7 but also, to ensure their itinerancy:*** *i.e.* unlike Cu^{3+}(I) holes the O^- holes are ***stable against re*** localization by self-trapping.

That the holes exist in the ***oxygen*** sublattice is, of course, consistent with the high sensitivity of T_c to the oxygen content; for the structure and integrity of the very bands in which the superconducting carriers (presumably pairs of oxygen holes) move is progressively degraded as oxygen atoms are lost from the ***b*** -axis sites.

We close this section with some remarks [41], motivated by the above considerations, on the possibility of optimizing and raising T_c, which compellingly suggest that the superconductivity observed in $YBa_2Cu_3O_{7-\delta}$ for $0 \leq \delta <0.5$ is based on ***oxygen holes*** , the concentration of which ($[O^-]$) is directly determined by δ, through:

$$[O^-] = 1-2\delta \tag{26}$$

Thus the empirical decrease in T_c as δ increases [42, 54] implies, ***via*** Eq.26, that T_c increases with increasing carrier concentration, $[O^-]$ - such as, in fact, predicted by both *BCS* and bipolaron theories. Attention is thus directed towards ways of ***optimizing*** $[O^-]$ for a given δ and towards ways of ***raising*** T_c ***above*** the present 90 *K* limit by compositional modifications - including substitutions of both cations and anions.

Consider first the optimization of T_c for a given δ. One possibility [41] of restoring T_c to the value appropriate to $\delta =0$, which is suggested by $La_{2-x}X_xCuO_4$, is to partially substitute the trivalent *Y* cation with a ***divalent*** element X^{2+}, for then overall electrical neutrality can be realized *via:*

$$\overset{3+}{Y}_{1-x}\ \overset{2+}{X}_{x}\ \overset{2+}{Ba}_{2}\ [\overset{2+}{Cu}(I)\ \overset{2+}{Cu}_{2}(II)]\ \overset{2-}{O}_{6+\delta-x}\cdot\ \overset{-}{O}_{1-(2\delta-x)}, \qquad (27)$$

whence to retrieve $T_c(\delta = 0)$ requires $x=2\delta$. In passing it is of interest to note that the opposite trend - namely a ***reduction*** in the value of T_c (for a particular δ -value) has, in fact, been observed [55] in *Pr* -doped $YBa_2Cu_3O_7$, in which the praseodymium, *Pr,* is ***tetra*** -valent, and, as such, causes a ***further*** reduction in the concentration of holes, ***over and above*** that appropriate to the δ -value of the material used [41].

Perhaps even more significant, in view of the well established ***insensitivity*** [56] of T_c to dopants which are ***iso*** valent with *Y,* is the fact that their presence does ***not*** result in any change in the hole concentration - as can be seen immediately from the above analysis.

An alternate possibility for raising T_c - which appears to enjoy some experimental support [57-59] - is by increasing $[O^-]$ above its present upper limit of unity, by the incorporation of ***fluorine*** into the lattice. Inspection of Eq.21 indicates [60] that there are ***three*** natural sites at which fluorine could be accommodated - namely the sites (δ) from which oxygen is preferentially lost (as δ increases from zero), *i.e.* the ***b*** -axis near-neighbours of the *Cu(I)* - and the two "perovskite" sites which are unoccupied by oxygen in $YBa_2Cu_3O_{7-\delta}$, namely those ($O_v(I)$) in the basal planes containing *Cu(I)* and those $(O_v(II))$ in the basal planes containing *Y.* It can easily be seen that in the case of total occupancy of the δ -sites by *F* :

$$[O^-] = 1-\delta. \qquad (28)$$

It should be stressed, however, that this value $[O^-]$ is ***not necessarily greater*** than that $(1-2\delta)$ of the unfluorinated material, since there is evidence [61] that δ itself ***increases*** upon fluorination. Total occupancy of the $O_v(I)$ and $O_v(II)$ sites increases $[O^-]$ by one and one, respectively - the maximum number per formula unit of each such site being unity. Accordingly,

$$[O^-]_{max} = 3, \qquad (29)$$

appropriate to the composition $YBa_2Cu_3F_2O_7$, which, it may be noted, is the ***maximum*** fluorine content apparently compatible with the occurrence of superconductivity [57]; the associated T_c is ***much higher*** ($\simeq 150$ *K*), as expected from the above considerations, in consequence of the higher hole concentration.

It is particularly interesting, and perhaps potentially significant, that occupancy of the $O_v(II)$ sites brings the hitherto "dormant" *Cu(II)* system into play - if in the first instance (following Eq. 19a) charge neutrality is considered to be maintained by the formation of $Cu^{3+}(II)$; for the number of *Cu(II)* per formula unit is ***twice*** that of *Cu(I),* thus ultimately entailing the possibility of higher concentrations of oxygen holes and hence higher T_c*'s!* Indeed just this possibility would appear to be realized in the case of the latest generation of cuprate superconductors

- namely those containing either ***bismuth*** or ***thallium***, to which we now turn our attention.

III.3 $CaSr_2Bi_2Cu_2O_{8+y}$ and $CaBa_2Tl_2Cu_2O_{8+y}$

Comparison of the crystal structures of the new materials, $CaSr_2Bi_2Cu_2O_8$ $(Ca_xSr_{3-x}Bi_2Cu_2O_{8+y})/CaBa_2Tl_2Cu_2O_8$ and $Ca_2Ba_2Tl_2Cu_3O_{10}$ with that of $YBa_2Cu_3O_7$ reveals [62] that the chain-forming *Cu(I)'s* of the latter are replaced by Bi^{3+}/Tl^{3+} ions, whose oxygen coordination is square planar - similar to that of the *Cu(II)'s* of the *Y* -material, from which oxygen loss is known to be significantly less than it is from the *O-Cu(I)-O* chains. Since, however, the trivalent *Bi/Tl* ions have ***closed*** shells they cannot play an electronically similar role to the *Cu(I)'s* in $YBa_2Cu_3O_{7-\delta}$, and attention is again focussed on the *Cu* ions; for *y=0 (i.e.* oxygen stoichiometry O_8) these *Cu* ions are **divalent**, having incomplete *d* -shells ($3d^9$) - as do the *Cu(II)* ions in $YBa_2Cu_3O_{7-\delta}$, for $0\leq\delta\leq1$. As explained in Section III.2, these Cu^{2+}(II) ions must be assumed to be Mott-insulating in order to reconcile the non-metallic nature of $YBa_2Cu_3O_6$ with the existence of incomplete shell Cu^{2+}(II) ions. For the new materials, however, such an assumption is ***empirically unnecessary*** since the materials are clearly ***metallic*** - as evidenced, for example, by the *T* -dependence of their normal state resistivity [63].

It must thus be inferred [62] that the superconductivity in the new materials is associated with a ***metallic subsystem based on*** Cu^{2+} ***ions*** ; it is to be stressed, however, that this conclusion ***is contingent*** upon the realization of metallic behaviour at *y=0: i.e.* that it is ***not**** essential to have *y>0*. At present, unfortunately, the experimental situation is not entirely clear and more work needs to be done to definitively establish the dependence on oxygen content. Indeed recent work [64] using angle resolved resonant photoemission claims evidence of a Fermi-liquid state with predominant *O 2p* character and of strongly localized states well away from the Fermi level, having both *Cu3d* and *O2p* character, such as might be anticipated ***for y>0,*** in accordance with the considerations of Section III.3 for $YBa_2Cu_3O_{7-\delta}$, at $\delta\leq0.5$. If, on the other hand, metallic behaviour is indeed realised ***for y=0,*** then one is forced to conclude that the nature of the superconductivity is here ***quite different*** from that in either $La_{2-x}X_xCuO_4$ or $YBa_2Cu_3O_{7-\delta}$, in that it occurs in a ***metallic*** *d* -band based on the *Cu* ions.

It is indeed tempting to attribute the greatly enhanced environmental and hydrolytic stability [65] of these new materials to this difference - coupled with the absence therein of any one-dimensional *O-Cu-O* chains from which oxygen might be easily lost, as happens in the case of the *Y* -material; for even under extreme reducing conditions where oxygen loss might occur from the (relatively more stable) square planar units, the carriers are ***not*** oxygen-based! These considerations motivate a reexamination of the situation, referred to in Section III.2, which obtains near δ =0.5 in samples of $YBa_2Cu_3O_{7-\delta}$ prepared using a gettering annealing technique [42].

III.4 $YBa_2Cu_3O_{7-\delta}$, Near δ=0.5, Revisited

As noted in Section III.2 it is possible to produce samples of $YBa_2Cu_3O_{7-\delta}$ which exhibit superconductivity near δ =0.5, characterized by a T_c ≃60 *K*. Whilst initially this phenomenon was found ***only*** in materials prepared using a relatively low temperature gettering technique, there is accumulating evidence of its existence in samples prepared otherwise [66-68]; the "plateau" structure in $T_c(\delta)$ near δ=0.5 so revealed must thus be taken seriously and attempts made to understand it. In this connection it must first be recalled that the earlier assumption that the *Cu(I)* ions - which at δ =0.5 are formally divalent - themselves constitute a Mott-insulating system (as does the Cu^{2+}(II) subsystem) was made on purely empirical grounds in order to reconcile the existence of incomplete Cu^{2+}(I) shells with the ***non*** -metallic nature of the material at δ >0.5, as indicated by early experiments; the ***metallic*** behaviour now revealed by the more recent work clearly indicates that there is ***no*** Mott-insulation within the Cu^{2+}(I) subsystem - the observed metallic behaviour being entirely in ***accord*** with elementary Band Theory, implying *U(I)<<W(I) c.f. U(II)>>W(II)*.

The observation of a plateau in T_c (≃60 *K*) - before T_c increases near δ =0.3, reaching 90 *K* at δ=0, - clearly reflects a certain (limited) stability of the superconductivity against change in oxygen content - such as would be anticipated if (again) a ***metallic d-band*** were involved, as opposed to oxygen holes. Since, however, the *Cu(II)* must still be assumed to remain Mott-insulating, in order to account for the non-metallic nature of $YBa_2Cu_3O_6$, any metallic *d* -band must be based on the *Cu(I)* ions. It is to be noted that the existence of such a band does not conflict with the non-metallic nature of $YBa_2Cu_3O_6$ since in this limit this band is ***full*** - the *Cu(I)* ions having the closed-shell configuration, $Cu^{1+}(I)(3d)^{10}$; in the opposite limit ($YBa_2Cu_3O_7$), on the other hand, the ½-filled *Cu(I)* *d* -band can be maintained by admitting electron transfer from the oxygen band to keep the concentration of Cu^{3+}'s at the value appropriate* to a ½-filled band. This transfer does, of course, generate holes (O^-) in the oxygen band - the number of which increases as the material is

---

* It will be recalled from the general discussion in Section II.1 that the existence of a metallic (½-filled) band means that the *Cu(I)* ions are divalent ***only on average;*** locally there must be valence fluctuations of the form:

$$2Cu(I)^{2+} = Cu(I)^{1+} + Cu(I)^{3+} \qquad (30)$$

- the occurrence of which is contingent on the inequality:

$$W(I) >> U(I), \qquad (31)$$

where *W* is the band-width and *U* is the dominant (on-site) contribution to the Coulomb potential energy associated with the altered-valent states (double-occupancies). In the case of a ½-filled orbitally non-degenerate *Cu* -band, there are at any instant

$$50\%\ Cu(I)^{2+}\text{'s},\ 25\%\ Cu(I)^{1+}\text{'s and } 25\%\ Cu(I)^{3+}\text{'s}. \qquad (32)$$

progressively oxidized above $YBa_2Cu_3O_{6.5}$, according to (where $0<\eta\leq 0.5$):

$$YBa_2Cu_3O_{6.5} \rightarrow YBa_2Cu_3O_{6.5+\eta} \leftrightarrow Y^{3+}Ba_2^{2+}[Cu^{<2+>}(I)\,Cu_2^{2+}(II)]O^{2-}_{6.5-\eta}O^{-}_{2\eta} \tag{33}$$

where, in accordance with Eq. 30, and the previous footnote, the notation $Cu^{<2+>}(I)$ denotes that the *Cu(I)'s* are divalent ***only on average,*** and ***locally*** should be interpreted as follows:

$$\overset{<2+>}{Cu(I)} \equiv \underset{0.25}{\overset{1+}{Cu(I)}} \cdot \underset{0.5}{\overset{2+}{Cu(I)}} \cdot \underset{0.25}{\overset{3+}{Cu(I)}} \tag{34}$$

The well-known sensitivity [69-72] of the superconductivity to ambient redox conditions near η =0.5 empirically indicates that these oxygen holes assume a dominating role in the superconduction mechanism in this limit. This could come about either by the holes ***themselves*** constituting a superconducting subsystem whose T_C - for some $\eta>\eta_C$ ($\simeq$0.25, empirically) - exceeds that of the *Cu(I) d* -band - ***or*** by providing an alternate pairing mechanism whose associated T_C ($\simeq$90 *K*) is higher than that ($\simeq$60 *K*) which already characterizes the *Cu(I) d* -band alone. In this latter connection a mechanism proposed some time ago by Fröhlich [73] is of particular interest, since it does ***not*** entail an Isotope Effect. Originally conceived with Transition Metals in mind, it is based on a screening of the *d* -band plasma by the *s* -electrons such that for long waves the frequency of the plasma becomes proportional to the wavenumber, which then accordingly represents a new "acoustic" branch which, like the usual (ion) acoustic branch, leads to the possibility of an attractive interaction between the *s* -electrons, and hence to superconductivity; slightly adapted - with the oxygen holes replacing the *s* -electrons of the Transition Metal - this mechanism merits serious consideration.

It should be noted that these ideas concerning the origin of the observed δ -dependence of T_C in $YBa_2Cu_3O_{7-\delta}$, in terms of a transition from pure *Cu(I)* -band conduction near δ =0.5 to one essentially dependent on oxygen holes near δ =0, are quite different and less problematic than that currently advanced [74], based on a transition from one-dimensional *Cu(I)* ***chain*** conduction near δ =0, to two-dimensional *Cu(II)* ***plane*** conduction near δ =0.5, for despite the appeal of such attempts with respect to, *e.g.,* the compatibility (within the context of the *BCS* theory) of the observed T_C ***plateau*** with the ***constant*** density of states associated with free particles in two-dimensions - they singularly fail to account for the ***non*** -metallic behaviour in the $YBa_2Cu_3O_6$ limit, where the *Cu(II)* are still divalent!

Finally, the following experimental investigations can be identified which would provide valuable checks on the ideas presented here:

1. ***Thermopower measurements*** on the normal states of new *Bi/Tl* materials and on $YBa_2Cu_3O_{7-\delta}$, for $0\leq \delta \leq 1$: differences in magnitudes and *T* -dependence are anticipated, according as whether or not there is oxygen hole conduction ***in addition*** to *Cu 3d* -band conduction; indeed, there are already indications [75] in $YBa_2Cu_3O_{7-\delta}$ of transport in ***more*** than one band near δ =0 .

2. ***Hydrolytic and deliquescent studies*** on $YBa_2Cu_3O_{7-\delta}$ near δ =0.5: these should indicate an enhanced stability of the superconductivity similar to that found in the *Bi/Tl* oxides if, ***in both cases***, a *Cu 3d* -band ***alone*** is involved.

3. ***T_c enhancement by oxidation*** : the already enhanced T_c's of $CaSr_2Bi_2Cu_2O_8/CaBa_2Tl_2Cu_2O_8$ (T_c <100 *K*), and particularly of $Ca_2Ba_2Tl_2Cu_3O_{10}$ (T_c ≃125 *K*), are consistent with the larger number of "*Cu(II)* -like" *Cu* ions (compared with the smaller number of *Cu(I)*'s in $YBa_2Cu_3O_7$ whose associated ***maximum*** T_c ≃90K). The increase in T_c from 60 *K* to 90 *K* realized in $YBa_2Cu_3O_{7-\delta}$ by oxidation ***above*** $O_{6.5}$, where the *Cu's* are formally divalent, suggests that attempts should be made to ***increase*** the oxygen stoichiometry in the new materials above the value O_8, where the *Cu's* are similarly divalent.

IV RESUMÉ AND OUTLOOK

The principal conclusion to be drawn from this Chapter on the relevance of ***Mott-insulation, Non-Stoichiometry and Altered Valence*** to the new high T_c superconducting materials is that both electron- electron and electron-lattice interactions play a dominant role in determining the electronic structure of the normal phase out of which the superconducting phase evolves. Specifically, the electron-electron interaction is responsible (under certain conditions) for the occurrence of ***Mott-insulation*** - a highly correlated state in which holes, which would otherwise be itinerant, remain localized on their "parent" *Cu* atoms; creation of potentially itinerant carriers out of this ground-state requires the formation of ***altered-valent*** cations, which is here realized either by ***doping*** or by departing from ***exact stoichiometry*** . The existence of such altered-valent cations is a necessary, but not necessarily sufficient, condition for conduction since, in consequence of the highly polar nature of the materials, significant interaction with their longitudinal optic modes must be anticipated which can, under certain conditions, result in a ***re*** -localization of these carriers into ***small-polaron*** eigenstates; experimentally, the occurrence of such "self-trapping" manifests itself as an ***activated*** (hopping) mobility *i.e.* in non-metallic-like transport.

In the case of $La_{2-x}X_xCuO_4$ and $YBa_2Cu_3O_{7-\delta}$, **metallic** -like behaviour is, nevertheless, ensured in consequence of the possibility of (low energy) charge-transfer excitations, whereby holes on the Cu^{3+} are effectively transferred to the oxygen system where, in consequence of the much greater band-widths, self-trapping does ***not*** occur; in the case of the *Y* -oxides such charge transfer is, in any case, ***essential*** to ensure the existence of a nonzero concentration of carriers in the limit $\delta \to 0$, where the lower Hubbard "band" of the *Cu(I)* system actually ***disappears*** as ***all*** the *Cu(I)*'s become trivalent*. At the same time, the ***impossibility*** of oxygen holes at δ >0.5 (where the *Cu(I)*'s acquire the closed shell configuration $Cu^{1+}(I)(3d)^{10}$) naturally accounts for the non-metallic behaviour found

---

* In $La_{2-x}X_xCuO_4$ this situation is not realized until *x=1*, which is ***far in excess*** of the actual value of *x* (≃0.15).

here, provided the "extra" electron on the Cu^{1+} (in the ***upper*** -Hubbard "band") is ***re*** localized, by self-trapping. In this connection it would be of interest to investigate if the same holds in $La_{2-x}X_xCuO_4$ under conditions analogous to δ >0.5 - *i.e.* when *X* is ***tetra-*** valent, so that preservation of overall charge neutrality requires the formation of Cu^{1+}; ***thorium*** might be tried in this respect, since the ionic radius of Th^{4+} is very close to that of La^{3+}.

To conclude we draw attention to a problem the solution of which could well entail a revision of the scenario developed in this Chapter. If conduction does occur *via* oxygen holes then it is necessary to reconcile this with the apparent disappearance of local magnetic moments as δ is reduced through 0.5, as evidenced by the ***absence*** of the anticipated ***Curie-Weiss*** behaviour in the paramagnetic susceptibility. Although for $\delta \geq 0.2$, the observed behaviour is consistent with the existence of a short-range, [2-dimensional], ***remnant*** of the long-range [3-dimensional] antiferromagnetic *(AFM)* order which obtains in the *Cu(II)* system at δ >0.6, the transition to ***Pauli-like behaviour**** at $\delta \leq 0.2$ [76] implies that here both *Cu(I)* ***and*** *Cu(II)* support ***metallic*** behaviour, and not just the *Cu(I)* as suggested in Section III.4; this requires that at δ <0.2 the polarization-induced reduction in *U(II)* is sufficient to ensure $U(II) \ll W(II)$ - contrary to the situation at δ=1. To account for the observed *p* -type behaviour it is then necessary to assume that the contribution of these *d* -bands to transport is ***dominated*** by that from the ***oxygen holes*** .

Unfortunately, the lack of agreement between available susceptbility data [77,78,79] does not, at present, permit further consideration of this crucial question, and it would be of great interest to now investigate the magnitude of the local moment associated with the ***strongly localized*** states (with both *Cu 3d* and *O 2p* character) recently detected [64] in $CaBi_2Sr_2Cu_2O_8$ (see Section III.3).

Finally, valuable information on the Mott-insulating *(Cu(II)) state near* δ =1 should be obtainable by analysing the structure of the *AFM* phase as a function of δ . For within this magnetic structure should be reflected the gradual transformation (Eq. 19b) of the non-magnetic Cu^{1+}(I) ions to Cu^{2+}(I), as the material is oxidized above $YBa_2Cu_3O_6$; indeed, the loss of AFM near δ=0.6 could well be due to a disruption of the long-range order amongst the Cu^{2+}(II) as the concentration of Cu^{2+}(I) increases.

---

* The same is found in the case of $La_{2-x}X_xCuO_4$.

REFERENCES

[1] B.T. Matthias, **Physics Today** 16, (1963) 21; **American Scientist** 58, (1970) 80.

[2] J.G. Bednorz and K.A. Müller, **Z. Phys. B.** 64, (1986) 189.

[3] C. Michel and B. Raveau, **Revue de Chimie Mineralé**, 21, (1984) 407, and references therein.

[4] M.K. Wu et al., **Phys. Rev. Letts.** 58, (1987) 908.

[5] J.R. Gaveler, **Appl. Phys. Letts.** 23, (1973) 480.

[6] S.S. Parkin et al., **Phys. Rev. Letts.** 60, (1988) 2539.

[7] A. Khurana, **Physics Today**, April (1987) 17.

[8] A. Khurana, **Physics Today**, July (1987) 17; see however, K.J. Leary et al., **Phys. Rev. Letts.** 59, (1987) 1236.

[9] P.W. Anderson, **Science** 235, (1987) 1196.

[10] P. Prelovsek et al., **J. Phys. C.**, 20L, (1987).

[11] P. Ganguly and C.N.R. Rao, **Mat. Res. Bull.** 8, (1973) 405.

[12] J. Hubbard, **Proc. Roy. Soc. A.** 276, (1963) 238.

[13] N.F. Mott, **Proc. Phys. Soc.** A 62, (1949) 416.

[14] K.F. Herzfeld, **Phys. Rev.** 29, (1927) 701.

[15] H. Fröhlich, 'Dielectric Instabilities - contribution to **"Ferroelectricity"**, p 9 (Ed. E.F. Weller), Elsevier Publ. Co., Amsterdam (1967).

[16] G.J. Hyland, **Phil. Mag.** 20, (1969) 837.

[17] D. Adler and J. Feinleib, **Phys. Rev. B.** 2, (1970) 3112 and **J. Solid State Chem.** 12, (1975) 332.

[18] S.K. Sinha, **MRS Bulletin**, June (1988) 24.

[19] N. Nücker et al., **Z. Phys. B.** 67, (1987) 9.

[20] E.J. Yoffa et al., **Phys. Rev. B.** 19, (1979) 1203.

[21] E.J. Yoffa and D. Adler, **Phys. Rev. B.** 12, (1975) 2260.

[22] H. Fröhlich, Transitions to the Metallic State - contribution to **"Quantum Theory of Atoms, Molecules and the Solid State"** p 465 (Ed. P.O. Lowdin), **Academic Press**, New York, (1966).

[23] J.D. Jorgensen et al., **Phys. Rev. Letts.** 58, (1987) 1024.

[24] P.W. Anderson, **Science** 235, (1987) 1196.

[25] S.A. Shaheen et al., **Phys. Rev. B.** 36, (1987) 7214.

[26] Y. Maeno et al., **Jap. J. Appl. Phys.** **26**, (1987) L402.

[27] N.P. Ong et al., **Phys. Rev. B.** **35**, (1987) 8807.

[28] M.W. Shafer et al., **Phys. Rev. B.** **36**, (1987) 4047.

[29] G.J. Hyland, **Phys. Stat. Solidi (b)** 144, (1987) 753.

[30] J.R. Cooper et al., **Phys. Rev. B.** 35, (1987) 8794.

[31] J.M. Tranquada et al, **Phys. Rev. B.** 35, (1987) 7187.

[32] E.E. Alp et al., **Phys. Rev. B.** 35, (1987) 7199.

[33] R.J. Cava et al., **Phys. Rev. B.** 58, (1987) 408.

[34] C.M. Varma et al., **Sol. State Commun.** 62, (1987) 681.

[35] V.J. Emery, **Phys. Rev. Letts.** 58, (1987) 2794.

[36] A. Fujimora et al., **Sol. State Commun.** 63, (1987) 857.

[37] N. Nücker et al., **Phys. Rev. B.** 37, (1988) 5158.

[38] R.C. Yu et al., **Phys. Rev. B.** 37, (1988) 7963.

[39] W.I.F. David et al., **Nature** 327, (1987) 310.

[40] J.D. Jorgensen et al., **Phys. Rev. B.** 36, (1987) 3608.

[41] G.J. Hyland, **Jap. J. Appl. Phys.** 27, (1988) L86.

[42] R.J. Cava et al., **Phys. Rev. B.** 36, (1987) 5719.

[43] H. Ishii et al., **Physica** 148B, (1987) 419.

[44] G.S. Grader et al., **Phys. Rev. B.** 38, (1988) 844.

[45] J. Hauck et al., **Z. Phys. B.** 67, (1988) 299.

[46] J.A. Yarmoff et al., **Phys. Rev. B.** 36, (1987) 3967.
[47] N. Nücker et al., **Phys. Rev. B.** 37, (1988) 5158.
[48] F. Baudelet et al., **Z. Phys. B.** 69, (1987) 141.
[49] A. Bianconi et al., **Z. Phys. B.** 67, (1987) 307.
[50] P. Steiner et al., **Z. Phys. B.** 69, (1988) 449.
[51] B. Lengeler et al., **Physica C** 153-155, (1988) 143.
[52] N. Stoffel et al., **Phys. Rev. B.** 36, (1987) 3986.
[53] I. Iwazumi et al., **Sol. State Commun.** 65, (1988), 213.
[54] P. Monod et al., **J. Physique,** 48 (1987) 1269.
[55] L. Soderholm et al., **Nature** 328, (1987) 604; Y. Dalichaouch et al, **Sol. State Commun.,** 65 (1987) 1001.
[56] T.J. Kistenmacher, **Sol. State Commun.** 65, (1988) 981.
[57] S.R. Ovshinsky et al, **Phys. Rev. Letts.** 58, (1987) 2579.
[58] R.N. Bhargava et al., **Phys. Rev. Letts.** 59, (1987) 1468.
[59] X-R Meng et al., **Sol. State Commun.** 64, (1987) 325.
[60] G.J. Hyland, **Jap. J. Appl. Phys.** 27, (1988) L598.
[61] P.K. Davies et al., **Sol. State Commun.** 64, (1987) 1441.
[62] G.J. Hyland, **Jap. J. Appl. Phys.** 27, (1988), No.8 (August).
[63] M.A. Subramanian et al., **Science,** Feb. (1988) 1015.
[64] T. Takahashi et al., **Nature** 334, (1988) 691.
[65] H. Maeda et al., **Jap. J. Appl. Phys.** 27, (1988) L209.
[66] Z.Z. Wang et al., **Phys. Rev. B.** 36, (1987) 7222.
[67] J.D. Jorgensen et al., **Physica C.** 153-155, (1988) 578.
[68] H. Ihara et al., **Physica C.** 153-155, (1988) 948.
[69] J.G. Thompson et al., **Mat. Res. Bull.** 22, (1987) 1715.
[70] Z. Dexin et al., **Sol. State Commun.** 65, (1988) 339.
[71] P. Narottam et al., **Appl. Phys. Letts.** 52, (1988) 323.
[72] J. Dominec et al., **Sol. State Commun.** 65 (1988) 373.
[73] H. Fröhlich, **J. Phys. C.** 1, (1968) 544.
[74] I.W. Chen et al., **Sol. State Commun.** 63, (1987) 997; C.N.R. Rao et al., **Mat. Res. Bull.** 23, (1988) 125.
[75] U. Gottwick et al, **Europhys. Letts.** 4 (1987) 1183.
[76] J.M. Tranquada et al., **Phys. Rev. B.** 38, (1988) 2477.
[77] Y. Nakazawa et al., **Jap. J. Appl. Phys.** 26, (1987) L796.
[78] H. Takagi et al., **Physica** 148B, (1987) 349.
[79] R.J. Cava et al, **Mat. Res. Symp. Proc.** 99, (1988) 19.

SOUND VELOCITY AND ELASTIC CONSTANTS IN OXIDE SUPERCONDUCTORS

R. Srinivasan

Department of Physics
Indian Institute of Technology, Madras 600036, India

INTRODUCTION

Studies on sound velocity and attenuation in superconductors are useful from the following points:

a) Some superconductors undergo a structural transformation at a temperature, T_S , well above the superconducting transition temperature T_c . A typical example is Nb_3Sn which undergoes a martensitic transformation before becoming superconducting. Changes in sound velocity and a peak in the attenuation at a certain temperature are indicators to a displacive phase transition at that temperature.

b) The superconducting phase transition at T_c is a second order transition which does not involve a change in density. The thermodynamics of the transition leads to the result

$$K_n - K_s = (K_n^2/4\pi)\ (\partial H_c/\partial p)^2_{T_c} \tag{1}$$

Here K_n and K_S are the bulk modulii in the normal and superconducting phases respectively at T_c , H_c is the thermodynamic critical field and p is the pressure. The bulk modulus in the superconducting phase will be less than the bulk modulus in the normal phase. Similar will be the behaviour of the other elastic modulii. So one should expect a slight decrease in the sound velocity as the sample becomes superconducting. This change will only amount to a few hundred parts per million.

c) The attenuation of ultrasonic waves in a metal in the normal phase is mainly caused by the interaction of acoustic phonons with the free electrons. In the superconducting phase the number of single particle excitations decreases rapidly as the temperature is reduced below T_c.

The acoustic attenuation coefficient, α_s in the superconducting state, decreases rapidly below T_c. The BCS expression for the temperature dependence is

$$\alpha_s / \alpha_n = 2/[1+\exp(2\Delta(T)/k_B T)] \tag{2}$$

Here α_n is the attenuation coefficient in the normal phase and $2\Delta(T)$ is the superconducting energy gap at temperature T. From a study of the temperature dependence of the ultrasonic attenuation in the superconducting phase there is a possibility of getting information about the energy gap parameter $2\Delta(0)$.

d) There could be other relaxation processes both above and below T_c. A thermally activated relaxation produces a peak in attenuation at a temperature T_M which should be a function of the frequency of the ultrasonic wave. The oxide superconductors exhibit a 'glassy' behaviour at low temperatures. In other glassy materials at low temperature (≈ 1 K) there is a characteristic dependence of the loss on temperature. This dependence is explained in terms of a model involving two level systems. It will be interesting to see if the oxide superconductors show a similar behaviour in ultrasonic attenuation at low temperature.

Experimental Methods

The experimental methods used in the study of ultrasonic velocity and attenuation can be broadly divided into two categories - (a) continuous wave and (b) pulse methods.

Among the continuous wave methods special mention must be made of resonance methods often used in such studies. In these methods the specimen in the shape of a rectangular bar or a sphere is excited to find the lowest resonance frequency. The shape of the resonance curve around this frequency is also traced. From the resonance frequency and the appropriate dimension of the sample the appropriate sound velocity is calculated. From the width at half maximum of the resonance curve the quality factor Q is obtained. The loss is related to Q^{-1}. Both the sound velocity and the loss are followed as a function of temperature. If the dimensions of the specimen are typically of the order of a few mm, the resonance frequency is of the order of a megahertz. The vibrating reed arrangement uses a specimen whose thickness is about a hundred micron. The excitation is at right angles to the length of the reed. In such a case the frequency is of the order of a kilo-hertz. The continuous wave methods yield the phase velocity of the sound wave.

In the pulse methods a pulse of ultrasonic waves of 10 to a few hundred MHz frequency is sent through the specimen. The echo of the pulse due to reflection at the end face of the specimen is received at the transmitter end. The time for such a round trip travel is measured and the velocity computed. By measuring the ratio of amplitudes of successive echoes the attenuation coefficient, α, can be determined. These methods give the group velocity of the waves.

With both methods a relative accuracy of 10 ppm in the sound velocity can be achieved.

Normally at these long wavelengths, or small wave numbers k, of the ultrasonic waves, the dependence of the angular frequency, ω, on k is expected to be linear. The phase velocity will then have the same value as the group velocity. However it has been pointed out [1] that in the perovskite structures there can be a soft optic mode coupled to the acoustic mode through a strong piezoelectric effect. This can give rise to a non-linear dispersion relation even for long wave length acoustic phonons. The phase velocity can then be different from the group velocity. The existence of such an effect in the oxide superconductors has, however, not been established.

In measuring the temperature dependence of the ultrasonic properties it will be ideal if the temperature of the specimen can be controlled and maintained constant at different values during measurement. In practice such a time consuming procedure is avoided and measurements are taken while the temperature of the specimen is slowly changed at a rate of less than 1 K/minute. The use of helium exchange gas in the specimen chamber facilitates thermal exchange and reduces the temperature gradient over the sample. However the oxide superconductors in the form of sintered pellets have a much lower thermal conductivity than metals. The lower thermal diffusivity will delay the attainment of thermal equilibrium throughout the sample. This effect can be taken care of at least partly by using as slow a cooling or warming of the specimen as possible, especially in large samples ($\approx$ 1 cm).

Effect of microstructure: In measuring the sound velocity use of single crystals is to be preferred. But it has been difficult to grow large sized single crystals of these materials. Only recently it has been possible to grow single crystals of a few millimetres in size and results on single crystals are just appearing in literature. Most of the measurements performed till now are on sintered pellets. The microstructure in these pellets can have certain effects. The density of a sintered pellet is less than the ideal value calculated on the basis of the unit cell parameters determined from a X ray powder diffractogram. This difference can vary from 5 to 30 %. In such a porous medium the velocity of sound waves can be different from that in the bulk material. The effect of porosity on attenuation can be even more drastic. If the grain size in the pellet is of the same order as the wavelength of the ultrasonic wave there can be an additional unknown contribution to the attenuation arising from the scattering of ultrasonic waves. The longer the wavelength of the ultrasonic wave in relation to the grain size the more representative is the attenuation to the bulk value.

There is another effect of microstructure. Sound velocity is anisotropic and the velocity measured in a sintered pellet is the value of the sound velocity averaged over the different orientations of the grains. The smaller the grain size the closer will the orientation approximate to a random distribution. On the other hand in large grained samples in which the grains are in the form of plates, the orientation of different grains will be clustered around the normal to the plates. The sound velocity measured will be closer to the value along the normal to the plates.

Microstructure will also have a role to play in hysteresis effects due to the release of internal strains between grains during thermal cycling.

In looking at the results on sintered pellets these different effects of the microstructure have to be borne in mind. Unfortunately data on the density of the pellet and the grain size have not been provided in many of the papers on sound velocity measurements.

One should also mention here that acoustic attenuation in yttrium barium copper oxide, $YBa_2Cu_3O_{7-y}$, will depend on the oxygen vacancies present. The oxygen deficiency parameter, y, is not stated in all the papers. The transition temperature T_c is usually given. However for a range of y values between 0 and 0.2 the T_c does not depend sensitively on y . So the transition temperature does not yield any information on the oxygen deficiency parameter except to say that y is less than 0.2.

Because of the above reasons one finds considerable differences in the results obtained on the same compound by different workers. In this review those observations which are in substantial agreement in the majority of the investigations will be collected first as we can have confidence that such effects indeed are present. Other observations in which discrepancies exist between different observers will be mentioned later. Further careful work on well characterised material is required to resolve the discrepancies.

DOPED LANTHANUM COPPER OXIDE COMPOUNDS

Lanthanum copper oxide, La_2CuO_4 , undergoes a structural transformation from a tetragonal to an orthorhombic structure at 533 K. The strontium and barium doped superconducting compounds are tetragonal at room temperature. The phase diagram of the tetragonal to orthorhombic transition in the strontium doped compound has been investigated by Moret et al.[2]. The transition from tetragonal to orthorhombic structure takes place in these materials below room temperature.

Sound velocity measurements on sintered pellets of these materials have been measured by different investigators using both pulse and continuous wave methods. Luthi et al. [3] studied the longitudinal and transverse sound velocities both in La_2CuO_4 and $La_{1.85}Sr_{0.15}CuO_4$ as a function of temperature using a phase comparison method in the frequency range of 5 to 15 MHz (Fig. 1). In contrast to the undoped compound in which the sound velocities did not show any rapid variation with temperature, both the longitudinal and transverse sound velocities in the doped compound showed a strong softening as the temperature decreased in the range 220 to 100 K. The softening was 11% for V_ℓ and 17% for V_t. This softening behaviour was frequency independent in the range investigated. They concluded that the softening is not due to any elastic relaxation process but must be attributed to a structural phase transition from tetragonal to the orthorhombic phase. The softening of the sound velocities was accompanied by peaks in the attenuation at 180 K. Stavola et al. [4] had observed the tetragonal to orthorhombic transition in this compound at 190 K. A similar variation of sound velocity was observed by Horie et al. [5] in $La_{1.86}Sr_{0.14}CuO_{4-y}$, using a pulse echo technique at a

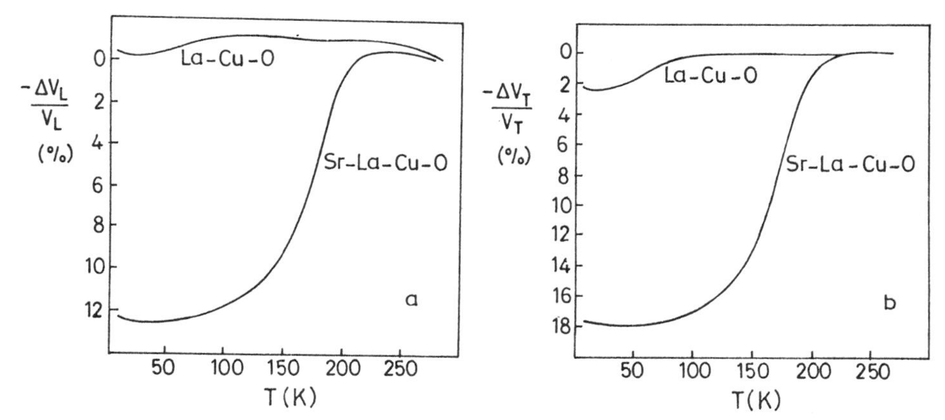

Fig. 1. Relative (a) longitudinal and (b) transverse sound velocities as a function of temperature in La_2CuO_4 and $La_{1.85}Sr_{0.15}CuO_4$ (from Luthi et al. [3]).

frequency of 10 MHz. However the attenuation curve of these authors differed from the results of Luthi et al. in showing two peaks, one at 200 K and a larger and sharper peak at 100 K. These authors speculate that the attenuation peaks at these two temperatures may be due to optical phonons. The sphere resonance method was used by Yoshizaki et al.[6] to study the structural transition by following the lowest resonance frequency as a function of temperature in a series of four strontium doped compounds with concentrations of strontium of 0.1, 0.15, 0.2 and 0.4. From the resonance frequency the corresponding elastic modulus was calculated and plotted as a function of temperature. The softening of the elastic modulus takes place for the first three concentrations, the total percentage change in the elastic modulus being the largest for $x = 0.15$. The structural transition temperatures so determined, plotted against concentration, fell on a straight line, which extra-polated backwards cut the temperature axis at the structural transition temperature for the un-doped compound. In the above examples the softening of the sound velocity was substantial (of the order of 10 to 20%). A measurement done by the pulse echo technique on $La_{1.85}Sr_{0.15}CuO_4$ at 10 MHz by Bishop et al.[7] also indicated a softening in the sound velocity from 250 to 120 K. But the percentage change in sound velocity was only 0.48%.

The longitudinal sound velocity in the barium doped compound with the composition $La_{1.85}Ba_{0.15}CuO_4$ was measured by Fossheim et al.[8] using a pulse method at 24 MHz. These authors also observed a sudden drop in the velocity starting from 200 to 240 K. This was accompanied by a sharp rise in attenuation which exhibited a broad peak at 100 K.

What happens to the sound velocity at the superconducting transition? Bishop et al.[7] noted that at T_c there was a slight softening of the sound velocity by 150 ppm (Fig. 2). These authors made an estimate of the softening to be expected from the thermodynamics of the phase transition using an approximate value for (dH_c/dp) and concluded that the softening

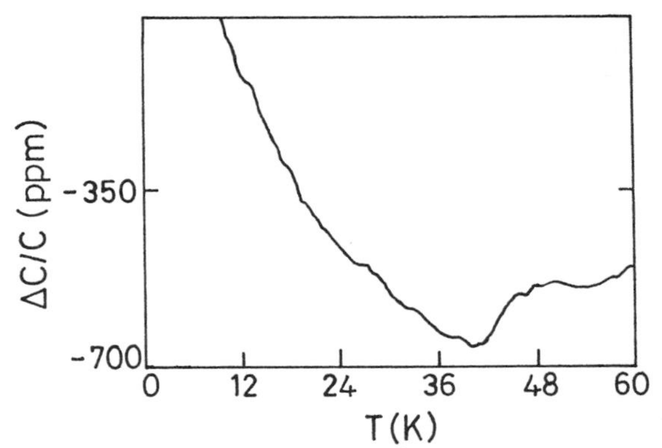

Fig. 2. Sound velocity change $\Delta c/c$ (ppm) as a function of temperature in $La_{1.85}Sr_{0.15}CuO_4$ near T_c (from Bishop et al.[7]).

observed is roughly in agreement with what is to be expected. However the other investigators have not observed such a softening. No change in attenuation at T_c was noticed by Bishop et al.[7].

All the authors quoted so far seem to observe an increase in sound velocity below the superconducting transition. One may expect a change in sound velocity below T_c due to the opening of the superconducting energy gap. But the observed variation appears to be orders of magnitude more than what is expected on this basis.

It is to be noted that all the authors mentioned above worked with frequencies from 0.5 MHz to 24 MHz. Esquinazi et al.[9] used the vibrating reed arrangement to study the resonance frequency and loss in the range 100 to 0.2 K. The resonance frequency was around 690 Hz. Fig. 3 shows the variation of sound velocity obtained by them. The velocity decreased from 100 K downwards and reached a minimum around 20 K. The total percentage variation in the sound velocity in this region is about 1.4 %. However no jump in the sound velocity at the transition temperature was noted by these authors. The sound velocity increased slightly below 20 K. The authors however noted that there was a distinct change in slope in the temperature dependence of the loss factor at 44 K (Fig. 4). The superconducting transition temperature of the compound was 40 K. Of greater interest is the fact that the loss varied as $T^{1.8}$ in the range 0.5 to 4 K and showed a plateau region thereafter upto 44 K. This plateau is similar to what is observed in many amorphous materials at such low frequencies and can be explained on the basis of a model based on two level systems (TLS). However such a model predicts a T^3 dependence of the loss at temperatures lower than the plateau region of temperatures. The observed temperature dependence is not in conformity with this prediction. The fact that the sound velocity also saturated at low temperatures instead of showing a maximum change in the plateau region of loss is in disagreement with the expectation on the TLS model.

Summarising, we see that the softening of the elastic constants in the range 220 to about 100 K is agreed upon by all observers and is

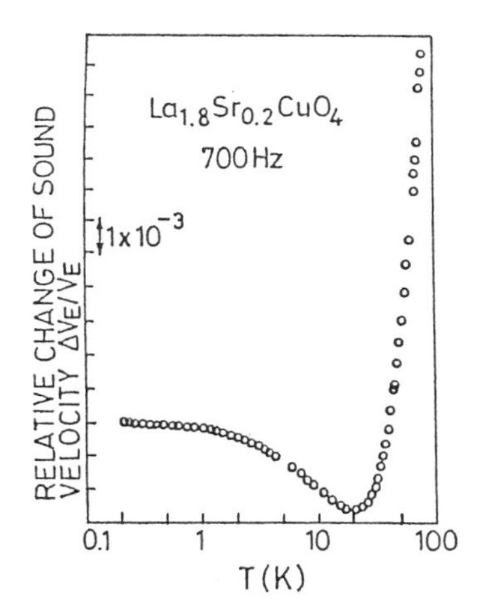

Fig. 3 Relative change in sound velocity as a function of temperature in $La_{1.8}Sr_{0.2}CuO_4$ (from Esquinazi et al.[9]).

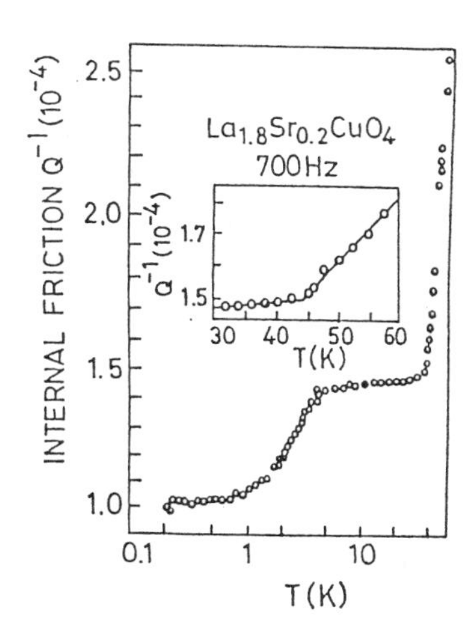

Fig. 4 Loss as a function of temperature in a sample of $La_{1.8}Sr_{0.2}CuO_4$. (from Esquinazi et al. [9])

clearly connected with the transition from tetragonal to orthorhombic structure. The open questions are (i) whether there is a softening of the sound velocity when the superconducting transition takes place in agreement with thermodynamic prediction and (ii) whether attenuation/loss at low temperatures is consistent with the TLS model.

COMPOUNDS BELONGING TO THE YTTRIUM BARIUM COPPER OXIDE FAMILY

The results on yttrium barium copper oxide are more confusing than the results on the doped lanthanum copper oxide compounds. $YBa_2Cu_3O_{7-y}$ has an orthorhombic structure at 300 K. Compounds with an oxygen deficiency y less than 0.2 are superconducting with a transition temperature T_c of about 91K. A further increase in y reduces the transition temperature, the compound with $y \geq 0.5$ not exhibiting any superconductivity at all. The oxygen stoichiometry is determined by the heat treatment around 900°C and the rate at which the sample is cooled from 900°C to room temperature.

Effect of microstructure: Muller et al.[10a] made sound velocity measurements at a frequency of 6 Hz by the pulse method. Their sample was a cylinder of the material annealed in oxygen at 770 K. The density of the sample was 80% of the ideal density. They observed a pronounced softening of the longitudinal sound velocity below 120 K while cooling the sample at 20 K/h. On warming the sample the temperature at which the velocity increased sharply was increased to about 220 K. The temperatures of softening while cooling and hardening while heating shifted when the rate of cooling or warming was increased to 60 K/h. While no sudden change in sound velocity was noticed at T_c , below 65 K a slight hardening of the sound velocity was observed. Muller et al.[10b] used a continuous wave technique at 5.6 MHz to measure the velocity in a compound, $Y_{1.2}Ba_{0.8}CuO_{4-y}$ which had a mixture of phases. In this compound they observed a monotonous increase of sound velocity as the temperature was lowered. By the application of a 6 T magnetic field they shifted the T_c to 72 K. They now observed a softening of the sound velocity by 460 ppm at T_c. Hysteresis effects were absent in the multiphase samples. The authors conclude that in the single phase sample hysteresis effects arise due to release of internal strains on cooling and warming.

Ewert et al. [11] studied the sound velocity and attenuation in sintered pellets of $YBa_2Cu_3O_{7-y}$ with different microstructures using a pulse echo method between 10 and 30 MHz. The first specimen had a density 78% of the ideal density. Electron microscope pictures showed large plate like grains almost normal to the c axis. The platelets had an average diameter of 50 μm and a thickness of 10 μm. The second pellet had a density 92% of the theoretical density and had grains of mean diameter 4 μm. These authors found that with a warming rate of 5 K/minute, strong hysteresis effects were observed with the first specimen. The sound velocity showed a softening of nearly 8% in the velocity in the range 100 to 220 K. The softening took place continuously over this temperature range on cooling while the reverse transformation took place from 150 to 220 K while warming. However when the measurements were repeated on sample 2, only a monotonic increase in sound velocity was observed as the temperature was reduced. Similar results were reported by Srinivasan et al.[12] and Ramachandran et al. [13] when they studied the longitudinal sound velocity in specimens of the compound having 63% and 85% of the ideal density (Fig.5). In the 63% compound the observed velocity of sound was very low (2200 m/s) while in the 85% compound the observed velocity of sound was 4400 m/s in agreement with the values of Ewert et al.

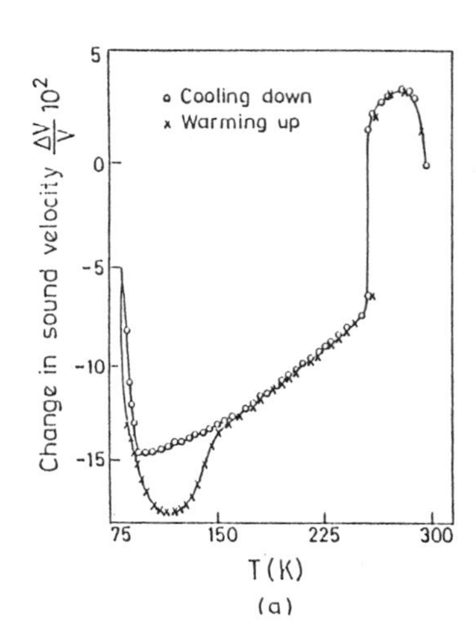

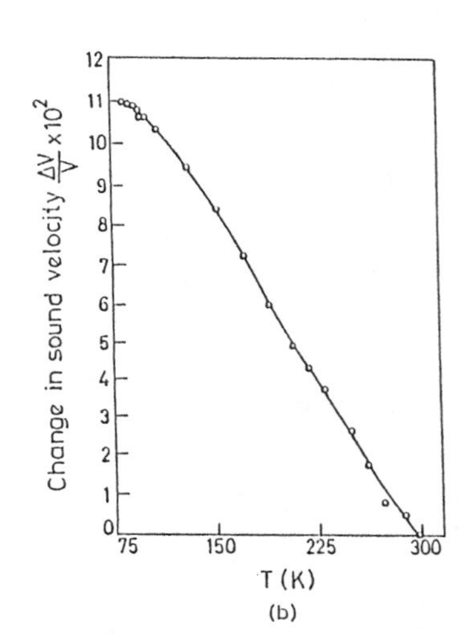

Fig. 5 Relative change in longitudinal sound velocity from its room temperature value in sintered pellets of $YBa_2Cu_3O_{7-y}$ with a density (a) 63% and (b) 85% of the ideal density (Ramachandran et al. [13]).

Is there a structural change around 230 K?

Srinivasan et al.[12] noted a steep drop in the velocity of sound around 230 K in the 63% dense sample. The shape of the temperature dependence of the velocity in the range 230 to 100 K was similar to the softening of the sound velocity in strontium doped lanthanum copper oxide discussed earlier. Even in the 83% dense sample a careful examination of the curve in Fig. 5 shows a small anomaly in slope around 230 K. They also saw a peak in attenuation around 225 K in both the samples. He Yusheng et al. [14] measured the ultrasonic attenuation in this material using a pulse echo technique at 10 MHz. Their single phase yttrium barium copper oxide had a T of only 85 K and so must have had an oxygen deficiency close to 0.2. They also made measurements on a sample of $Y_3Ba_5Cu_4NbO_{18-y}$ which existed in two phases and had a T_c of 77.3 K. Their attenuation results showed two clear peaks at 250 K and 160 K. A third peak was also seen above T_c. They observed hysteresis in the attenuation at 235 and 160 K. They have stated that their differential scanning calorimetry results showed distinct peaks at these two temperatures. In the Nb compound they observed a distinct dip in the sound velocity at 225 K. Similar measurements were made by Hori et al.[15]. They observed two small attenuation peaks at 245 and 160 K. Cannelli et al.[16] observed the loss factor as a function of temperature using a resonance technique at 1.8 to 6.5 kHz. They found that the resonance frequency showed an inflection at 235 K and the loss factor showed a spike. A small peak was seen in the loss factor at 165 K which shifted with frequency and may therefore be attributed to a thermally activated relaxation process. Using a 15 MHz pulse echo method Xu et al.[17] studied the temperature dependence of the attenuation coefficient in a sintered pellet having 78% of the ideal density. They observed a maximum in the attenuation coefficient at 252 K. Calemczuk et al.[18] have determined the Young's modulus and internal friction of a specimen of $YBa_2Cu_3O_{6.9}$ using the vibrating reed technique at a frequency of 2 kHz. They found a forty per cent change in the Young's modulus and a change in loss factor by one order of magnitude in the range 200-240 K.

These different observations appear to indicate some structural change around 230 to 250 K. Srinivasan et al.[19] measured the lattice parameters by X ray diffraction of a sample of $YBa_2Cu_3O_{7-y}$ prepared under identical conditions to the sample used for sound velocity measurements. Fig. 6 shows the variation of the lattice parameters c, a and b as a function of temperature. A clear dip in the c lattice parameter is observed at 230 K. Recently Laegried et al.[20] showed from high resolution specific heat measurements a small anomaly in the specific heat around this region which is reminiscent of a first order phase transformation. Calemczuk et al.[18] also measured the specific heat and resistivity of the sample and found evidence for a first order phase transition at 230 K. There are other properties which also have been seen to show small anomalies in this region. The transformation seen at this temperature could arise from the ordering of oxygen vacancies. However it should be stated that no anomaly in the c lattice parameter has been observed in the X ray diffraction data of by Horn et al.[21]. On the other hand Francois et al.[22] observed a small anomaly in the value of (a-b)/(a+b) at 230 K. Srinivasan et al.[19] stated that the nature of the anomaly might depend on the type of oxygen annealing treatment the sample received.

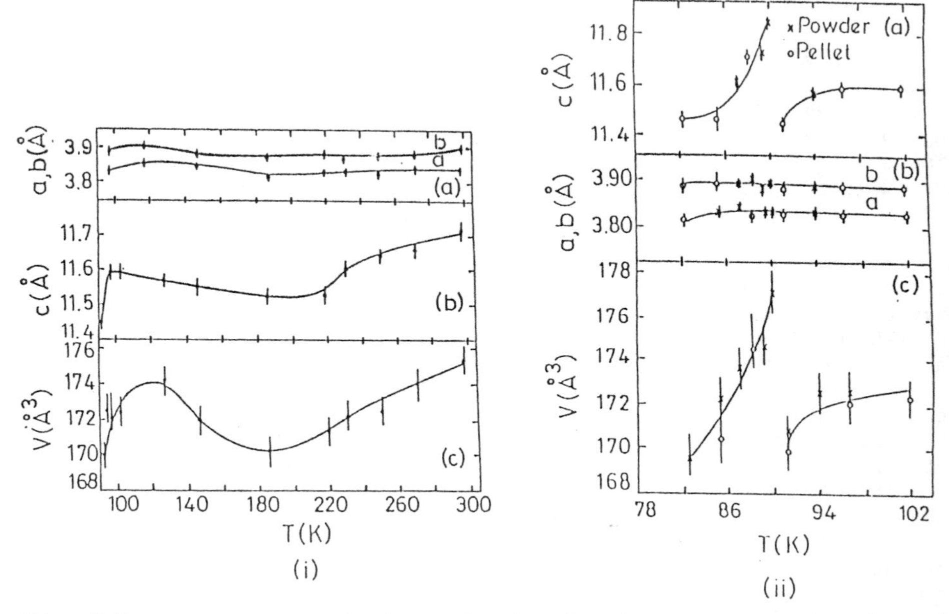

Fig. 6 Temperature variation of the lattice parameters a,b and c and the cell volume V in a sample of $YBa_2Cu_3O_{7-y}$ as a function of temperature (i) in the range 91-300 K and (ii) near T_c (from Srinivasan et al.[19]).

<u>Sound velocity change at T_c:</u> Ewert et al.[11] and Srinivasan et al.[12] noted a steep increase in the sound velocity after the superconducting transition is passed on the low density samples they used. The change was appreciable and amounts to a few per cent. But in the denser samples they used, the sound velocity increased monotonically as the temperature was lowered and there was no abrupt change at T_c. No drop in the sound velocity at T_c was found by Bishop et al.[23]. This was in agreement with the small pressure coefficient of T_c observed in this material. Migliori et al. [24] used a resonance method and found a jump in the resonance frequency at T by 0.4 % in a superconducting $YBa_2Cu_3O_{7-y}$ sample. As stated in the introduction one should expect a drop in the sound velocity when the material becomes superconducting. Also an estimate based on $\partial T_c / \partial p$ indicates that such a change should be about 50 ppm. So the observed behaviour is contradictory to expectation.

Two samples of yttrium barium copper oxide were studied by Suzuki et al [25]. The first sample was superconducting and had an oxygen deficiency of 0.1. The second sample had an oxygen deficiency of 0.8, was tetragonal and did not show superconductivity. They observed a small change in slope of the sound velocity at 90 K in the superconducting sample. By applying a magnetic field of 8 T they shifted the T_c to 72 K. But the change in slope at 90 K persisted. Such an anomaly at 90 K not linked with T_c can arise due to some structural deformation at 90 K. Duran et al.[26] also observed a similar change in slope at the transition temperature of a sample of yttrium barium copper oxide. The change in slope was more noticeable in a sample of europium barium copper oxide. The loss factor in their vibrating reed experiment also showed a well defined peak at T_c both in the yttrium and the europium compounds. They state that this behaviour resembles what one would observe in a charge density wave transition.

In their X ray diffraction studies Srinivasan et al [19] observed an anomalous change in c lattice parameter near T_c which was parallelled by the variation in the sound velocity observed by them on the 63% dense sample. The X ray diffraction results of Horn et al.[21] and the neutron diffraction data of Francois et al.[22] show a small change in (a-b)/(a+b). Srinivasan et al.[19] had stated that the details of the change in the lattice parameters near T_c may depend on the details of the oxygen annealing process.

Contrary to the above behaviour in the yttrium and europium compounds, Brown et al.[27] found a sharp decrease in the resonance frequency of a sample of $GdBa_2Cu_3O_{7-y}$ by 70 ppm at T_c. They used a vibrating reed technique at a frequency of 320 kHz. They found this decrease in agreement with the value estimated on the basis of the thermodynamics of the superconducting transition.

Variation of sound velocity below T_c: Bishop et al [23] found an increase in the sound velocity below T_c by about 1000 ppm in yttrium barium copper oxide. This was much larger than what one would expect on the basis of the opening of the superconducting energy gap. Migliori et al [24] found that the sound velocity kept increasing as the temperature was reduced to 40 K. The observed increase was larger than that observed by Bishop et al. An increase in sound velocity of 3.5% down to 4.2 K was also noted by Suzuki et al. [25] in their sample of yttrium barium copper oxide. A hardening of the sound velocity below T_c both in the yttrium and europium compounds was also found by Duran et al.[26]. A similar result was obtained by Brown et al. [27] in the gadolinium compound.

Variation of attenuation or loss: Almond et al.[28] noted a large background attenuation of 13 dB/cm in a sintered pellet of $YBa_2Cu_3O_{7-y}$. Superposed on this background a small variation of attenuation of about 1 dB/cm was observed. Notable result was the linear increase in attenuation with increasing temperature below 65 K. The attenuation reached a peak value at 65 K and showed thermal hysteresis. Since the electron mean free path is small in these materials these authors believe that the linear attenuation below T_c is due to neutral spin excitations. Cannelli et al. [16] found a decrease in the loss factor below T_c in a sample of $YBa_2Cu_3O_{7-y}$. Assuming the loss to arise from the BCS mechanism these authors fitted their results to the formula (2). Fig. 7 shows the results. From the fit these authors deduced a value for the superconducting energy gap of 41 meV. Single particle tunneling measurements give energy gaps of this approximate magnitude (for example see Prakash Fortunata Rajam et al.[29]).

Esquinazi et al.[30] have measured the Young's modulus, velocity and the loss in a sample of $YBa_2Cu_3O_{7-y}$ below 100 K with a vibrating reed experiment and frequencies in the range 1 to 2 kHz. Between 0.15 to 1 K they found the velocity to vary linearly with the logarithm of the temperature with a slope C of 7.5×10^{-5}. The variation of loss with

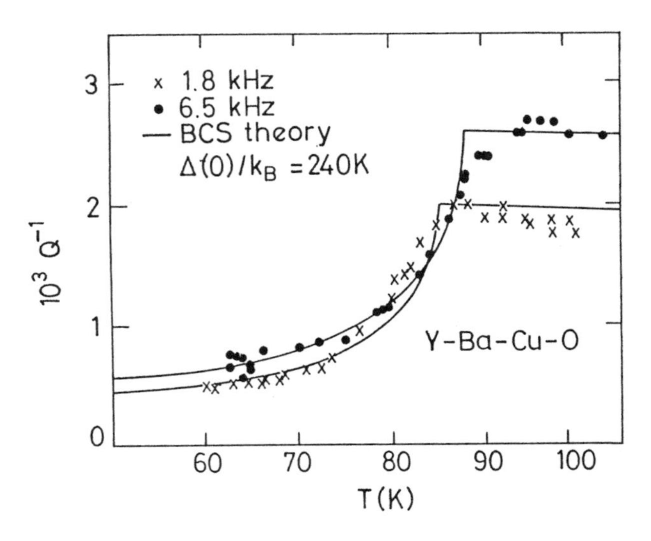

Fig. 7 Fit of BCS theory to the experimental data on loss in a sample of yttrium barium copper oxide (from Cannelli et al. [16]).

temperature is shown in Fig. 8. One observes a plateau region in the loss between 1 and 5 K. On a two level model the plateau value of the loss should be $\pi C/2$. The value of C deduced from the plateau value of the loss is 9.5×10^{-5} which agrees fairly well with the value deduced from the temperature dependence of the velocity. From their results the authors conclude that tunnelling entities exist in these materials and they are perhaps related to oxygen vacancies. Xu et al.[17] studied the variation of attenuation in a sintered pellet using the pulse echo technique at a frequency of 15 MHz. They observed a peak in the attenuation at a temperature slightly below T_c and a change in attenuation below this peak which varied as T^n where n was 2.1 above 42 K and 1.6 below 42 K. Thus these results are in contradiction to the results of Esquinazi et al.[31]. Xu et al. believe that their results indicate a Heavy Fermion like behaviour in yttrium barium copper oxide.

<u>Work on single crystal:</u> Only one paper on a single crystal of yttrium barium copper oxide has come to the notice of the author of this review. Saint Paul et al.[31] have studied the ultrasonic velocities along the c axis and in the ab plane as a function of temperature in a single crystal of $YBa_2Cu_3O_{7-y}$ of dimension 0.5 x 2.5 x 2.5 mm^3. They used a frequency between 10 and 200 MHz. They measured the velocities of the longitudinal and pseudoshear mode along the c axis and the longitudinal mode along the ab plane. Fig. 9 shows the results. In the case of waves along the c axis they found a total change in velocity of about 4% in the temperature range 190 to 40 K. In the case of the longitudinal wave in the ab plane the total increase in the sound velocity is only about a tenth as large. The increase in slope of the velocity-temperature curve in sintered specimens as the sample is cooled through T_c is not observed in the single crystal. Neither was any anomalous change in attenuation

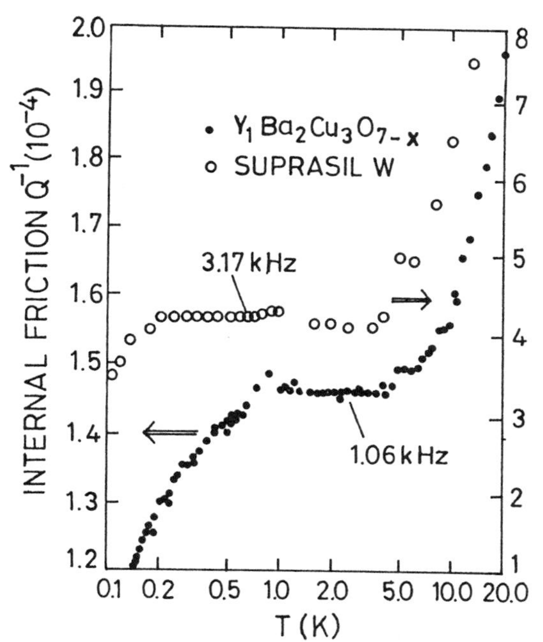

Fig. 8 Internal friction as a function of temperature in a sample of $YBa_2Cu_3O_{7-y}$ (from Esquinazi et al. [29]).

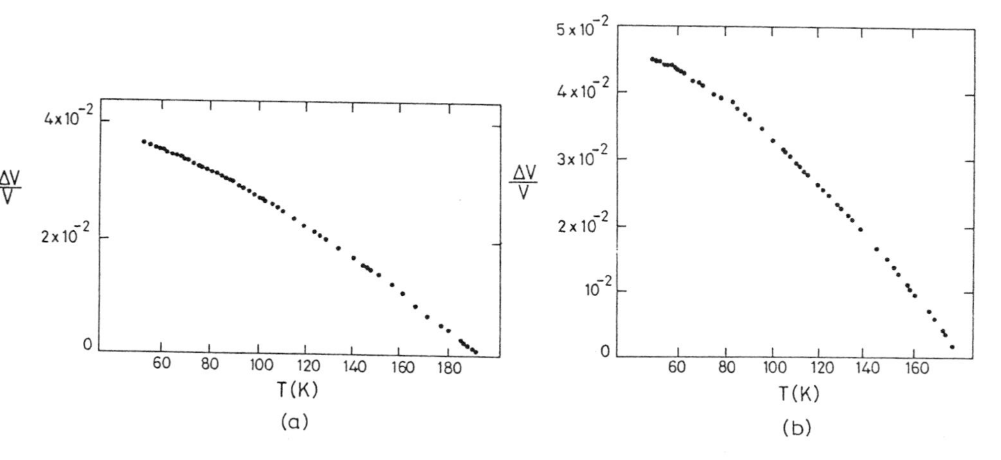

Fig. 9 Temperature dependence of the sound velocity of (a) the longitudinal mode at 15 MHz and (b) the pseudo-shear mode at 190 MHz along the c axis in a single crystal of $YBa_2Cu_3O_{7-y}$ (from Saint Paul et al. [31])

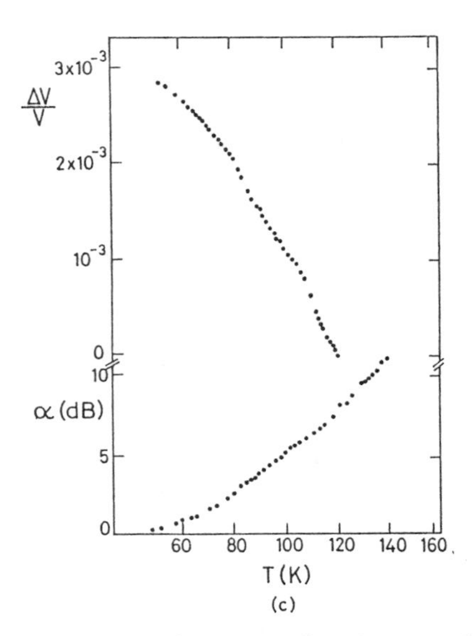

Fig. 9c Temperature dependence of the sound velocity of a longitudinal wave at 210 MHz in the ab plane of the same single crystal as above (from Saint Paul et al. [31]).

seen in the ab plane. The attenuation decreases monotonically till 40 K. Thus the results on a single crystal appear to differ from other measurements on sintered specimens.

In summary the experiments on sintered specimens of $YBa_2Cu_3O_{7-y}$ indicate the following:

(a) Microstructure seems to play a more important role in the yttrium compound than in the lanthanum compounds.

(b) There is evidence for some structural change at 230 K in agreement with the measurements on other properties of these compounds.

(c) Below T_c there is a slight hardening of the sound velocity, and there is evidence for a change in slope dV/dT at T_c. However measurements on single crystals are required to verify if this effect exists.

(d) Loss measurement at very low temperature at low frequency appears to give some support to the two level model.

The situation in yttrium-barium-copper-oxide is quite confusing. There are still many questions to be answered. Further measurements on good quality single crystals are required to resolve many questions.

Acknowledgment: The author is thankful to Mr. G.A. Ramdass, Dr. V.Sankaranarayanan and Dr. V. Ramachandran for their help in the preparation of this article.

REFERENCES

[1] V. Muller, D. Maurer, Ch. Roth, C. Hucho, D. Winau, K. de Groot, H. Eickenbusch and R.Schollhorn, Physica C, 153-155 (1988) 280.

[2] R. Moret, J.P. Pouget, and G. Collin, Europhys. Lett. 4 (1987) 365.

[3] B. Luthi, B. Wolf, T. Kim, W. Grill and B. Renker, Jap. J. Appl. Phys. 26 (1987) 127.

[4] M. Stavola, R.J. Cava, and E.A. Rietman, Phys. Rev. Letters 58 (1987) 1571.

[5] Y. Horie, T. Fukami and S. Mase, Solid State Commun. 63 (1987) 653.

[6] R. Yoshizaki, T. Hikata, T. Han, T. Iwazumi, H. Sawada, T. Sakudo, T. Suzuki, and E. Matsura, Jap. J. Appl. Phys. 26 (1987) 1129.

[7] D.J. Bishop, P.L. Gammel, A.P. Ramirez, R.J. Cava, B. Batlogg, and E.A. Rietman, Phys. Rev. B 35 (1987) 8788.

[8] K. Fossheim, T. Laegried, E. Sandvold, F. Vassenden, K.A. Muller, and J.G. Bednorz, Phys. Rev. Letters 58 (1987) 1143.

[9] P. Esquinazi, J. Luzuriaga, C. Duran, D.A. Esparza and C. D'Ovidio, Phys. Rev. B 36 (1987) 2316.

[10] a) V. Muller, K. de Groot, D. Maurer, Ch. Roth, K.H. Rieder, E. Eickenbusch and R. Schollhorn, Jap. J. Appl. Phys. 26 (1987) 2139.
b) V. Muller, K. de Groot, D. Maurer, Ch. Roth and K.H. Rieder, presented at the Yamada Conference, Sendai, (1987).

[11] S. Ewert, S. Guo, P. Lemmens, F. Stellmach, J. Wynants, G. Arlt, Dorothee Bonnenberg, H. Kliem, A. Comberg and Helga Passing, Solid State Commun. 64 (1987) 1153.

[12] R. Srinivasan, V. Ramachandran, A.T. Seshadri and G.Ananda Ramadass, Pramana J. Phys. 29 (1987) L603.

[13] V. Ramachandran , G.A. Ramadass, and R. Srinivasan, Physica C, 153-155 (1988) 278.

[14] He Yusheng, Zhang Baiwen, Lin Sihan, Xiang Jiong, Lou Yongming, and Chen Haoming, J. Phys. F. Met. Phys. 17 (1987) L243.

[15] Y. Hori, Y. Terashi, H. Fukuda, T. Fukami and S. Mase, Solid State Commun. 64 (1987) 501.

[16] G. Cannelli, R. Cantelli, F. Cordero, G.A. Costa, M. Ferretti and G.L. Olcese, Phys. Rev. B 36 (1987) 8907.

[17] M.F. Xu, H.P. Baum, A. Schenstrom, Bimal K. Sharma, Moise Levy, K.J. Sun, L.E. Toth, S.A. Wolf and D.U. Gubser, Phys. Rev. B 37 (1988) 3675.

[18] R. Calemczuk, E. Bonjour, J. Y. Henry, L. Forro, C. Ayache, M.J.M. Jurgens, J. Rossat-mignod, B. Barbara, P. Burlet, M. Couach, A.F. Khoder and B. Salce, Physica C153-155 (1988) 960.

[19] R. Srinivasan, K.S. Girirajan and V. Ganesan, Vijàyashree Radhakrishnan and G.V. Subba Rao, Phys. Rev. B 38 (1988) 889.

[20] T. Laegried, K. Fossheim, E. Sandvold, S. Julsrud, Nature, 330 (1987) 637.

[21] P.M. Horne, D.T. Keane, G.A. Held, J.L. Jordan-Sweet, D.L. Kaiser, F. Holtzberg and T.M. Rice, Phys. Rev. Letters, 59 (1987) 2772.

[22] M. Francois, A. Junod, K. Yvon, P. Fischer, J.J. Capponi, P. Strobel, M. Marezio and A.W. Hewat, Physica C, 153-155 (1988) 962.

[23] D.J. Bishop, A.P. Ramirez, P.L. Gammel, B. Batlogg, E.A. Rietman, R.J. Cava and A.J. Mills, Phys. Rev. B 36 (1987) 2408.

[24] A. Migliori, Ting Chen, B. Alavi and G. Gruner, Solid State Comm. 63 (1987) 827.

[25] M. Suzuki, Y. Okuda, I. Iwasa, A.J. Ikushima, T. Takabatake, Y. Nakazawa, and M. Ishikawa, Physica C, 153-155 (1988) 266.

[26] C. Duran, P. Esquinazi, C. Fainstein, and M. Nunez Regueiro, Solid State Commun. 65 (1988) 957.
[27] S.E. Brown, A. Migliori and Z. Fisk, Solid State Commun. 65 (1988) 483.
[28] D.P. Almond, E.F. Lambson, G.A. Saunders, and Wang Hong, J. Phys. F. Met. Phys. 17 (1987) L 261.
[29] Prakash Fortunata Rajam, C.K. Subramanian and R. Srinivasan, Rev. of Solid State Science, 2 (1988) 339.
[30] P. Esquinazi, C. Duran, C. Fainstein and M. Nunez Reguiero, Phys. Rev. B 37 (1988) 545.
[31] M. Saint Paul, J.L. Tholence, P. Monceau, H. Noel, J.C. Levet, M. Potel, P. Gougeon, and J.J. Capponi, Solid State Commun. 66 (1988) 641.

X-RAY PHOTOELECTRON SPECTROSCOPIC STUDIES OF HIGH- Tc OXIDE SUPERCONDUCTORS

B.D. Padalia and P.K. Mehta
Department of Physics, I.I.T. Bombay 400 076, India

INTRODUCTION

The field of superconductivity has been of considerable interest ever since the phenomenon of superconductivity (zero resistance state) in Hg at low temperatures below 4.2K was first discovered by Kammerlingh Onnes in 1911 [1]. Lateron, several metals and alloys exhibiting superconductivity were identified [2]. An alloy, Nb_3Ge exhibited superconductivity upto 23.2K which was the highest transition temperature (T_c) achieved in metals and alloys [3]. In the meantime, the celebrated BCS theory of superconductivity based on the interaction of electrons and phonons was developed [4]. These can be called as Low T_c Superconductors.

The search for high T_c superconducting materials was going on for several years. In 1986, the unprecedented discovery of superconductivity in a multiphase La-Ba-Cu-O system with T_c in the range of 30-40K announced by Bednorz and Müller created world wide excitement [5]. This was soon followed by the spectacular discovery of superconductivity above liquid nitrogen temperature in Y-Ba-Cu-O system with $T_c \sim 90K$, stimulating spurt of activity and interest in the physical properties and technological potentials of these materials [6]. Subsequently other rare earth based compounds of composition, $RBa_2Cu_3O_{7-\delta}$ (R = rare earths, except Ce, Pr and Tb) with $T_c \sim 90K$ were synthesized [7,8]. Intensive search for other oxide superconductors without rare earths led to the synthesis of new Bi-based materials ($T_c \sim 105K$) and those of the Tl based oxide superconductors with $T_c \sim 110K$ [9,10]. Further, there are reports of achieving higher T_c values [11-13]. Attempts are also being made to synthesize new superconducting materials without Cu as a constituent [14]. Vigorous search for new materials is likely to continue till room temperature superconductivity ($T_c \sim 300K$) is achieved.

$YBa_2Cu_3O_{7-\delta}$ compound is the high temperature superconducting phase in Y-Ba-Cu-O system. The orthorhombic structure ($\delta \leq 0.15$) exhibits superconductivity with T_c in the vicinity of 90K where as the tetragonal compound ($\delta \geq 0.5$) is non superconducting. A reversible orthorhombic to tetragonal phase transition occurs at elevated temperatures [15]. Fig.1a shows lattice structure of $YBa_2Cu_3O_{7-\delta}$ compound in orthorhombic form (a $\neq$ b $\neq$ c). For the purpose of comparison, the tetragonal structure of $YBa_2Cu_3O_{7-\delta}$ (a = b $\neq$ c) is shown in Fig.1b. A perusal of the figures indicates that there are three distorted perovskite unit cells stacked along the c-axis. The Cu atoms are at the corners and O atoms are on the cube edges. Ba and Y are the ordered metal ions located at the body centered positions of these unit cells in the sequence Ba-Y-Ba along c-axis. There are two inequivalent Cu sites represented by Cu1 and Cu2. O2 - Cu2 - O3 form planar sheets perpendicular to the c-axis. Since Y plane has no oxygen atoms, the Cu2 planes adjacent to Y planes are slightly distorted due to pulling of O atoms towards Y. An interesting feature of these structures is the presence of O1 - Cu1 - O1 linear chains along the b-direction in the orthorhombic structure which are absent in the case of tetragonal structure (Fig.1b). The Cu1 chains are weakly coupled to Cu2 planes through O4. However, the Cu1 - O4 bonds are strong and short. It is noted that in both the orthorhombic and tetragonal structures, a-axis at Cu1 site is completely devoid of oxygen. Out of possible nine oxygen sites, only seven are nearly occupied. The XRD patterns for the orthorhombic ($\delta \sim 0.15$) and tetragonal ($\delta \geq 0.5$) forms of $YBa_2Cu_3O_{7-\delta}$ are shown in Fig.1c and 1d.

The oxygen vacancy in the defect perovskite layered oxidic compounds, $YBa_2Cu_3O_{7-\delta}$ leads to charge anomaly. If formal valencies of 3+, 2+ and 2- are assigned to Y, Ba and O, respectively in $YBa_2Cu_3O_{7-\delta}$, Cu can have non integral valence depending on the oxygen vacancy concentration parameter, δ . For example, if $\delta = 0$, the valence of Cu is 2.33 (Cu^{2+} and Cu^{3+} mixed valence) while it is 2+ for $\delta = 0.5$ (tetragonal structure). Alternately, if Y, Ba and Cu are in 3+, 2+ and 2+ valence states, respectively then for charge balancing some of the oxygen atoms should be in O^{1-} valence state. Further, if we assume Cu in 2+ valence state in $YBa_2Cu_3O_{7-\delta}$, the charge neutrality condition suggests ($1 - 2\delta$) holes per unit cell. This implies that a knowledge of the oxygen vacancy concentration, δ can be utilized for estimating the valence of constituent elements in $YBa_2Cu_3O_{7-\delta}$ compounds. Since δ also governs the transition temperature, T_c, a correlation between superconductivity (T_c) and the valence state of the constituents in $YBa_2Cu_3O_{7-\delta}$ is desirable.

In order to understand the mechanism of superconductivity in the high T_c superconducting (SC) materials, several theories different from the traditional BCS approach, have been proposed [16]. Although, the importance of the electron correlations in the high T_c SC materials is generally accepted, the origin of the electron-electron attraction giving rise to Cooper pairing is still an open puzzle. The fundamental problem regarding the mechanism for superconductivity in high T_c SC materials has stimulated extensive experimental work employing a variety of techniques. Further, this has enhanced interest in the experimental studies of the electronic structure (ground state properties) of the high T_c SC materials and its relation with superconductivity.

The electronic structure of the high T_c oxide superconductors has been extensively investigated by various high energy spectroscopic techniques. These include X-ray Photoelectron Spectroscopy (XPS), Ultra-violet Photoelectron Spectroscopy (UPS),1 Auger Electron Spectroscopy (AES), X-ray Emission Spectroscopy (XES), X-ray Absorption Spectroscopy (XAS), Bremsstrahlung Isochromat Spectroscopy (BIS) and electron Energy Loss Spectroscopy (EELS). Both the laboratory and the synchrotron radiation sources have been utilized for electron excitations. In XAS, data have been collected using X-ray Absorption Near Edge Structure (XANES) as well as the Extended X-ray Absorption Fine Structure (EXAFS). An elegant review of these high energy spectroscopies for the study of high T_c oxide superconductors has recently been given by Wendin [17].

The valence states of Cu and O in $YBa_2Cu_3O_{7-\delta}$ compound have received much attention. This is considered to be important to understand the Cu-O interaction and thereby to lend support to the theoretical models of superconductivity based on Cu 3d and O 2p. Since XPS has provided precise information on the valence state of a variety of materials including mixed valence rare-earth based intermetallics [18,19], this technique has been employed extensively for the studies of Cu 2p and O 1s core levels in high T_c SC $YBa_2Cu_3O_{7-\delta}$. XPS valence band and core level studies of $YBa_2Cu_3O_{7-\delta}$ have been carried out by various experimental groups with the aim of relating the valence state information with the superconducting state. However, it appears that interpretations of results by different research groups lead to different conclusions and controversy.

In this chapter, discussion has been confined to XPS technique and its applications to the studies of valence and core levels in $YBa_2Cu_3O_{7-\delta}$ and other high T_c SC compounds. Salient features of the XPS technique are discussed and its strength and limitations are outlined. An attempt is made to provide an overview of the work done on high T_c SC compounds using XPS technique and its role in the understanding of the mechanism of superconductivity is pointed out.

SALIENT FEATURES OF XPS

XPS is a powerful tool for studying the electronic structure of solids and surfaces. In this technique, a sample is bombarded with monochromatic X-ray photons and the kinetic energy distribution of the photoejected electrons from the sample is analysed. Historically the birth of XPS is attributed to Robinson and Rawlinson who in 1914 observed a line spectrum due to emission of electrons from a sample irradiated with X-rays [20]. There was no significant progress in the development of XPS technique for several years as the research was limited by the technology of that period. It was around 1955 when Siegbahn and his co-workers at Uppsala decided to adapt the apparatus designed for high resolution β -ray spectroscopy to the study of X-ray photoelectrons and there was rapid increase in activity [21, 22]. Several reports were published during 1964 and 1970 [23-26] illustrating the potentialities of XPS in the study of core level Binding Energies (BE) of all the elements in the periodic table other than hydrogen and helium. The XPS technique was promptly employed for the studies of changes in

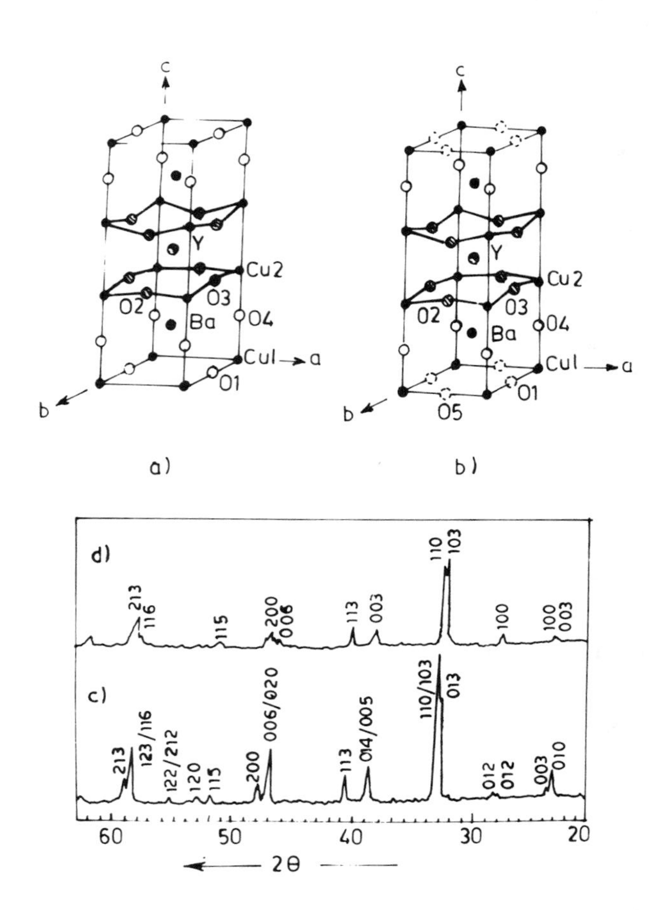

Fig.1 : Lattice structure of $YBa_2Cu_3O_{7-\delta}$ compound a) orthorhombic structure ($\delta \sim 0.15$) and b) tetragonal structure ($\delta \gtrsim 0.5$); XRD patterns for c) orthorhombic ($\delta \sim 0.15$) form and d) tetragonal ($\delta \gtrsim 0.5$) form.

BE (shifts) due to chemical environment and was popularily known as Electron Spectroscopy for Chemical Analysis (ESCA). The ability of XPS to detect carbon, oxygen and other light elements and their different chemical species in a variety of materials expanded the range of applications of this method [21,22]. Further, realization of the fact that XPS is essentially a nondestructive method and requires small quantities of samples in the form of powder, single crystal or thin film attracted much attention of scientists and technologists. This led to the design and fabrication of commercial photoelectron spectrometers with ultrahigh vacuum (UHV), sample handling and improved data acquisition facilities [27,28]. Soon, the ESCA (or XPS) machines became available in several research laboratories and industries. With the advent of commercial UHV photoelectron spectrometers, XPS emerged as an important technique for the studies of electronic structure of solids and surfaces with typical analysis depth of the order of ∼ 20 A [29,30]. Further, importance of the XPS technique has been established in the study of valence fluctuation phenomenon observed in rare earth based intermetallics and alloys [18,19,31]. The choice of XPS technique is dictated by the fact that the characteristic probing time of XPS is $\sim 10^{-16}$ sec which is faster than the valence fluctuation time ($\sim 10^{-13}$ sec). XPS provides signatures of both the valences in a mixed valent system. Since, the high T_c oxide superconductors are also highly correlated systems, the XPS technique provides reliable information on valence and other parameters in these compounds.

Basic Principle

The basic event in XPS is the ejection of an electron either from the core level or valence band in a solid by an incident X-ray photon($h\nu$). The process of photo-emission is illustrated in Fig.2. The energetics of photoemission process for a solid sample are governed by the relation:

$$E_{kin} = h\nu - E_B - \phi_{sp} \qquad (1)$$

where E_B and E_{kin} are, respectively the Binding energy and the Kinetic energy of the ejected electron, ϕ_{sp} is the work function of the spectrometer material. Momentum conservation requires that a term related to recoil be included. However, this term is negligible and normally not included [22].

Equation (1) applies to simple photoemission process in which the photoelectron does not suffer further energy loss and the system after photoejection returns to its lowest energy final state. Here it is assumed that the photoelectrons on ejection escape from the thin surface layer, equal to or less than their elastic mean free path. The mean free path is energy dependent and ranges from 20 A° for ∼ 1.5 keV photons (X-rays) to nearly 5 A° for ∼ 40 eV ultraviolet photons. The photoelectrons are assumed to escape with the kinetic energy, E_{kin} imparted to them on ejection from atoms in the sample. In practice, many photoelectrons are scattered inelastically which gives rise to a tail on the low kinetic energy side of each photoelectron peak. Plasma losses due to collective oscillations of the valence electrons may occur which produce peaks at lower BE than the primary

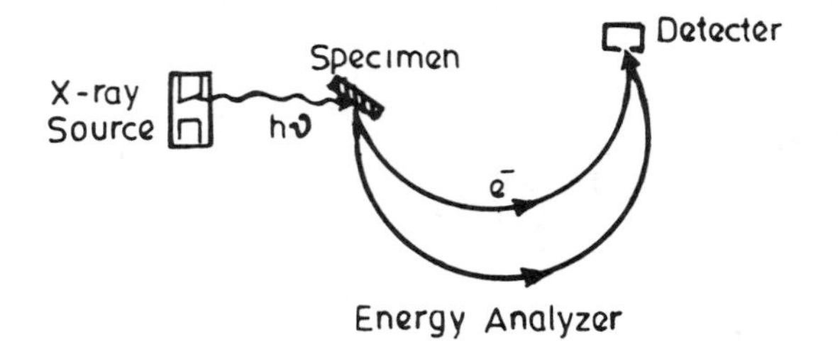

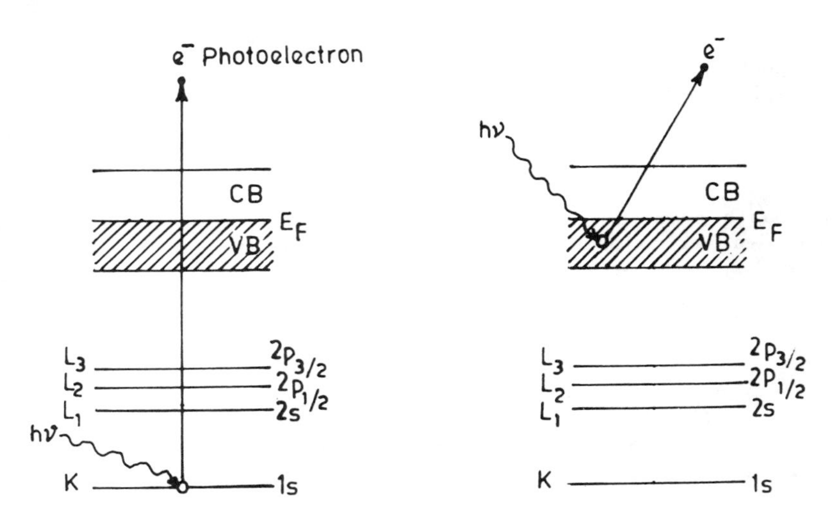

Fig.2 : Process of photoemission; core level XPS and Valence Band (VB) XPS.

photoelectron peaks. Shake up and shake off satellites associated with the outer electron excitation on production of a core hole state (two-hole-one-electron levels) may also be visible in the spectra. The photoelectron spectrum may be further complicated by the presence of X-ray induced Auger electron spectra (AES). Artefacts formed during cleaning the sample by Argon ion etching may also be present in the X-ray photoelectron spectra. An understanding of all these possibilities is necessary for an unambiguous interpretation of the spectra. For convenience of comparison, basic principles of various competitive processes are illustrated in Fig.3. It is obvious from Fig.2 that XPS is a single step process while AES and XRF (X-ray Fluorescence Spectroscopy) shown in Fig.3 are two-step processes. Further, AES is dominant in the case of light elements (atomic number, $Z \leq 12$) and also when incident beam energies are less than 10 keV. Normally the energies involved in AES are less than 2.5 keV and at least three electronic states and two electrons take part in Auger process. The kinetic energies of the Auger electrons are independent of the excitation source energies. This provides a method of identification for the Auger peaks in the X-ray induced photoelectron spectra. For this purpose, a twin anode (MgK_α and AlK_α) facility is provided in the ESCA machine.

Determination of Binding Energies

Fig.4 illustrates the basic principle of determining the BE of a solid sample in XPS spectrometer. Accurate determination of binding energies of the valence and core levels in a solid sample is made with respect to Fermi level, E_F which is taken as zero of the energy scale. For metals E_F is distinct and can be located precisely. In the case of semiconductors and insulators, the sample looses electrons continuously on X-ray bombardment. The sample may become positively charged and the photoelectron peak shifts towards higher BE side. The Fermi Edge in the valence band spectrum of an insulator or semiconductor is rather difficult to detect from the spectrum. In these cases, the BE is inferred from the kinetic energy and from the previously determined work function, ϕ_{sp} of the spectrometer. In the photoelectron spectra of semiconductors and insulators, the charging correction is generally done by recording C 1s peak (BE = 285 eV) or by depositing thin film of gold on sample surface and recording Au $4f_{7/2}$ peak at BE 83.7 eV. Normally the spectra are recorded as total electron intensity (counts per sec) against BE in eV. The intensity scale is calibrated by standard procedure and it is linear in photoelectron current but the zero is chosen arbitrarily for convenience.

EXPERIMENTAL

XPS measurements of valence and core levels are now generally made with commercial ESCA machines with UHV. Several such machines are available and their comparative performance has been reviewed by some investigators [27,28,32]. Combination of various suitable techniques such as UPS, XPS, AES, etc. in the same system is an attractive feature of these machines. Since each technique has its own relative advantages, these can be operated in the single or combined

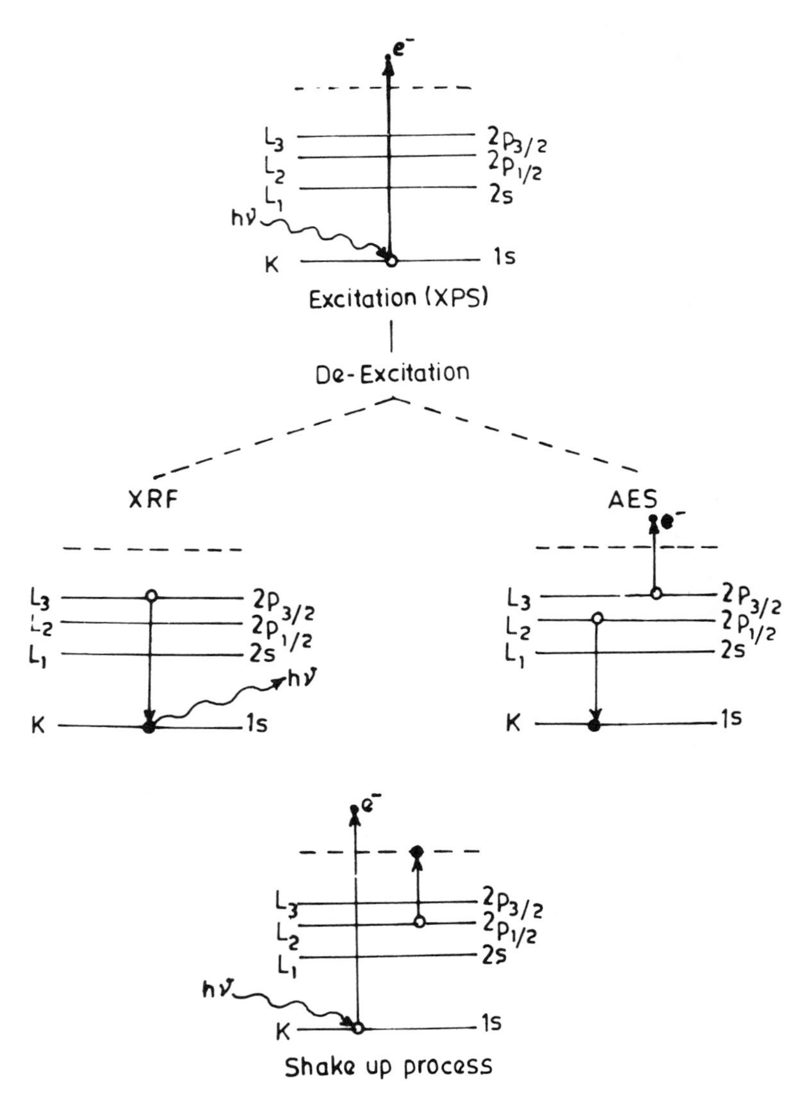

Fig.3 : Competitive processes following the photoemission process : The de-exitation may take place either by emission of a photon (XRF) or by radiationless transition (AES).

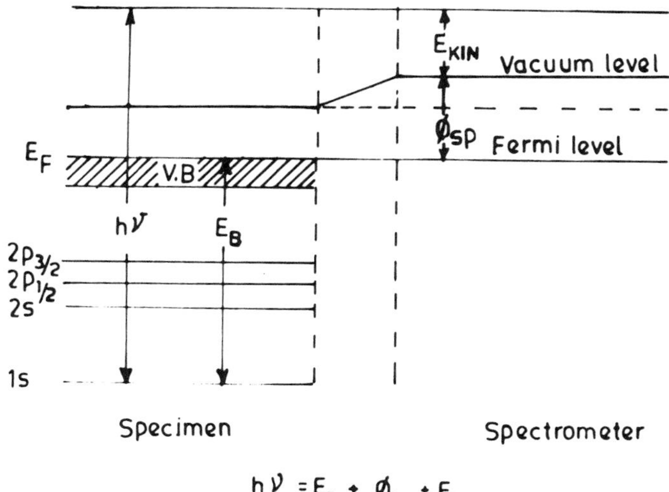

$$h\nu = E_B + \phi_{SP} + E_{KIN}$$

Fig.4 : Energy level of a solid specimen in electrical equilibrium with the spectrometer. The Fermi level, E_F of the sample and spectrometer are aligned in thermodynamic equilibrium. E_{Kin} is the Kinetic energy of the photoelectron with respect to electron spectrometer. $h\nu$ is the energy of the incident photon, E_B is the BE of the electron and ϕ_{sp} is the work function of the spectrometer (For details see [24]).

mode. Such a multiple - technique facility is employed for complete examination of the solids and surfaces, under nearly the same experimental conditions. Besides, a more recent innovation is the development of Angular Dispersion Electron Spectrometer (ADES) with UHV for studying the energy distribution of electrons at various angles of emission [24].

A schematic block diagram of a photoelectron spectrometer (VG ESCA 3-II) is shown in Fig.5. The UHV system, X-ray source, analyser chamber, sample preparation chamber, sample probe, sample handling mechanism, signal detector, etc. are the essential components of the instrument. The X-ray tube has twin anode of Al and Mg. The K_α line of Al with energy 1486.6 eV has a half width (FWHM) of nearly 0.9 eV while the FWHM of MgK_α line of energy 1253.6 eV is $\sim$0.8 eV. The electron energy analyser which is the heart of the instrument is of electrostatic type. This is easier to shield from stray megnetic fields than the magnetic type used earlier by Siegbahn et al [17]. In VG ESCA 3 (II) machine, a retarding field hemispherical analyser is used while in some other ESCA machines a cylindrical mirror analyser is preferred. Electrons ejected from a sample by X-rays and permitted to pass through the analyser are detected by channel electron multiplier at the exit slit of the analyser. The output pulses are amplified and fed to the rate meter. The rate meter counts these pulses and in turn feeds it to the recorder. The Y axis is calibrated for intensity and the X-axis for kinetic energy (or BE). Alternately, the amplified signals

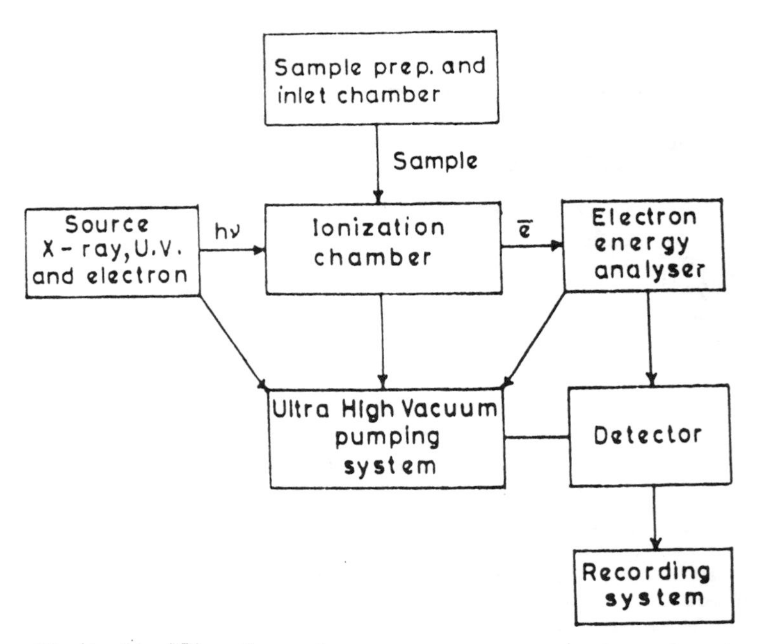

Fig.5 : Block diagram of an electron spectrometer (see for details the Ph.D. Thesis of V. Prabhawalkar, I.I.T. Bombay, 1982).

are fed directly into the on-line computer.

An important feature of the machine design is the provision of a separate sample preparation chamber. This chamber is provided with insitu evaporation unit, argon ion gun, scraper, low and high temperature controls, gas inlet valves, and the sample transfer mechanism from sample preparation chamber to the analyser chamber without breaking the vacuum. Depending upon the requirements, there are ports where other facilities such as quadrupole mass spectrometer, fracture stage etc. can be fitted to the instrument.

Generally, calibration of the spectrometer is done using Au $4f_{7/2}$ peak at 83.7 eV as standard before recording the spectra. A typical example is shown in Fig.6 by recording the spectra of evaporated gold sample using AlK_α as excitation source and the analyser energy of 20 eV. The half width (FWHM) of the $4f_{7/2}$ peak of gold which is a measure of instrumental resolution is $\sim$ 1 eV. For high resolution work, a monochromator is used which may provide a resolution of the order of $\sim$ 0.3 eV. Since in the process of monochromatization there is enormous loss of intensity, a synchrotron radiation source with intensity several orders of magnitude higher than that of a laboratory source of X-rays is preferred.

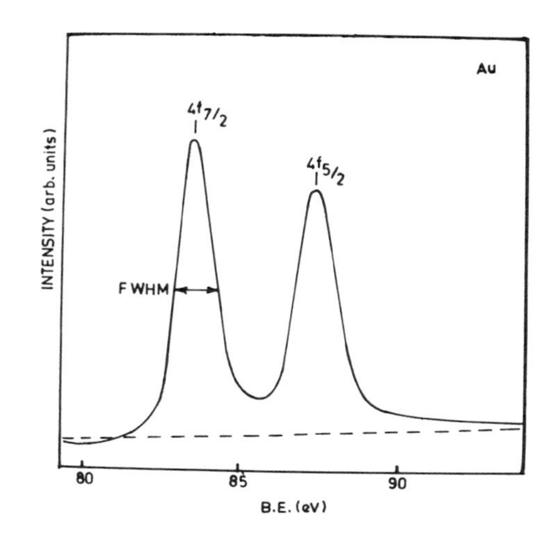

Fig.6 : XPS 4f core level spectra of pure evaporated Au. The Full Width at Half Maximum (FWHM) of Au $4f_{7/2}$ line at 83.7 eV is $\sim$1.00 eV.

X-RAY PHOTOELECTRON SPECTRA

XPS is a surface sensitive technique and it has the ability to identify the chemical states of the atoms (including the adsorbed atoms) on the sample surface. The XPS signals arise from the sample surfaces and the sampling depths (or escape depth or mean free path of electrons) are in the range of 5-25 A^o for metals and 15-40 A^o for oxides. Normally, the starting point is to take a wide energy survey spectrum of the sample surface in the 'as received' condition and identify the elements present through the measurement of their binding energies (BE). Fig.7 shows a typical XPS 1000 eV scan of a $YBa_2Cu_3O_{7-\delta}$ sample (without surface cleaning) using AlK_α radiation as excitation source. The spectrum has been recorded under UHV condition (pressure better than 10^{-9} torr). The spectrum exhibits valence band region, core levels of Y, Ba, Cu and O and a region of broad Auger transitions. The presence of a strong C 1s peak in the spectrum indicates that the sample surface is contaminated. Before proceeding further for recording the smaller energy scans (say of 30 eV) of valence band region and those of individual core levels, the surface of the sample is cleaned and made contamination free.

The XPS valence band spectra of $YBa_2Cu_3O_{7-\delta}$ compound have been recorded by several groups [33-39]. Fig.8 shows a sample spectrum of the valence band region of a clean $YBa_2Cu_3O_{7-\delta}$ sample using monochromatised AlK_α radiation (resolution $\sim$0.65 eV) at 300K and at 10K [33]. The spectral features at 300K resemble closely with those recorded at 10K. A broad peak centered at $\sim$4 eV below the Fermi level, E_F with a width of nearly 4 eV is evident. The intensity of the peak at E_F is insignificant indicating that density of states (DOS) at E_F is low. On the other hand, the independent electron band structure calculations [40] predict a narrow peak close to E_F (between E_F and $\sim$2 eV below E_F) and higher

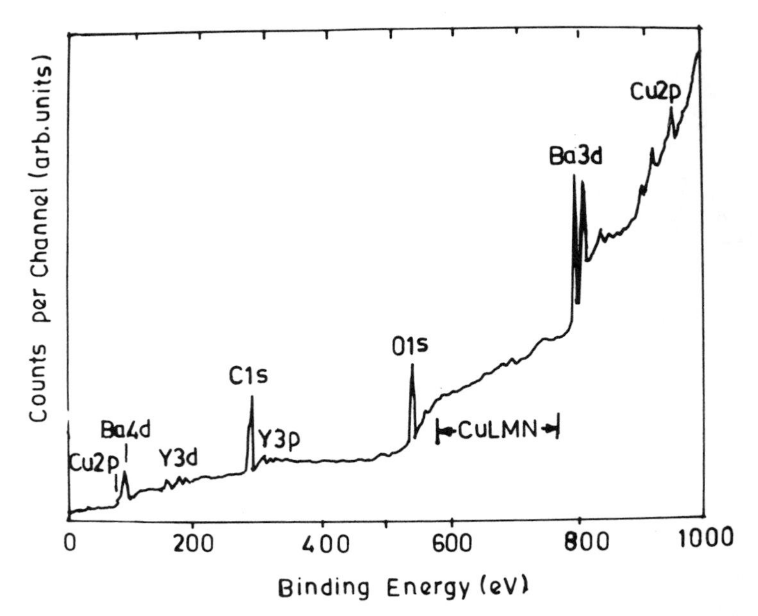

Fig.7 : A typical XPS 1000 eV scan of a $YBa_2Cu_3O_{7-\delta}$ 'as received' sample using AlK_α radiation (1486.6 eV) as excitation source. The spectra also exhibits contribution from CuLMN Auger process.

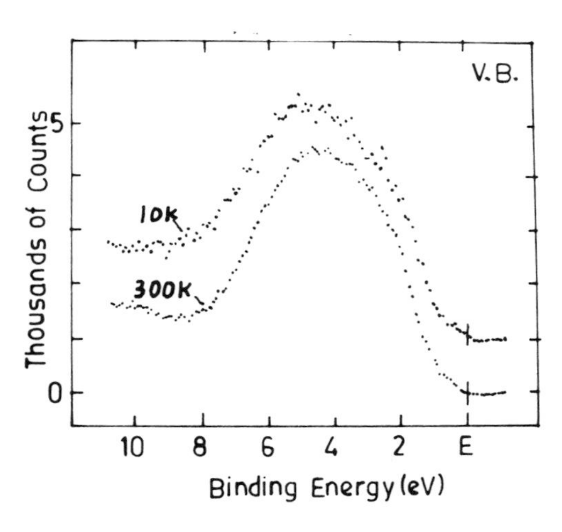

Fig.8 : Valence band region of a clean $YBa_2Cu_3O_{7-\delta}$ sample at 300K and 10K using monochromatised Alk_α radiation [33].

DOS at E_F. A comparison of the experimental results with the theoretical calculations suggests the importance of localisation and correlation effects on the Cu 3d band and O 2p hole in the high T_c SC samples. It is known that a direct comparison between XPS valence bands and DOS calculations in the case of high T_c SC compounds with strong electronic correlations is not meaningful. However, the XPS valence band of samples with negligible electron-electron correlation, are related to DOS calculations.

It is expected that valence band region studied with tunable photon energy from a storage ring would provide better agreement with the DOS calculations. For this purpose, an excellent result of Ming Tang et al [38] on valence band spectra of a single crystal of $YBa_2Cu_3O_{7-\delta}$ taken with an angle resolved electron energy analyser using different photon energies emitted by a storage ring is illustrated in Fig.9. The spectra exhibit two clearly resolved feature at BE $\sim$2.5 eV and $\sim$1.5 eV below E_F. These are associated with mainly O 2p and Cu 3d derived levels, respectively. The same two structures are also observed for the sintered samples by other investigators. The changes in the intensity ratio of the structures are explained on the basis of photoionization cross-sections. It is noted that the photoionization cross-section of O 2p levels falls while that of Cu 3d levels remains nearly the same as the energy is varied from 30 eV to 120 eV.

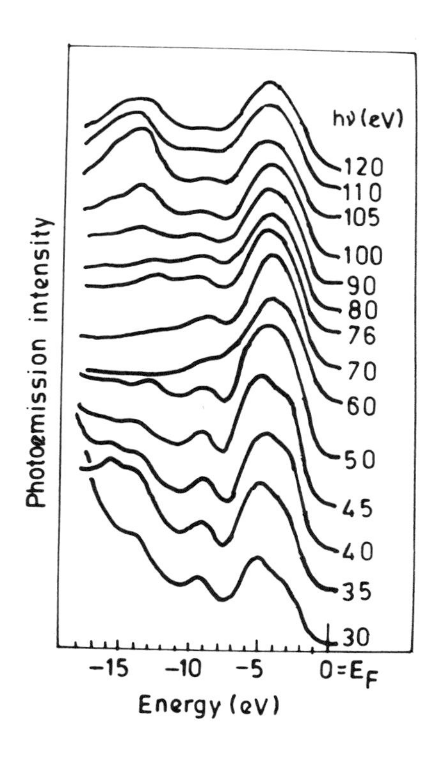

Fig.9 : Valence band spectra of a single crystal of $YBa_2Cu_3O_{7-\delta}$ taken with an angle resolved energy analyser using different photon energies [38].

The spectra also exhibits structures at ~9.5 eV and ~12.5 eV below E_F. The feature at ~ 9.5 eV was a subject of considerable controversy. Some investigators attributed the 9.5 eV structure to contaminants while some others regarded it as a satellite of the O 2p derived peak at 2.5 eV below E_F. This problem appears to be solved by the clean work of Ming Tang et al [38] who observed the same feature at ~9.5 eV below E_F in all the insitu cleaved surfaces of single crystal $YBa_2Cu_3O_{7-\delta}$ under UHV conditions. Following Wendin's explanation [17], the 9.5 eV structure has been ascribed to a satellite of the primary O 2p related photoemission peak at ~ 2.5 eV below E_F. The structure observed at ~12.5 eV below E_F in the valence band region of $YBa_2Cu_3O_{7-\delta}$ has also been attributed to a satellite associated with Cu 3d derived photoemission peak at ~ 4.5 eV below E_F. It is noted that the satellite structures observed at ~ 9.5 eV and ~ 12.5 eV below E_F in the valence band region of $YBa_2Cu_3O_{7-\delta}$ can not be accounted for by the one-electron band structure calculations [17,38]. The photoemission studies provide an evidence for strong correlations [41,42]. The chemical origin of these satellites is confirmed by resonance photoemission studies [43]. This suggests that the theoretical formulation used to explain these structures in the valence band spectra of the high T_c oxide superconductors should include correlation effects.

It is evident from the valence band spectra recorded at 300K and 10K (T_c ~ 90K) that there is no significant change in the spectral features due to variation of temperature (Fig.8). This implies that there are no specific features which can be associated with superconductivity. Further, Fuggle et al [44] have noticed that the valence band spectra of $YBa_2Cu_3O_{7-\delta}$ are similar to those recorded for La-Sr-Cu-O system. Some decrease in the intensity in the energy range between 3.5 and 7.5 eV below E_F has been detected by Sarma et al [36] in the XPS and UPS valence band spectra of $YBa_2Cu_3O_{7-\delta}$ while cooling the sample from 300K to 80K. They have also noticed a simultaneous significant increase in the intensity in the energy range of ~ 8 eV at 80K in the valence band spectra recorded with He(II) excitation source. These changes in the intensity are attributed to dimerization of oxygen ions at low temperatures. This is an interesting result. Support to this has been obtained from the study of O 1s XPS in sintered samples of $YBa_2Cu_3O_{7-\delta}$ compounds [33,36]. However, if the result could be reproduced by studying clean insitu cleaved single crystal of $YBa_2Cu_3O_{7-\delta}$ under UHV condition using tunable photon energy, it is likely to lead to the understanding of mechanism of superconductivity in high T_c oxide superconductors. It should be mentioned that the available experimental data on valence bands and theoretical calculations have not yet succeeded in relating the mechanism of superconductivity with the electronic structure of the high T_c superconductors. It is, therefore natural to turn to the XPS core level studies of these superconductors.

XPS Core Level Spectra

It is noted that the valence band XPS studies indicate strong hybridization between O 2p and the Cu 3d valence level in $YBa_2Cu_3O_{7-\delta}$ superconductors. This is reflected in the XPS core level spectra of these compounds. The XPS core level binding energy shifts are related to charge transfer between the ligand and the metal atoms. The effects of hybridization are observed in the line shapes and satellite structures. The valence states of the constituents are generally determined by comparison with the standard compounds in which the valence states are well established. The determination of valence state (chemical valence or formal valence) by the above mentioned finger-print method is rather not possible if a strong satellite structure appears in the XPS spectra. It is realized that strong configuration mixing in the initial and final states in the photoemission process complicates the problem and a direct determination of the valence state using finger-print method is not considered to be appropriate. It appears that attempts to assign formal valency to Cu by direct method of comparison have led to the existing controversy regarding the valence of Cu in $YBa_2Cu_3O_{7-\delta}$ compounds.

Fig.10 shows the XPS $2p_{3/2}$ core level spectra of Cu in oxygen annealed $YBa_2Cu_3O_{7-\delta}$ ($\delta \sim 0.07$) with $T_c \sim 90K$ recorded at 300K and 80K. The $YBa_2Cu_3O_{7-\delta}$ samples were prepared by standard ceramic technique and the single phase formation was checked by X-ray diffraction method [45]. The samples were mounted onto probe tip of spectrometer and cleaned gently by an alumina scraper. The spectra were recorded with a X-Y recorder using AlK_α radiation as an excitation source. The vacuum in the analyser chamber was maintained better than 10^{-9} torr. The instrumental resolution was $\sim$1 eV.

The spectra show two peaks at binding energies (BE) $\sim$ 933 eV and $\sim$ 942 eV. The intense peak at BE $\sim$ 933 eV is the main Cu $2p_{3/2}$ XPS signal while the associated feature at BE $\sim$ 942 eV corresponds to a satellite structure. It is apparent from Fig.10 that the spectral features exhibit no significant change on cooling from 300K to 80K. However, some investigators [33,36,46] have detected changes in the shapes of spectra of the SC compounds on cooling below T_c. Increase in the relative intensities of the main Cu $2p_{3/2}$ peak with respect to the

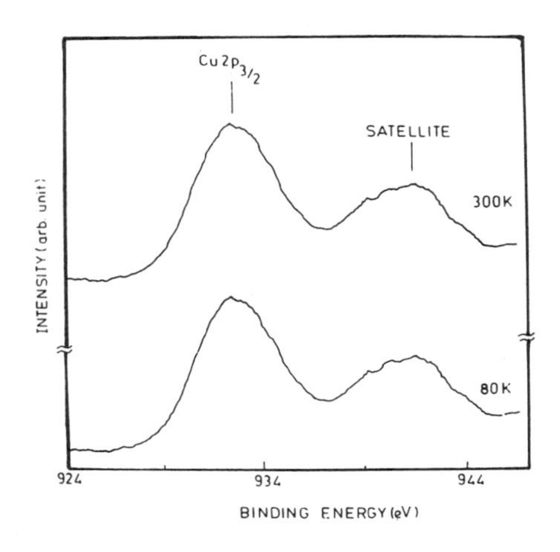

Fig.10 : XPS $2p_{3/2}$ core level spectra of Cu in oxygen annealed $YBa_2Cu_3O_{7-\delta}$ ($\delta \sim 0.07$) at 300K and 80K.

satellite structure have been observed. This is considered to be due to dimerization of oxygen. It is noted that the spectra for the $YBa_2Cu_3O_{7-\delta}$ samples resemble closely with those recorded for the $La_{2-x}Sr_xCuO_4$ at 300K and 80K [47]. Further, the simulated spectra of the Cu 2p region of $La_{2-x}Sr_xCuO_4$ at 300K and 80K look very similar to the experimental XPS Cu $2p_{3/2}$ spectra of the same sample indicating reasonable success of the configuration interaction approach.

Several research groups have recorded the XPS Cu $2p_{3/2}$ spectra for the high temperature oxide superconductors and compared the same with reference compounds in which the valence of Cu is well established. Fig.11 shows the XPS $2p_{3/2}$ core level spectra of Cu in the superconducting compound, $YBa_2Cu_3O_{7-\delta}$ and the reference compound, CuO (Cu^{2+}). It is obvious that the full width at half maximum (FWHM) of Cu $2p_{3/2}$ peak for the superconducting sample is nearly the same as that of CuO. It is rather difficult to understand the unusual broadening of the Cu $2p_{3/2}$ peak in CuO and $YBa_2Cu_3O_{7-\delta}$ ($T \sim 90K$). Further, the binding energy (BE) of the Cu $2p_{3/2}$ peak position in CuO is nearly the same as that of $YBa_2Cu_3O_{7-\delta}$. This comparison under normal condition leads to the conclusion that Cu in $YBa_2Cu_3O_{7-\delta}$ is in 2+ valence state. However, assignment of formal valence to Cu in the high T_c oxide superconductors is rather questionable as has been stated earlier in this chapter.

It is noted that the main $2p_{3/2}$ peak of Cu in $YBa_2Cu_3O_{7-\delta}$ samples is broad and asymmetric. This, perhaps motivated Ihara et al [48] to deconvolute the broad 2p peak into three narrow peaks assigned to Cu^{1+}, Cu^{2+} and Cu^{3+} with intensity ratio 1:2:1. The presence of Cu^{2+} and Cu^{3+} are favoured by the charge neutrality condition. Further, some theoretical models require the presence of Cu^{3+} for the occurance of high T_c. However, there appear to be several problems with the convolution of Cu $2p_{3/2}$ peak by Ihara et al. First of all, there are no apparent shoulder like structures in the Cu 2p peak to provide

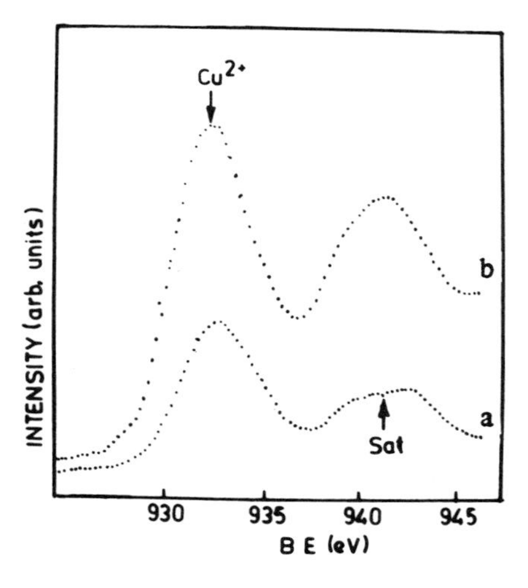

Fig.11 : XPS $2p_{3/2}$ core - level spectra of Cu in superconducting compound a) $YBa_2Cu_3O_{7-\delta}$ and b) CuO.

a hint that the broad peak represents the convolution of three narrow peaks. Secondly, the width of $2p_{3/2}$ peak for the superconducting sample is nearly the same as that of divalent CuO. Therefore, there is no convincing justification for such a deconvolution.

Enormous amount of work has been carried out on the determination of valence of Cu in the High T_c oxide superconductors. This was partly motivated by the requirements of the theoretical models based on Cu 3d and O 2p states. It is surprising to note that the interpretations of the spectra given by different groups are different and often lead to controversy regarding the valence of Cu. For example, in the case of $YBa_2Cu_3O_{7-\delta}$ compounds, Ihara et al [48] assign XPS $2p_{3/2}$ main peak to an admixture of Cu^{1+}, Cu^{2+} and Cu^{3+} valencies while Horn et al [49] attribute the same peak in the same compounds to Cu^{2+} valence state. Further, Sarma et al [36] and Sasaki et al [50] report Cu in Cu^{1+} and Cu^{2+} mixed valence states where as Steiner et al [51], Garcia-Alvarado et al [52] and Eichenbusch et al [53] detect the presence of both Cu^{2+} and Cu^{3+} states. There are several other investigators [46,48,54] who have found evidence of Cu^{3+} besides the Cu^{1+} and Cu^{2+} valence states in $YBa_2Cu_3O_{7-\delta}$ compounds. Table 1 summarises the assignments made by different groups to the XPS Cu $2p_{3/2}$ peak in the $YBa_2Cu_3O_{7-\delta}$ compounds. It is obvious that the attempts made to assign Cu valence state [46,48,51-54] are controversial and the role played by mixing of valencies of Cu in the high T_c oxide superconductors is not clear.

A typical example of the procedure followed for the assignment of valence states to XPS Cu $2p_{3/2}$ spectra of $YBa_2Cu_3O_{7-\delta}$ compounds, by Steiner et al [55] is given here. In Fig.12, the spectra of reference compounds CuO (Cu^{2+}), La_2CuO_4 (Cu^{2+}, the base structure for superconducting compounds) and $NaCuO_2$ (Cu^{3+}) are compared with the binding energies and widths of the XPS Cu $2p_{3/2}$ peak and the associate satellite structure in $YBa_2Cu_3O_{7-\delta}$ compounds. The BE and the FWHM of Cu $2p_{3/2}$ main peak for CuO (Cu^{2+}) are nearly the same as those of La_2CuO_4 and $YBa_2Cu_3O_{7-\delta}$ compounds. The BE and width of the main Cu $2p_{3/2}$ peak for $NaCuO_2$ in which Cu is in trivalent state, are significantly different from those of CuO, La_2CuO_4 and $YBa_2Cu_3O_{7-\delta}$. Comparison of the spectra in Fig.12 leads to the conclusion that Cu in $YBa_2Cu_3O_{7-\delta}$ is predominantly in 2+ state. This conclusion is supported by the comparison of the shape and intensity of the satellite structure associated with the main Cu $2p_{3/2}$ peak in CuO, La_2CuO_4 and $YBa_2Cu_3O_{7-\delta}$ compounds. However, the possibility of the presence of a small Cu^{3+} component is not ruled out due to the unusually large width of the $2p_{3/2}$ peak in $YBa_2Cu_3O_{7-\delta}$ compounds. A more recent work of Steiner et al [51] on the measurements of the XPS 2p core level spectra of $YBa_2Cu_3O_{7-\delta}$ samples as a function of annealing temperatures under 1 bar flowing oxygen has revealed interesting formation on the valence of Cu. A small Cu^{3+} signal has been detected in the Cu $2p_{3/2}$ spectra of $YBa_2Cu_3O_{7-\delta}$ samples annealed in the temperature range 350°C and 550°C. The Cu^{3+} concentration saturates at about 10% below 500°C. Above 500°C the Cu^{3+} concentration decreases to zero. Between 600°C and 800°C, the Cu exists in nearly Cu^{2+} state. Above 800°C, a Cu^{1+} signal appears and it increases with increasing annealing temperature. In all these cases the samples after annealing were quenched to room temperature. These are, indeed interesting results and require confirmation by other

Table 1 : Summary of the results reported on XPS Cu 3d and O 1s core level spectra of high T_c oxide superconductors

XPS Cu $2p_{3/2}$ spectra		O 1s spectra assignments	Reference
Main peaks and assignments	Satellite structure and assignment		
$YBa_2Cu_3O_{7-\delta}$ Cu^{2+} No evidence of Cu^{3+}	-	Presence of oxygen holes	[49]
Peak at BE ~933 eV due to 'well screened' $3d^{10}$ states Reduction of Cu^{2+} to Cu^{1+} below T_c	Structure at BE~942 eV due to 'poorly screened $3d^9$ states	Peak at BE 528.7 eV assigned to O^{2-} ions At BE 531.3 eV due to contaminant species At BE 533.3 eV assigned to absorbed oxygen dimers.	[36]
Peak at BE 933.6 eV at room temperature assigned to pure Cu^{2+} Reduction of the Cu^{2+} to Cu^{1+} state on heating	Structure at BE 940-945 eV due to $3d^9$ final state configuration Intensity decreases on heating	Peak at BE~529 eV due to O4-site oxygen At BE 531.2 eV due to O2-site oxygen At BE 532.4 eV due to O1-site oxygen	[50]
Deconvolution into three peaks Peak at BE 931.5 eV assigned to Cu^{1+} At BE 932.6 eV to Cu^{2+} At BE 934.2 eV to Cu^{3+}	-	-	[48]
Peak at BE 933.8 eV due to $3d^{10}$ final state Majority of Cu^{2+} at 300K At low temperature ~80K increase in FWHM interpreted as presence of Cu^{3+} and Cu^{1+} in addition to Cu^{2+} state	Structure at BE ~942 eV due to $3d^9$ final state	-	[46]

Table 1 (Contd.)

XPS Cu $2p_{3/2}$ spectra		O 1s spectra assignments	References
Main peaks and assignments	Satellite structure and assignment		
Assignments to Cu^{1+}, Cu^{2+} and Cu^{3+} No direct evidence of Cu^{3+}	Structure at BE ~943 eV due to charge transfer from oxygen into empty Cu d-state	Peak at BE 528.2 eV due to pyramidal configuration At BE 531.1 eV due to square planar configuration At BE 533.2 eV due to octahedral configuration	[54]
Peak at BE 932.8 eV due to Cu^{2+} At BE 934.8 eV due to Cu^{3+} (Cu^{3+} is 36%)	-	-	[52]
Peak at BE 933.3 eV due to Cu^{2+} shoulder at BE 934.9 eV due to Cu^{3+}	Structure at BE 943.3 eV	Peak at BE 528.9 eV due to transition metal oxide At BE 531.7 eV due to main group element oxides	[53]
Peak due to $3d^{10}\underline{L}$ configuration	Structure due to $3d^9$ configuration	-	[42]
Peak at BE 933.1 eV due to Cu^{2+}	Structure at BE ~943 eV	Peaks at BE ~529 eV and ~531 eV due to nonequivalent oxygen atoms	[59]
Peak at BE ~933 eV identified as Cu $3d^{10}$ O $2p^5$ 'well screened' final state a broadening is observed on cooling in $3d^{10}$ O $2p^5$ final state	Structure located between BE 940 and 945 eV due to the Cu $3d^9$ O $2p^6$ 'poorly screened' final state	Peak at BE 528.9 eV assigned to oxide Origin of peak at BE ~531 eV is not understood Peak at BE ~533 eV assigned to O_2^{2-} incipient dimers	[33]

Table 1 (Contd.)

XPS Cu $2p_{3/2}$ spectra		O 1s spectra assignments	References
Main peaks and assignments	Satellite structure and assignment		
Peak at BE~933 eV assigned to $2\underline{p}_{3/2}\ 3d^{10}\underline{L}$ final state where $\underline{L}$ refers to ligand hole and $2\underline{p}_{3/2}$ refers to hole in the $2p_{3/2}$ level	Structure at BE ~942 eV assigned to $2\underline{p}_{3/2}\ 3d^{9}$ final state	Peak at BE 528.9 eV assigned to O^{2-} ions At BE 530.9 eV to oxygen atoms with holes induced by $3d^{9}\underline{L}$ configuration	[37]
Peak at BE ~933 eV assigned to $2\underline{p}_{3/2}\ 3d^{10}\underline{L}$ final state At BE 934.9 eV due to $\vert 3d^{9}LK\rangle$ configuration in ground state (can be regarded as particular case of Cu^{3+}) where K is a conduction electron	Structure at BE ~943 eV assigned to $2\underline{p}_{3/2}\ 3d^{9}$ final state	Peak at BE ~529 eV assigned to O^{2-} At BE ~531 eV to O^{1-}	[58]
Cu^{2+} and Cu^{3+} (~10%) definite Cu^{3+} for very slow cooled sample	-	Holes on the oxygen sites in the basal plane	[51]
Cu^{2+} state	-	-	[34] and [55]
$La_{2-x}Sr_xCuO_4$			
Deconvoluted into three peaks Peak at BE 933.0 eV assigned to Cu^{1+} At BE 934.3 eV to Cu^{2+} At BE 935.7 eV to Cu^{3+}	-	-	[89]

Table 1 (Contd.)

XPS Cu $2p_{3/2}$ spectra		O 1s spectra assignments	References
Main peaks and assignments	Satellite structure and assignment		
Peak at BE 933.2 eV attributed to $2\underline{p}\ 3d^{10}\ \underline{L}$-like final states No Cu^{3+}	Structure at BE ~941 eV attributed to $2\underline{p}\ 3d^{9}$ - like multiplet of final states	-	[41]
Possibility of Cu^{2+}/Cu^{3+} charge fluctuation	-	Peak at BE 528.8 eV main line At BE 531.3 eV satellite	[90]
Peak at BE~933 eV due to 'well screened' $3d^{10}$ state Cu^{1+} Absence of Cu^{3+}	Structure at BE ~942 eV due to 'poorly screened' $3d^{9}$ state	Peak at BE 529 eV due to O^{2-} oxide species At BE 531.5 eV can be due to O^{1-} species and/or contaminant like $(OH)^{-}$ or $(CO_3)^{2-}$ At BE ~533 eV attributed to peroxo-like O_2^{2-} species	[91]
Bi-Ca-Sr-Cu-O			
Basically Cu^{2+} state	-	Single O 1s peak	[92]
Cu^{2+} state	-	-	[93]
Tl-Ca-Ba-Cu-O			
Cu^{1+} and Cu^{2+} states	-	Presence of oxygen holes and their dimers	[95]

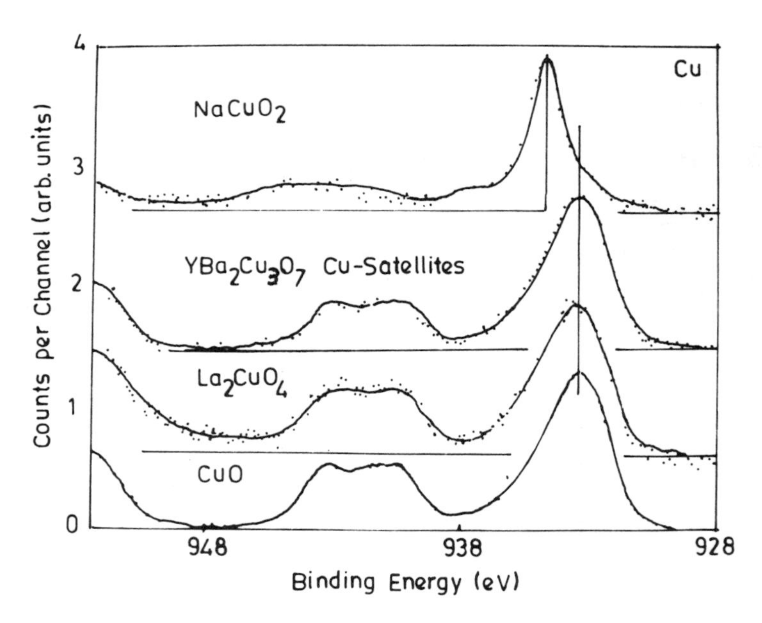

Fig.12 : XPS $2p_{3/2}$ spectra of Cu in SC compound $YBa_2Cu_3O_{7-\delta}$ and reference compounds CuO, La_2CuO_4 and $NaCuO_2$ [55].

groups equipped with excellent sample preparation and high resolution XPS facilities, preferably with single crystals using synchrotron radiation source. It is equally important to relate these results with high T_c superconductivity mechanism.

It should, however be emphasized that $KCuO_2$ and $NaCuO_2$ compounds themselves are a subject of controversy regarding the formal valence state 3+ to Cu. The possibility of defect structure with complicated Cu valence state or oxygen dimerization in $KCuO_2$ ($NaCuO_2$) can not be ruled out. The system requires careful investigations. It is not surprising if a similarity appears in the Cu valence state in $KCuO_2$ + 2CuO mixture and $YBa_2Cu_3O_{7-\delta}$ wherein 66% of Cu is in passive Cu^{2} state (similar to CuO).

Since CuO and $YBa_2Cu_3O_{7-\delta}$ compounds are highly correlated electronic systems, their XPS 2p spectra exhibit satellite structures associated with the main XPS Cu 2p peaks. The origin of satellite structures in the XPS Cu 2p spectra of divalent copper compounds was explained earlier by Larsson [56] and Van der Laan et al [57] in terms of final state energies diminished by coulomb interactions between the valence electrons and the core hole. It was noted that the intensity of the satellite provides information on the relative position of the Cu 3d and ligand valence hole and the strength of hybridization. On the basis of their experience with the interatomic intermediate valence (IIV) oxide, NiO, Bianconi et al [37] realized that the description of

$YBa_2Cu_3O_{7-\delta}$ ($\delta = 0.5$) is similar to NiO. This means that $YBa_2Cu_3O_{7-\delta}$ ($\delta = 0.5$) can be described by mixing of many configurations $3d^9$ and $3d^{10}$ $\underline{L}$ where $\underline{L}$ is a hole in the oxygen derived band also called ligand-hole. The superconducting state in $YBa_2Cu_3O_{7-\delta}$ ($\delta < 0.5$) can be assigned to the intinerant $3d^9$ $\underline{L}$ configuration. The holes in oxygen band are coupled with the localized Cu 3d elements giving 3d $\underline{L}$ manybody configurations. The ground state, $|i\rangle$ can be obtained by mixing these configurations. Still what is in reality the spectroscopic form of the ground state needs to be understood. Assignments of the initial and final states associated with the XPS Cu 2p spectra of $YBa_2Cu_3O_{7-\delta}$ have been made by some investigators with slight variations [40,51,58] and these are included in Table 1.

The main Cu $2p_{3/2}$ peak at BE $\sim$933 eV (Fig.12) has been assigned to the transition associated with $\underline{2p}_{3/2}$ $3d^{10}$ $\underline{L}$ final states while the satellite structure at BE $\sim$ 942 eV is attributed to $\underline{2p}_{3/2}$ $3d^9$ final states. In these notations, $\underline{2p}_{3/2}$ refers to a hole in the $2p_{3/2}$ core level of Cu and $\underline{L}$ is a hole in the O-Cu ligand. It is evident that the final states involve an electron transfer from oxygen 2p orbital to Cu 3d band. The configuration mixing, therefore involves ligand-metal charge transfer and leads to hybridization. Finally, whether we call $3d^9\underline{L}$ as Cu^{2+} or Cu^{3+} is still an open question.

The presence of the $3d^9\underline{L}$ in the ground state indicates that holes in the oxygen 2p band are coupled with the localized Cu 3d electrons giving $3d^9\underline{L}$ many body configuration. This should reflect in the XPS O 1s core level spectrum of $YBa_2Cu_3O_{7-\delta}$ samples. Fig.13 shows XPS O 1s spectrum in the 'as received' oxygen annealed $YBa_2Cu_3O_{7-\delta}$ sample. The deconvoluted spectra (solid lines) show peaks at $\sim$529 eV and $\sim$531 eV. The intensity of the peak at BE = 531.3 eV is large compared to the peak at BE $\sim$ 529 eV. The peak at $\sim$529 eV is due to O^{2-} in bulk oxide while the peak appearing at $\sim$ 531 eV is difficult to analyse. If we examine the structure of $YBa_2Cu_3O_{7-\delta}$, we find inequivalent O-sites in the CuO_2 planes and Cu-O chains. Further, as stated earlier XPS is a surface sensitive technique and therefore, the surface contamination

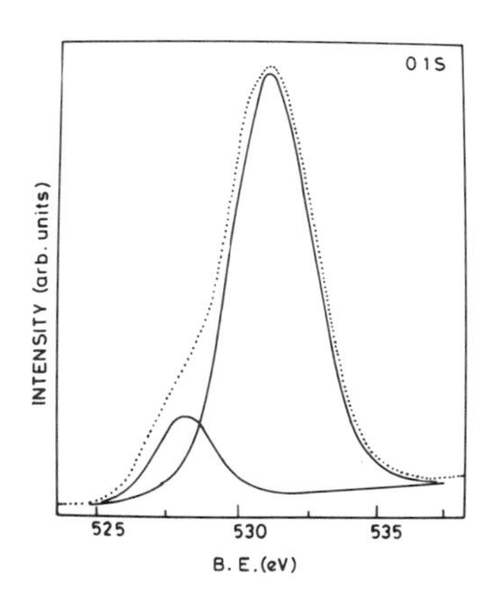

Fig.13 : XPS O 1s spectrum in the 'as received' oxygen annealed $YBa_2Cu_3O_{7-\delta}$ sample. The deconvoluted spectra are shown by solid curves.

affects the O 1s spectrum. A perusal of the O 1s spectrum reported by different groups provides an idea of the quality of the sample analysed.

Bianconi et al [37] have reported XPS O 1s spectrum of a freshly scraped $YBa_2Cu_3O_{7-\delta}$ sample and have observed a peak $\sim$529 eV and another intense peak at 530.9 eV. They observed reduction in the intensity of 530.9 eV peak and simultaneous increase in the intensity of $\sim$529 eV peak by keeping the sample in vacuum and also by ion etching the sample surface. This indicates extreme surface sensitivity of the $\sim$531 eV peak.

Further, Fig.14 shows XPS O 1s spectra of $YBa_2Cu_3O_{7-\delta}$ sample [33]; the spectra were recorded before and after scraping the sample. It is evident from the figure that the $\sim$529 eV peak grows in intensity and the $\sim$531 eV peak intensity decreases on scraping. This provides a

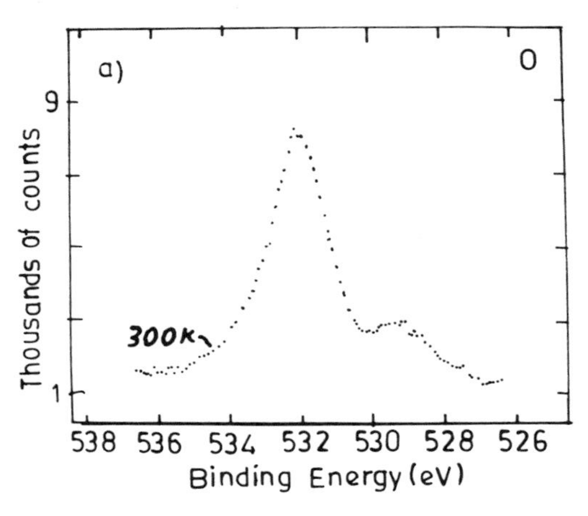

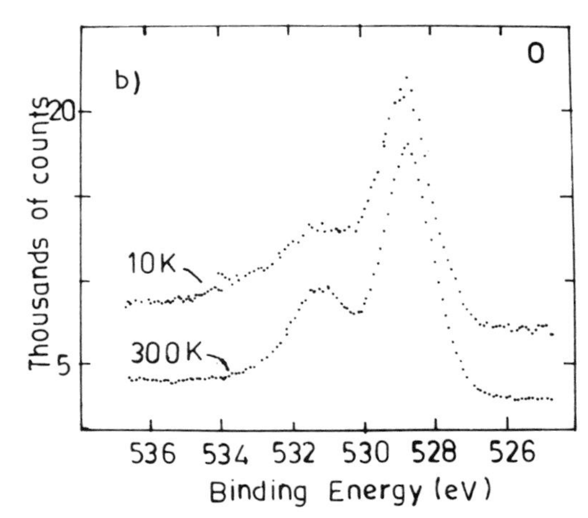

Fig.14 : XPS O 1s spectra of $YBa_2Cu_3O_{7-\delta}$ sample a) before scraping and b) after scraping [33].

strong evidence of the fact that the majority of the ~531 eV peak intensity is due to chemisorption etc. There is still evidence of some intensity, at ~531 eV which appears to be due to the presence of O^{1-} species (or holes induced by $3d^9\underline{L}$ configuration). It is realized that a small fraction (13%) of lattice oxygen should be in O^{1-} state in $YBa_2Cu_3O_{7-\delta}$ (~0.07) to satisfy charge neutrality condition.

Some investigators have observed an additional structure in the XPS O 1s spectra of $La_{1-x}Sr_xCuO_4$ and $YBa_2Cu_3O_{7-\delta}$ at ~533 eV and have associated the origin to peroxo-like O_2^{2-} species [33,36]. It is noted that intensity of O_2^{2-} like species increases on cooling the sample (300K to 80K). The structure appearing at ~531 eV has been assigned to O_2^{2-} incipient dimers [33,36]. This feature becomes distinct on cooling the sample to 10K. However, further systematic and high resolution work on XPS O 1s is required to decide in favour of dimerization.

Although XPS $2p_{3/2}$ and O 1s spectra in $YBa_2Cu_3O_{7-\delta}$ compounds have been a subject to considerable discussion, Ba 3d and 4d core levels are equally important. Fig.15 shows Ba XPS $3d_{5/2}$ peak in oxygen annealed $YBa_2Cu_3O_{7-\delta}$ compounds recorded at ~300K and ~80K [60]. A comparison of the XPS $3d_{5/2}$ spectra of Ba in the superconducting samples with those recorded for the non superconducting Y_2BaCuO_5 and $BaCO_3$ compounds suggests that Ba in $YBa_2Cu_3O_{7-\delta}$ compounds retains the 2+ valence state. The Ba $3d_{5/2}$ peak at ~780 eV for the $YBa_2Cu_3O_{7-\delta}$ compound when cooled to 80K shows an asymmetry in the line profile. The deconvoluted spectra

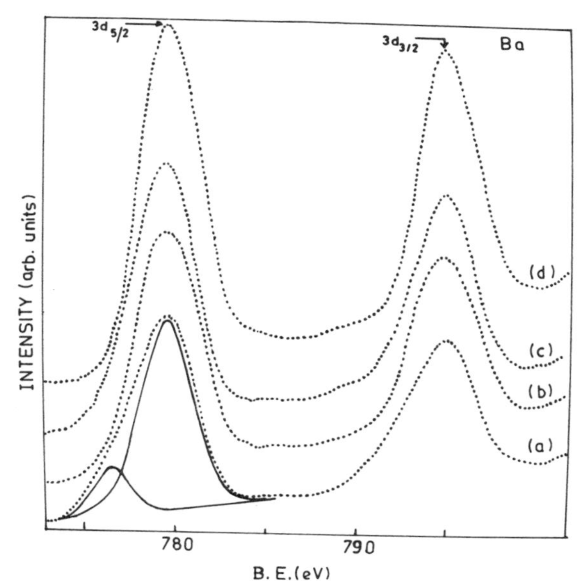

Fig.15 : XPS Ba 3d spectra in oxygen annealed $YBa_2Cu_3O_{7-\delta}$ a) at ~300K and b) ~80K and in those of reference compounds c) Y_2BaCuO_5 and d) $BaCO_3$.

are shown by solid curves (Fig.15). Similar asymmetry in the XPS Ba $3d_{5/2}$ spectra of $YBa_2Cu_3O_{7-\delta}$ has been observed by Steiner et al [51] and Fjellvag et al [59]. The asymmetry has been considered to be due to (i) Ba^{2+} ions coordination to O^{2-} ions and O^{2-} vacancies and (ii) in-equivalent crystallographic sites and defect structures.

Earlier in this chapter, we have mentioned that Steiner et al [51] annealed $YBa_2Cu_3O_{7-\delta}$ samples in the temperature range 350°C and 550°C and observed a small Cu^{3+} signal in the XPS Cu $2p_{3/2}$ spectra (Fig.16). This is associated with distinct changes in the Ba XPS 4d spectra (Fig.17). The XPS Ba 4d spectra for samples of $YBa_2Cu_3O_{7-\delta}$ annealed at 1000°C and 420°C at 1 bar O_2 for 15th are shown in Fig.17a and 17b, respectively. The difference spectra are shown in Fig.17c. It is obvious that shoulder like pattern in the Ba $4d_{5/2}$ peak (Fig.17b) is associated with the appearance of a weak signal (assigned to Cu^{3+}) in the XPS Cu $2p_{3/2}$ spectra (Fig.16). This structure saturates at ~30% and above 600°C intensity of this structure decreases continuously to zero.

In order to see if the asymmetry observed in the XPS Ba $3d_{5/2}$ spectra of oxygen annealed $YBa_2Cu_3O_{7-\delta}$ compounds is a characteristic feature of the high T_c SC samples, we have recorded the spectra of air annealed ($T_c \sim 90K$) and nitrogen annealed ($T_c < 77K$) samples where in oxygen stoichiometry is changed. Further, the XPS $3d_{5/2}$ spectra of Ba in the $YBa_2Cu_3O_{7-\delta}$ replacing Y ions by magnetic Gd ions were

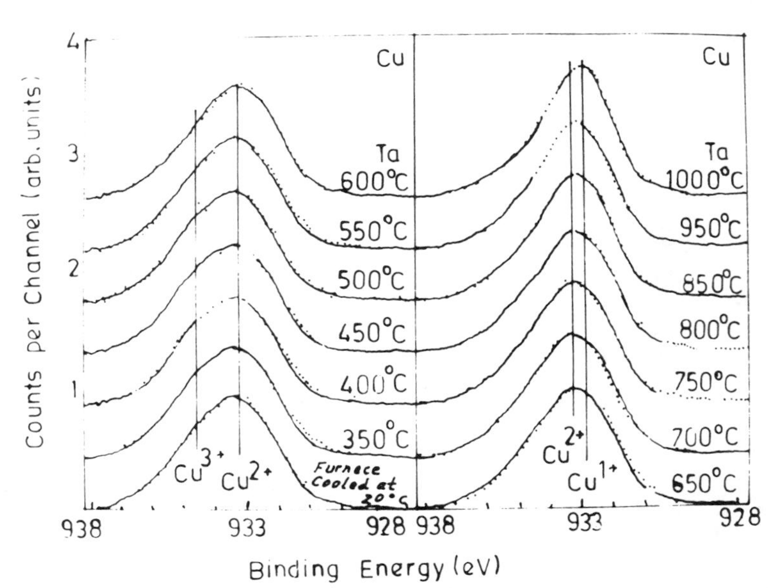

Fig.16 : Measurements of the XPS Cu $2p_{3/2}$ core level spectra of $YBa_2Cu_3O_{7-\delta}$ samples as a function of annealing temperatures (Ta) under 1 bar flowing oxygen in the temperature range, 350°C to 650°C.

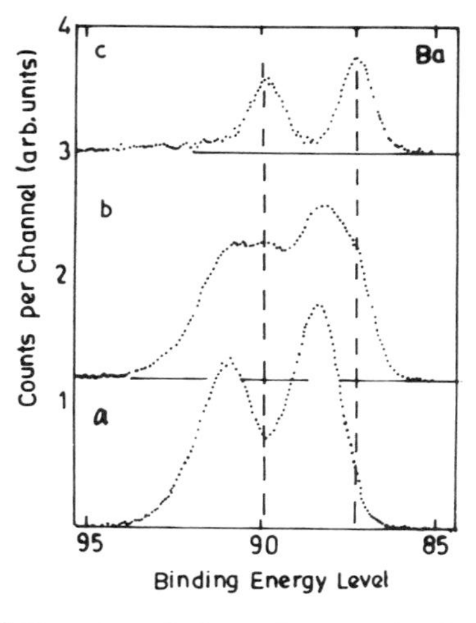

Fig.17 : XPS Ba 4d spectra for samples of $YBa_2Cu_3O_{7-\delta}$ annealed at a) 1000°C, b) 420°C and c) the difference spectra of a - 0.7b.

also recorded and are included in Fig.18. It is observed that no asymmetry appears in the XPS Ba 3d spectra for the air and N_2 annealed $YBa_2Cu_3O_{7-\delta}$ samples as well as for oxygen annealed $GdBa_2Cu_3O_{7-\delta}$. Further, the Ba 3d spectra recorded for the non-superconducting Y_2BaCuO_5 sample also show asymmetry. It, therefore appears that the observed asymmetry in the XPS 3d core levels of Ba in the oxygen annealed $YBa_2Cu_3O_{7-\delta}$ samples is not associated with the superconducting behaviour [60]. However, relation with the superconductivity of the Ba(1) component in the XPS Ba 4d spectra corresponding to the appearance of a small component of Cu^{3+} signal reported by Steiner et al [51] in the XPS

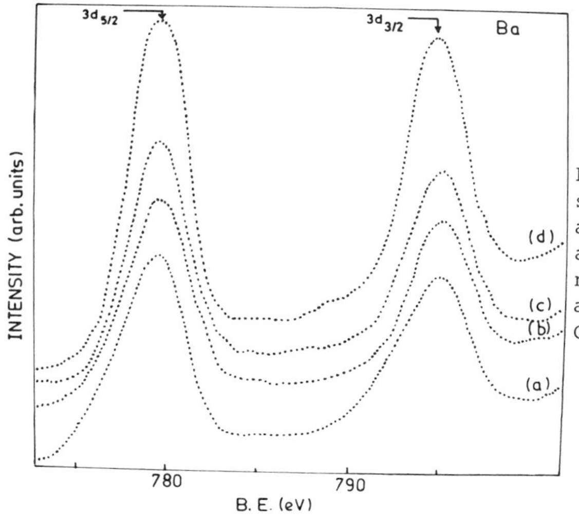

Fig.18 : XPS 3d core level spectra of Ba in a) oxygen annealed $YBa_2Cu_3O_{7-\delta}$, b) air annealed $YBa_2Cu_3O_{7-\delta}$, c) nitrogen annealed $YBa_2Cu_3O_{7-\delta}$ and d) oxygen annealed $GdBa_2Cu_3O_{7-\delta}$.

Cu $2p_{3/2}$ spectra of $YBa_2Cu_3O_{7-\delta}$ samples with controlled annealing temperatures is not clear.

Fig.19 shows the XPS spectra of Y in $YBa_2Cu_3O_{7-\delta}$ compound at 300K and 10K [33]. Two distinct peaks corresponding to spin-orbit split components $3d_{5/2}$ and $3d_{3/2}$ are evident. The spectra do not exhibit any significant change on cooling from 300K to 10K except a clearly detectable narrowing of the line at 10K may be due to decrease in phonon broadening. The binding energies of the Y 3d peaks for $YBa_2Cu_3O_{7-\delta}$ sample show no shift with respect to Y_2O_3 indicating that Y in the superconducting sample $YBa_2Cu_3O_{7-\delta}$ is in 3+ state.

Measurements of the XPS core level binding energies of Y,Ba,Cu and O in pure $YBa_2Cu_3O_{7-\delta}$ compounds indicate that Y is in 3+ and Ba is in 2+ valence states. The valence states of Cu and O are difficult to assign due to strong hybridization between Cu 3d and O 2p derived levels. The assignments are associated with the configurations $\underline{2p}\ 3d^9$ $3d^9\underline{L}$ and $3d^{10}\underline{L}$ in Cu while oxygen assignments $1\underline{s}\ 2p^6\ 3d^{10}\underline{L}$, $1s\ 2p^6\ 3d^9$ as well as $1\underline{s}\ 2p^5\ (3d^{10} + 3d^{10}\underline{L})$ are tentatively proposed [17].

Effects of Annealing, Quenching and Substitution

Since oxygen and copper sites in $YBa_2Cu_3O_{7-\delta}$ compounds are considered to play important role in the mechanism of superconductivity, it is desirable to study the effects of annealing, quenching and ionic substitutions on T_c. $YBa_2Cu_3O_{7-\delta}$ ($\delta \leq 0.15$) compound with defect orthorhombic structure ($T_c \sim 90K$) when quenched from elevated temperature ($> 800°C$) transforms to tetragonal structure ($\delta \geq 0.5$) and does not show superconducting behaviour [45,51]. It is observed that in the orthorhombic, $YBa_2Cu_3O_{7-\delta}$ compound, Cu plays central role in superconductivity and partial substitution of Cu by any cation lowers T_c. Suppression of T_c occurs over a wide magnitude depending on the substituting ions. For example 5% substitution of Ti lowers T_c down to

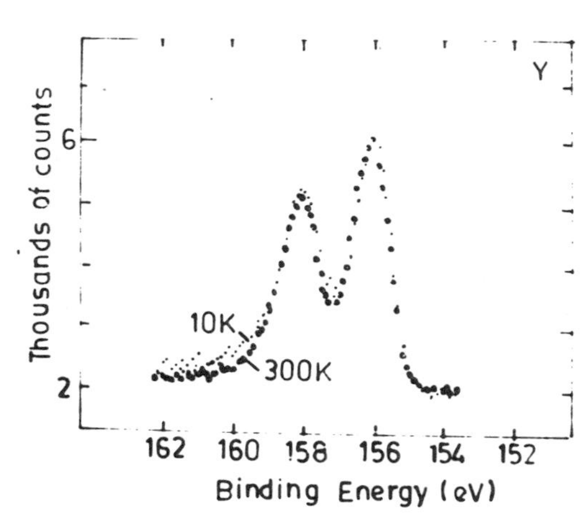

Fig.19 : XPS 3d spectra of Y in $YBa_2Cu_3O_{7-\delta}$ compound at 300K and 10K [33].

76K, Fe suppresses T_c to ∽ 40K and Zn almost kills the T_c [61-78]. It is noted that orthorhombic to tetragonal phase transition occurs in the substituted compounds. The effect of ionic substitutions on oxygen site is equally interesting. Although T_c remains unaffected on S substitution at oxygen sites, the Cl substitution suppresses T_c from 90K to 72K due to lowering of hole concentration [70]. F substitution at O sites provides an interesting situation for charge balancing in $YBa_2Cu_3O_{7-\delta}$ compound. The results of substitutions have shown wide range of variations in T_c mainly due to the method of sample preparation [78-82]. Further substitution of Sr at Ba site in $YBa_2Cu_3O_{7-\delta}$ compound lowers T_c while T_c remains almost unaffected on replacing Y by any lanthanide ion except Ce, Pr and Tb [7,8,83-85]. Interesting results have emerged from the annealing, quenching and selective substitutions in $YBa_2Cu_3O_{7-\delta}$. Some typical examples of the XPS core level measurements carried out by us as well as by others are discussed here.

Oxygen annealed and quenched samples of $YBa_2Cu_3O_{7-\delta}$ compound were prepared [77]. Like the pure oxygen annealed $YBa_2Cu_3O_{7-\delta}$ compound, Fe substituted $YBa_2(Cu_{1-x}Fe_x)_3O_{7-\delta}$ samples with $x \leq 8\%$ compositions were also prepared [76]. F substituted samples were prepared by taking air annealed pure $YBa_2Cu_3O_{7-\delta}$ in powder form along with NH_4F in proportionate quantities to form the green compacted pellet. The preparation details of Fe and F substituted samples are discussed elsewhere [76-78]. The samples were characterized by X-ray diffraction (XRD) method using CuK_α radiation. Fig.20 shows XRD patterns of the Fe and F substituted $YBa_2Cu_3O_{7-\delta}$ compounds. The results indicate that the oxygen annealed pure $YBa_2Cu_3O_{7-\delta}$ compound has orthorhombic (Pmmm) structure as is evident from the split line at $2\theta \sim 33°$. The Fe and F substituted samples show merging of the characteristic orthorhombic/tetragonal split-line in the vicinity of $2\theta \sim 33°$. The orthorhombic structure is driven towards tetragonality on substitution. The quenched sample has tetragonal (P4/mmm) structure. From resistivity

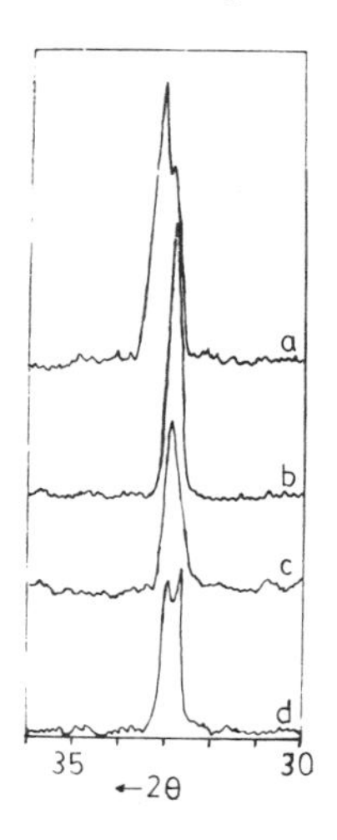

Fig.20 : XRD patterns for $YBa_2Cu_3O_{7-\delta}$ compounds a) O_2 annealed, b) 8% Fe substituted, c) 20% F substituted and d) quenched.

measurements on the oxygen annealed pure $YBa_2Cu_3O_{7-\delta}$ compound, T_c was found to be ∽ 90K and the quenched smaple did not show T_c down to 4.2K. The Fe and F substituted $YBa_2Cu_3O_{7-\delta}$ samples do not exhibit Meissner Effect down to 77K.

Fig.21 shows the $2p_{3/2}$ core level photoelectron spectra of Cu in the pure, quenched, Fe and F substituted $YBa_2Cu_3O_{7-\delta}$ samples. The figure includes the XPS Cu $2p_{3/2}$ spectra of Cu_2O (Cu^{1+}) and CuO (Cu^{2+}) compounds for the purpose of comparison. An asymmetry is evident from the spectra of Fe and F substituted samples. The deconvoluted spectra (solid lines) exhibit two peaks at BE ∽ 931 eV and ∽ 933 eV. The peak at BE ∽ 933 eV appears at the same BE position as that observed for Cu^{2+} in CuO. The peak at BE ∽ 933 eV is assigned to $2p_{3/2}$ $3d^{10}$ $\underline{L}$ final state configuration based on the discussion earlier in this chapter. The satellite structure which shows some variations in its line shape on substitution, is attributed to $2p_{3/2}$ $3d^9$ final states. An interesting feature of the XPS Cu $2p_{3/2}$ spectra for the Fe and F substituted $YBa_2Cu_3O_{7-\delta}$ samples is the appearance of an additional weak structure at BE ∽ 931 eV. This feature is absent in the XPS Cu $2p_{3/2}$ spectra of pure $YBa_2Cu_3O_{7-\delta}$ compound. The BE of this structure corresponds to Cu^{1+} in Cu_2O. This weak peak has, therefore been tentatively assigned to Cu^{1+}.

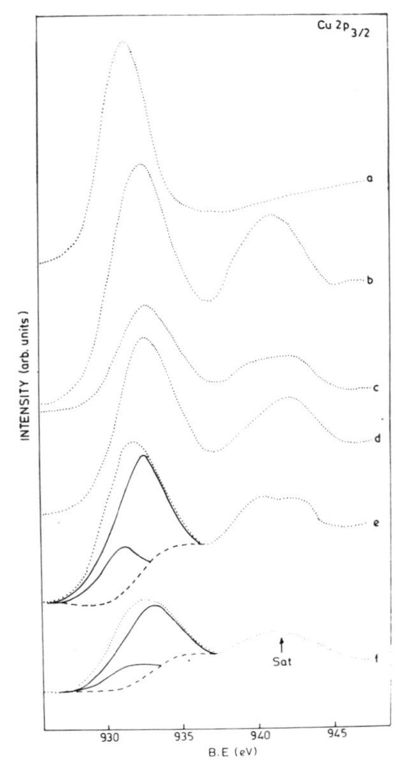

Fig.21 : Cu 2p core level XPS of a) Cu_2O, b) CuO, c) O_2 annealed, d) quenched, e) 8% Fe substituted and f) F substituted $YBa_2Cu_3O_{7-\delta}$ compounds.

XPS Fe $2p_{3/2}$ spectra for the $YBa_2(Cu_{1-x}Fe_x)_3O_{7-\delta}$ samples indicate that Fe in the Fe substituted samples is in $3+$ state. On the basis of charge neutrality condition, presence of Fe^{3+} leads to the possibility of Cu^{1+} in these Fe substituted samples. Further support to the assignment of the additional weak peak to Cu^{1+} is obtained from the studies of F-substituted samples. F being in monovalent (1-) state could lead to decrease in the effective anion valence in the F-substituted $YBa_2Cu_3O_{7-\delta}$ compounds, thus forcing part of Cu to be in Cu^{1+} state. Appearance of Cu^{1+} in the quenched $YBa_2Cu_3O_{7-\delta}$ compounds due to depletion of oxygen from oxygen sites in understandable. Further, the observation of Cu^{1+} in the Fe and F substituted as well as quenched $YBa_2Cu_3O_{7-\delta}$ samples is supported by the XANES studies on the same samples [76,77].

It is now known that substitutions of lathanide ions (except Ce, Pr and Tb) at Y-site in $YBa_2Cu_3O_{7-\delta}$ compounds do not affect T_c [7,8]. Still it was thought to be of interest to see the effect of Ce, Gd and Yb substitutions at Y site on the X-ray induced photoelectron spectra. The choice of Ce, Gd and Yb is dictacted by the fact that Ce at the beginning of the lanthanide series can exist in Ce^{3+} $(4f^1)$, Ce^{4+} $(4f^0)$ or mixed valent (Ce^{3+} and Ce^{4+}), Gd in the middle of the series with $4f^7$ configuration remains in pure 3+ state while towards the end of the series Yb may also exist in divalent $(4f^{14})$, trivalent $(4f^{13})$ or mixed valent (Yb^{2+} and Yb^{3+}) state. For this purpose, polycrystalline samples of $GdBa_2Cu_3O_{7-\delta}$ and $Y_{1-x}R_xBa_2Cu_3O_{7-\delta}$ with $0 \leq x \leq 0.1$ for R = Ce and $0 \leq x \leq 1$ for R = Yb were prepared by standard ceramic technique. The XRD patterns indicated typical single phase orthorhombic (Pmmm) structure except for 10% Ce substituted samples in which $BaCeO_3$ impurity phase was visible. Except for 10% Ce substituted sample, strong Meissner effect at ~77K was exhibited by the samples. The XPS Cu $2p_{3/2}$ spectra of the Ce, Gd and Yb substituted samples was similar to those reported for the pure $YBa_2Cu_3O_{7-\delta}$ compound. XPS Ce 3d spectra of substituted Ce compounds resemble closely with those of CeO_2, indicating Ce in mixed valent state (Ce^{3+} and Ce^{4+}) in the substituted samples. The presence of part of Ce in 3+ state leads to a strong possibility of a fraction of Ce going to Ba site in $(Y_{1-x}Ce_x)Ba_2Cu_3O_{7-\delta}$ on the basis of significantly larger ionic radius of Ce 3+ (1.11 Å) compared to other trivalent lanthanide ions. Further support to this contention is lent by earlier studies of partial lanthanide ion substitutions on to the Ba site [84,85], leading to suppression of T_c below 77K.

Although a large number of papers have been published on substitutions [61-85], lthere are not many reports on the XPS studies. However, an interesting pattern appears to emerge from the XPS studies of substituted compounds [76-78]. The XPS, XRD and T_c measurements on the Fe and F substituted $YBa_2Cu_3O_{7-\delta}$ samples show interesting correlations. The appearance of Cu^{1+} in XPS Cu $2p_{3/2}$ spectra of Fe and F substituted $YBa_2Cu_3O_{7-\delta}$ compounds is associated with the lowering of T_c and also with merging of the characteristic orthorhombic/tetragonal split line in XRD pattern at $2\theta \sim 33°$. This is an important correlation, first reported by Prakash et al [76].

More XPS studies of samples in which suitable ions are substituted at Y-,Ba-,Cu- and O- sites in $YBa_2Cu_3O_{7-\delta}$ are necessary to understand the relation of substitutions with T_c and crystal structure. XPS studies of the Fermi level position in the substituted $YBa_2Cu_3O_{7-\delta}$ compounds are likely to provide valuable information. Further, studies of various ion substitutions at oxygen anion site by XPS method are of special importance for obtaining insight regarding the electronic structure of Cu-chains in the substituted samples which enhanced or suppressed T_c.

Effects of Lowering Temperature

Some interesting changes in the spectral features due to the effect of lowering the temperature of the superconducting samples below T_c on their XPS valence band and core level spectra have been detected. When the SC samples are cooled from 300K to temperatures below T_c, the intensity of the satellite peak ($2\underline{p}_{3/2}$ $3d^9$) associated with the XPS Cu $2p_{3/2}$ main peak in $YBa_2Cu_3O_{7-\delta}$ compounds decreases. The O 1s spectral intensities are also affected on cooling the SC samples below T_c. Possibilities of oxygen dimerization at low temperatures and its relation with T_c are conceived. It appears that the XPS studies of O 1s spectra at low temperatures may provide intricate clues about the holes in the O 2p derived band and the pairing of correlated holes. This may lead to the understanding of mechanism of superconductivity in high T_c oxide superconductors.

However, the major problem with the XPS studies of high T_c SC materials is the presence of contaminants on the sample surface. The nature of these high T_c materials does not permit the use of argon ion beam for insitu cleaning of the surfaces. At present, an alumina or diamond scraper is used to clean the surfaces gently. This process does not ensure perfect cleanliness of the surfaces. Since XPS is a surface sensitive technique, it can detect the contaminants even if they are present in traces [86,87]. In the case of high T_c SC materials, XPS O 1s spectra can be regarded as an important monitor of surface contamination [88]. If we examine the XPS O 1s core level spectra reported by various groups, it is obvious that the surface conditions of the high T_c samples examined in different laboratories are not the same. This may be a probable reason for the existing discrepancy in the reported XPS results on the high T_c oxide superconductors.

It may be added that the XPS Cu $2p_{3/2}$ and O 1s core level spectral features for the superconducting La-Ba-Cu-O system[41,89-91]Bi-Ca-Sr-Cu-O system [92-94] and Tl-Ba-Ca-Cu-O system [95] appear similar to those observed for the Y-Ba-Cu-O system. The interpretations of the XPS results regarding the valence states of Cu and O in these systems have not led to new conclusions. Due to the different surface conditions of the superconducting samples in question, the XPS results reported so far, are not unambiguous and the controversy persists.

CONCLUSIONS

Although XPS technique has undoubtedly provided valuable data on the valence bands and core levels of high T_c SC materials success has not yet been achieved in relating the mechanism of superconductivity with the electronic structure. However, the following important guide lines have emerged from such studies.

i) The theoretical models for high T_c SC materials should include in their formulation, the strong electron-correlations.
ii) A deeper insight into the oxygen vacancies (holes) is required.
iii) XPS O 1s spectral features for contamination free and well characterised samples at low temperatures need careful investigations.
iv) XPS valence band and core level studies of clean single crystal surfaces in UHV condition with tunable photon energy at different temperatures and annealing conditions are likely to provide a distinct pattern.
v) Studies of the selectively substituted samples and correlations of XPS results with crystal structure and T_c variations may lead to a better understanding of the mechanism of superconductivity.
vi) In dynamic mode, formal valence states either to Cu or oxygen in a superconducting system need not be assigned. The configuration mixing approach appears to be more reasonable, and
vii) Intensive efforts are required to find definite relationship between the electronic structure and superconductivity.

ACKNOWLEDGEMENTS

We would like to express our sincere thanks to the publishers (American Physical Society and Springer-Verlag) and the authors (Steiner et al, Dauth et al and Ming Tang et al) for permission to reprint some published figures in the present article. Thanks are also due to Dr.Om Prakash, an active member of our research team on the high T_c oxide superconductors for inspiration, discussion and valuable suggestions. Financial assistance provided by the Department of Science and Technology (Programme Management Board), Government of India (Project No.TSG/10/018/88-R2) is thankfully acknowledged.

REFERENCES

[1] H.K. Onnes, Akad. Van Wetenschappen (Amsterdam) 14, (1911) 113,818.
[2] A.V. Narlikar and S.N. Ekbote, South Asian Publishers, New Delhi (1983).
[3] J.R. Gavaler, Appl. Phys. Lett. 23, (1973) 480.
[4] J. Bardeen, L.N. Cooper and J.R. Schrieffer, Phys. Rev. 106, (1957) 162; 108, (1957) 1175.
[5] J.G. Bednorz and K.A. Müller, Z. Phys. B64, (1986) 189.
[6] M.K. Wu, J.R. Ashburn, C.J. Torng, P.H. Hor, R.L. Meng, L. Gao, Z.J. Huang, Y.Q. Wang and C.W. Chu, Phys. Rev. Lett. 58, (1987) 908.
[7] Y. Le Page, T. Siegrist, S.A. Sunshine, L.F. Schneemeyer, D.W. Murphy, S.M. Zahurak, J.V. Waszczak, W.R. Mckinnon, J.M. Tarascon, G.W. Hull and L.H. Greene, Phys. Rev. B36, (1987) 3617.
[8] T. Itoh, M. Uzawa and H. Uchikawa, Phys. Lett. A129, (1988) 67.
[9] C. Michel, M. Hervieu, M.M. Borel, A. Grandin, F. Deslandes, J. Provost and B. Raveau, Z. Phys. B68, (1987) 421.

[10] Z.Z. Sheng, A.M. Hermann, A. El Ali, C. Almasan, J. Estrada, T. Datta and R.J. Matson, Phys. Rev. Lett. 60, (1988) 937.
[11] Proc. X-ray and Inner-Shell Processes, Paris, France, Sept. 14-18 (Ed. P. Lagarde, F.J. Wuilleumier and J.P. Briand) : Journal De Physique 48, (1987).
[12] Proc. Int. Conf. on High Temp. Superconductors and Materials and Mechanisms of superconductivity, Interlaken, Switzerland, Feb. 28 - March 4 (Ed. J. Müller and J.L. Olsen) : Physica C153-155, (1988).
[13] Proc. 5th Int. Conf. on X-ray Absorption Fine Structure, Seattle, Washington, Aug. 21-26 (1988).
[14] R.J. Cava, B. Batlogg, J.J. Krajewski, R. Farrow, L.W. Rupp Jr., A.E. White, K. Short, W.F. Peck and J. Kometani, Nature 332, (1988) 814.
[15] P.K. Gallagher, H.M. O'Bryan, S.A. Sunshine and D.W. Murphy, Mat. Res. Bull. 22, (1987) 995.
[16] P. Flude, physica C-153-155, (1988) 1769 and references therein.
[17] G. Wendin, Journal De Physique 48, (1987) C9-1157 and references therein.
[18] B.D. Padalia and Bhakt Darshan, Indian J. Phys. 60B, (1986) 128.
[19] B.D. Padalia and Sujata Patil, Proc. Indo-U.S. Workshop on Advanced Tech. for Microstructural Characterization, Bombay, India, January 11-13 (1988).
[20] H. Robinson and W.F. Rawlinson, Philos. Mag. 28, (1914) 277.
[21] K. Siegbahn, C. Nordling, A. Fahlman, R. Nordberg, K. Hamrin, J. Hedman, G. Johansson, T. Bergmark, S.E. Karlsson, I. Lindgren and B. Lindberg, ESCA : Atomic, Molecular and Solid State Structure studied by Means of Electron Spectroscopy, Nova Acta Regiae Soc. Sci. Upsaliensis, Ser. IV, Vol. 20, (Ed. Almqvist and Wiksells) Stockholm (1967).
[22] K. Siegbahn, C. Nordling, G. Johansson, J. Hedman, P.F. Heden, K. Hamrin, U. Gelius, T. Bergmark, L.O. Werme, R. Manne and Y. Baer, ESCA Applied to Free Molecules, North-Holland, Amsterdam, (1969).
[23] S.B.M. Hagstrom and C.S. Fadley, X-ray Spectroscopy (Ed. L.V. Azaroff), McGraw-Hill, New York, (1974).
[24] C.S. Fadley, Electron Spectroscopy, Theory, Techniques and Applications, Vol.2 (Eds. C.R. Brundle and A.D. Baker), Pergamon Press (1978).
[25] Proc. Int. Conf. on Electron Spectroscopy, California, U.S.A. (Ed. D.A. Shirley), North Holland (1972).
[26] K. Siegbahn, J. Electr. Spectr. 5, (1974) 3.
[27] T.A. Carlson, Photoelectron and Auger Spectroscopy, Plenum Press, New York (1975).
[28] B.D. Padalia, DAE Proc. Struct. Properties Correlation and Instr. Techniques in Mat. Sci., Rourkela (1977) 99.
[29] C.R. Brundle, J. Vac. Sci. Technol. 11, (1974) 212.
[30] B.D. Padalia, Proc. DAE Nucl. Phys. and Solid State Phys. Symposium, Poona, India, 20A, (1977) 132.
[31] J.M. Lawrence, P.S. Riseborough and R.D. Parks, Reports on Progress in Physics 44, (1981) 1.
[32] W.M. Riggs and M.J. Parker, Methods of Surface Analysis (Ed. A.W. Czanderna), Elsevier Scientific Publ. Co., (1975) 103.

[33] B. Dauth, T. Kachel, P. Sen, K. Fischer and M. Campagna, Z. Phys. B68, (1987) 407.
[34] P. Steiner, V. Kinsinger, I. Sander, B. Siegwart, S. Hüfner and C. Politis, Z. Phys. B67, (1987) 19.
[35] P.D. Johnson, S.L. Qiu, L. Jiang, M.W. Ruchman, M. Strongin, S.L. Hulbert, R.F. Garrett, B. Sinkovic, N.V. Smith, R.J. Cava, C.S. Jee, D. Nichols, E. Kaczanowicz, R.E. Salomon and J.E. Crow, Phys. Rev. B35, (1987) 8811.
[36] D.D. Sarma, K. Sreedhar, P. Ganguly and C.N.R. Rao, Phys. Rev. B36, (1987) 2371.
[37] A. Bianconi, A. Clozza, A.C. Castellano, S.D. Longa, M. De Santis, A. Di Cicco, K. Garg, P. Delogu, A. Gargano, R. Giorgi, P. Lagarde, A.M. Flank and A. Marcelli, Proc. Adriatico Research Conference on High Temperature Superconductors, Trieste, Italy, July 6-8, (Ed. Y. Lu, M. Tosi and E. Tosatti), World Scientific, Singapore (1987).
[38] M. Tang, N.G. Stoffel, Q.B. Chen, D. LaGraffe, P.A. Morris, W.A. Bonner, G. Margaritondo and M. Onellion, Phys. Rev. B38, (1988) 897.
[39] M.H. Frommer, Phys. Rev. B38, (1988) 2444.
[40] L.F. Matheiss and D.R. Hamann, Solid State Commun. 63, (1987) 395.
[41] N. Nücker, J. Fink, B. Renker, D. Ewert, C. Politis, P.J.W. Weijs and J.C. Fuggle, Z. Phys. B67, (1987) 9.
[42] A. Fujimori, E. Takayama-Muromachi, Y. Uchida and B. Okai, Phys. Rev. B35, (1987) 8814.
[43] R.L. Kurtz, R.L. Stockbauer, D. Mueller, A. Shih, L.E. Toth, M. Osofsky and S.A. Wolf, Phys. Rev. B35, (1987) 8818.
[44] J.C. Fuggle, P.J.W. Weijs, R. Schoorl, G.A. Sawatzky, J. Fink, N. Nücker, P.J. Durham and W.M. Temmerman, Phys. Rev. B37, (1988) 123.
[45] R. Prakash, O. Prakash and N.S. Tavare, Pramana- J. Phys. 30, (1988) L597.
[46] D.H. Kim, D.D. Berkley, A.M. Goldman, R.K. Schulze and M.L. Mecartney, Phys. Rev. B37, (1988) 9745.
[47] D.D. Sarma, Phys. Rev. B37, (1988) 7948.
[48] H. Ihara, M. Hirabayashi, N. Terada, Y. Kimura, K. Senzaki, M. Akimoto, K. Bushida, F. Kawashima and R. Uzuka, Jpn. J. Appl. Phys. 26, (1987) L460.
[49] S. Horn, J. Cai, S.A. Shaheen, Y. Jeon, M. Croft, C.L. Chang and M.L. den Boer, Phys. Rev. B36, (1987) 3895.
[50] T.A. Sasaki, Y. Baba, N. Masaki and I. Takano, Jpn. J. Appl. Phys. 26, (1987) L1569.
[51] P. Steiner, S. Hüfner, V. Kinsinger, J. Sander, B. Siegwart, H. Schmitt, R. Schulz, S. Junk, G. Schwitzgebel, A. Gold, C. Politis, H.P. Müller, R. Hoppe, S. Kemmler-Sack and C. Kunz, Z. Phys. B69, (1988) 449.
[52] F. Garcia-Alvarado, E. Moran, M. Vallet, J.M. Gonzalez-Calbet, M.A. Alario, M.T. Perez-Frias, J.L. Vicent, S. Ferrer, E. Garcia-Michel and M.C. Asensio, Solid State Commun., 63, (1987) 507.
[53] H. Eickenbusch, W. Paulus, R. Schollhorn and R. Schlogl, Mat. Res. Bull. 22, (1987) 1505.
[54] H. Watanabe, K. Ikeda, H. Miki and K. Ishida, Jpn. J. Appl. Phys. 27, (1988) L783.
[55] P. Steiner, V. Kinsinger, I. Sander, B. Siegwart, S. Hüfner, C. Politis, R. Hoppe and H.P. Müller, Z. Phys. B67, (1987) 497.

[56] S. Larsson, Chem. Phys. Lett. 40, (1976) 362.
[57] G. van der Laan, C. Westra, C. Haas and G.A. Sawatzky, Phys. Rev. B23, (1981) 4369.
[58] T. Gourieux, G. Krill,M. Maurer, M.F. Ravet, A. Menny, H. Tolentino and A. Fontaine, Phys. Rev. B37, (1988) 7516.
[59] A. Fjellvag, P. Karen, A. Kjekshus and J.K. Grepstad, Solid State Commun. 64, (1987) 917.
[60] B.D. Padalia, P.K. Mehta, S.R. Reddy, O. Prakash and N. Venkataramani, Mod. Phys. Lett. B2, (1988) 1177.
[61] N. Venkataramani, K. Muraleedharan, S.N. Bhatia, A. Dutta, O. Prakash and C.M. Srivastava, Pramana- J. Phys. 30, (1988) L455.
[62] S.B. Oseroff, D.C. Vier, J.F. Smyth, C.T. Salling, S. Schultz, Y. Dalichaouch, B.W. Lee, M.B. Maple, Z. Fisk, J.D. Thompson, J.L. Smith and E. Zirngiebl, Solid State Commun. 64, (1987) 241.
[63] J.M. Tarascon, P. Barboux, P.F. Miceli, L.H. Greene, G.W. Hull, M. Eibschutz and S.A. Sunshine, Phys. Rev. B37, (1988) 7458.
[64] T. Siegrist, L.F. Schneemeyer, J.V. Waszczuk, N.P. Singh, R.L. Opila B. Batlogg, L.W. Rupp and D.W. Murphy, Phys. Rev. B36, (1987) 8365.
[65] Y. Maeno, T. Tomita, M. Kyogoku, S. Awaji, Y. Aoyaki, K. Hoshino, A. Minami and T. Fujita, Nature 328, (1987) 512.
[66] G. Xiao, M.Z. Cieplak, A. Gavrin, F.H. Streitz, A. Bakshai and C.L. Chein, Phys. Rev. Lett. 60, (1988) 1446.
[67] B.R. Zhao, Y.H. Shi,Y.Y. Zhao and Lin Li, Phys. Rev. B38, (1988) 2486.
[68] E. Takayama-Muromachi, Y. Uchida and K. Kato, Jpn. J. Appl. Phys. 26, (1987) L2087.
[69] I. Felner, I. Nowik and Y. Yeshurun, Phys. Rev. B36, (1987) 3923.
[70] C.V.N. Rao, B. Jayaram, S.K. Agarwal and A.V. Narlikar, Physica C152, (1988) 479.
[71] Z. Yong, Z. Han, S. Shifang, S. Zhenpeng, C. Zuyao and Z. Qirui, Solid State Commun. 67, (1988) 31.
[72] T. Sugita, M. Yabuuchi, K. Murase, H. Okabayashi, K. Gamo and S. Namba, Solid State Commun. 67, (1988) 95.
[73] B.D. Dunlap, J.D. Jorgensen, W.K. Kwok, C.W. Kimball, J.L. Matykiewicz, H. Lee and C.U. Segre, Physica C153-155, (1988) 1100.
[74] R. Gomez, S. Aburto, V. Marquina, M.L. Marquina, M. Jimenez, C. Quintanar, T. Akachi, R. Escudero, R.A. Barrio and D. Rios-Jara, Physica C153-155, (1988) 1557.
[75] J.L. Dormann, S.C. Bhargava, C. Djega-Mariadassou, J. Jove, O. Gorochov, R. Suryanarayan and H. Pankowska, Physica C153-155, (1988) 1553.
[76] O. Prakash, P.K. Mehta, S.R. Reddy and B.D. Padalia, Rev. Solid State Sci., 2, (1988) 249.
[77] B.D. Padalia, P.K. Mehta and O. Prakash, Proc. Int. Conf. on Hi T_c SC, Jaipur (Ed.K.B. Garg) Oxford and IBH Publ. Comp., New Delhi (1988) in press.
[78] P.K. Mehta, S.R. Reddy, D.T. Adroja, O. Prakash and B.D. Padalia, Proc. Int. Conf. on Hi T_c SC, Jaipur (Ed.K.B. Garg) Oxford and IBH Publ. Comp., New Delhi (1988) in press.
[79] S.R. Ovshinsky, R.T. Young, D.D. Allred, G. De Maggio and G.A. Vande Leeden, Phys. Rev. Lett. 58, (1987) 2579.
[80] H.H. Wang, A.M. Kini, H.C.I. Kao, E.H. Appelman, A.R. Thompson, R.E. Botto, K.D. Carlson, J.M. Williams, M.Y. Chen, J.A. Schlueter, B.D. Gates, S.L. Hallenbeck and A.M. Despotes, Inorganic Chemistry 27, (1988) 6.

[81] P.K. Davies, J.A. Stuart, D. White, C. Lee, P.M. Chaikin, M.J. Naughton, R.C. Yu and R.L. Ehrenkaufer, Solid State Commun. 64, (1987) 1441.
[82] N.P. Bansal, A.L. Sandkuhl and D.E. Farrell, Appl. Phys. Lett. 52, (1988) 838.
[83] P.H. Hoγ, R.L. Meng, Y.Q. Wang, L. Gao, Z.J. Huang, J. Bechtold, K. Forster and C.W. Chu, Phys. Rev. Lett. 58, (1987) 1891.
[84] K. Zhang, B. Dabrowski, C.U. Segre, D.G. Hinks, I.K. Schuller, J.D. Jorgensen and M. Slaski, J. Phys. C20, (1987) L935.
[85] K. Yoji, N. Takazhi, H. Tetsuo, N. Tetsuo and F. Tetsuo, Jpn. J. Phys. 26, (1987) L2089.
[86] B.D. Padalia, J.K. Gimzewski, S. Affrossman, W.C. Lang. L.M. Watson and D.J. Fabian, Surface Science 61, (1976) 468.
[87] J.M. Gimzewski, B.D. Padalia, S. Affrossman, L.M. Watson and D.J. Fabian, Surface Science 62, (1977) 386.
[88] M. Grioni, J.C. Fuggle, P.J.W. Weijs, J.B. Goedkoop, G. Rossi, F. Schaefers, J. Fink and N. Nücker, Journal De Physique 48, (1987) C9-1189.
[89] H. Ihara, M. Hirabayashi, N. Terada, Y. Kimura, K. Senzaki and M. Tokumoto, Jpn. J. Appl. Phys. 26, (1987) L463.
[90] P. Steiner, J. Albers, V. Kinsinger, I. Sander, B. Siegwart, S. Hüfner and C. Politis, Z. Phys. B66, (1987) 275.
[91] D.D. Sarma and C.N.R. Rao, J. Phys. C20, (1987) L659.
[92] E.G. Michel, J. Alvarez, M.C. Asensio, R. Miranda, J. Ibanez, G. Peral, J.L. Vicent, F. Garcia, E. Moran and M.A. Alario-Franco, Phys. Rev. B38, (1988) 5146.
[93] S. Kohiki, T. Wada, S. Kawashima, H. Takagi, S. Uchida and S. Tanaka, Phys. Rev. B38, (1988) 7051.
[94] Z.X. Shen, P.A.P. Lindberg, I. Lindau, W.E. Spicer, C.B. Eom and T.H. Geballe, Phys. Rev. B38, (1988) 7152.
[95] A.K. Ganguli, K.S. Nanjunda Swamy, G.N. Subbanna, A.M. Umarji, S.V. Bhat and C.N.R. Rao, Solid State Commun. 67, (1988) 39.

SYNTHESIS OF HIGH-Tc OXIDE SUPERCONDUCTORS IN THE Y-, Bi- AND Tl- SYSTEMS: THE ROLE OF CHEMISTRY

R.M. Iyer and J.V. Yakhmi

Chemical Group, Bhabha Atomic Research Centre, Bombay-400085, India

I. INTRODUCTION

After the pioneering work of Bednorz and Muller [1] there has been an explosion of worldwide interest in superconductivity, particularly in oxide superconductors. This has resulted in quick discovery of superconductivity above liquid nitrogen temperature first in Y-Ba-Cu-O system [2] in early 1987, and later, in Bi-Ca-Sr-Cu-O system [3] and Tl-Ca-Ba-Cu-O system [4] in early 1988 and more recently in Ba-K-Bi-O [5] The underlying superconducting compound in the Y-system, $YBa_2Cu_3O_{7-x}$ was first identified by Cava et al [6]. This compound exists in two modifications : tetragonal phase (x ~ 1) can be obtained by quenching from high temperatures whereas a superconducting compound (x ~ 0) with an orthorhombic distorted oxygen-deficient perovskite structure is obtained by slow-cooling in air (or oxygen). $YBa_2Cu_3O_{\sim 7}$ is characterized by a very narrow homogeneity range with respect to the metallic constituents [7]. For this reason, synthesis of bulk polycrystalline samples of $YBa_2Cu_3O_{7-x}$ superconducting phase devoid of any minor secondary-phase impurities is a challenge.

The discovery of high-temperature superconductivity in both Bi-Ca-Sr-Cu-O and Tl-Ca-Ba-Cu-O systems has provided a further stimulus to the studies in the field of high-T_c oxides. However, the more complex chemical nature of these quinary systems produces an array of phases with different compositions and superconducting properties, and also leads to intergrowth of different phases during synthesis. The crystal structures of the superconducting phases, in these two systems are similar to each other [3,4] with each system containing three different superconducting phases with varying Tc's. For instance, the Tc's of $Tl_2Ca_2Ba_2Cu_3O_{10}$ and $Bi_2Ca_2Sr_2Cu_3O_{10}$ (the 2223 phases) are 125K and 110K respectively, whereas $Tl_2CaBa_2Cu_2O_8$ and $Bi_2CaSr_2Cu_2O_8$ (the 2122 phases) become superconducting below ~106K and 85K, respectively. The Tc-values for calcium-free compounds $Tl_2Ba_2CuO_6$ and $Bi_2Sr_2CuO_6$ (2021 phases) are somewhat uncertain and are reported to be 6K-80K and 8K-20K, respectively[8], depending on the preparation conditions.

As is the case for any technology based on materials, one of the most crucial aspects of research in high-Tc oxides is the synthesis and processing of high-quality single-phase materials. Eventhough superconductivity can be observed when only a small portion of the sample is superconducting, synthesis of single-phase material is of paramount importance for any meaningful scientific measurement and for most practical applications. A detailed phase diagram is, of course, an invaluable guide to materials synthesis. In the case of Y-Ba-Cu-O system a number of preliminary phase diagrams have now been established which are in general agreement[7]. However, reliable and authentic phase diagrams for Bi- and Tl-system are not available. The usefulness of these high-temperature superconducting oxides in practical devices is critically dependent on certain crucial parameters such as Tc, Jc and Hc_2 characterizing the superconducting state as well as on bulk mechanical properties of these ceramic materials. These parameters, in turn are dependent on the micro- and macroscopic homogeneity towards which chemistry plays a major role. The processing conditions for these multicomponent oxides are, for instance, dependent on the reactivity of individual component oxides which are widely different. Control of reactivity of the individual components is thus an important factor in determining the formation of these high-Tc oxides. In what follows, we make an attempt to elaborate on the role of chemistry in the synthesis of single-phase Y-,Bi- and Tl- based superconducting oxides. Crucial factors in the synthesis of superconducting oxides using the powder metallurgical and the solution chemistry methods are elaborated by taking the synthesis of Y-Ba-Cu-O as an example in section II. A large variety of studies have been reported only on $YBa_2Cu_3O_{7-x}$ and related substitutional compounds. Hence, sections II to VII deal with important chemical aspects such as reactivity, oxygen stoichiometry, ion-sorption and fluorine incorporation, etc. Special problems pertaining to the synthesis of Bi-and Tl-based superconducting oxides necessitate the use of a matrix reaction method, which is detailed exclusively in section VIII.

II. PROCESSING METHODS FOR BULK MATERIALS

II.A. Solid State Reaction Method

Solid-state reaction or powder metallurgy is a commonly adopted synthesis route for producing multicomponent oxides which involves mixing pulverizing, calcination, compacting and sintering of component oxides/ carbonates/oxalates/nitrates. This method has been employed widely, due to its simplicity, to produce superconducting oxides. For $YBa_2Cu_3O_{7-x}$ (Tc≈ 90K). typical starting mixtures are $BaCO_3$, Y_2O_3 and CuO powders, although many alternate materials have also been used such as $Ba(OH)_2$, $Ba(NO_3)_2$, BaO_2, BaF_2, $Y_2(CO_3)_3$, $CuCO_3$ and $Cu(NO_3)_2$. $BaCO_3$ has the advantage that it has a stable composition although one must ensure complete removal of CO_2 during calcination steps. It should be noted that $BaCO_3$ itself decomposes only above 1300K whereas when it is admixed with Y_2O_3 and

CuO the decomposition occurs at much lower temperatures.

For small batches (5-50 gm) the constituent oxides (or their precursors) are mixed in appropriate proportion in an agate mortar and then fired (calcinated) in alumina crucibles between 1075K and 1225K for 1-3 days in air. For larger batches, the components are mixed as slurry using ball mills, dried and then calcinated. As in typical ceramic processing techniques, the product is finally ground for homogenising and pelletised under pressure (50 to 500 MPa) prior to sintering. This step is repeated a number of times to ensure reasonable homogeneity of the reactants. With each intermittent grinding the quality of the end-product is expected to improve. However, there is a possibility of impurity intrusion from the container and grinding media. Certain organic binders like polyvinyl alcohol are sometimes used in small proportions to provide additional strength to green pellets. Sintering of these pellets at high temperatures leads to densification and facilitates reaction of components. Since $YBa_2Cu_3O_{7-x}$ melts congruently at $\gtrsim$1275K, the sintering temperature is kept between 1175-1235K. Before sintering, it is, however, necessary to heat the pellets gradually to $\sim$575K in order to drive off any moisture and traces of organic binder. Subsequently, the temperature is raised quickly to the sintering temperature and kept at it for several hours. An oxygen/air flow over the pellets, during sintering, is helpful to remove the occluded gases and thus aids in minimising porosity. $YBa_2Cu_3O_{7-x}$ pellets thus sintered are not superconducting unless annealed and slow-cooled in oxygen.

Another solid-state method [9] available for the synthesis of superconducting ceramics like $RBa_2Cu_3O_{7-x}$ is the high-temperature oxidation of a Y(RE)BaCu alloy precursor. However, due to the wide variation in the reactivity of the individual metals towards oxygen and the consequent lattice expansion upon oxide formation the end-product is not expected to be very homogeneous and free from strain-induced microcracks.

II.B. Solution Chemistry Routes for Synthesis

Conventional solid-state route for synthesis of superconducting oxides suffers from certain practical problems such as compositional inhomogeneity, lack of control over distribution of shape and size of particles (morphology) and impurity pick-up during repeated pulverization. This makes it a challenging task to use the solid state route to synthesize high-purity ceramic superconductors with high-density and reproducible microstructure. On the other hand, solution chemistry methods offer greater control of homogeneity, stoichiometry, particle-size and purity. This is because solution chemistry routes allow mixing of constituent materials at molecular levels thus precluding the necessity to go for repeated grinding and milling.

Solution chemistry routes usually cover co-precipitation method, sol-gel process and freeze-drying method.

The co-precipitation method involves simultaneous precipitation of two or more insoluble species from the solution of a mixture of metal salts as a result of interaction of the ions in the solution. In a multicomponent solution the rates of precipitation of different species could be different, giving rise to chemical inhomogeneity in the co-precipitate. It requires special skills in solution chemistry to choose the right precipitant, pH and other parameters to achieve the simultaneous separation of all the insoluble species from the solution. For instance, co-precipitation of mixed citrates, oxalates or hydroxide carbonates of Y, Ba and Cu have been reported for the synthesis of $YBa_2Cu_3O_{7-x}$. We, in our laboratory have adopted a partial co-precipitation (semi-wet) route for synthesis of high-quality single-phase $YBa_2Cu_3O_{7-x}$ samples. In this method, Y and Cu are co-precipitated as oxalates and dry barium acetate powder is mixed with this co-precipitate to get over the problem of preferential leaching-out of Ba in a complete co-precipitation method. After calcination, the mixtures are pelletised and sintered at 1175K for 72 hours in air. These pellets are reground, pelletised and resintered at 1225K in oxygen for 18 hours and then furnace-cooled. It was possible to obtain Tc (R=o) in the region of 100K following this procedure. One particular sample, in fact, showed a zero-resistance state at a temperature as high as 106K [10] (Fig.1). It must be mentioned, however, that the coprecipitation process involves the use of alkalis or other precipitating agents which if present in small quantities as impurities in the product could be deleterious. Precaution should also be taken against a non-uniform or incomplete precipitation and to avoid losses of the precipitate by leaching out during washing and ingress of impurities during the precipitation process.

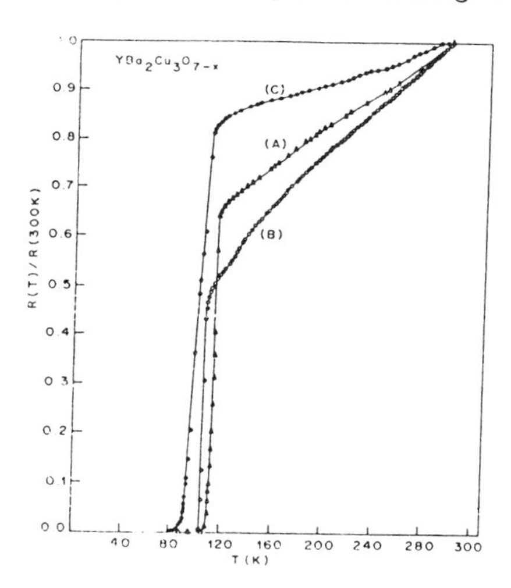

Fig 1. Normalised resistance for $YBa_2Cu_3O_{7-x}$: (A) furnace cooled from 1225K; (B) soaked for 4 hrs. in O_2 at 1225K and slow cooled; and (c) soaked for 1 hr in O_2 at 1225K and slow-cooled.

In the sol-gel process, a "sol" is gelated by the flocculation of the sol particles, which essentially raises the viscosity of the sol in a gradual way, eventually, converting it into a homogeneous solid product with no concentration gradient. The term "sol" is used to describe stable dispersions of ultrafine particles in a given solvent. Sols of oxides can be prepared by using aqueous or non-aqueous solvents. The sol-gel route requires the application of principles of colloid chemistry or polymerization chemistry depending on whether the sol particles are synthesized by dispersing hydrous oxide precipitate or in situ in a non-aqueous solvent. Details of solution chemistry for this process leading to the synthesis of oxide superconductors have been described in a number of reports[11,12]. It is worth noting, however, that the sol-gel process is comparatively expensive and time-consuming and cannot avoid impurity ingress due to the very nature of the gelation process although homogeneity control is much superior.

Freeze-drying is a relatively less familiar solution chemistry method that has been employed towards the synthesis of $YBa_2Cu_3O_{7-x}$. This method exploits the advantage of a solution technique but avoids the disadvantages of co-precipitation and sol-gel techniques. Typically a multicomponent solution of soluble salts (nitrates or acetates) is sprayed into liquid nitrogen, thus freezing it instantaneously which ensures homogeneity. The frozen beads of solution are dried, calcined and heat-treated as in conventional ceramic processing[13]. $YBa_2Cu_3O_{7-x}$ prepared thus has higher density and should, in principle, exhibit sharper resistivity drops and larger Meissner effects. However, this method has not been developed and employed extensively due to certain inherent problems such as occurrence of "melting" before drying in which the frozen beads coalesce. Additionally, sprayed material requires special care during calcination since it tends to become hygroscopic after freeze-drying.

III. CHEMICAL REACTIVITY CONSIDERATIONS

Container materials commonly employed during the high-temperature synthesis of ceramics are found to be reactive with $YBa_2Cu_3O_{7-x}$ at the synthesis temperatures (1075K-1225K) exceptions being silver and gold. The reactivity towards crucible material like alumina increases with temperature and extent of contact. It appears that the best way to avoid contact with container material is to either wrap the green pellet in an inert metal foil or to produce a lining of the ceramic powder of the same material to avoid contact with the crucible while sintering. For thin-film processing, reactivity with substrates is a severe problem and necessitates shortest possible processing times and as low temperatures as possible.

In addition to the reversible uptake of oxygen, $YBa_2Cu_3O_7$ gets degraded by reaction with ambient moisture and CO_2. Reaction with H_2O liberates oxygen and decomposes the material into $Ba(OH)_2$, Y_2BaCuO_5 and CuO. Of course, this reaction is strongly dependent on temperature and surface area. Recent studies conducted in our laboratory have shown that $YBa_2Cu_3O_{7-x}$ is reasonably stable in mixed-solvent media over long periods[14].

Although $YBa_2Cu_3O_{7-x}$ even in powder form may be handled in air without significant changes in properties, long-term stability of this material requires protection. There have been reports regarding the deterioration in superconducting characteristics of $YBa_2Cu_3O_{7-x}$ due to ageing and environmental changes. Structurally well-ordered material has been found to degrade in air at ambient temperature and to exhibit basal-plane faulting[15]. A neutron diffraction study conducted in our labs showed the effect of ageing in $YBa_2Cu_3O_{7-x}$ in the form of significant changes in line-width and diffuse scattering background when the samples were kept in open environment upto one week[16]. The deterioration in structural parameters was, however, much reduced when the samples were kept sealed in quartz. It is believed that this degradation in structural integrity of $YBa_2Cu_3O_{7-x}$ upon ageing is due to the absorption of moisture which could promote slow crack growth and strain fields. This can lead to structural distortions due to partial release of excess oxygen and its migration to the grain-boundary region.

IV. OXYGEN STOICHIOMETRY IN $YBa_2Cu_3O_{7-x}$

The distorted oxygen deficient perovskite structure of $YBa_2Cu_3O_{7-x}$ derives itself from the ideal perovskite structure by the presence of ordered vacancies in the oxygen sublattice[17] resulting in an orthorhombic symmetry with the space group Pmmm. Due to the ordering of the oxygen vacancies, two third of the copper atoms are surrounded by only five oxygen atoms and form a two-dimensional network of CuO_5 square-pyramids in the a-b plane linked through corner-shared oxygens. The remaining one-third of the copper atoms have square planar coordination (CuO_4) in the b-c planes and form quasi-one-dimensional chains linked through corner-shared oxygens along the b-axis. A conversion from the orthorhombic to the tetragonal structure can, however, occur either due to a disordering of the oxygen atoms, say, by quenching the material from high temperature, or by heating in vacuum, or in an inert gas which results in the creation of a tetragonal structure for $YBa_2Cu_3O_{\sim 6}$ which is not superconducting. The effect of oxygen stoichiometry on the electronic properties is dramatic. By slow-cooling the material in an oxygen atmosphere the superconducting orthorhombic state, ($YBa_2Cu_3O_{\sim 7}$) can be stabilized. As mentioned earlier, the oxygen-content in the orthorhombic phase can thus be maximised which improves the superconductivity characteristics. A remarkable feature of $YBa_2Cu_3O_{7-x}$ structure is that the oxygen diffusion kinetics are quite slow at room temperature, allowing stability of samples over the entire range $0.0 \leq x \leq 1.0$. Materials prepared or sintered above 1175K, generally, have $x \approx 0.7$ and are tetragonal (not superconducting) and must be annealed at 675-775K in oxygen to obtain the highest oxygen-content and Tc. Although films, powders and porous ceramic samples readily equilibrate with oxygen in this temperature range, dense ceramics and large single-crystals require longer times for oxygen to diffuse in.

The existing information indicates that the decomposition of $YBa_2Cu_3O_{7-x}$ starts at T = 675K at Po_2 = 1 bar and that the compound melts incongruently. Growth of large single crystals from a stoichiometric melt could become possible if it were possible to change the phase diagram at higher oxygen pressure so that the compound melts congruently. Karpinski et al[18] have made systematic thermodynamic investigation of the three-dimensional P-T-x phase diagram of $YBa_2Cu_3O_{7\pm x}$-O_2 system upto 2200 bar and 1475K. In particular, they have shown that under 100 bar oxygen, it is possible to avoid the orthorhombic-tetragonal phase transion upto the temperature of 1275K.

Isobaric thermogravimetry experiments in air and pure oxygen atmosphere were carried out on $YBa_2Cu_3O_{7-x}$ in our laboratory[19] using a variety of heating and cooling rates. The data obtained was complemented by the x-ray diffraction analysis of the samples of varying oxygen stoichiometry obtained by heating this compound to different temperatures in argon, followed by cooling to room temperature. It was concluded that the compound $YBa_2Cu_3O_{7-x}$ on heating in air (or oxygen) loses oxygen and tansforms to a tetragonal phase after the oxygen content falls below 6.5 atoms per formula weight of the compound. The superconducting orthorhombic phase exhibits a homogeneity range of composition upto x = 0.5. For oxygen content less than 6.5 atoms per formula weight of the compound, existence of a biphasic region containing both orthorhombic and tetragonal phases, could be established. The tetragonal phase, formed from the oxygen rich $YBa_2Cu_3O_{\sim 7}$ orthorhombic phase by the loss of oxygen, also exhibited a wide homogeneity range and did not break up into constituent oxides upto the composition $YBa_2Cu_3O_{5.92}$. An important conclusion of our TG studies was that samples heated at moderately high heating rates, when cooled very slowly could pick up almost all the oxygen they had lost in the heating cycle (Fig.2). Following on this, by adopting a very gradual programmed cooling procedure in oxygen environment it was demonstrated during our studies[20,21] that a very sharp superconducting transition ($\Delta Tc \lesssim 1K$) can be obtained, while retaining the high Tc-value ($\simeq 105K$) (Fig.1) of the samples prepared by a semi-wet route, as described earlier. For this purpose, pellets sintered earlier at 1175K for 72 hrs in air were re-sintered at 1225K for 3 hrs in air and 1 hr in oxygen after which they were slow-cooled in flowing oxygen at a rate of 0.1K/min upto 1125K, then at 2K/min to 675K, followed by furnace-cooling to room temperature.

Estimation of oxygen-content per formula unit in $YBa_2Cu_3O_{7-x}$ and other high-Tc oxides is an important parameter and has generally been

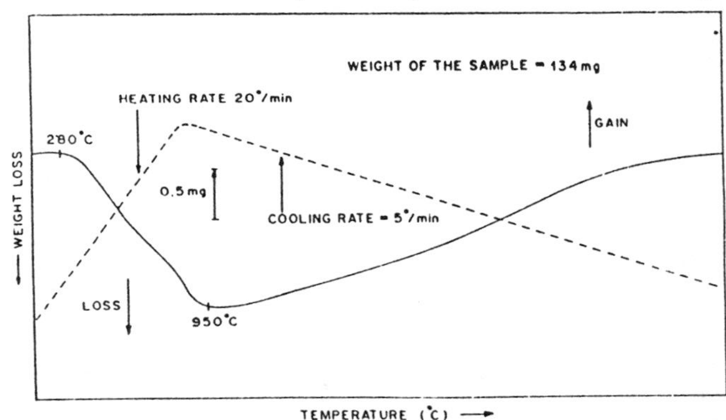

Fig 2. Influence of cooling rate on oxygen recovery by $YBa_2Cu_3O_{7-x}$

estimated by thermogravimetric(TG) methods, iodometric titration methods or inert gas fusion technique. TG method, however, is only an indirect one since one has to assume the initial oxygen composition if an oxidising atmosphere (air, O_2) is used or the final composition if a reducing atmosphere (Ar, H_2) is used. Carbon monoxide produced by the fusion of the specimen in graphite crucible is estimated by infrared measurements in the inert gas fusion method for evaluation of net oxygen content directly. However, this method is also prone to errors since it does not have sufficient precision to distinguish between close oxygen stoichiometries and also because the oxygen content determined is sensitive to the presence of other oxygen-bearing species such as moisture or unreacted oxide. According to detailed studies conducted in our labs[22] the modified Appleman's iodometric method gives a very high precision in determining oxygen content of Y-Ba-Cu-O superconductors since it involves just a single titration by which the Cu present in the specimen as Cu (III) is exclusively estimated by dissolving the sample in cold HBr with subsequent addition of KI. The addition of citrate results in complexing copper in Cu(II) state and the iodine liberated due to reduction of Cu(III) $\rightarrow$ Cu(II) is estimated. Oxygen stoichiometry could be determined to third decimal place, e.g. $YBa_2Cu_3O_{6.885\pm0.003}$. Precision of the method can be appreciated from the values obtained on repeat analysis from the same batch, the values being 6.887, 6.881, 6.886, 6.884 and 6.888.

V. PROCESSING FOR HIGHER Jc

Both Tc and Hc_2 are satisfactorily high for the high-temperature oxide superconductors but the critical current density (Jc) which is also a crucial parameter for applications is undesirably low. Although oriented thin films and single crystals have been made with Jc = 4×10^6 A/cm^2 (at 77K in zero field)[23], Jc is usually $\approx$ 200A/cm^2 at 77K (in zero field) for bulk sintered samples, and it drops further under applied magnetic fields. Jc, however, is not an intrinsic characteristic of superconducting materials and can be strongly modified by structural features such as defects, grain boundaries and impurities. The magnetic field dependence of Jc is thought to arise from the low-Jc weak links between high-Jc grains. These weak-links are generally attributed to structural or compositional deviation at interfaces, or to the presence of non-superconducting or insulating impurities giving rise to poor connectivity between grains. New processing techniques such as melt textured growth[24] which employ partial or complete melting followed by controlled recrystallisation have demonstrated that in the case of $YBa_2Cu_3O_{7-x}$ this leads to dramatic improvement in Jc (1.7×10^4 A/cm^2 at 77K and H = O) primarily due to alignment of the grains. We have recently shown that by substituting small amounts of gadolinium ($\simeq$ 0.2 at %) in place of yttrium, the transport-Jc is nearly doubled, going up from ~250A/cm^2 at 77K (H=O) for the pure compound to ~450A/cm^2 at 77K (H=O)[25]. Simultaneously, it was observed that the normal-state resistivity and the transition-width (ΔTc) also decreased considerably as a result of substitution of 0.2 at% Gd. (Fig.3) The improvement in superconducting characteristics arises probably because Gd-addition acts as a scavenger of insulating phases which may be present as intergranular material, thus enhancing the transport across the grains. This is an area worth exploring further since the conditions under which impurity phases (or non--superconducting phases) get formed at the grain boundaries are not understood at present. Methods to mitigate the formation of weak-links will probably emerge through finer control of the chemistry at the processing stages.

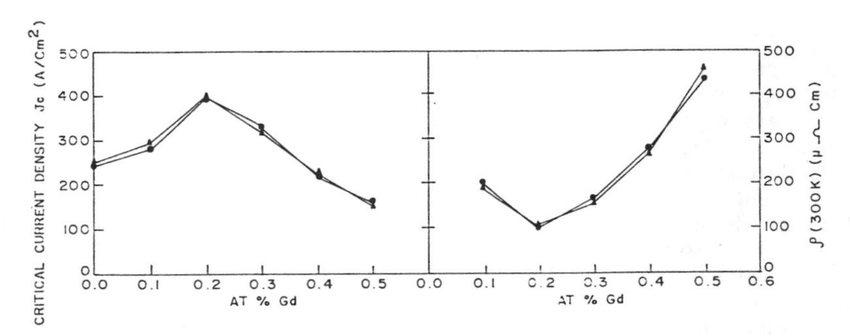

Fig 3. Effect of Gd-substitution on (a) transport-J_c, and, (b) room temperature resistivity.

Another question which can be addressed at this stage is whether high-purity Y_2O_3 is really needed for synthesizing $YBa_2Cu_3O_{7-x}$ of acceptable superconducting characteristics. It is a fact that most rare-earth elements (excepting Ce, Pr and Tb) when substituted partially or fully for Y do not deteriorate the superconducting properties of $YBa_2Cu_3O_{7-x}$ appreciably[26]. Preparation of 99.95% pure Y_2O_3 involves initial solvent extraction followed by an ion-exchange process, the latter step being a major escalant in the cost. Recent efforts in our laboratory[27]have shown that polycrystalline $YBa_2Cu_3O_{7-x}$ superconducting material of acceptable quality (Tc$\simeq$ 95K, transport-Jc $\simeq$250 A/Cm^2 at 77K) can be synthesized from 93% pure Y_2O_3 itself which can be prepared in a cost-effective way from the 60% Y_2O_3-concentrate by using the solvent extraction process alone.

VI. ION-SORPTION BEHAVIOUR

Many mixed oxides having different crystal structures are known to possess ion-exchange property. It is useful to explore the ion-exchange and /or ion-sorption behaviour, if any, of the high-Tc oxides because it would then provide a route to introduce in them other ions of interest through a solution medium at specific sites starting with a single-phase oxide. It could thus lead to the preparation of oxides having different compositions without resorting to high-temperature techniques. In this context, ion-sorption behaviour and stability of $YBa_2Cu_3O_{7-x}$ has been studied in detail in our labs both in aqueous and mixed solvent media[14]. When $YBa_2Cu_3O_{7-x}$ is in contact with aqueous solution, Ba is leached out extensively, resulting in the breakdown of its structure. However, in a mixed-solvent (e.g. methanol-water mixture) containing a minimum of

0.1M of dissolved salt not only the leaching of Ba is significantly reduced due to solubility considerations, the structure of $YBa_2Cu_3O_{7-x}$ too, remains in tact and its superconductng behaviour is not affected. Further experiments with mixed-solvents containing dissolved $CaCl_2$ showed that Ca-ions could be incorporated into the oxide-matrix. The amount of Ca incorporated depended on the nature of the mixed-solvent used. Mg-ion has also been incorporated into $YBa_2Cu_3O_{7-x}$ and $Bi_2CaSr_2Cu_2O_{8+x}$ through the solution medium. Such studies open up new vistas in the field of structural chemistry and inorganic ion-exchangers.

VII. FLUORINE-INCORPORATION IN $YBa_2Cu_3O_{7-x}$

Since the first reports of Tc-enhancement[28] in fluorine-doped 123 compound, many groups have attempted to incorporate fluorine into $YBa_2Cu_3O_{7-x}$ although a majority of them could not observe any enhanced superconducting properties in F-doped material. The conflicting results are due to the inherent difficulties during synthesis since BaF_2, CuF_2 or YF_3 are difficult to decompose and to incorporate into the perovskite structure. During a systematic study conducted in our labs it has been shown that as long as Ba forms a part of the lattice, attempts to incorporate fluorine into $YBa_2Cu_3O_7$ at elevated temperatures invariably lead to preferential leaching out of BaF_2. For instance, the reaction of $YBa_2Cu_3O_{6.5+\delta}$ with NH_4HF_2 at 725K yields BaF_2, CuO and Y_2O_3 as reaction products as evidenced from the XRD patterns of the product. Separate attempts to prepare fluorinated compound ($YBa_2Cu_3O_{6.0+\delta}F$) from reaction of YOF, $BaCuO_2$ and CuO at 1200K also lead to the formation of BaF_2, CuO and Y_2O_3 only as reaction products[29]. It is also noticed that BaF_2 by itself cannot act as fluorinating agent even at its melting point.

Incorporation of fluorine into the $YBa_2Cu_3O_{6.5+\delta}$ orthorhombic superconducting oxide, has been subsequently accomplished in our labs by a reaction of this compound with NH_4HF_2 at sufficiently low temperatures, viz. 600K. TG/DTA experiments on mixtures of $YBa_2Cu_3O_{6.5+\delta}$ and NH_4HF_2 in different ratios have confirmed that reaction between the two occurs and the weight loss levels off at $T \geqslant 600K$. The weight loss can be accounted for in terms of fluorine substitution at oxygen sites. The product was found to be orthorhombic and chemical analysis revealed that depending on the amount of NH_4HF_2 for the reaction, upto four fluorine atoms can be incorporated into the lattice without affecting the structure[30]. However, the fluorinated compound on heating to 1000K showed the emergence of BaF_2 lines in XRD patterns whose strength exactly matched the amount of fluorine incorporated. In view of this, evaluation of Tc by resistivity measurements on sintered pellets was not feasible. Superconductivity of the fluorinated compound was established using microwave absorption technique on powder samples. No enhancement of Tc was observed upon fluorination.

VIII. MATRIX REACTION METHOD

Conventional powder-metallurgical approach where a mixture of component oxides, or a co-precipitate, is heated together to form a single-phase multicomponent oxide does not yield the best results when one goes over from the quaternary Y-Ba-Cu-oxide to the quinary ones Bi-Sr-Ca-Cu-O and Tl-Ba-Ca-Cu-O mainly due to the diverse nature of the reactivity of the component oxides which leads to the formation of unwanted ternary or quaternary oxide phases depending on the rate of interdiffusion of individual components (cations) at any given reaction temperature. Raising the reaction temperature generally leads to preferential losses of the more volatile components even before the reaction has been completed, leading to micro-inhomogeneity and presence of unreacted oxides. A practical solution to this problem is to reduce the number of reacting components of diverse reactivity and independent character. This is conveniently achieved when one employs what is known as matrix reaction method, a technique we have employed extensively for synthesizing both Bi- and Tl-based superconducting oxides. Typically, instead of reacting a mixture of component oxides in a single step, one chooses to prepare a matrix by reacting only two, or at most three, components (say CuO + SrO + CaO) at first. Guidelines for the formation of a uniform matrix and conditions for this reaction are obtained from TG/DTA analyses, and X-ray diffraction analysis of the end-product. This matrix e.g. Sr-Ca-Cu-oxide or Ba-Ca-Cu-oxide is then reacted with the remaining last component (Bi_2O_3 or Tl_2O_3) which is relatively more volatile. Specific instances of this method are provided in subsequent sections. Distinct advantage of the matrix reaction method is towards incorporation of substituents which are desirable but difficult to achieve in a conventional method. For instance, our attempts to substitute Gd or other rare-earth elements in $YBa_2Cu_3O_{7-x}$ have shown that the reactivity of rare-earth oxides with Y-Ba-Cu-O system is rather sluggish, thus requiring much longer periods of processing and sintering to achieve rare-earth incorporation. However, long heating periods could be detrimental since most superconducting phases are metastable and possess a narrow range of stability in the phase diagram. We have been able to substitute K-ions in Tl-sites in Tl-Ca-Ba-Cu-oxide [31,32] and Pb-ions in Bi-sites in Bi-Ca-Sr-Cu-oxide[33] using the matrix reaction method with exceptional success apart from using this technique in the synthesis of single-phase Bi-2122 and Tl-2122 superconducting oxides[32,34].

VIII A. Single-Phase $Bi_2CaSr_2Cu_2O_8$ by Matrix Reaction

Many investigators[35,36] have stated that single-phase Bi-2122 is not formed easily when the constituent oxides/carbonates are reacted in the stoichiometric ratio but can be obtained from Cu-rich compositions. This invariably leads to materials containing small amounts of unreacted Bi_2O_3 and CuO. We have applied the matrix reaction method for the synthesis of single phase Bi-2122 superconducting phase with T_c (R=O) of 81K[34]. The method of synthesis consisted of reacting together Bi_2O_3 and a base matrix of $CaSr_2Cu_2O_5$. The matrix was first made by mixing thoroughly appropriate amounts of $CaCO_3$, $SrCO_3$ and CuO and heating in an alumina boat at 1235K for 72 hours with intermediate grindings. Stoichiometric amounts of precursor matrix and Bi_2O_3 were mixed to give a nominal composition of $Bi_2CaSr_2Cu_2O_8$, pelletised and heated in air at 1200K for 3-5 minutes till the mixture turned completely black. The reacted powder was cooled, ground well and repelletised to be heated at 1200K for another 3 minutes after which the pellet was quickly withdrawn from the furnace in order to quench it. Finally, this pellet was annealed at 1125 K for 2 hours and furnace-cooled to room temperature. XRD analysis of the compound thus prepared did not show the presence of any parasitic phases confirming its single-phase nature. Rietveld profile refinement of the room-temperature neutron diffraction pattern was interpreted in terms of a Fmmm space group[34].In the light of current thinking on the structure of Bi-2122, we have recently re-examined the neutron profile refinement according to A2aa/Amaa structures. The modified structure suggests closely-packed near-linear BiO_2 moieties stacked parallel to a-axis (Fig.4). The contacts between oxygens of neighbouring BiO_2 groups are unusually short ($\sim$ 1Å)[37].The zero electrical resistance state as well as the diamagnetic-onset is reached at 81K for the compound thus prepared and there is absolutely no trace of 110K (Bi-2223) phase both in terms of an electrical resistivity drop or a magnetisation kink, thus underlining the importance of the matrix reaction technique over conventional single-step sintering method commonly employed, for production of single-phase samples.

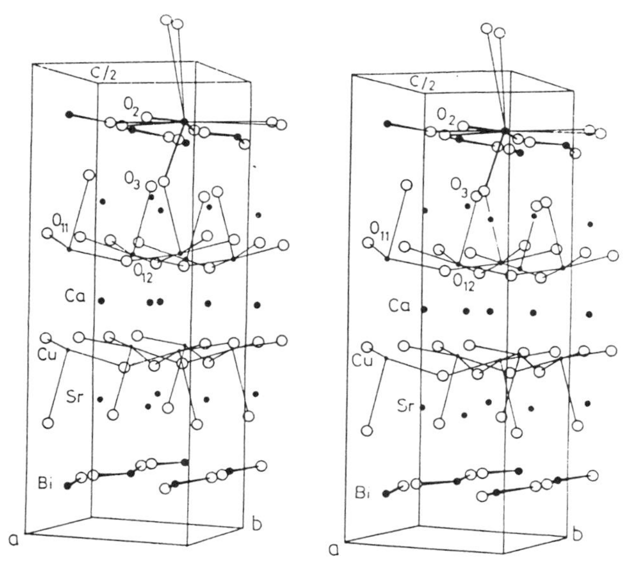

Fig 4. A stereoscopic diagram of Bi-2122 structure as per the refined A2aa/Amaa model showing the packing and the coordination of bismuth.

VIII B. Stabilising the $Bi_2Ca_2Sr_2Cu_3O_{10}$ Phase through Pb-substitution

While it is possible to isolate the 85K phase ($Bi_2CaSr_2Cu_2O_8$) in pure form, the 110K phase (i.e. $Bi_2Ca_2Sr_2Cu_3O_{10}$) cannot be isolated due to the intergrowth of 85K phase[35].This could be due to difficulties relating to the kinetics of the reaction or to the thermodynamical equilibrium of phases[36]. Consequently, although samples with a major proportion of 110K-phase have been prepared by sintering nominal Bi-1112 or Bi-4334 mixtures for long periods which show a large drop in resistance above 100K, zero-resistance state is achieved in all these samples at 80K-85K only due to a foot in the R vs T curves arising from the intergrowth of Bi-2122 phase[3]. Relative proportions of the 85K and 110K phases can be changed by adopting different heat treatments but it has been a challenge to isolate the 110K-phase in bulk. Superconducting Bi-Ca-Sr-Cu-O films with R=O at ~100K have, of course, been obtained using sputtering and evaporation deposition method[38]. Partial substitution of Bi by Pb has been shown to promote the formation of the 110K-phase (Bi-2223). Superconducting onset at 120K and R=O state at 105K have been observed[39] for $Bi_{0.7}Pb_{0.3}Sr_1Ca_1Cu_{1.8}O_{\delta}$ samples prepared by solid state reaction of component salts. Both R=O state and a sharp magnetization drop have also been reported at 108K for $Bi_{3.2}Pb_{0.8}Ca_5Sr_4Cu_7O_x$ samples prepared by sintering of component oxides for long periods yielding 2223 as a major phase[40].

We have used the matrix reaction method, for the first time, to facilitate incorporation of Pb towards stabilising the Bi-2223 phase. Our attempts have met with exceptional success since the matrix-route has yielded samples with the best possible connectivity resulting in a zero-resistance state at 120K[33]. For this purpose, the matrix $Ca_2Sr_2Cu_3O_x$ was prepared by reacting $CaCO_3$, $SrCO_3$ and CuO at 1235K. This matrix was reacted with $Pb(Ac)_2$ and Bi_2O_3 powders at 1175K for 3-5 minutes, pelletised and annealed for long intervals at ~1135K. Thin slices from the bulk, when heated to their melting point for 1-2 minutes, showed a zero-resistance state at 120K, after proper annealing. The X-ray diffraction patterns of samples thus prepared showed the presence of Bi-2223 and Bi-2122 phases in 70:30 proportion, approximately.

VIII C. Synthesis of 125K Tl-Ca-Ba-Cu Oxide Superconductor

The conventional powder-metallurgical approach, namely, mixing and heating of the constituent oxides Tl_2O_3, CaO, BaO and CuO was not found to be feasible in the synthesis of Tl-Ca-Ba-Cu-O superconductors due to the high volatility of Tl_2O_3 at the reaction temperature at which the compound is expected to be formed. Extreme precautions must be taken since Tl_2O_3 has high vapour pressures above 975K apart from being highly toxic. Hence, synthesis of Tl-based oxides is carrried out in fumehoods under protective environment. Tl_2O_3 decomposes above 1125K to Tl_2O which melts at 575K itself, thus providing high-reactivity in the liquid-phase. The problem of Tl-loss can be significantly curbed by using the matrix reaction method. In order to prepare the 125K Tl-Ca-Ba-Cu-O superconductor we synthesized the matrix $CaBaCu_2O_x$ by reacting $CaCO_3$, $BaCO_3$ and CuO in the ratio 1:1:2 at 1195K for 72 hours with several intermediate grindings. Stoichiometric ratios of Tl_2O_3 and $CaBaCu_2O_x$ matrix were mixed, ground and pelletised. The superconducting bulk phase with Tc=125K (R=O) could be obtained by adopting the following heat-treatment protocol[41]: the pellet was introduced into a furnace preheated to 1200K and held there for 5-8 minutes in flowing oxygen containing controlled amounts of water-vapour. The pellet was then quenched to room

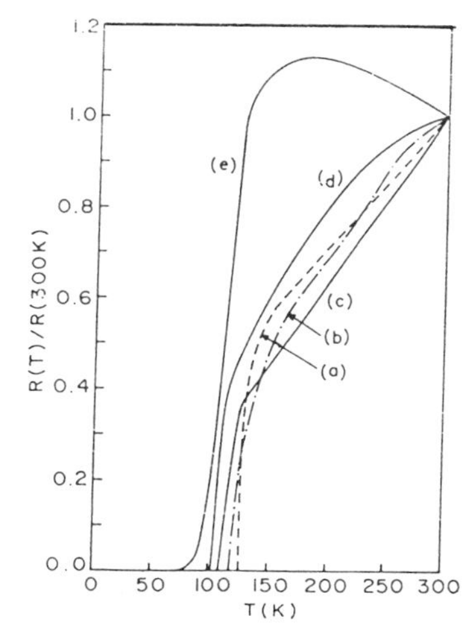

Fig 5. Resistivity of Tl-Ca-Ba-Cu-O of nominal composition : (a) 1112; (b) 1112 with 70% K; (c) 2122; (d) 2122 with 50% K; and (e) 2122 argon-annealed .

temperature by withdrawing it from the furnace. It was then annealed at 1200K in flowing oxygen and furnace-cooled by switching off the furnace. X-ray diffraction analysis showed that the resulting sample was multiphase and most of the diffraction lines could be identified as a mixture of Tl-2223 ($Tl_2Ca_2Ba_2Cu_3O_{10}$) and Tl-2122 ($Tl_2CaBa_2Cu_2O_8$) as reported[42]by Hazen et al and had a sharp resistance drop leading to a zero-resistance state at 125K (Fig.5). Presence of water vapour in controlled amounts during synthesis appears to be a must for obtaining R=O state at 125K since X-ray diffraction analysis of samples prepared by following the above protocol in dry oxygen gas contained only Tl-2122 phase and had a lower Tc of 106K (R=O). An important observation that was made during the initial quick-reaction step at 1200K was that the pellets begin to melt and densify even in 5 minutes residence time. Relatively high value of transport-Jc (250 A/cm^2 at 77K) for polycrystalline bulk samples obtained by the above process could be related to the high Tc as well as to the densification of the material. The last mentioned step of annealing for about 2 hours at 1200K and furnace cooling also appears to be crucial for attainment of high values of Tc since although the X-ray diffraction patterns of the samples prepared without going through this step looked similar to that of 125K-phase, they had a much lower Tc ($\sim$ 77K-80K). This could be either due to an insufficient oxygen pick-up in such samples or due to possible conversion of Tl^{1+} to Tl^{3+} state at quenching stage, thereby converting Cu^{3+} to Cu^{2+}. Stability of the superconducting phase in Tl-Ca-Ba-Cu-O appears to be very sensitive to the heat treatment protocols, as concluded from our detailed studies. An increase in residence time or temperature can adversely affect the Tcf (R=O) value. For example, at a processing temperature of 1210K, Tcf came down from 125K to 114K. Similarly longer residence times(10-15 minutes) at 1200K resulted in reduction of Tcf from 125K to 105K and below. Prolonged heating at 1200K also led to the emergence of matrix lines in the XRD patterns, indicating preferential loss of Tl from the sample. There are also some indications that prolonged storage of 125K material in humid environment results in deterioration in Tcf which can be avoided by storing the samples either at 475K or under partial vacuum at room temperature.

Bulk superconductivity at 125K (Tcf) was first achieved by Parkin et al[43]in samples prepared by reacting mixtures of Tl_2O_3, CaO, BaO and CuO at 1150K for 3 hours in pellet form in sealed quartz tubes after wrapping them in gold foil. They, however, observed that the largest percentage of the 125K phase (Tl-2223) was obtained only if the starting composition was $Tl_1Ca_3Ba_1Cu_3$ whereas starting composition $Tl_2Ca_2Ba_2Cu_3$, in fact, gave rise to the 108K-phase (Tl-2122) predominantly. Even by using the quick-reaction matrix route Hazen et al[42]could obtain only Tl-2122 as a major superconducting phase (and no trace of the higher-Tc Tl-2223 phase) starting from $Tl_2Ca_2Ba_2Cu_3$ nominal cation ratio. The 125K phase could be obtained by Hazen et al[42] only from $Tl_2Ca_2BaCu_3$ and $Tl_2Ca_3Ba_1Cu_3$ nominal compositions using the matrix route. In our laboratory, we have successfully adopted a variation of the method described earlier[41]with which bulk samples of 125K- majority phase of consistent quality can be produced if the $CaBaCu_2O_x$ matrix powder is first equilibrated with water vapour at ambient temperature before mixing with Tl_2O_3 and pelletization. The pellet is then completely wrapped in a foil of Pd-Ag alloy before heat treatment. The holding time at 1200K has been optimized in this case to 15 minutes and the rest of the synthesis protocol remains as explained above. However, in the absence of moisture equilibration, the product obtained is predominantly Tl-2122 phase (Tc=106K) only.

VIII D. Synthesis of Single-phase Tl-2122 (Tc=108K)

2122 nominal composition is known to yield majority Tl-2223 phase. Cu-rich nominal compositions do yield majority Tl-2122 phase but leave behind some unreacted CaO and CuO. Detailed experiments conducted in our labs have shown that the formation of single-phase Tl-2122 (Tc = 108K) is facilitated by minimising the presence of moisture during a matrix reaction procedure[32].The matrix $CaBa_2Cu_2O_5$ was prepared by reacting a mixture of $BaCO_3$, $CaCO_3$ and CuO at 1120K for 24 hours followed by heating at 1200K for 48 hours with several intermittent grindings. The matrix and Tl_2O_3 powders were mixed in the stoichiometric 2122 ($Tl_2CaBa_2Cu_2O_x$) composition and reacted according to the quick reaction method. Care was taken to avoid exposure to any moisture during synthesis (by doing the processing in a dry box) since our experience is that the presence of any moisture promotes the growth of Tl-2223 phase as explained in earlier sections. Since the synthesis of large samples (> 5gm) needed for, say, powder neutron diffraction studies necessitates longer reaction periods (> 10 minutes) the pellets of (matrix+Tl_2O_3) mixture were wrapped in a Pd-Ag alloy foil (or a Ni-foil) to avoid Tl-losses before reacting at 1200K for 15 minutes. X-ray diffraction pattern of this sample confirmed that it was single-phase $Tl_2CaBa_2Cu_2O_x$ in agreement with reported X-ray pattern and did not contain lines corresponding to any impurity phase.

The superconducting transition temperature corresponding to zero-electrical resistance state was confirmed to be 108K (Fig.5). The Rietveld profile refinement of the room-temperature neutron diffraction pattern, based on the space group I4/mmm quickly converged to an R-value (R_p) of 2.21 %, and indicated that the sample is single phase Tl-2122 (Fig.6).

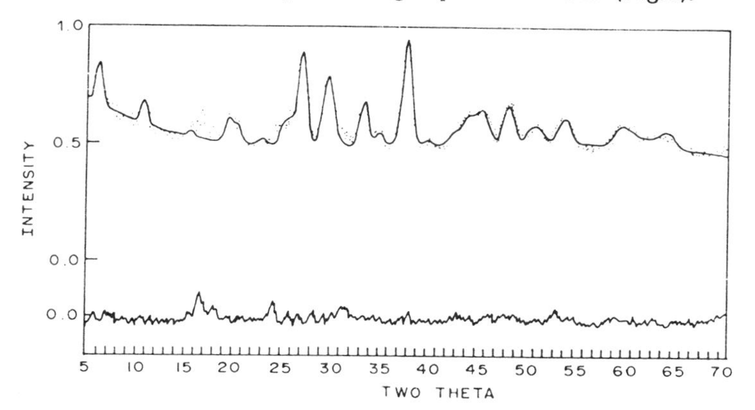

Fig 6. Calculated and observed (dots) neutron diffraction patterns of $Tl_2CaBa_2Cu_2O_x$ taken at room temperature. The difference pattern is shown at the bottom of the figure.

VIII E. Substitution of K at Tl-sites in Tl-Ca-Ba-Cu-O

It is known that facile substitution of thallium by potassium is possible in many compounds of thallium including the oxides. Therefore, successful attempts have been made in our laboratories to substitute K in Tl-sites in Tl-Ca-Ba-Cu-O system, and the superconducting behaviour and stability of Tl(K)-2122 and Tl(K)-2223 structures have been evaluated. One of the objectives of this study was to evaluate how much K-substitution at Tl-sites can be tolerated without affecting Tc.

Potassium was first sought to be incorporated by first preheating a pellet made of an intimate mixture of K_2CO_3 and CuO to 1175K which was then ground and mixed with the base matrix, $CaBaCu_2O_x$, pelletised and heated at 1185K for 30 minutes in oxygen. The cooled mass was then ground and mixed with appropriate mass of Tl_2O_3 to yield $Tl_{1-x}K_xCaBaCu_2O_y$, and made into pellets. These pellets wrapped in Ni foil were heated at 1200K for 5-8 minutes in flowing oxygen according to the method described earlier for synthesis of pure Tl-2122. X-ray diffraction analysis of samples containing 0%, 50%, 70% and 100% K as replacement for Tl, showed that XRD patterns of K-free and 50% K loaded samples were nearly identical and 50% K-substituted sample i.e. $Tl_{.5}K_{.5}Ca_1Ba_1Cu_2O_x$ did not show any of the prominent lines of the base matrix or of K_2O/K_2CO_3. However, for higher loading of K, matrix lines started appearing. It was observed that superconducting transition temperature Tcf(R=O) did not deteriorate appreciably upon K-substitution. Whereas K-free $TlCaBaCu_2O_x$ (nominal) compound gave a Tcf = 125K, Tcf value for the 50%K- and 70%K-substituted samples came down only marginally to 123K and 118K, respectively (Fig.5) [31]. The 90%K-substituted samples showed metallic behaviour below 200K with no evidence of superconductivity down to 20K. One important conclusion, drawn from our studies is that presence of optimum amount of water vapour during synthesis has a much more crucial role than anything else in the stabilisation of the higher-Tc Tl-2223 phase even in the K-substituted samples. Synthesis of 50% K- and 70% K-substituted samples conducted in the absence of any moisture led to the attainment of R=O state only at ~102K, despite the fact that all other parameters during the synthesis protocol were kept as before.

In order to establish the substitution of K in Tl-sites unequivocally we decided to undertake a neutron-diffraction analysis. For this purpose, synthesis of a 50% K- substituted Tl-2122 ($Tl_1K_1Ca_1Ba_2Cu_2O_y$) was attempted using the matrix reaction method as described before, starting from stoichiometric proportions of the matrix $K_1Ca_1Ba_2Cu_2O_x$ and Tl_2O_3. X-ray diffraction pattern of the sample thus prepared showed lines corresponding to 2122-phase in addition to evidence for formation of 2223-phase. It is believed that formation of some 2223-phase is unavoidable in this case since it is impossible to exclude moisture during synthesis due to the highly hygroscopic nature of K_2O/K_2CO_3. The inherent moisture could promote the growth of 2223-phase. The superconducting resistive transition (R=O) for this $Tl_1K_1Ca_1Ba_2Cu_2O_x$ (nominal) sample was measured to be 103K (Fig.5). Rietveld profile refinement of the room temperature neutron diffraction pattern of the compound showed that the cell parameters do not change appreciably upon K-substitution [32]. The occupancy factor shows that nearly 50% of the original Tl-sites in Tl-2122 are now occupied by K-ions in $Tl_1K_1Ca_1Ba_2Cu_2O_y$. Substitution of K^{1+} in Tl-sites also manifests itself in complete elimination of the mismatch of Ba-sheets and 0(2)-sheets along c-axis due to creation of a deficit of net positive charge in the Tl-double layers consequent upon K-substitution.

IX. PROBLEMS OF CURRENT INTEREST

It is generally believed that substitution of Pb for Bi stabilises the 110-K (Bi-2223) and promotes its growth. There have been a number of successful attempts in many laboratories in this regard. The general conclusion is that Pb-substitution enhances the relative proportion of Bi-2223 phase vis-a-vis the lower-Tc Bi-2122 phase but there has been no success thus far in synthesizing single-phase bulk samples of Bi-2223. The enhancement of 2223 fraction consequent upon Pb-substitution has led to the observation of sharp resistive onsets (no foot) with R=O state above 100K in many cases, the highest Tcf (R=O) reported being 110K. For the first time, using the matrix reaction technique, we have been able to achieve very encouraging results viz. a zero resistance state at 120K for $Bi_{1.6}Pb_{0.4}Ca_2Ba_2Cu_3O_x$ although the sample is still multiphasic [33]. What is remarkable is that our preliminary studies using the matrix reaction method have led to synthesis of $Bi_5Ca_4Sr_4Cu_8$ (nominal composition with no Pb) samples with a sharp (80%) drop in resistance at 120K indicating that in Bi-Ca-Sr-Cu-O system one can hope to achieve bulk superconductivity at ≃ 120K, substantiating the diamagnetic onsets observed at this temperature by some groups. The exact role of Pb-substitution, however, remains to be elucidated.

Another point of interest is the valence state of Tl in Tl-Ca-Ba-Cu-O system. Most researchers presume Tl to be in +3 state, although there is some speculation that both Tl^{1+} and Tl^{3+} states coexist. Our neutron--diffraction experiments have unequivocally proved that upto 50% Tl-sites can be occupied by K-ions in Tl-2122 structure and that Tl is not exclusively in +3 state in this compound. Any alteration in the Tl^{1+} : Tl^{3+} ratio should also reflect, due to internal redox chemistry, in the Cu^{2+} : Cu^{3+} balance, which should affect the superconducting behaviour, as is the general belief. A direct consequence of this argument would be that complete (100%) replacement of Tl by K, if possible, should affect the superconducting characteristics drastically. Preliminary attempts made in our laboratory in this direction indicate that K-equivalents of Tl-2122 or Tl-2223 phases are not superconducting down to 20K[44].In our view the ideal method to define Tl^{1+}:Tl^{3+} ratio would be NMR since the two nuclei of Tl are known to be strongly NMR-active. However, the large line-width problems in solid-state NMR of Tl has so far prevented such an elucidation. An entirely different approach to understand the Tl-valence would be to try and alter the Tl^{1+} : Tl^{3+} balance internally, say in Tl-2122. Recent experiments conducted in our laboratory on $Tl_2CaBa_2Cu_2O_8$ (Tc = 108K) show that the resistive ΔTc can deteriorate from 5K to 30K by annealing the compound in argon (Fig.5). Re-oxygenation of the inert-gas annealed samples not only restores the sharpness of the transition, but also enhances the Tcf from 108K to 114K. These effects are entirely reproducible upon repeat-cycling[45].If one presumes that argon-annealing essentially alters the Tl^{1+} : Tl^{3+} ratio, then our studies show that it is crucial to optimise this ratio for obtaining the best superconducting characteristics. On the other hand, inert gas annealing may be altering the oxygen-stoichiometry and thus, indirectly, affecting the superconducting behaviour. Efforts are continuing in our laboratory to assess these competing possibilities.

An important aspect to consider is whether so much Tl as present in double-layers is really needed for superconductivity in Tl-Ca-Ba-Cu-O. Over a limited range, Tl-deficiency does not appear to alter the superconducting behaviour of Tl-based oxides. In fact, fairly high Tc's ($\approx$ 100K) have been obtained even for single Tl-layer structures[46].Our recent studies indicates that Tl-2122 (108K-phase) when partially depleted of Tl, in fact, shows improved superconducting characteristics (Tc = 140K) (Fig.7).

The matrix reaction technique has provided a vehicle to synthesize complex structures and its success is now well established. It is expected that the use of this technique would provide avenues to synthesize new superconducting compounds as is evident from the recent reports[47] of discovery of a new family of cuprate superconductors $Pb_2Sr_2ACu_3O_{8+\delta}$ (A=Ln, or Ln+Sr or Ca) exploiting this method. Current trends in thinking to find new superconducting compositions have primarily centred on the complex chemistry and the multivalent nature of certain elements like Bi,Pb and Tl. An understanding of solid-state chemistry of these ions coupled with the elegance of the matrix reaction method is bound to give a fresh

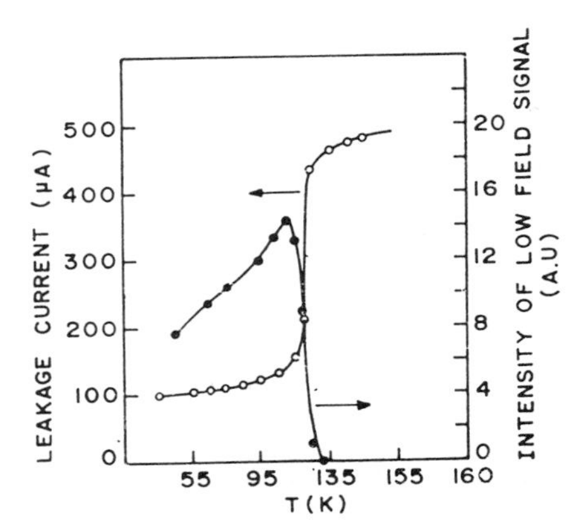

Fig 7. Evidence of superconductivity at 140K from leakage current measurements on Tl-depleted Tl-2122 samples.

impetus to the discovery of new compounds and structures relevant to a better understanding of high-Tc phenomenon.

ACKNOWLEDGEMENTS

The authors are grateful to a large number of colleagues at Bhabha Atomic Research Centre for help and collaboration at different stages. What has been reported above is essentially based on a joint program of research on high-temperature superconductors. Special thanks are due to Drs. A. Sequeira, M.D. Sastry, G.M. Phatak, U.R.K. Rao, S.R. Dharwadkar, B. Venkataramani, I.K. Gopalakrishnan, P.V.P.S.S. Sastry, H. Rajagopal and K. Gangadharan. Patience of Mr. P. Ramadas in typing the manuscript is gratefully acknowledged.

REFERENCES

[1] J.G. Bednorz and K.A. Muller, Z. Phys. B 64 (1986) 189.

[2] M.K. Wu, J.R. Ashburn, C.J. Torng, P.H. Hor, R.L. Meng, L. Gao, Z.J. Huang, Y.Q. Wang and C.W. Chu, Phys. Rev. Lett. 58 (1987) 908.

[3] H. Maeda, Y. Tanaka, M. Fukutomi and T. Asano, Jap. J. Appl. Phys. 27 (1988) L 209.

[4] Z.Z. Sheng and A.M. Herman, Nature 332 (1988) 55; Nature 332 (1988) 138.

[5] R.J. Cava, B. Batlogg, J.J. Krajewski, R. Farrow, L.W. Rupp Jr., A.E. White, K. Short, W.F. Peck and T. Kometani, Nature 332 (1988) 814.
[6] R.J. Cava, B. Batlogg, R.V. VanDover, D.W. Murphy, S. Sunshine, T. Siegrist, J.P. Remeika, E.A. Rietman, S. Zahurak and G.P. Espinosa, Phys. Rev. Lett. 58 (1987) 1676.
[7] K.G. Frase and D.R. Clarke, Adv. Ceram. Mater. 2 (1987) 295.
[8] C.C. Torardi, M.A. Subramanian, J.C. Calabrese, J. Gopalakrishnan, E.M. McCarron, K.J. Morrissey, T.R. Askew, R.B. Flippen, U. Chowdhury and A.W. Sleight, Phys. Rev. B. 38 (1988) 225.
[9] G.J. Yurek, J.B. VanderSande, W.X. Wang, D.A. Rudman, Y. Zhang and M.M. Mathiesen, Metall. Trans. A 18 (1987) 1813.
[10] I.K. Gopalakrishnan, J.V. Yakhmi and R.M. Iyer, Nature 327 (1987) 604.
[11] G. Kordas et al, Mater. Lett. 5 (11&12) (1987) 417.
[12] P. Barboux, J.M. Tarascon, F. Shokoohi, B.J. Wilkens and C.L. Schwartz, Bellcore, NJ, USA. To be published
[13] S.M. Johnson, M.I. Gusman, D.J. Rowcliffe, T.H. Geballe and J.Z. Sun, Adv. Ceram. Mater. 2 (1987) 337.
[14] B. Venkataramani, R.M. Kadam, Y. Babu, I.K. Gopalakrishnan, P.V.P.S.S. Sastry, P.V. Ravindran, G.M. Phatak, M.D. Sastry and R.M. Iyer, Solid State Comm. 67 (1988) 397.
[15] B.G. Hyde, J.G. Thompson, R.W. Withers, J.G. Fitzgerald, A.M. Stewart, D.J.M. Bevan, J.S. Anderson, J. Bitmead and M.S. Peterson, Nature 327 (1987) 402.
[16] A. Sequeira, H. Rajagopal and J.V. Yakhmi, Solid State Comm. 65 (1988) 991.
[17] F. Beech, S. Miraglia, A. Santoro and R.S. Roth, Phys. Rev. B. 35 (1987) 8778.
[18] J. Karpinski, K. Conder and E. Kaldis, Physica C. 153-155 (1988) 401.
[19] S.R. Dharwadkar, V.S. Jakkal, J.V. Yakhmi, I.K. Gopalakrishnan and R.M. Iyer, Solid State Comm. 64 (1987) 1429.
[20] J.V. Yakhmi, I.K. Gopalakrishnan, M.A. Vaidya and R.M. Iyer, Pramana 29 (1988) L 597.
[21] I.K. Gopalakrishnan, J.V. Yakhmi, M.A. Vaidya and R.M. Iyer, Appl. Phys. Lett. 51 (1987) 1367.
[22] T.S. Krishnamoorthy, N. Mahadevan, S.S. Desai and C.C. Dias, Pramana (in press).
[23] P. Choudhari, R.H. Koch, R.B. Laibowitz, T.R. McGuire and R.J. Gambino Phys. Rev. Lett. 58 (1987) 2684.
[24] S. Jin et al Phys. Rev. B 37 (1988) 7850.
[25] P.V.P.S.S. Sastry, I.K. Gopalakrishnan, J.V. Yakhmi

and R.M. Iyer, Appl. Phys. Lett. 52 (1988) 1447.
[26] S.E. Brown, J.D. Thompson, J.O. Willis, R.M. Aikin, E. Zirngieble, J.L. Smith, Z. Fisk and R.B. Schwarz, Phys. Rev. B 36 (1987) 2298.
[27] P.V.P.S.S. Sastry, I.K. Gopalakrishnan and S.L. Mishra, Ind. J. Technol. (in press)
[28] S.R. Ovshinsky, R.T. Young, D.D. Allred, G. DeMaggio and G.A. Vander Leden, Phys. Rev. Lett. 58 (1987) 2579.
[29] A.K. Tyagi, S.J. Patwe, U.R.K. Rao and R.M. Iyer, Solid State Comm. 65 (1988) 1149.
[30] U.R.K. Rao, A.K. Tyagi, S.J. Patwe, R.M. Iyer, M.D. Sastry, R.M. Kadam, Y. Babu and A.G.I. Dalvi, Solid State Comm. 67 (1988) 385.
[31] R.M. Iyer, P.V.P.S.S. Sastry, G.M. Phatak, I.K. Gopalakrishnan, K. Gangadharan and M.D. Sastry, Physica C 152 (1988) 505.
[32] A. Sequeira, H. Rajagopal, I.K. Gopalakrishnan, P.V.P.S.S. Sastry, G.M. Phatak, J.V. Yakhmi and R.M. Iyer, Physica C 156 (1988) 599.
[33] P.V.P.S.S. Sastry et al. To be published.
[34] P.V.P.S.S. Sastry, I.K. Gopalakrishnan, A. Sequeira, H. Rajagopal, K. Gangadharan, G.M. Phatak and R.M. Iyer, Physica C 156 (1988) 230.
[35] R.L. Meng et al, Mod. Phys. Lett. B (to appear).
[36] J.M. Tarascon, Y. LePage, P. Barboux, B.G. Bagley, L.H. Greene, W.R. McKinnon, G.W. Hull, M. Giroud and D.M. Hwang, Phys. Rev. B 37 (1988) 9382.
[37] A. Sequeira, H. Rajagopal and J.V. Yakhmi, Physica C (Submitted).
[38] Y. Yoshitake, T. Satoh, Y. Kubo and H. Igarashi, Jap. J. Appl. Phys.. 27 (1988) L 1089.
[39] U. Balachandran et al, Physica C 156 (1988) 649.
[40] B.W. Statt, Z. Wang, M.J.G. Lee, J.V. Yakhmi, P.C. DeCamargo, J.F. Major and J.W. Rutter, Physca C 156 (1988) 251.
[41] I.K. Gopalakrishnan, P.V.P.S.S. Sastry, K. Gangadharan, G.M. Phatak, J.V.. Yakhmi and R.M. Iyer, Appl. Phys. Lett. 53 (1988) 414.
[42] R.M. Hazen, L.W. Finger, R.J. Angel, C.T. Prewitt, N.L. Ross, C.G. Hadidiacos, P.J. Heaney, D.R. Veblen, Z.Z. Sheng, A. ElAli and A.M. Herman, Phys. Rev. Lett. 60 (1988) 1657.
[43] S.S.P. Parkin, V.Y. Lee, E.M. Engler, A.I. Nazzal, T.C. Huang, G. Gorman, R. Savoy and R. Beyers, Phys. Rev. Lett. 60 (1988) 2539.
[44] To be published.
[45] To be published.
[46] B. Morosin, D.S. Ginley, J.E. Schirber and E.L. Venturini, Physica C 156 (1988) 587.
[47] R.J. Cava et al, to be published in Nature.

SUBSTITUTIONAL STUDIES ON HIGH TEMPERATURE SUPERCONDUCTORS

A.V. Narlikar, C.V. Narasimha Rao and S.K. Agarwal

National Physical Laboratory, Hillside Road, New Delhi-12, India

HIGH Tc SYSTEMS - RELEVANCE OF SUBSTITUTIONS

A great deal of interest in the studies of high temperature superconductors is currently focussed on the substitutional effects in them. Presently there exist five prominent high Tc systems, namely the four cuprates formed with La (Tc=30K to 40K), Y (Tc of about 90K), Bi(Tc=80K to 110K) and Tl (Tc=100K to 125K) and a non-copper based system (Tc of about 30K). The significance of substitutional studies can be readily understood from the fact that the very discovery of these high temperature superconductors owes to various substitutions and dopings carried out in the related low Tc or even nonsuperconducting systems. Thus,there seems every hope that investigations of substitutional effects might yet pave the way for further enhancement of Tc, the discovery of new high-Tc phases and possibly also go a long way in unfolding the mechanism responsible for high temperature superconductivity.

Bednorz and Müller's[1] pioneering discovery of La-based cuprates containing Ba, possessing the Tc of about 30K, was the result of a systematic and logical process involving substitutions. Their thinking in part was essentially based on Cooper pair formation due to possible stronger electron-phonon coupling in oxides, and ensuing Jahn-Teller distortion giving rise to polarons. Although these theoretical ideas perhaps may not be so strongly asserted now, the approach they followed to enhance Tc was more empirical based taking full advantage of the mixed valence character of Cu[2]. Their reasoning led them to partial substitution of La^{3+} by Ba^{2+} in La_2CuO_4 whereby they could change the ratio of Cu^{3+} /Cu^{2+} in a controlled way. Thus, in contrast to the nonsuperconducting metallic perovskite $LaCuO_3$(of ABO_3 type) which had only Cu^{3+} and the insulating perovskite La_2CuO_4(having the perovskite related K_2NiF_4 crystal structure-Fig.1a) having only Cu^{2+}, the substituted material $La_{1.85}Ba_{.15}CuO_4$ became metallic,exhibiting superconductivity close to 30K. Tc was raised to 35K when Ba was replaced by yet another alkaline-earth metal Sr[3]. The marginal increase in Tc may be ascribed

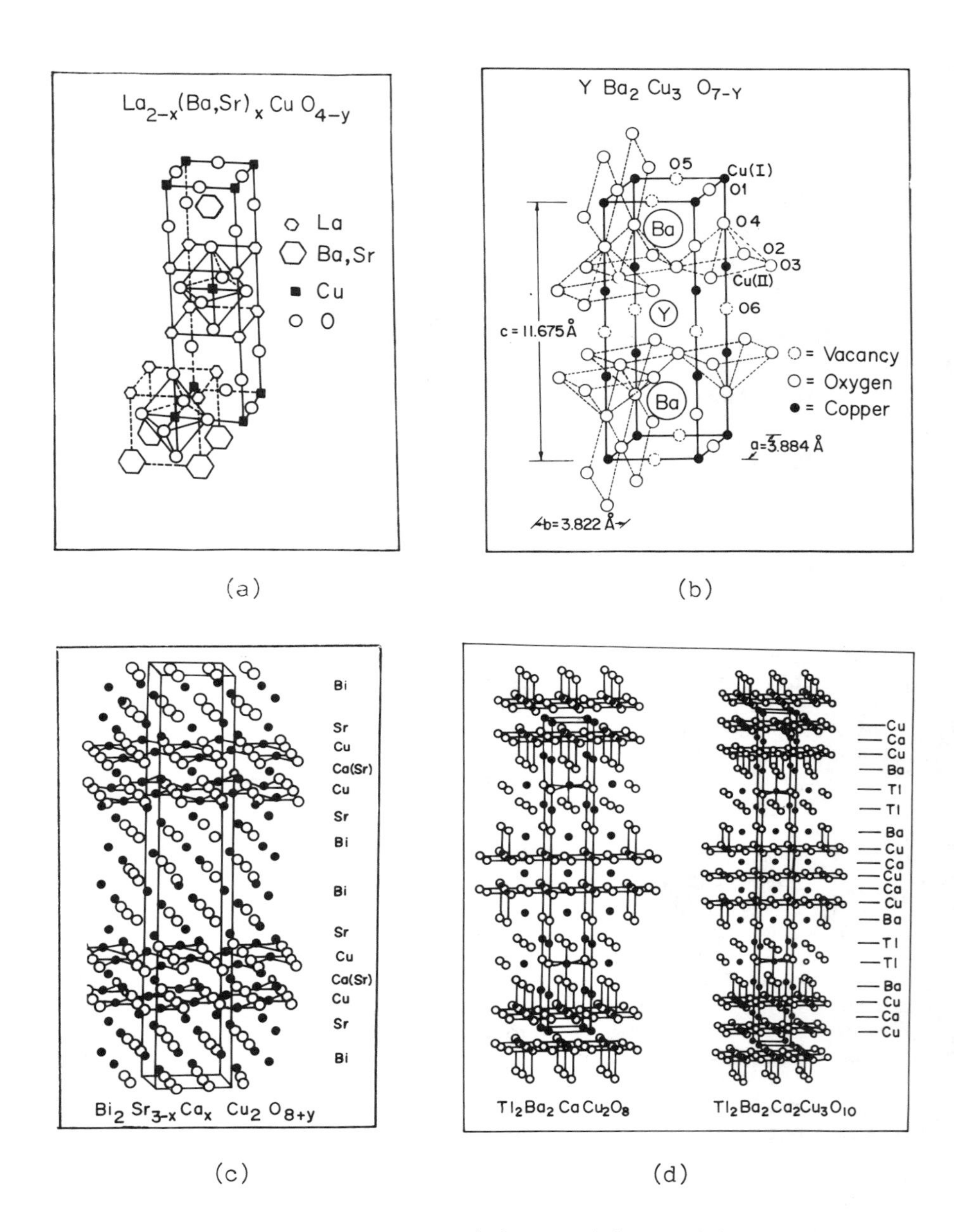

(a) (b)

(c) (d)

Fig.1 : Crystal structures of (a) La-, (b) Y-, (c) Bi- and (d) Tl-based superconducting cuprates.

to a smaller ionic size of Sr as compared to that of Ba. The substitutional effects in the lanthanum based 2-1-4 system are to be described later.

The advent of the yttrium based superconductor with Tc of about 90K[4] was again essentially the result of substitutional studies involving replacement of La by Y in La-Ba-Cu-O system. The crystal structure of the new compound so formed was, however, different. The high pressure experiments had revealed a significantly large increase in the Tc-onset of Sr doped lanthanum cuprate. The smaller ionic radius of yttrium substituting for lanthanum possibly led to the internal chemical pressure in the crystal lattice resulting in Tc enhancement. The high Tc phase was a triple layer perovskite of ABO_3 type,namely $BaCuO_3YCuO_3BaCuO_3$,but having oxygen deficiency; the final composition being $YBa_2Cu_3O_{7-y}$, commonly referred to as the 1-2-3 compound. The crystal structure is depicted in Fig.1b. The oxygen stoichiometry plays a particularly dominant role in controlling Tc of this class of compounds. Relatively marginal decrease in oxygen concentration from 7 to about 6.4 quenches superconductivity completely and transforms the crystal structure from orthorhombic to tetragonal. This aspect as will be seen later in this chapter significantly complicates the substitutional studies especially when the dopants are aliovalent.

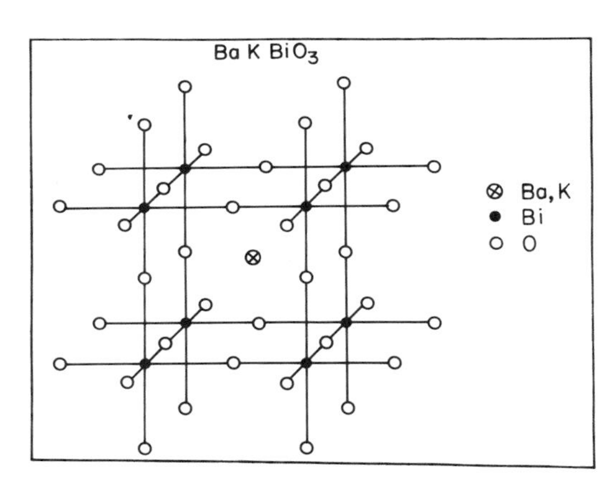

Fig.2: Crystal structure of $(Ba,K)BiO_3$

Both bismuth and thallium based cuprates are analogous to the so called Aurivillius family of oxides [5] and possess perovskite related structures. Michel et al.[6] had found the Tc of Bi-Sr-Cu-O compounds to range from 7K to 22K. This is suggestive of La being substituted by Bi in the La-Sr cuprate, but the resulting structure is different from K_2NiF_4. Later Maeda et al.[7] reported a considerably enhanced Tc of about 100K for Bi-(Ca,Sr)-Cu-O. Depending on the nominal composition and the heat treatments followed, the Tc of this system is found to vary from 80K to 110K. Thallium based systems are similar and Tl-Ba-Cu-O and Tl-(Ca,Ba)-Cu-O respectively exhibit superconductivity at about 80K and 100K- 125K [8,9]. Recently, Tl-(Ca,Sr)-Cu-O compounds have also been reported [10] with twin superconducting transitions at about 20K and 70K. The crystal structures of $Bi_2Ca_1Sr_2Cu_2O_z$ (2-1-2-2 compound) and of

Tl-based 2-1-2-2 and 2-2-2-3 compounds formed with Ba, depicted in Figs. 1c and 1d, are distinctly different from either 2-1-4 or 1-2-3 system. In contrast to the latter the Tc of Bi and Tl based cuprates are relatively insensitive to the oxygen stoichiometry.

Turning finally to the noncuprate systems, the perovskites $BaBiO_3$ and $BaPbO_3$ were respectively known to be insulating and metallic with the latter becoming superconducting at 0.5K. The partial substitution of Bi for Pb in the latter, i.e. $BaPb_{1-x}Bi_xO_3$, made the system metallic and superconducting with x=0.3 giving the optimum Tc of about 13K [11]. Interestingly, when Ba is partially replaced by K, i.e. $Ba_{1-x}K_xBiO_3$, for x=0.4 the Tc was increased significantly to about 30K[12]. Its crystal structure, a cubic perovskite, is shown in Fig.2. As with Bi- and Tl-cuprates the critical temperature is found to depend sensitively on the composition and heat treaments.

GENERAL CONSIDERATIONS OF SUBSTITUTIONS

One of the prominent ways in which the four high Tc cuprates differ from the non-copper system is that while the latter possesses a cubic crystal structure the former four are orthorhombic or tetragonal having a large c-parameter. Interestingly, as may be seen from Figs.1a to 1d all the cuprates possess highly conducting two dimensional Cu-O networks and in the 1-2-3 system additional one dimensional Cu-O chains are present along the b-axis. High temperature superconductivity is believed to be linked with these Cu-O networks. The low dimensionality in general leads to anisotropy of various properties, in particular the range of coherence ξ, the normal state resistivity ρ_n and the related parameters such as upper and lower critical fields Bc_2 and Bc_1 and the critical current density Jc when measured along and perpendicular to the a-b plane. As will be brought out later, any substitution which directly or indirectly interferes with these Cu-O networks tend to seriously affect superconducting properties of the material. In this respect these networks seem akin to the linear orthogonal transition metal chains of the conventional A-15 superconductors whose integrity has long since been known to be crucial for their Tc. Compared to the Cu-sites, substitutions at other cation sites, in general, have lesser effect on Tc, their role primarily being to provide the characteristic crystal structures possessing the Cu-O networks. The new superconducting phases, possibly of higher Tc, might result through different crystal structures incorporating substitutions at these cation sites. This contention seems to receive credence as all the existing high Tc cuprates have essentially evolved through non-Cu site substitutions in the related structures, as mentioned earlier.

In contrast to cuprates, the cubic crystal structure of the non-copper system excludes the two dimensional metal-oxygen sublattice and consequently, the low dimensionality and associated anisotropies are absent in this system. Instead, the presence of three dimensional sublattice of BiO_3 seems important for high Tc, and substitutions at the Bi-site deleteriously affect superconductivity.

Broadly speaking the site occupancy in substitutional studies is primarily controlled by three factors : (1) the ionic radii, (2) the valence state and (3) the coordination number of the on-site cation. The valence state and the ionic radii corresponding to different coordinations of various cations forming the parent high Tc systems are given in Table-1. As may be seen from the table, the cations Ba and K, in non-Cu system possess an optimum 12-fold coordination while the similar cations in cuprates, namely La, Y, Bi, Tl, Ba, Sr and Ca, all exist in lower coordinations. This aspect presumably manifests in making the conducting Cu-O networks in the latter low dimensional. In general, complete substitution of a dopant is feasible only when it is isovalent and for the same coordination its radius matches within about 15% to that of the on-site cation (cf.Table-1). The aliovalent substitutions, on the other hand, tend to be only partial.

Table-1: Valence state, coordination number and ionic radius of various cations forming HTSCs.

Material	Cation	Valence state	Existing Co-ordination number	Corresponding ionic size (Å)
$La_2(Sr)CuO_4$	La(Sr)	3 +	9	1.20
	Cu	2 +	6	.73
$Y_1Ba_2Cu_3O_7$	Y	3 +	8	1.02
	Ba	2 +	10	1.52
	Cu(I)	2 +	4	·62
	Cu(II)	2 +	5	·65
$Bi_2CaSr_2Cu_2O_x$	Bi(Tl)	3 +	6	1.02 (.68)
	Ca	2 +	8	1.12
	Sr(Ba)	2 +	9	1.30 (1.4)
	Cu	2 +	5	.65
$Ba(K)BiO_3$	Ba(K)	2 +	12	1.60
	Bi	3 +	6	1.02

Complete or partial substitutions of any of the constituents of the parental high Tc systems are in general attempted with the three broad objectives: (i) to replace toxic, less abundant or expensive components with less hazardous, more abundant or cheaper ones without making any compromise on superconducting properties, (ii) to study any anomalous changes in Tc which might give some insight into the mechanism of superconductivity and (iii) to explore the possibility of forming new superconducting phases of still higher Tc. The aforesaid systems have been explored and particularly extensive data are available for 2-1-4 and 1-2-3 cuprates, where one or more of the constituent sites have been doped by various isovalent and aliovalent cations, compatible with different parameters given in Table-1. The site occupancy need not necessarily be unique and predetermined and is normally controlled by the valence state of the dopant and the process parameters. The aliovalent cation substitutions in the parental stoichiometry inevitably lead to changes in the oxidation state of Cu and/or in the oxygen stoichiometry. The former is determined by the environment of the on-site cation, its valence state in relation to that of the dopant. A change in the oxygen stoichiometry in 1-2-3 system can bring about orthorhombic to tetragonal (O-T) transformation. There are instances where such a transformation resulting from substitutions suppresses Tc rapidly,and also contrarily in certain cases superconductivity persists even in the tetragonal state.

Tentatively, one can chalk out some of the prominent features associated with substitutions. As mentioned earlier, in the case of cuprate systems it is the low dimensional Cu-O network which is most relevant and the way the superconductivity gets influenced by dopings is essentially the manifestation of how the Cu-O networks would respond to substitutions at various sites. The networks can in general get distorted by (1) cation substitution directly at the Cu-sites, (2) anion substitution at the O-sites and (3) the aliovalent cation substitution at the non-Cu sites. Both cation and anion substitutions at these network sites, as will be seen, in general lead to lowering of Tc through host of reasons like, localization of carriers, reduction in the effective carrier concentration, obstruction of hopping process etc., and the effect is found to become more significant when the average spacing between dopants in the network matches with the range of coherence in the a-b plane. When the substituted dopant at the Cu-sites or at the non-Cu sites is aliovalent, it tends to disturb the oxygen stoichiometry in the Cu-O network between Ba-O planes of the 1-2-3 compound and lead to O-T change and Tc depression. In all these situations where the gradual incorporation of dopants disturbs or adulterates the Cu-O networks and Tc is lowered, the nature of resistance-temperature (R-T) curve is found to turn from metallic to semiconductor-like, characterised by a negative dR/dT. Isovalent substitutions of comparable ionic radii at the non-Cu sites, on the other hand, generally have relatively lesser effect on Tc and the R-T curves continue to depict metallicity. The Tc variation resulting from such substitutions may be ascribed to ensuing small changes in the lattice parameters.

In high temperature superconductors perhaps no other aspect has been studied more extensively than substitutions -both in copper and non-copper based compounds and the data available are indeed profuse. Clearly, it might be futile and seemingly unfeasible to review each and every report, but instead our attempt will be to focus the attention on some of the salient features emerging from substitutional studies. Detailed discussion of crystal structures and chemistry of these materials are included in the other contributions of the present volume.

LANTHANUM BASED CUPRATES

Undoped and stoichiometric La_2CuO_4 is orthorhombic at ambient temperatures with a=5.360 Å, b=5.402 Å and c=13.143 Å [13]. It undergoes O-T transformation at 530K. It does not become superconducting and is known to be an antiferromagnetic insulator. The material is considered to be in two dimensional antiferromagnetically ordered quantum spin fluid state with spins having no measurable time averaged magnetic moment[14] . Small departure from the O-stoichiometry in both ways,i.e. O_{4+} or

O_{4-} transforms the material from insulating to superconducting[13,15,16] although the latter have also been reported to be insulating[14]. The former state can be achieved through high temperature-high pressure oxygenation. The excess oxygen is understood[17] to have realised in the form of a superoxide ion O_2^-. which, in simple terms, oxidizes the divalent Cu and creates excess holes that are responsible for conduction and superconductivity. Alternatively, the effect of excess oxygen can be viewed as a consequence of lanthanum deficiency, i.e. $La_{2-x}CuO_4$. Depletion in the O-stoichiometry, on the other hand, transforms Cu^{2+} to Cu^{1+} which again gives rise to mobile holes.

The creation of mobile holes is the essence of making La_2CuO_4 superconducting which can be brought about by various mono and divalent substitutions at the La-site.

Substitutions at the La-site

Historically, the substitutions at the La-site are important as the field of HTSCs was evolved when the trivalent La was partially substituted by the divalent Ba[1], followed by Sr[3]. Since then both these substitutions have profusely been studied (Table-2). Also given in the table are the results of other substitutions Ca, Na and Pb. The divalent cations Ba, Sr and Ca, for concentrations $x<0.05$ are found to retain the orthorhombic distortion, but for $x \geqslant 0.1$ the structure is found to be tetragonal at the ambient temperature. The O-T transformation

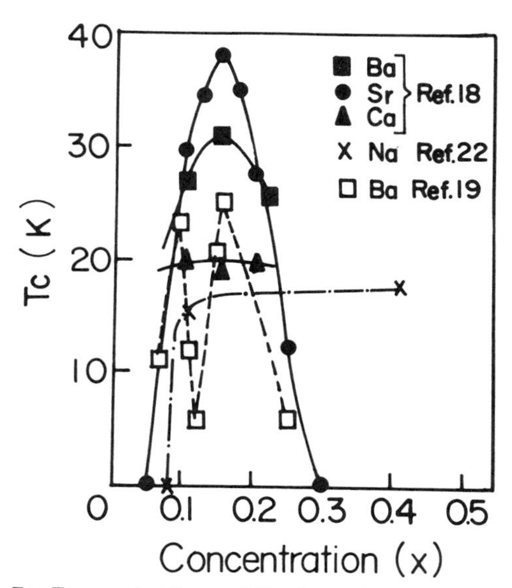

Fig.3: Tc variation with dopant concentration

temperature of 530K is rapidly lowered with increasing x as the detailed neutron diffraction studies at various temperatures have shown that for both Ba and Sr doped samples with x=0.15 the transformation occurs at about 180K. As may be seen from the table, Pb and Na substitutions have

Table-2: Lattice parameters, and critical temperatures for various partial substitutions at La-site in 2-1-4 system.

M_X $La_{2-X}M_X CuO_{4-Y}$	a (Å)	b (Å)	c (Å)	Volume (Å^3)	Tc mid (K)	Ref.
Ba .05	3.787x$\sqrt{2}$	3.801 x $\sqrt{2}$	13.192	189.89	NS	18, 19
.10	3.785	Tetragonal	13.244	189.74	27.0	
.15	3.784	"	13.285	190.22	31.5	
.20	3.784	"	13.323	190.77	25.0	
Sr .05	3.783x$\sqrt{2}$	3.810x$\sqrt{2}$	13.163	189.72	NS	18, 20
.10	3.783	Tetragonal	13.209	189.04	29.5	
.15	3.773	"	13.224	188.25	38.0	
.20	3.777	"	13.231	188.75	32.0	
Ca .05	3.781x$\sqrt{2}$	3.803x$\sqrt{2}$	13.149	189.07	NS	18, 22
.10	3.780	Tetragonal	13.158	188.01	20.0	
.15	3.779	"	13.166	188.02	19.5	
.20	3.790	"	13.159		22.5 (onset)	
Na .08		Orthorhombic		—	NS	21
.11		Tetragonal		—	16	
.41		"		—	18	
Pb .10	3.801	Tetragonal	13.191	190.58	NS	22
.30	3.792	"	13.158	189.10	NS	

NS: not superconducting

analogous effects although the former substitution does not lead to superconductivity.

Fig.3 shows Tc variation with concentration x of Ba, Sr, Ca and Na as reported by Ohishi et al.[18] and Markert et al.[21]. As may be seen, whereas the divalent substitutions yield the peak Tc at x=0.15 the monovalent Na depicts a broad maximum,possibly around x=0.4. Recent results of Ba substitution[19] essentially agree with the former, but interestingly they report an additional peak at x=0.09 and a minimum value of 5K for x=0.12. The XRD at the ambient temperature however shows no anomalies. Of the three divalent substitutions Sr results in relatively higher Tc. It seems surprising that Tc resulting from Ca substitution is lower than from Ba or Sr substitutions whereas the ionic size of La matches more closely with Ca than with Ba or Sr. The oxygen content in the three samples show[18] that the oxygen vacancies are produced more rapidly with Ca substitution. It is argued that the charge compensation in the Ca-substituted samples takes place via vacancy formation while with the other two substitutions this occurs through increase in the oxidation state of Cu^{2+}. Further, because of the smaller size of Ca at the La-site the oxygen ions tend to come closer together and due to their mutual repulsion more anion vacancies are created. Turning to Na substitution, in general one expects that half of the monovalent substitution should suffice in producing essentially the same effect as a divalent cation. The broad maximum at the nominal composition of x=0.4 may be ascribed to hygroscopic nature of the doped samples.

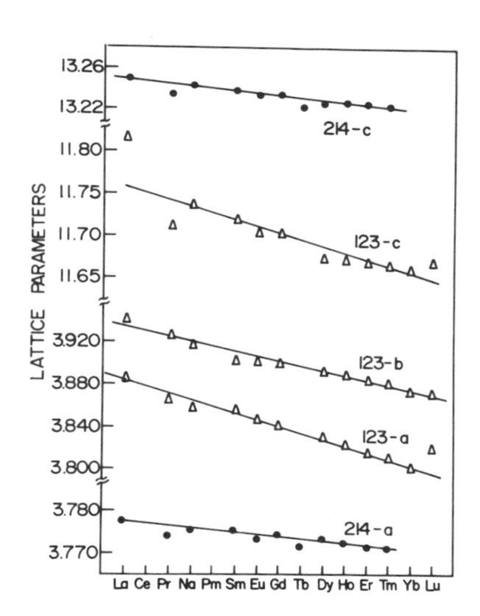

Fig.4: Lanthanide contraction in La-and Y-systems.

Taking La-Sr-Cu-O and La-Ba-Cu-O as parental systems host of substitutions have been explored where La and Cu are partly replaced by iso- and aliovalent substitutions. Tables-3 and 4 summarise some of the available data. Complete substitution of Cu by Ag has not been successful[23], though a partial replacement[24] improves Tc of the La-Ba cuprate. All other substitutions are found to lower Tc, but the effect is

Table-3: Various substitutions at La and Cu-sites in $La_{1.85}Sr_{.15}CuO_{4-y}$.

Dopant in	Tc (Onset) (K)	Tc (R=0) (K)	μ_{eff} μB	Tc Variation (K/at %)	Normal state behaviour near Tc	Ref.
$La_{1.85}Sr_{.15}CuO_{4-Y}$	41-37	38-28	—	—	Metallic	21, 27, 32 34, 37
at La-site						
$(La_{1.8}Y_{0.5})$	32	12	—	−4	Metallic	27
$(La_{1.45}Nd_{.4})$ ...	30	9	3.9/Nd	−1	Semicond.	28
at Cu-site						
... $(Cu_{.99}Cd_{.01})$	37[a]	—	—	−1	—	32
.. $(Cu_{.99}Hg_{.01})$	37[a]	—	—	−1	—	32
$(Cu_{.99}Fe_{.01})$	26[a]	—	3.0/Fe	—	—	33
$(Cu_{.95}Fe_{.05})$	NS	—	3.7/Fe	—	—	33
$(Cu_{.98}Ni_{.02})$	25	20	—	—	Semicond.	35
$(Cu_{.97}Ni_{.03})$	17	9	—	−8	Semicond.	35
$(Cu_{.925}Ni_{.075})$	NS	—	0.21/Cu	—	—	34
$(Cu_{.99}Zn_{.01})$	32	28	—	−10	Metallic	37
$(Cu_{.98}Zn_{.02})$	8	5	—	—	Metallic	37
$(Cu_{.975}Zn_{.025})$	17	10	—	−10	Metallic	34
$(Cu_{.925}Zn_{.075})$	NS	—	0.16/Cu	—	—	34
$(Cu_{.99}Zn_{.01})$	29[a]	—	—	−10	—	32

Table-4: Various substitutions at La and Cu-sites in $La_{1.8}Sr_{.2}CuO_{4-y}$ and at Cu-site in La-Ba-Cu-O.

Dopant in	Tc Onset (K)	Reference	Dopant in	Tc Onset (K)	Reference
$La_{1.8}Sr_{.2}CuO_{4-Y}$	38	26, 36	At Cu - Site		
At La - Site			$La_{1.8}Sr_{0.2}(Cu_{.95}Ti_{0.5})$	NS	29
$(La_{1.7}Pr_{.1})Sr_{2}CuO_{4-Y}$	36[a]	25, 29	$(Cu_{.99}V_{.01})$	NS	"
$Nd_{.1}$	35[a]	"	$(Cu_{.99}Cr_{.01})$	23	"
$Sm_{.1}$	32[a]	"	$(Cu_{.98}Mn_{.02})$	32	"
$Eu_{.1}$	33[a]	"	$(Cu_{.95}Fe_{.05})$	NS	"
$Gd_{.1}$	30[a]	"	$(Cu_{.95}Co_{.05})$	NS	"
$Tb_{.1}$	30[a]	"	$(Cu_{.95}Ni_{.05})$	23	"
$Dy_{.1}$	32[a]	"	$(Cu_{.99}Zn_{.01})$	29	36
$Ho_{.1}$	33[a]	"	$(Cu_{.98}Zn_{.02})$	19	36
$Er_{.1}$	35[a]	"	$La_{1.8}Sr_{.2}AgO_{4-Y}$	NS	23
$Tm_{.1}$	35[a]	"			
$Yb_{.1}$	40[a]	26			
$La_{1.9}Ba_{.1}CuO_{4-Y}$	40-30	21, 24	$La_{1.8}Ba_{.2}CuO_{4-Y}$	37	30
$(Cu_{.95}Ag_{.05})$	34	24	$(Cu_{.975}Li_{.025})$	19	30
$(Cu_{.90}Ag_{.10})$	35	24	$(Cu_{.95}Li_{.05})$	15	30
			$La_{1.8}Ba_{.2}AgO_{4-Y}$	NS	23

a) inductively measured

more deleterious when the doped cation is at the Cu-site [25-37]. Despite their magnetic moment, the partial lanthanide substitutions at the La-site have shown only marginal effect on Tc[25-29]. Efforts pertaining to complete replacement of most of the rare-earths for La have failed possibly because of their preference to 6- and 8-fold coordinations (Table-5) against the 9-fold coordination of La in the 2-1-4 system. On the other hand Y and Nd can exist in the 9-fold coordination but owing to their smaller ionic size they are expected to give rise to more anion vacancies as it is seen in the case of Ca substituted samples. Partially lanthanide doped samples exhibit the so called "lanthanide contraction",i.e., a gradual decrease in the a- and c-parameters [25]as the dopant changes from Nd to Yb (Fig.4). The paramagnetic susceptibility and Mossbauer absorption studies of Eu doped samples have confirmed the trivalent state of Eu and no anomalies have been noticed in respect of both the isomer shift or Debye-Waller factor across the superconducting transition[26].

Substitutions at the Cu- and O-sites

Several different substitutions have been tried for the Cu-site in both La-Sr and La-Ba systems (Tables-3 and 4). Li substitution for Cu in the La-Ba cuprate [30] has a moderate effect on Tc-onset, however the same substitution in pure La compound yielded no effect, the material remained nonsuperconducting[30,31]. Cd and Hg[32] substitutions for Cu have resulted in negligible effect on Tc which owing to their compatible ionic size are expected to preferably occupy the La-site. In the case of 3d elements irrespective of whether the dopant cation is magnetic or nonmagnetic, superconductivity is found to be quenched at the 5% level of substitution. Most of the substitutions at the Cu-site invariably lead to the negative dR/dT.

With Fe, 3% substitution quenches superconductivity and 5% doped sample, through Mossbauer studies[33], have indicated a magnetic order below 15K. In contrast to Fe, Ni at low concentrations does not reveal magnetic moment[34]. At very low concentrations Tc depression with x is slower which however becomes rapid and quenches superconductivity at 4%[35]. This perhaps indicates that in the Ni doped 2-1-4 compound the magnetic moment has some role in killing superconductivity.

Decrease in Tc is most rapid for Zn substitution[32, 34,36,37] and in all cases the rate is about 10K/% or more. Narasimha Rao et al.[36] have studied Zn substitution in detail. With increasing x the R-T curves turn from metallic to semiconducting like, accompanied by decrease in Tc as shown in Fig.5. The rates of depression are 9K and 11K for Tc-onset and Tc(R=0) respectively (Fig.6). The over all decrease of Tc with increasing Zn concentration can be understood in terms of its unique electronic structure. In contrast to divalent Cu which exists in $3d^9$ configuration, the d-shell of divalent Zn is completely filled, i.e. $3d^{10}$. As a direct consequence, there is a diminished overlap between the d-orbitals of Zn and p-orbitals of O. Moreover, unlike Cu, the Zn is in a nonfluctuating divalent state and as such serves as an obstacle for charge carriers to move during conduction. The changeover to the negative dR/dT with increasing Zn concentration may be ascribed to either decrease in hole concentration or to on-site localization. About 2 to 2.5% of Zn is sufficient to destroy superconductivity completely[36,37]. Interestingly, at this concentration the average spacing between randomly distributed Zn cations at Cu-sites seems to match with the range of coherence of about 24Å in Cu-O planes and thus the coherence of the superconducting state gets destroyed by the dopant.

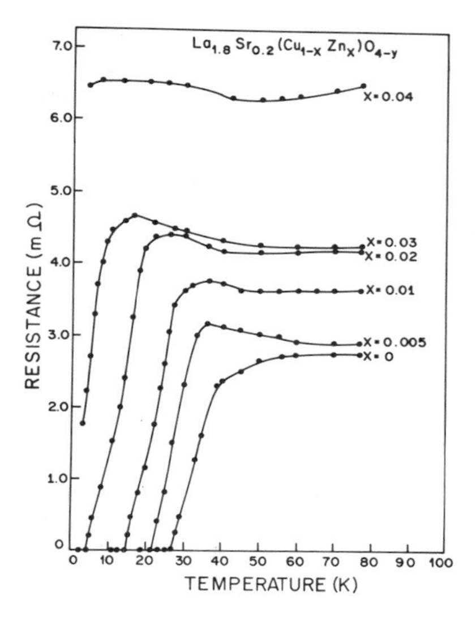

Fig.5: R vs T curves for different x values

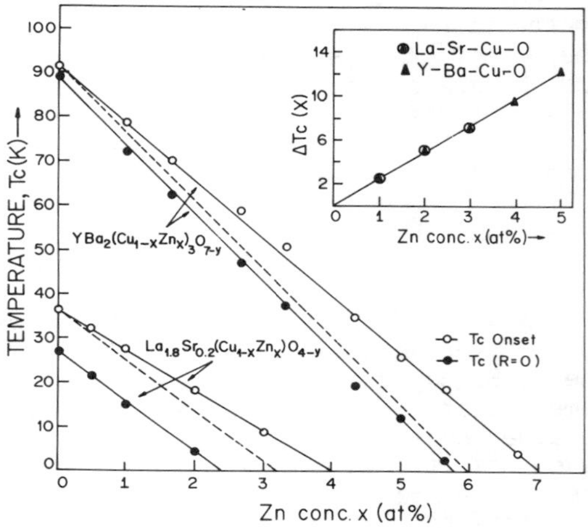

Fig.6: Tc vs x

Only limited data are available about the effect of anion substitutions in 2-1-4 system. Interesting result[38] is that La_2CuO_{4-y} which did not exhibit superconductivity showed Tc of 35K when the sample was treated with fluorine gas at 350°C. The effect was however confined to the sample surface. The fluorine ions presumably because of their higher electronegativity give rise to excess hole density which is responsible for superconductivity. The fluorination, as will be seen in the next section, has been reported to produce a much greater effect in the Y-based 1-2-3 cuprate.

YTTRIUM BASED CUPRATES

The crystal structure of 1-2-3 compound $YBa_2Cu_3O_{7-y}$ comprises (Fig.1b) of intergrown cation-oxygen planar stacks with the exception of yttrium plane which is completely devoid of oxygen giving it the 8-fold, instead of 12-fold coordination (Table-1) characteristic of the perovskite structure. The local charge neutrality is ensured by the adjacent CuO_2 planes having surplus anion content. The structure accommodates excess anion vacancies at sites located farthest away from the highly charged Y-plane,i.e.,between the two Ba-O planes. This results in two crystallographically distinct sites for Cu, Cu(I) and Cu(II). Depending on the oxygen content, the former can have 2-,4- and even 6-fold coordinations while the latter can exist only in the square pyramidal 5-fold coordination. The O_7 compound gives a 4-fold coordination to Cu(I) site leading to CuO-chains (CuO_4 ribbons) and Ba acquires 10-fold coordination. The structure as well as superconductivity are sensitive to O-stoichiometry as for y=1 it is fully tetragonal and insulating while for y=0 it is orthorhombic and superconducting.

For superconductivity, the effective copper valence is important which is controlled by oxygen stoichiometry. Isovalent substitutions, in general, do not cause any significant change in either, but aliovalent substitutions can change either of them or both, depending on the on-site cation.

Substitutions at the Y-site

The cations which have been most extensively studied for the substitution at the Y-site are from the lanthanide(Ln) family. Data pertaining to their valence state, coordination and the ionic radius are given in Table-5. All the lanthanides fulfill the criteria for substitution for Y and indeed their complete replacement has been possible without any significant change in the parental Tc of about 90K[39-44]; the exceptions are however Ce, Pr and Tb. Ce and Tb form multiphases[45] which are more stable than their corresponding 1-2-3 structure,though their partial replacement has been possible with decrease in Tc. Pr is of particular interest as it can completely substitute for Y with the ensuing structure remaining isomorphic to 1-2-3, but the material is neither metallic nor superconducting[46].

Table-6 gives various parameters of 1-2-3 samples with the Y-site completely substituted by lanthanides. As may be seen, they have the orthorhombic distortion and their lattice parameters exhibit the lanthanide contraction[40,43,47] as shown in Fig.4. Formation of 1-2-3 compound of La depends on the process parameters and the highest value reported to date is 92K by Wada et al.[48]. Tc values observed on different Ln substitutions are mutually close, in the range of 88K to 94K. Their other properties are also essentially similar. The gap coefficient is found to be around 3.2 which is close to the BCS weak coupling limit and the gap is temperature dependent. Cu-O vibrational modes as revealed by Raman studies of these materials formed with Y, Eu or Gd are found to be identical[49]. Pressure studies carried out on 1-2-3 samples of Y, Gd, Er and Yb have shown nearly the same rate of Tc increase (0.15K/kbar) and the presence of localised magnetic moment appears to have practically no influence on Tc under pressure[50]. Closeness of various properties resulting from isovalent substitutions suggests that the Y-site is only of secondary importance for high temperature superconductivity; it only helps in realising the characteristic crystal structure.

The values of the upper critical field $Bc_2(0)$ are found to cluster around 160 T, though a small systematic decrease is indicated from Nd to Tm[51]. The parameter linked with Bc_2 is the normal state resistivity ρ_n,which also shows a gradual decrease. Magnetic susceptibility in the normal state consists of temperature independent part (Pauli-like) χ_o plus a temperature dependent Curie-Weiss term $\chi(T)$ Compounds formed with heavy rare-earths, e.g. from Gd to Tm, have shown ideal Curie-Weiss

Table-5: Data for Y-site substitution in 1-2-3.

Element	Valence state	Co-ordination number	Ionic radius
Y	3+	6	0.892
La		8	1.02
		9	1.10
La	3+	6	1.06
		7	1.10
		8	1.18
		9	1.22
		10	1.28
		12	1.32
Ce	3+	6	1.034
		8	1.040
		12	1.290
	4+	6	0.80
		8	0.97
Pr	3+	6	1.01
		8	1.14
	4+	6	0.78
		8	0.99
Nd	3+	6	0.995
		8	1.12
		9	1.09
Pm	3+	6	0.979
Sm	3+	6	0.964
		8	1.09
Eu	2+	6	1.17
		8	1.25
	3+	6	0.95
		8	1.07

Element	Valence state	Co-ordination number	Ionic radius
Gd	3+	6	0.938
		8	1.06
Tb	3+	6	0.923
		8	1.04
	4+	6	0.76
		8	0.88
Dy	3+	6	0.908
		8	1.03
Ho	3+	6	0.894
		8	1.02
Er	3+	6	0.881
		8	1.00
Tm	3+	6	0.869
		8	0.99
Yb	3+	6	0.858
		8	0.980
Lu	3+	6	0.848
		8	0.97
Sc	3+	6	0.73
		8	0.87
U	4+	6	0.97
		7	0.98
		8	1.00
		9	1.05
Th	4+	6	1.00
		8	1.06
		9	1.09

Table-6: Various parameters of (1-2-3) with Ln at Y-site.

Dopant	a (Å)	b (Å)	c (Å)	$g\sqrt{J(J+1)}$ μB/atom	P_{eff} μB/atom	θK (K)	Tc (K) (R=0)	NSR (300K) μΩcm	Xo 10^{-4} cm^3/mol.	Tm (K)	Ref.
Y	3.823	3.887	11.657	0	–	4.41	92.0	2500	3.41	<.5	40,43,52 58,65
La	3.885	3.938	11.817	0	–	25.4	92.0	–	0.99	–	48, 52
Ce	–	–	–	2.54	2.16	22.6	–	–	10.03	–	52
Pr	3.864	3.926	11.709	3^+ 3.58 4^+ 2.54	2.67	-5.25	NS	–	0.99	–	40,52,65
Nd	3.855	3.914	11.736	3.62	3.10	-22.60	92.0	4200	10.80	0.52	40,43,44 52, 58
Sm	3.855	3.899	11.721	0.84	0.70	-17.90	88.3	2800	11.80	0.61	40,43,44 52,58
Eu	3.845	3.901	11.704	0	(5.6)	–	93.7	2500	–	<.5	40,43,44 58
Gd	3.840	3.899	11.703	7.94	8.00	-12.70	9.22	1650	0	2.25	40,43,44 52, 58
Tb	–	–	–	9.72	8.65	-45.60	35.0	–	0	–	41, 52
Dy	3.828	3.889	11.668	10.63	10.70	-12.10	91.2	1550	0	0.90	40,43,44 52,58
Ho	3.822	3.888	11.670	10.60	9.70	-10.30	92.2	1200	0	0.17	40,43,44 52,58
Er	3.815	3.885	11.659	9.59	9.30	-12.30	9.15	1880	0	0.553	40,43,44 52,58
Tm	3.810	3.882	11.656	7.57	7.90	-13.00	91.2	1110	0	< .5	40,43,44 52,58
Yb	3.799	3.873	11.650	4.54	4.50	-2.40	85.6	1000	25.50	< .5	40,43,44 52,58
Lu	3.802	3.870	11.645	0	–	18.5	88.2	–	0	–	40,43,52 63

behaviour with insignificant Pauli susceptibility. On the other hand the light rare-earths have yielded a nonzero Pauli susceptibility which may be due to the van Vleck contribution [52]. As may be seen from Table-6 in many instances the magnetic moment of the lanthanide ion in 1-2-3 compound above Tc as estimated from the Curie-Weiss behaviour has a value slightly lower than the free ion value which has been interpreted as due to 'some unidentified diamagnetic shielding effect' above Tc[53]. Interestingly, the possibility of extra high Tc phases at 230K and above have been indicated in 1-2-3 system through partial resistance drops and inverse a.c.Josephson studies [54] as well as from R=0 at T=260K observed in Er(1-2-3) system[55]. Unidentified diamagnetic shielding effect might originate from the presence of such extra high Tc phases.

The fact that Tc is insensitive to the presence of various magnetic rare-earth ions suggests that the local magnetic moments at the Y-site have little or no effect on superconducting state. Mossbauer studies of Eu and Gd in their (1-2-3) structures confirm the trivalent state of these ions and give direct evidence that there is no interaction between the localised 4f electrons and the conduction electrons[56-58]. Specific heat measurements have shown sharp peaks at very low temperatures for 1-2-3 compounds formed with various rare-earths, Gd, Dy, Sm, Nd and Er (Table-6) which are attributed to their magnetic ordering, coexisting with superconductivity[41,59]. However,the magnetic order was not found for Y, Eu, Ho, Tm and Yb down to 0.5K[59]. But, subsequently the magnetic order was detected in Ho (1-2-3) at 0.17K.[60]. Curie-Weiss temperature is negative in these systems indicative of its antiferromagnetic nature which is substantiated by neutron scattering measurements. The origin of magnetic ordering is attributed either to the dipole-dipole interaction[61] or to the RKKY-like interaction involving electrons some distance away from the Fermi level[62].

As mentioned earlier, although a complete substitution of Pr is found feasible, superconductivity has been observed only for a partial substitution up to about x=0.6[64], the rate of Tc depression being about 2K/at.%.[64,65]. The reason for the Tc depression is attributed to the valence state of Pr which is close to 4+. In order to ensure the local charge neutrality, substitution of Pr at the Y-site favours a reduction in the effective Cu-valence from 2.33 to 2.2 rather than increase in the oxygen content, which is supported by the measured oxygen stoichiometry [64,65]. Partial substitutions by Sc, In, and Tl (Table-7) in appreciable concentrations have yielded insignificant changes in Tc[66,67]. A partial replacement (x=0.2) by Ca showed a small decrease in Tc which is attributed to creation of extra oxygen vacancies[58,68]. Similarly, a partial substitution of aliovalent Ce^{4+} has been found to lower Tc and for $x > 0.05$ unidentified phases appear[69].

Substitutions at the Ba-site

Substitutions at the Ba-site,in general, have a larger effect in lowering Tc. As to the likely candidates, the lanthanide cations substituting for Y, when present in excess of stoichiometry have a tendency to occupy the Ba-site(cf.Table-9). Various rare-earth substitutions have been attempted at the Ba-site of Y(Ln)-based 1-2-3 compounds. Relatively small change in Tc and invariance of orthorhombic

Table-7: Partial substitutions for Y-site in 1-2-3 system.

Dopant	a (Å)	b (Å)	c (Å)	Tc(R=O) (K)	NSR behaviour	Remarks	Ref.
$Y_{.8}Pr_{.2}$	3.828	3.890	11.673	68	Metallic	(i) Reduction in Cu valence from 2.33-2.20 with X	64,65
$Y_{.5}Pr_{.5}$	3.840	3.895	11.675	14	Semicond.	(ii) Cu(I)-O(4) bond length increases with X	
Y_0Pr_1	3.863	3.910	11.694	NS	"		
$Y_{.9}Sc_{.1}$	—	—	—	90	Metallic	(i) No solubility in Y (ii) Multiphases present	66
$Y_{.6}In_{.4}$	—	—	—	89.2	Metallic	(i) Multiphases present	67
$Y_{.4}Tl_{.6}$	—	—	—	85.0	Metallic	(ii) Probably not going in to the lattice	
$Y_{.9}Ce_{.1}$	3.821	3.890	11.678	89	Metallic	(i) Ce in 4+ state	69
$Y_{.8}Ce_{.2}$	3.823	3.888	11.681	83	Metallic	(ii) Multiphases present for $x > 0.05$	
$Y_{.8}Ca_{.2}$	3.851	3.862	11.730	79	—	(i) Induces O-T for $X > .2$ (ii) Extra O vacancies created	68,72

lattice parameters found after partial substitution (10%) of Eu or Gd for Ba (Table-8) in the Y-compound [70] however cast some doubt about the occupancy of the Ba-site. La and Pr, on the other hand, owing to their relatively bigger ionic size, are more inclined to substitute for Ba. As may be seen from the Table-9, both La and Pr, beyond 15% substitution, induce O-T change but superconductivity persists at a reduced Tc in the tetragonal state[70-72]. Similarly, a partial substitution of Nd for Ba is also possible due to its bigger ionic size and its tendency for a higher coordination(cf.Table-5). All these substitutions beyond 10-15% are found to induce the O-T transformation(cf.Table-9). The lighter rare-earths have more solubility than heavier ones. Interestingly, this solubility is comparatively more in Ln 1-2-3 than in the Y-compound. This is expected due to a smaller ionic size of lanthanides compared to Ba and substitution of Ln for Ba squeezes the unit cell which is not favourable for 1-2-3 formation and thus precludes a higher solubility for Ba in Y-compound. The unit cell of 1-2-3 formed with most of Ln for Y is comparatively larger and this enlargement takes care of the squeezing effect resulting from Ln substitutions at the Ba-site. Due to this factor the solubility of various Ln for Ba is relatively more in Ln-based 1-2-3 system than in Y-(1-2-3) compound.

Beyond the oxygen content O_7,the material is already tetragonal and with increasing doping the oxygen content rises and Tc continues to decrease. Substitutions of other rare-earths at the Ba-site in Ln-(Ba,Ln)-Cu-O have shown essentially similar effects (Table-9): Tc depression, O-T change and increase in O-content[70,73,74]. These studies of Ln-substitutions at the Ba-site give credence to the belief that it is not the crystal structure but the oxygen stoichiometry which is more important for the occurrence of superconductivity.

Turning to non-lanthanide substitutions[75-85], the cations from alkali and alkaline-earth series such as Na, K, Cs, Rb, Mg, Ca, Sr as well as other elements like Zr, Sn, Pb etc. have been partly substituted for Ba and the resulting properties investigated (Table-8). The valence state, the possible coordination number and the corresponding ionic size of some of these cations are given in Table-10. Most of the alkali and alkaline-earth metals, as seen from the table, exist in 10-fold coordination and their corresponding radii are quite close to that of Ba. Substitutions of the alkali metals Na and K in appreciable concentrations did not affect Tc and for $x > 0.3$, X-ray diffraction showed unidentified phases[75-77]. Rb and Cs substitutions also showed similar effect. These elements because of their differed valence state are probably soluble only in small concentrations and at higher concentrations they seem to form multiphases instead. Amongst the alkaline earth family Mg substitution is not favoured because of its lower coordination and a smaller ionic size in comparison to Ba. Attempts made to substitute Mg for Ba have yielded multiphases. In the case of Sr substitution, Ono et.al.[78] and Veal et al.[79] reported a linear Tc depression with increasing Sr concentration x and the oxygen concentration was first found to decrease for low values of x and then increase for higher x. The substitution was found to stabilize the tetragonal phase at lower temperatures. These observations are consistent with those of Oda et al.[80] who could succeed in full replacement of Ba by Sr and found a Tc of 81K for the sample in the tetragonal state with O_{7+y}(Table-9). For small x, Sr continues to exist in the 10-fold coordination provided by

Table-8: Partial substitutions for Ba-site in 1-2-3 system.

Dopant	a (Å)	b (Å)	c (Å)	Tc onset (K)	Tc(R=O) (K)	Remarks	Ref.
$Ba_{1.8}Eu_{.2}$	3.831	3.882	11.645	90	84		70
$Ba_{1.6}Eu_{.4}$	3.829	3.884	11.651	90	77		
$Ba_{1.8}Gd_{.2}$	3.828	3.889	11.667	90	84		
$Ba_{1.8}Mg_{.2}$	—	—	—	87	77	X > .2 multiphase	76
$Ba_{1}Mg_{1}$	3.84	3.890	11.64			X < .7 no change in Tc.	
$Ba_{1.6}Na_{.4}$	3.84	3.89	11.64		77	unidentified phases present	76
$Ba_{1.5}K_{.5}$	3.83	3.89	11.64	93	90	Tc is insensitive	75,76
$Ba_{1}K_{1}$	—	—	—	93	90	X >.3 multiphases are present	
$Ba_{1.8}Rb_{.2}$	—	—	—	95	90	multiphases present	76,77
$Ba_{1.5}Rb_{.5}$	3.83	3.89	11.68	—	82		
$Ba_{1.8}Cs_{.2}$	—	—	—	96	90		
$Ba_{1.5}Cs_{.5}$	3.83	3.89	11.65	—	82		
$Ba_{1.9}Ca_{.1}$			V= 173.21	—	86	Tc decreases linearly 2nd phase appears X>.25. O vacancies increase.	72
$Ba_{1.8}Ca_{.2}$			V=172.94	—	82		
$Ba_{1.5}Sr_{.5}$	(b-a)=0.020		C=11.665	—	85	Linear & non-linear Tc variation reported.	78,79
$Ba_{1}Sr_{1}$		0.029	11.500	—	82-87	a little change in O content.	
$Ba_{.75}Sr_{1.25}$		0.002		—	80		
$Ba_{1.8}Pb_{.2}$	—	—	—	95	90	density increase is significant	
$Ba_{1.5}Zr_{.5}$	—	—	—	81	68	$BaZrO_3$ increases with X	76,77
$Ba_{1.8}Sn_{.2}$	—	—	—	95	85		85,86

Table-9: Data of tetragonal superconductors formed with various substitutions at Ba- and Y-sites in Ln(1-2-3) and Y(1-2-3) compounds.

Compound	a (Å)	b (Å)	c (Å)	Tc onset (K)	Tc(R=0) (K)	7+Y (Z)	Ref.
Nd Ba_2 Cu_3 O_Z	3.871	3.914	11.756	88	77	7.04	73
($Ba_{1.8}$ $Nd_{.2}$)	3.890	Tetragonal	11.696	54	33	7.10	
($Ba_{1.7}$ $Nd_{.3}$)	3.890	"	11.661	50	14	7.14	
($Ba_{1.5}$ $Nd_{.5}$)	3.876	"	11.649	—	—	7.30	
Sm Ba_2 Cu_3 O_Z	3.858	3.910	11.741	92	82	7.01	70,73
($Ba_{1.8}$ $Sm_{.2}$)	3.881	Tetragonal	11.654	47	29	7.09	
($Ba_{1.6}$ $Sm_{.4}$)	3.872	"	11.615	—	—	7.19	
Eu Ba_2 Cu_3 O_Z	3.844	3.904	11.709	92	88	7.01	73
($Ba_{1.8}$ $Eu_{.2}$)	3.873	Tetragonal	11.631	56	28	7.18	
($Ba_{1.6}$ $Eu_{.4}$)	3.873	"	11.619	43	13	"	
Y ($Ba_{1.8}$ $Pr_{.2}$) Cu_3O_Z	3.831	3.882	11.645	67	56	—	70
($Ba_{1.6}$ $Pr_{.4}$)	3.861	Tetragonal	11.582	37	25	—	
Y ($Ba_{1.8}La_{.2}$)Cu_3O_Z	3.824	3.885	11.650	90	82	7.10	71,72
($Ba_{1.7}$ $La_{.3}$)	—	Tetragonal	—	62	54		
($Ba_{1.5}$ $La_{.5}$)	3.858	"	11.554	43	26	7.19	
(Ba_1 La_1)	3.859	"	11.555	NS	NS	7.56	
($Nd_{.8}$ $Ca_{.2}$)$Ba_2Cu_3O_Z$	3.890	Tetragonal	11.763	76	68	—	74
Eu ($Ba_{1.85}K_{.15}$)Cu_3O_Z	3.863	Tetragonal	11.600	—	79	6.66	75
($Ba_{1.5}$ $K_{.5}$)	"	"	"	—	65	6.54	
Y Sr_2 Cu_3 O_Z	—	Tetragonal	—	86	81	7+y	80

the host lattice and due to smaller ionic size of Sr, the cubo-octahedron is expected to shrink and, for the reasons mentioned in the case of 2-1-4 system containing Ca, more oxygen vacancies will be created. Since the ionic size of Sr for the 12-fold coordination agrees better with that of Ba in 10-fold coordination, for higher Sr content the cubo-octahedron adjusts itself to the former situation and in doing so acquires more oxygen. The same, however, does not hold for Ca whose ionic size is comparatively smaller and only partial substitution has been found feasible. Tc decreases linearly upto x=0.25, beyond which multiphases have been reported [81].

Both Sr and Ca substitutions are of further interest because they have given indications of extra high Tc phases. The samples with nominal compositions of both 1-2-3 and 2-2-3 when substituted partially by Sr and Ca (for Ba) in appreciable concentrations have shown[82] noticeable resistance drops starting from room temperature down to 240K (Fig.7). These results were corroborated by Ihara et al.[83] when they found resistance drops at temperatures as high as 328K in a multiphase system of Y-Ba-Sr-Cu-O. Inverse a.c. Josephson studies[54] and microwave absorption experiments[84] on Sr-substituted samples have substantiated the possibility of extra high Tc phases near room temperature. The interconnectivity of such phases is however found to be unstable with respect to thermal cycling. These substitutions possibly give rise to superstructures in the grain boundary regions having extra high Tc.

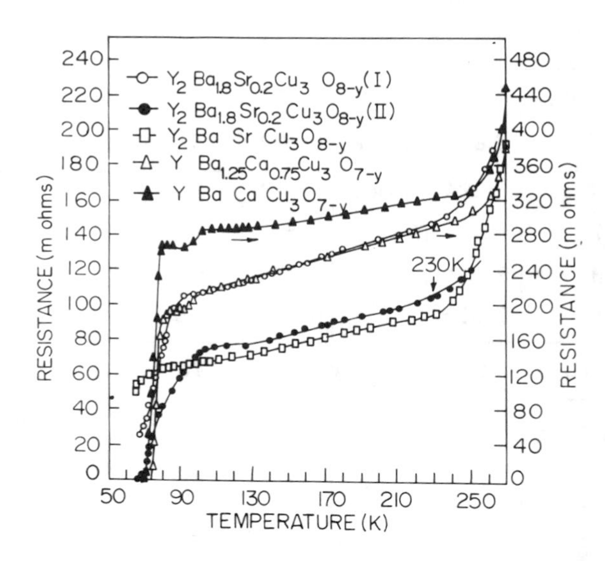

Fig.7: R vs T curves in Sr and Ca substituted Y-system

Sn substitution in 1-2-3 system has an advantage since it can serve as a Mössbauer probe. The effect of Sn substitution (x=0.2 for Ba) is found to reduce Tc (R=0) to 82K and to make the transition width ΔTc

Table-10: Data for cations at Ba-site in 1-2-3.

Element	Valence state	Co-ordination number	Ionic size (Å)
Mg	2+	4	0.49
		6	0.72
		8	0.89
Ca	2+	6	1.00
		7	1.07
		8	1.12
		9	1.18
		10	1.28
		12	1.35
Sr	2+	6	1.16
		7	1.21
		8	1.25
		9	1.31
		10	1.32
		12	1.44
Ba	2+	6	1.36
		7	1.39
		8	1.42
		9	1.47
		10	1.52
		12	1.60
Na	1+	4	0.99
		5	1.00
		6	1.02
		7	1.13
		8	1.16
		9	1.24
K	1+	6	1.38
		7	1.46
		8	1.51
		9	1.55
		10	1.59
		12	1.60

Element	Valence state	Co-ordination number	Ionic size (Å)
Zr	4+	6	0.72
		7	0.78
		8	0.84
Hf	4+	6	0.71
		8	0.83
Bi	3+	5	0.99
		6	1.02
		8	1.11
	5+	6	0.74
Pb	2+	4	0.94
		6	1.18
		8	1.24
		9	1.33
		11	1.39
		12	1.49
	4+	6	0.78
		8	0.94
Tl	1+	6	1.50
		8	1.60
		12	1.76
	3+	6	0.68
		8	1.00
In	3+	6	0.79
		8	0.92
Cs	1+	6	1.70
		9	1.78
		10	1.81
		12	1.88
Rb	1+	6	1.49
		7	1.56
		8	1.60
		12	1.73

Table-11: Data for cations at Cu-sites in 1-2-3.

Element	Valence state	Co-ordination number	Ionic size (Å)
Ti	2+	6	0.86
	3+	6	0.67
	4+	5	0.53
		6	0.61
V	2+	6	0.79
	3+	6	0.64
	4+	6	0.59
	5+	4	0.36
Cr	2+	6	0.73
	3+	6	0.62
	4+	4	0.44
		6	0.55
Mn	2+	6	0.67
		8	0.93
	3+	5	0.58
		6	0.58
	4+	6	0.54
Fe	2+	4	0.63
		6	0.77
	3+	4	0.49
		6	0.55
Co	2+	6	0.65
	3+	6	0.53
Ni	2+	6	0.70
	3+	6	0.56
Cu	1+	2	0.46
		6	0.96
	2+	4	0.62
		5	0.65
		6	0.73

Element	Valence state	Co-ordination number	Ionic size (Å)
Zn	2+	4	0.60
		5	0.68
		6	0.75
Ag	1+	2	0.67
		4	1.02
		5	1.12
		6	1.15
		7	1.24
		8	1.30
	2+	6	0.89
	3+	4	0.65
Nb	2+	6	0.71
	3+	6	0.70
	4+	6	0.69
	5+	4	0.32
Ta	3+	6	0.67
	4+	6	0.66
	5+	6	0.64
W	4+	6	0.65
	6+	4	0.41
		6	0.58
Li	1+	4	0.59
		6	0.74
Al	3+	4	0.39
		5	0.48
		6	0.53
Ga	3+	4	0.47
		5	0.55
		6	0.62

larger[85]. Mössbauer spectra have been fitted in two ways: (i) two single line components and (ii) one quadrupole doublet with a singlet, indicating that Sn occupies two crystallographically inequivalent sites in 1-2-3 lattice. Across the superconducting transition, one of the components and the quadrupole doublet exhibit temperature variation whereas the second component as well as the singlet are found to be comparatively insensitive. The latter is attributed to the Y-site occupancy of Sn^{2+} while the former to that of Ba-site. Substitutions of Zr[86] and Pb[77] for x=0.2 have yieled no change in Tc. Higher Zr substitutions have led to multiphases consisting of nonsuperconducting barium zirconate and other superconducting phases such as 1-4-5, 3-3-4[87] as well as the parent 1-2-3 phase.

Substitutions at the Cu-sites

The problem of partial replacement of Cu has been studied most exhaustively in 1-2-3 system. The situation is complicated by the presence of two chemically inequivalent sites for Cu, Cu(I) and Cu(II) and the dopant ion occupies either one of the two sites or both, depending on its preferential coordination number. Since the ionic radii of all the 3d elements are close to that of Cu they can be substituted for the latter. Table-11 gives the possible valence states, coordination number and their corresponding ionic radii of various cations which have been substituted for Cu. Since the available data are so profuse and varying only some interesting features of different cation substitutions will be presented briefly.

(i) Iron, cobalt and nickel:

For $x < 0.02$ (i.e. 2%), iron and cobalt substitutions are found to retain the orthorhombic distortion with little change in Tc, but for 3% or more, the O-T transformation is found to occur resulting in a rapid depression (5-6K/% for Fe and about 7K/% of Co) in Tc[88-91]. 20% of Fe substitution is found to make the material nonsuperconducting[92]. The Curie-Weiss behaviour indicates that both Fe and Co have magnetic moment of about $3.4\mu_B$ /atom. Since the free ion value of Co^{3+} for S=2 is about $4.9\mu_B$ and for S=1, it is $2.83\mu_B$, the observed value may be ascribed to the possibility of an intermediate spin state. Similarly, the magnetic moment of Fe does not agree either with the high spin ($5.9\mu_B$) or with the low spin ($2\mu_B$) configuration. Mössbauer studies[89] on Fe doped samples showed three kinds of doublets corresponding to Fe(I), Fe(II) and Fe(III). The former two respectively correspond to Cu(I) and Cu(II) while Fe(III), to Cu(I) in the 6-fold coordination. The Isomer shift values of Fe(III) indicate it to be in the 3+ state, while of Fe(I) and Fe(II) suggest that they could be either in the localised 3+ state or in the metallic state. The Mössbauer studies of Qui et al.[93] on 6% Fe doped sample indicated the coexistence of the magnetic ordering of Fe and superconductivity at 4.2K. With increasing Co concentration its magnetic moment is found to decrease. For 33% substitution, the magnetic moment is $2.79\mu_B$ which suggests that for higher Co concentrations the magnetic properties are controlled by 3+ state (S=1) on the Cu(I) site[94]. As can be seen from Table-11, Co exists only in the 6-fold coordination and as a consequence it would prefer the Cu(I) site with both O(1) and O(5) sites occupied and thereby promote an increase in the oxygen stoichiometry (see Table-12 which enlists the tetragonal superconductors resulting from

Table-12: Tetragonal superconductors with dopings at Cu-sites in 1-2-3.

Dopant	a (Å)	c (Å)	Tc onset (K)	Magnetic moment (μB/atom)	Tc depression (K/at%)	Oxygen content	Reference
Fe 3%	3.860	11.683	76	4.7	–	6.92	
Fe 10%	3.867	11.681	43–38	3.7	6	6.96	88–92
Fe 15%	3.864	11.590	20	3.6	5	–	
Fe 20%	–	–	NS	–	–	–	
Co 3%	3.861	11.681	73	4.1		7.05	
Co 10%	3.871	11.680	22	3.4	7	7.09	88,89,94
Co 33%	3.890	11.660	NS	2.79		7.20	
Al 7%	3.862	11.673	85	–	2	6.91	89,104
Al 10%	3.863	11.672	73	–	4	6.99	
Ga 6%	3.864	11.670	85	–	4	6.94	99,105
Ga 10%	3.861	11.696	66–62	–		6.97	

Table-13: Data for 1-2-3 compounds with various partial substitutions at Cu-sites.

Dopant	a (Å)	b (Å)	c (Å)	T_c onset (K)	T_c (R=0) (K)	Tc depression (K/at%)	Metallic/ Semicond.	7–Y	Reference
Fe 2%	3.83	3.87	11.66	90	–	<1	M	–	
Co 2%	3.83	3.883	11.667	93	–	–	M	–	89,91
Ni 2%	3.822	3.892	11.674	87	82	4	M	7.00	95,96
Ni 10%	3.827	3.889	11.671	70	65	2.3	Semicond.	6.88	100,101
Ni 20%	3.822	3.880	11.610	67	60	1.2	Semicond.	6.94	
Zn 2%	3.819	3.890	11.670	66	58	13–15	M	6.83	
Zn 3%	–	–	–	57		12–13	M		91
Zn 4%	3.820	3.890	11.670	42	28	13–15	Semicond.	6.83	97–101
Zn 6%	3.821	3.891	11.670	15	<24	13–15	Semicond.	6.84	
Al 3%	3.849	3.895	11.657	92.85	66	1–4	M		89,99
Ga 2%	3.848	3.874	11.689	87	–	–	–	6.933	104,105
Ti 10%	–	–	–	84	75	<2	M		
Cr 10%	–	–	–	90	84	<1	M		
Mn 10%	–	–	–	–	79	1	Semicond.		
Ag 10%	–	–	–	78	60	3	M		
Mg 2%	3.820	3.890	11.664	–	65	12	–		88,101
Li 5%	3.818	3.888	11.649	–	78	2	–		106,108
Nb 2%	3.808	3.873	11.622	–	92	no change	–		
W 2%	–	–	–	–	90	no change	–		
Si 10%	3.821	3.886	11.632	–	89	no change	–		

substitutions at Cu-sites). Since, the rate of Tc decrease with two substitutions is nearly the same, it is reasonable to conclude that there is no drastic difference between the two sites as regards to suppression of Tc.

As may be seen from Table-13, unlike Co and Fe, Ni seems to suppress Tc at a faster rate at low concentrations, less than 2%, but the lattice parameters remain almost invariant upto 10 to 20% of substitution[91,95]. The normal state resistivity is found to be metallic upto 3 to 4% of substitution beyond which it turns semiconductor-like[96]. The Curie-Weiss behaviour is consistent with the divalent state of Ni giving an effective magnetic moment of $3.2\,\mu_B$ [97]. This is consistent with the fact that Ni substitution does not lead to increase in the O-stoichiometry and cause the O-T transformation, which occurs for aliovalent substitutions at the Cu(I) site. As a result, Ni prefers Cu(II) site, in conformity with other reports[89,91].

(ii) Zinc

As with 2-1-4 system the Tc of 1-2-3 compounds is most vulnerable to Zn substitution; the rate of Tc depression being 13 to 15K/% and less than 6% of substitution quenches superconductivity[36,98-101]. The reasons for this are as discussed in the case of 2-1-4 system, and the quenching concentration is consistent with the reported coherence length of about 15Å in the a-b plane.

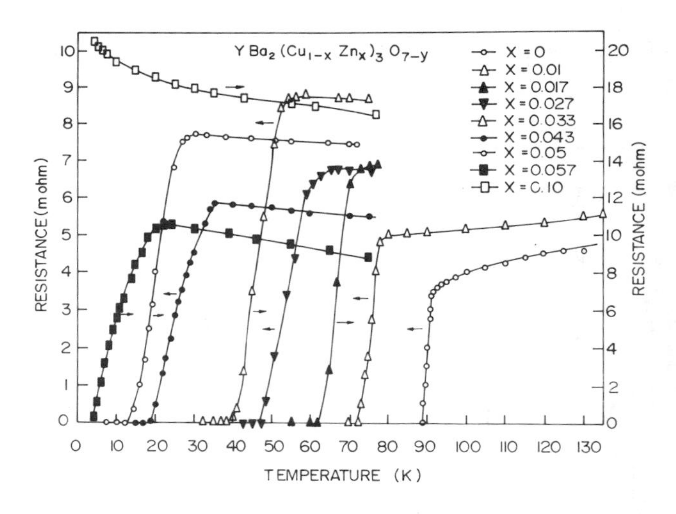

Fig.8: R vs T curves for Zn substituted 1-2-3 system

As may be seen from Fig.8, with increasing Zn concentration, the normal state conduction gradually changes from metallic to semiconductor-like which may be due to localization of charge carriers or decrease in the carrier density. As may be seen from Table-11, Zn can prefer all the three coordinations 4, 5 and 6 in its divalent state and as such both Cu(I) and Cu(II) sites seem equally preferable. This receives credence from neutron diffraction studies[102] which have indicated that Zn occupies both Cu(I) and Cu(II) sites. Interestingly, not only Tc but also the transition width vary almost identically with Zn concentration, which suggests[103] that the mechanism responsible for superconductivity in both 30K and 90K systems are unlikely to be radically different, although the former shows a weak isotope effect and the latter, its near absence.

(iii) Aluminium and Gallium

Both Al[89,104], and Ga[99,105] induce O-T transformation at about 5 to 6% of substitution (Table-12). The rate of Tc depression is about 2 to 4K/%. Surprisingly, upto about 10% concentration there is no change in the oxygen stoichiometry and the ensuing O-T transformation may be attributed to the local disorder in the Cu(I)-O plane. Both cations exist in trivalent state and their ionic size, in the 6-fold coordination, is close to that of divalent Cu. Consequently, they tend to create local disorder at Cu(I) site and a part of O(5) sites gets occupied.

As regards to other substitutions at Cu-sites, Table-13 presents the data obtained for limited substitutions of Ti, Cr, Mn, Nb, W, Ag, Li, Mg and Si[88,101,106-108]. For 2% substitution of Nb, W and Si there is no change in Tc, suggesting that these cations perhaps do not enter the lattice. The same situation seems to hold also for Cr. For Ti and Mn, the Tc depression is about 1K-2K/% of substitution. In the case of Ag, Tc is found to decrease systematically and the resistance behaviour turning to semiconductor-like[106]. On the other hand Cahen et al.[107] have found Ag to be insoluble for large concentrations.

Substitution at the O-sites

The prime interest in the anion substitution is to explore the possibility whether high temperature superconductivity can at all exist in compounds other than oxides. However, even with partial substitutions the situation as regards to Tc seems far from being transparent. Attempts have been made to replace oxygen partially by fluorine and sulphur. With the former three approaches have been tried: (i) solid state diffusion using fluorides as starting materials,(ii) heat-treatments of superconducting oxide(1-2-3 system) in the fluorine atmosphere and (iii) implantation with fluorine ions. Ovshinsky et al.[109] and Gupta et al.[110] respectively reported extra high Tc superconducting phases with Tc-onsets at 200K to 300K (Fig.9) formed using the first and the third method of preparation. The figure also shows the extra high Tc data reported by Xian-Ren et al.[111] obtained again by ion implantation. The first method has been repeated by several workers, but in most of the cases, at best, only a marginal increase in Tc has been detected and the material is found to contain unreacted BaF_2,CuO along with some unidentified phases[112-114]. The solubility of F is less than 1% and owing to its very high electronegativity it readily forms fluorides with other cations. The second method too has not been much of a success as, depending on the fluorine treatment the material turns out to be either nonsuperconducting or exhibits no change in the original Tc of about 90K. The chemical analysis shows a little or no fluorine in the system[115].

If incorporation of F takes place it is expected to have varied effects on Tc depending on its occupancy. For instance, if it replaces oxygen in planes or chains it should lead to Tc depression or quenching, accompanied by semiconducting like behaviour. If,instead, the vacant O-sites are occupied by F or if it goes preferentially to the grain boundary regions, its effect would be to populate the oxygen-hole density whereby Tc may increase. Essentially the same contentions are drawn by Hsu et al.[116] by considering the changes in the band structures resulting from fluorination.

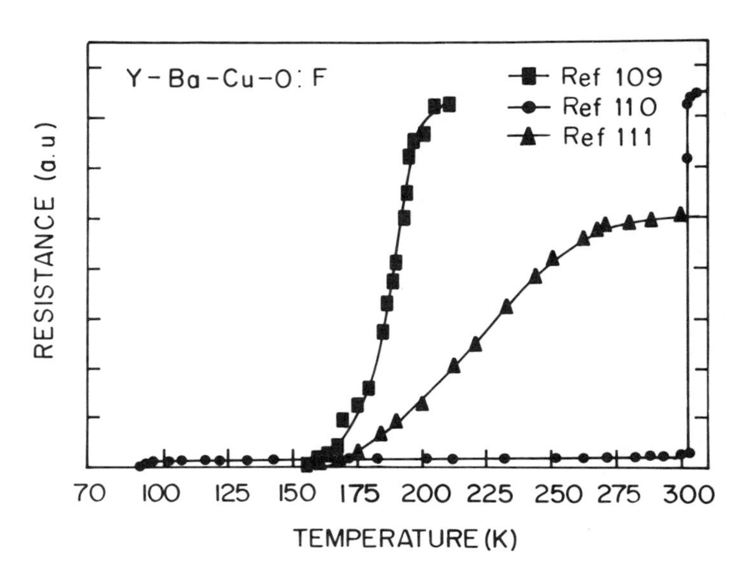

Fig.9: Extra high Tc of fluorinated 1-2-3 samples

The Tc-onset is reported to be insensitive to sulphur substitution carried out by solid state reaction, using CuS and Cu_2S. The samples formed are found to contain multiphases along with the original 1-2-3 phase[117-119]. On the other hand, Taylor et al.[120] found Tc to increase to 108K for 10% of CuS substitution, but the transition came down to 90K with two thermal cyclings. Heating below 100°C under H_2S atmosphere is found to cause swelling of the unit cell without any change in Tc.On heating at higher temperatures, the transition gradually disappears and the material is found to become semiconducting-like[117].

BISMUTH AND THALLIUM BASED CUPRATES

Both these systems constitute the rare-earth free cuprates having two superconducting phases represented by 2-1-2-2 and 2-2-2-3 with the constituents:Bi(Tl), Ca, Sr(Ba) and Cu, also known as low and high Tc phases. For the Bi-system they correspond to about 80K and 110K while for the Tl-system their Tc values are 110K and 125K. However, there are no unambiguous claims of high Tc single phase while the low Tc phase in substantial volume fraction can be readily realised. The presence of stacking defects and impurities in the reacted sample makes estimation of the oxygen content unreliable. The precise relation between Tc and the oxygen content remains to be established though it is generally believed that this dependence is not so sensitive as in the rare-earth based cuprates. Moreover, the depletion of Sr and Ca in Bi-cuprate and of Tl and Ca in the Tl-cuprate have also been reported. With such uncertainties, substitutional studies in these compounds are yet to give definite conclusions relating to the role of the Cu-O networks and of nonconducting cations but in certain cases they have paved the way towards stabilization of the high Tc phase.

Pb substitution has been extensively studied[121-123] in $Bi_{1-x}Pb_xCaSrCu_2O_z$ (1-1-1-2) and also in $Bi_{2-x}Pb_xCa_2Sr_2Cu\ O_z$ (2-2-2-3). The positive features of Pb substitution are: (1) effective copper valence is increased with doping, (2) Pb and Bi belong to the same family and form the layer structure and (3) Pb lowers the melting point of the host compound which is found to be advantageous for promoting the high Tc phase[124]. Partial replacement of Bi by Pb for x=0.3 in 1-1-1-2 compound,as reported by many groups, increases the volume fraction of the high Tc phase(Tc=110K) which gets enhanced by increasing the annealing time(at 845°C),although the XRD data show the system to remain as multiphase[121]. The 2-2-2-3 system is also found to respond in nearly identical fashion to Pb-substitution; for x=0.25 the Tc-onset and Tc(R=0) are 120K and 105K respectively[122]. Similarly, the addition of Pb in 2-2-2-3 is also found to favour the formation of high Tc phase[125].

The 2-1-2-2 compound which is known to have a tetragonal subcell with a modulated structure along the b-axis shows orthorhombic symmetry and insulating behaviour on complete replacement of Ca by Y,Nd and Eu. The insulating behaviour in $Bi_2LnSr_2Cu_2O_z$ is explained on the basis of oxygen stoichiometry and reduction in the valence state of Cu from 2.15 to 1.875[126]. Interestingly, the complete substitution of Ca by Ba in 2-1-2-2 is reported to be semiconducting while in 2-2-2-3 it is still superconducting with Tc-onset of about 94K[127]. A partial substitution of Y for Ca (40%) retains superconductivity with diminished Tc and the normal state resistivity showing the metallic behaviour. Tc decrease may be ascribed to a reduction in carrier (hole) concentration, in conformity with the reduced valence state of Cu while the metallic nature of the normal state resistivity suggests that the Cu-O networks are not significantly affected. With 50% substitution, the material shows no superconductivity and the resistivity behaviour turns semiconductor-like. Full replacement,as revealed by susceptibility measurements, is found to exhibit a spin-glass like behaviour at 13K[128].

Substitutions at the Bi-site have also been tried in 4-3-3-4. Partial substitutions of Eu and Gd (upto 7%) are found to increase Tc marginally but for higher doping levels the Tc is reduced. Partial substitutions of Pb and K (5-12%) show the presence of high Tc phase[129]. La when substitutes partially for Bi is found to lower Tc[130]. As regards to substitutions at the Cu-site, Zn is found to depress Tc at a rate of about 6K/% which is much lower than for 2-1-4 and 1-2-3 systems. Upto about 3% of substitution the normal state resistivity remains metallic which for higher concentrations rapidly turns to semiconductor-like[131]. In addition, host of Ln have been substituted partially for Sr in low Tc (20K) 2-2-1 and 1-1-1 oxides of Bi-Sr-Cu; Tc is found to be relatively insensitive to La,Nd and Pr but is lowered with other substitutions[132,133].

Substitutional studies have been extended to Tl-based systems and complete replacement of Tl by In and Pb has been found unfruitful. 50% Tl replacement by Bi in 2-2-2-3 is found to be superconducting with the Tc-onset of above 100K. Complete substitution of Ca by Mg and 50% substitution of Sr by Pb in 2-1-2-2 are found to be metallic while a complete replacement of Ba by Sr shows semiconducting character with higher resistivity down to 60K[127,134]. More recently, however Tl-Ca-Sr-Cu-O has been found superconducting at 20K in the nominal composition of 2-2-2-3 and at 70K in 2-2-1-3[10]. 50% substitution of Pb for Tl in Tl-Ca-Sr-Cu-O is found to form the low Tc phase in 2-1-2-2 composition and the high Tc phase in 2-2-2-3 with Tc(R=0) at 82K and 115K respectively[135].

NON-COPPER SYSTEM

As mentioned in the beginning, Sleight et al.[11] and Cava et al.[12] found superconductivity in $Ba(Pb_{1-x}Bi_x)O_3$ and $(Ba_{1-x}K_x)BiO_3$, respectively known as BPBO and BKBO, with the former having Tc of 13K for x=0.25 and the latter about 28K for x=0.4. The results of substitutional studies in both of them are yet to be reported. Their normal state resistivity is semiconductor-like,indicative of low carrier density[136,137]. Tc of both systems is sensitive to x. In the BPBO the material becomes nonsuperconducting for $x > 0.3$ and shows a considerable increase in the room temperature resistivity. Tunnelling measurements indicate BPBO to be a strongly coupled superconductor rather than a Bose condensate of charged bipolarons. The gap coefficient, on the other hand, matches with the BCS weak coupling value of 3.5[136]. By X-ray and neutron diffraction techniques, a series of phase transformations have been found from orthorhombic-tetragonal-orthorhombic-monoclinic with increasing Bi concentration[138-140]. The first transformation occurs for $x > 0.05$ while the second, for $x > 0.3$ and the third for $x > 0.9$. Oda et al.[141] have however shown that the orthorhombic phase can be stabilized over a wider concentration range of x upto 0.9 by suitable change in the process parameters. This way, the Tc for x=0.25,in the orthorhombic state is found to be 10K, whereas for the same concentration, in the tetragonal state ,it is found to be 12K.

The BKBO for x=0.4 appears altogether different and simple in structure, namely a cubic perovskite at room temperature[12,142]. Tc changes from 24K to 28K as x is increased from 0.35 to 0.45. For smaller x there is no superconductivity while for larger x the material exhibits superconductivity and the presence of multiphases[137,142]. The high Tc phase,corresponding to x=0.4 is found to remain cubic in the temperature range of 14K to 300K, with the lattice parameter increasing linearly from 4.274Å to 4.286Å, as revealed by neutron diffraction[143]. Interestingly, in contrast to cuprate superconductors, the BKBO is found to show a pronounced isotope effect with respect to oxygen; the numerical exponent being close to 0.4 in agreement with the conventional phononic superconductivity.

CONCLUDING COMMENTS

To sum-up, in this chapter we have attempted to present and review the salient features associated with host of different substitutions at various sites of recently discovered high temperature superconductors. In the broad sense, the understanding evolved of the role of different cations and anions, substituting fully or partly at various sites of 2-1-4 and 1-2-3 lattices is quite noteworthy. The situation however does not appear to be so transparent in the case of non-rare-earth and non-copper based systems which are very sensitive to the processing conditions and the parennial problem is of realising high Tc single phases. The indications available seem to give sufficient credence to the belief that substitutional studies hold the key of realising phases with much higher Tc values than hitherto possible.

Acknowledgements: The authors would like to thank Professor S.K. Joshi, the Director of their laboratory for his keen interest and encouragement.

REFERENCES

[1] J.G. Bednorz and K.A. Müller, Z.Phys. B64, (1986) 187.
[2] M.K.Wu, J.R.Ashburn, C.J. Torng, P.H. Hor, R.L. Meng, L. Gao, Z.J. Huang, Y.Q. Wang and C.W. Chu, Phys.Rev. Lett., 58 (1987) 908.
[3] H. Takagi, S. Uchida, K. Kitazawa and S. Tanaka, Jap.J. Appl.Phys., 26 (1987) L123.
[4] N. Nguyen, C.Michel, F. Studer and B. Raveau, Mat.Chem. Chem., 7 (1982) 413.
[5] B. Aurivillius, Arkiv Kemi, 1 (1950) 463.
[6] C. Michel, M. Hervieu, M.M. Borel, A. Grandin, F. Deslandes, J. Provost and B. Raveau, Z.Physik, B68 (1987) 421.
[7] H. Maeda, Y. Tanaka, M. Fukutomi and T. Asano, Jap.J. Appl.Phys., 27 (1988) L209.
[8] Z.Z. Sheng and A.M. Hermann, Nature, 33 (1988) 55, 138.
[9] R.M. Hazen, L.W. Finger, R.J. Angel, C.T. Prewitt, N.L. Ross, C.G. Hadidiacos, P.J. Heaney, D.R. Veblen, Z.Z. Sheng, A.E. Ali and A.M. Hermann, Phys.Rev.Lett., 60 (1988) 1657.
[10] Z.Z. Sheng, A.M. Hermann, D.C. Vier, S. Schultz, S.B. Oseroff, D.G. George and R.M. Hazen, Phys.Rev. B., 38 (1988) 7074.
[11] A.W. Sleight, J.L. Gillson and P.E. Bierstedt, Sol. State Commu., 17 (1975) 27.
[12] R.J. Cava, B. Batlogg, J.J. Krajewski, R. Farrow, L.W. Rupp Jr., A.E. White, K. Short, W.F. Peck and T. Kometani, Nature, 332 (1988) 814.
[13] J.E. Schirber, B. Morosin, R.M. Merrill, P.F.Hlava,E.L. Venturini, J.F. Kwak, P.J. Nigrey, R.J. Baughman and D.S. Ginley, Physica C, 152 (1988) 121.
[14] M. Oda, Y. Hidaka, M. Suzuki, Y. Enomoto, T. Murakami K. Yamada and Y. Endoh, Sol. State Commu., 67 (1988), 257.

[15] J. Beille, B. Chevalier, G. Demazeau, F. Deslandes, J. Etourneau, O. Laborde, C. Michel, P. Lejay, J. Provost, B. Raveau, A. Sulpice, J.L. Tholence and R. Tournier, Physica B, 146 (1987) 307.
[16] R. Yoshizaki, H. Sawada, T. Iwazumi and H. Ikeda, Sol. State Commu.,65 (1988) 1539.
[17] J.W. Rogers,Jr., N.D. Shinn, J.E. Schriber, E.L. Venturini, D.S. Ginley and B. Morosin, Phys.Rev. B, 38 (1988) 5021.
[18] K. Ohishi, M. Kikuchi, Y. Syono, N. Kobayashi , T. Sasaoka, T. Matsuhira, Y. Muto and H. Yamauchi, Jap.J. Appl.Phys., 27 (1988) L1449.
[19] A.R. Moodenbaugh, Y. Xu, M. Suenaga, T.J. Folkerts and R.N. Shelton, Phys.Rev.B 38 (1988) 4596.
[20] P. Przyshupski, J. Igalson, J. Rauhszkiewicz and T. Skoskiewicz, Phys.Rev.B 36 (1987) 743.
[21] J.T. Markert, C.L. Seaman, H. Zhou and M.B. Maple, Sol. State Commu., 66 (1988) 387.
[22] D.U. Gubser, R.A. Hein, S.H. Lawrence, M.S. Osofsky, D.J. Schrodt, L.E. Toth and S.A. Wolf, Phys.Rev.B, 35 (1987) 5350.
[23] K.K. Pan, H. Mathias, C.M. Rey, W.G. Moalton, H.K. Ng, L.R. Testardi and Y.L. Wang, Phys.Lett. A, 125 (1987) 147.
[24] Y. Saito, T. Noji, A. Endo, N. Matsuzaki, M. Katsumata and N. Higuchi, Jap.J.Appl.Phys., 26 (1987) L834.
[25] K. Kishio, K. Kitazawa, T. Hasegawa, M. Aoki, K. Fueki, S. Uchida and S. Tanaka, Jap.J.Appl.Phys., 26 (1987) L391.
[26] A.K. Grover, S.K. Dhar, P.L. Paulose, V. Nagarajan, E.V. Sampathkumaran and R. Nagarajan, Sol.State Commu., 63 (1987) 1003.
[27] G.M. Phatak, A.M. Umarji, J.V. Yakhmi, L.C. Gupta, K. Gangadharan, R.M. Iyer and R. Vijayaraghavan, Sol.State Commu., 10 (1987) 905.
[28] G.W. Crabtree, W.K. Kwok, A. Umezawa, L. Soderholm, L. Morss and E.E. Alp, Phys.Rev. B, 36 (1987) 5258
[29] T. Hasegawa, K. Kishio, M. Aoki, N. Ooba, K. Kitazawa, K. Fueki, S.Uchida and S. Tanaka, Jap.J.Appl.Phys.,26, (1987) L337.
[30] M.A. Kastner, R.J. Birgeneau, C.Y. Chen, Y.M. Chiang, D.R. Gabbe, H.P. Jenssen, T. Junk, C.J. Peters, P.J. Picone, T. Thio, T.R. Thurston and H.L. Tuller, Phys. Rev. B, 37 (1988) 111.
[31] C.N.R. Rao, Chemistry of Oxide Superconductors (Ed. C.N.R. Rao), Blackwell Sci.Pub., (1988) p. 2.
[32] K. Remschnig, P. Rogl, R. Eibler, G. Hilscher, N. Pillmayr, H. Kirchmayr and E. Bauer, Physica C, 153-155 (1988) 906.
[33] J.M. Matykiewicz, C.W. Kimball, J. Giapintzakis, A.E. Dwight, M.B. Brodsky, B.D. Dunlap, M. Slaski and F.Y. Fradin, Phys.Lett. A, 124 (1987) 453.
[34] J.M. Tarascon, L.H. Greene, P. Barboux, W.R. McKinnon, G.W. Hull, T.P. Orlando, K.A. Delin, S. Foner and E.J. McNiff,Jr., Phys.Rev. B, 36 (1987) 8393.

[35] W. Kang, H.J. Schulz, D. Jerome, S.S.P. Parkin, J.M. Bassat and Ph. Odier, Phys.Rev. B, 37 (1988) 5132.
[36] C.V. Narasimha Rao, B. Jayaram, S.K. Agarwal and A.V. Narlikar, Physica C, 152 (1988) 479.
[37] G. Xiao, A. Bakhshai, M.Z. Cieplak, Z. Tesanovic and C.L. Chien, to be published.
[38] B.M. Tissue, K.M. Cirillo, J.C. Wright, M. Daeumling and D.C. Larbalestier, Sol.State Commu., 65 (1988) 51.
[39] P.H. Hor, R.L. Meng, Y.Q. Wang, L. Gao, Z.J. Huang, J. Bechtold, K.Forster and C.W. Chu, Phys.Rev.Lett., 58 (1987) 1891.
[40] J.M. Tarascon, W.R. McKinnon, L.H. Greene, G.W. Hull and E.M. Vogel, Phys.Rev. B, 36 (1987) 226.
[41] S.E. Brown, J.D. Thompson, J.O. Wills, R.M. Aikin, E. Zirngiebl, J.L. Smith, Z. Fisk and R.B. Schwarz, Phys. Rev. B, 36 (1987) 2298.
[42] G.V. Subbarao, Chemistry of Oxide Superconductors (Ed. C.N.R. Rao), Blackwell Sci.Pub. (1988) p.65.
[43] T. Kaneko, H. Toyodo, H. Fujita, Y. Oda, T. Kohara, K. Ueda, Y. Yamada, I. Nakada and K. Asayama, Jap.J.Appl. Phys.,26 (1987) L1956.
[44] G. Xiao, F.H. Streitz, A. Gavrin and C.L. Chien, Sol. State Commu.,63 (1987) 817.
[45] L.F.Schneemmeyer, J.V. Waszezak, S.M. Zahorak, R.B. van Dover and T. Siegrist, Mat.Res.Bull.,22 (1987) 1467.
[46] A. Matsuda, K. Kinoshita, T. Ishii, H. Shibata, T. Watanabe and T. Yamada, Phys.Rev. B, 38 (1988) 2910.
[47] M.A. Alario-Franco, E.Moran-Miguelez, R. Saez-Puche, F. Garcia-Alvarado, U. Amador, M. Barahona, F. Fernandez, M.T.Perez-Fria and J.L. Vicent, Mat.Res. Bull., 23 (1988) 313.
[48] T.Wada, N. Suzuki, T. Maeda, A. Maeda, S. Uchida, K. Uchinokura and S. Tanaka, Appl.Phys.Lett. 52 (1988) 1989.
[49] G.A. Kourouklis, A. Jayaraman, B. Batlogg, R.J. Cava, M. Stavola, D.M. Krol, E.A. Rietman and L.F. Schneemeyer, Phys.Rev. B, 36 (1987) 8320.
[50] H.A. Borges, R. Kwok, J.D. Thompson, G.L. Wells, J.L. Smith, Z. Fisk and D.E. Peterson, Phys.Rev. B , 36 (1987) 2404.
[51] T.P.Orlando, K.A. Delin, S. Foner, E.J. McNiff,Jr, J.M. Tarascon, L.H. Greene, W.R. McKinnon and G.W. Hull, Hull, Phys.Rev. B, 36 (1987) 2394.
[52] H. Zhou, X.L. Seaman, Y. Dalithaouch, B.W. Lee, K.N. Yang, R.R. Hake, M.B. Maple, R.P. Guertin and M.V. Kuric, Physica C, 152 (1988) 322.
[53] F. Zuo, B.R. Patton, D.L. Cox, S.I. Lee, Y. Song, J.P. Golben, X.D. Chen, S.Y. Lee, Y. Cao, Y. Lu, J.R. Gaines J.C. Garland and A.J. Epstein, Phys.Rev. B, 36 (1987) 3603.
[54] A.K. Gupta, B. Jayaram, S.K. Agarwal, A. Gupta and A.V. Narlikar, Phase Transitions, 10 (1987) 29.
[55] P. Ayyub, P. Guptasarma, A.K. Rajarajan, L.C. Gupta, R. Vijayaraghavan and M.S. Multani, J.Phys.C:Solid State Phys., 20 (1987) L673.

[56] M. Eibschutz, D.W. Murphy, S. Sunshine, L.G. Van Uitert S.N. Zahurak and W.H. Grodkiewicz, Phys.Rev.B , 35 (1987) 8714.
[57] P. Boolchand, R.N. Enzweiler, I. Zitkovsky, R.L. Meng, P.H. Hor, C.W. Chu and C.Y. Huang, Sol.State Commu., 63 (1987) 521.
[58] E.E. Alp, L. Soderholm, G.K. Shenoy, D.G. Hinks, D.W. Capone II, K. Zhang and B.D. Dunlap, Phys.Rev.B 36 (1987) 8910.
[59] B.W. Lee, J.M. Ferreira, Y. Dalichaouch, M.S. Torikachvili, K.N. Yang and M.B. Maple, Phys Rev.B, 37 (1988) 2368.
[60] B.D. Dunlap, M. Slaski, D.G. Hinks, L. Soderholm, M. Beno, K. Zhang, C.Segre, G.W. Crabtree, W.K. Kwok, S.K. Malik, I.K. Schuller, J.D. Jorgensen and Z. Sungaila, J.Mag.Mag.Mat., 68 (1987) L139.
[61] B.D. Dunlap, M. Slaski, Z. Sungaila, D.G. Hinks, K. Zhang, C. Segre, S.K. Malik, E.E. Alp, Phys.Rev. B, 37 (1988) 592.
[62] S.H. Liu, Phys.Rev.B, 37 (1988) 7470.
[63] A. Oota, Y. Sasaki, M. Ohkubo and T. Hioki, Jap.J.Appl. Phys., 27 (1988) L1425.
[64] K. Kinoshita, A. Matsuda, H. Shibata, T. Ishii, T. Watanabe and T. Yamada, Jap.J.Appl.Phys., 27 (1988) L1642.
[65] B. Okai, M. Kosuge, H. Nozaki, K. Takahashi, and M. Ohta, Jap.J.Appl.Phys., 27 (1988) L41.
[66] G. Svensson, Z. Hegedus, L. Wang and O. Rapp, Physica C 153-155 (1988) 864.
[67] A.K. Bhattacharya and K.K. Singh, Physica C, 152 (1988) 283.
[68] M. Kosuge, B. Okai, K. Takahashi and M. Ohta, Jap.J. Appl.Phys., 27 (1988) L1022.
[69] R. Liang, Y. Inaguma, Y. Takagi, T. Nakamura, Jap.J. Appl.Phys., 26 (1987) L1150.
[70] A. Suzuki, E.V. Sampathkumaran, K. Kohn, T. Shibuya, A. Tohdake and M. Ishikawa, Jap.J.Appl.Phys., 27 (1988) L792.
[71] M.R. Harrison, I.C.S.T. Hegedus, W.G. Freeman, R. Jones P.P. Edwards, W.I.F. David and C.C. Wilson, Chemistry of Oxide Superconductors (Ed.C.N.R. Rao) Blackwell Sci. Pub. (1988) p.137.
[72] A. Tokiwa, Y. Syono, M. Kikuchi, R. Suzuki, T. Kajitani N. Kobayashi, T. Sasaki, O. Nakatsu and Y. Muto, Jap.J. Appl.Phys., 27 (1988) L1009.
[73] S. Li, E.A. Hayri, K.V. Ramanujachary, M. Greenblatt, Phys.Rev.B, 38 (1988) 2450.
[74] H. Uwe, T. Sakudo, H. Asano, T.Han, K. Yagi, R. Harada, M. Iha and Y. Yokoyama, Jap.J.Appl.Phys.,27 (1988) L577
[75] I. Felner, M. Kowitt, Y. Lehavi, L. Ben-dor, Y. Wolfus, B. Barbara and I. Nowik, Physica C, 153-155 (1988) 898.
[76] Y. Matsumoto, T. Abe, M. Tanaka, T. Tazawa and E. Sato, Mat.Res.Bull., 23 (1988) 1241.
[77] A.S. Bhalla, R. Roy and L.E. Cross, Chemistry of Oxide Superconductors (Ed. C.N.R. Rao), Blackwell Sci.Pub.,

(1988) p.76.
[78] A. Ono, T. Tanaka, H. Nozaki and Y. Ishikawa, Jap.J. Appl.Phys., 26 (1987) L1687.
[79] B.W. Beal, W.K. Kowak, A. Umezawa, G.W. Crabtree, J.D. Jorgensen, J.W. Downey, L.J. Nowicki, A.W. Mitchell, A.P. Paulikas and C.H. Sowers, Appl.Phys.Lett., 51 (1987) 279.
[80] M. Oda, T. Murakami, Y. Enomoto and M. Suzuki, Jap.J. Appl.Phys., 26 (1987) L804.
[81] Y. Zhao, H. Zhang, T. Zhang, S.F. Sun, Z.Y. Chen and Q.R. Zhang, Physica C, 152 (1988) 513.
[82] B. Jayaram, S.K. Agarwal, A. Gupta and A.V. Narlikar, Sol.State Commu., 63 (1987) 713.
[83] H. Ihara, N. Teroda, M. Jo, M. Hirabayashi,M. Tokumora, M. Kimura, T. Matsubara and R. Sagise, Jap.J.Appl.Phys. 26 (1987) 169.
[84] S.M. Bhagat, Int.Workshop on HTSC, Srinagar (India), May, 1988.
[85] R.P. Sharma, K.G. Prasad, B. Jayaram, S.K. Agarwal, A. Gupta and A.V. Narlikar, Phys.Lett. A 128 (1988) 217.
[86] B. Jayaram, S.K. Agarwal, K.C. Nagpal, A. Gupta and A.V. Narlikar, Mat.Res.Bull., 23 (1988) 701.
[87] S.B. Qadri, L.E. Toth, M. Osofski, S. Lawrence, D.U. Gubser and S.A. Wolf, Phys.Rev. B, 35 (1987) 7235.
[88] G. Xiao, F.H. Streitz, A. Garvin, Y.W. Du and C.L. Chien, Phys.Rev.B, 35 (1987) 8782.
[89] J.M. Tarascon, P. Barboux, P.F. Miceli, L.H. Greene, G.W. Hull, M. Eibschutz and S.A. Sunshine, Phys.Rev.B, 37 (1988) 7458.
[90] Y. Maeno, M. Kato, Y. Aoki and T. Fujita, Jap.J.Appl. Phys., 26 (1987) L1982.
[91] E. Takayama-Muromachi, Y. Uchida and K. Kato, Jap.J. Appl.Phys., 26 (1987) L2087.
[92] X. Zhang, S. Labroo, P. Hill, N. Ali, E. Funk and J.R. Gains,Jr., Phys.Lett. A, 130, (1988) 311.
[93] Z.Q. Qiu, Y.W. Du, H. Tang, J.C. Walker, W.A. Bryden, K. Moorjani, J.Mag.Mag.Mat., 69 (1987) L221.
[94] Y. Kimishima, S. Hirabayashi and N. Miyata, Jap.J.Appl. Phys., 27 (1988) L1123.
[95] J.F. Bringley, T.M. Chen, B.A. Averill, K.M. Wong and S.J. Poon, Phys.rev.B, 38 (1988) 2432.
[96] Z. Yong, Z. Han, S. Shifang, S. Zhenpeng, C. Zuyao, Sol.State Commu., 67 (1988) 31.
[97] K. Westerholt, M. Arndt, H.J. Wuller, H. Bach and P. Stauche, Physics C, 153-155 (1988) 862.
[98] B. Jayaram, S.K. Agarwal, C.V. Narasimha Rao and A.V. Narlikar, Phys.Rev.B, 38 (1988) 2903.
[99] G. Xiao, M.Z. Cieplak, D. Musser, A. Gavrin, F.H. Streitz, C.L. Chien, J.J. Rhyne and J.A. Gotaas, Nature, 332 (1988) 238.
[100] Y. Shimakawa, Y. Kubo, K. Utsumi, Y. Takeda and M. Takano, Jap.J.Appl.Phys., 27 (1988) L1071.
[101] M.F. Yan, W.W. Rhodes and P.K. Gallagher, J.Appl.Phys., 63 (1988) 821.
[102] T. Kajitani, K. Kusaba, M. Kikuchi, Y. Syono and M.

Hirabayashi, Jap.J.Appl.Phys., 27 (1988) L354.
[103] C.V. Narasimha Rao, S.K. Agarwal, B. Jayaram and A.V. Narlikar, Pramana-J.Phys., 31 (1988) L323
[104] P.B. Kirby, M.R. Harrison, W.G. Freeman, I. Samuel and M.J. Haines, Phys.Rev. B, 36 (1987) 8315.
[105] M. Hiratani, Y. Ito, K. Miyauchi and T. Kudo, Jap.J. Appl.Phys., 26 (1987) L1997.
[106] C.V. Tomy, A.M. Umarji, D.T. Adroja and S.K. Malik Sol.State Commu., 64 (1987) 889.
[107] D. Cohen, Z. Moisi and M. Schwartz , Mat.Res.Bull., 22 (1987) 1581.
[108] M. Kuwabara and N. Kusaka, Jap.J.Appl.Phys., 27 (1988) L1504.
[109] S.R. Ovshinsky, R.T. Young, D.D. Allred, G. DeMaggio and G.A. Van der Leeden, Phys.Rev.Lett., 58 (1987) 2579
[110] R.P. Gupta, W.S. Khokle, G.S. Virdi, B.C. Pathak, B. Jayaram, R.C. Dubey, S.K. Agarwal and A.V. Narlikar, Pramana-J.Phys., 30 (1988) L71.
[111] M. Xian-Ren, R. Yan-Ru, L. Ming-Zhu, T. Qing-Yun, L. Zhen-Jin, S. Li-Hua, D. Wei-Qing, F. Min-Hua, M. Qing-Yun, L. Chang-Jiang, L. Xiu-Hai, Q. Guan-Liang and C. Mou-Yuan, Sol.State Commu., 64 (1987) 325.
[112] A.K. Tyagi, S.J. Patwe, U.R.K. Rao and R.M. Iyer, Sol. State Commu., 65 (1988) 1149.
[113] R. Sugise, H. Ihara, Y. Yokoyama, T. Okumura, M. Hirabayashi, N. Terada, M. Jo, N. Koshizuka and I. Hayashida, Jap.J.Appl.Phys., 27 (1988) L1254.
[114] N.P. Bansal and A.L. Sandkuhl, Appl.Phys.Lett., 52 (1988) 838.
[115] K.M. Cirillo and J.C. Wright, J. Seuntjns, M. Dacumling and D.C. Larbalestier,Sol.State Phys., 66 (1988) 1237.
[116] W.Y. Hsu, R.V. Kasowski and F. Herman, Phys.Rev.B, 37 (1988) 5824.
[117] I. Matsubara, H. Tanigawa, T. Ogura and S. Kose, Jap.J. Appl.Phys., 27 (1988) L1080.
[118] D. Bhattacharya, S.K. Ghatak, T.K. Dey, P. Pramanik, K.L. Chopra, M. Bhatnagar and D.K. Pandya, Sol.State Commu. 66 (1988) 961.
[119] I. Felner, I. Nowik and Y. Yeshurun, Phys.Rev.B, 36 (1987) 3923.
[120] K.N.R. Taylor, D.N. Matthews and G.R. Russel, J.Crys. Growth, 85 (1988) 628.
[121] E. Yanagisawa, D.R. Dietderich, H. Kamakara, K. Togano, H. Maeda and K. Takahashi, Jap.J.Appl.Phys., 27 (1988) L1460.
[122] N. Murayama, E. Sudo, M. Awano, K. Kani and Y. Torii, Jap.J.Appl.Phys., 27 (1988) L1629.
[123] H. Mazaki, M. Takano, J. Takada, K. Oda, H. Kitaguchi, Y. Imura, Y.Ikeda, Y. Tomii and T. Kubozoe, Jap.J.Appl. Phys., 27 (1988) L1639.
[124] H. Nobumasa, K. Shimizu, Y.Kitano and T. Kawai, Jap.J. Appl.Phys., 27 (1988) L1669.
[125] M. Mizuno, H. Endo, J. Tsuchiya, N. Kijima, A. Sumiyama and Y. Oguri, Jap.J.Appl.Phys., 27 (1988) L1225.
[126] N. Fukushima, H. Niu and K. Ando, Jap.J.Appl.Phys., 27

(1988) L790.
[127] C.N.R. Rao,Chemistry of Oxide Superconductors,(Ed. C.N. R. Rao), Blackwell Sci. Pub. (1988), p.183.
[128] R. Yoshizaki, Y. Saito, Y. Abe and H. Ikeda, Physica C, 152 (1988) 408.
[129] E.V. Sampathkumaran, N. Ikeda, K. Kohn and R. Vijayaraghavan, J.Phys.F:Met.Phys., 18 (1988) L163.
[130] N. Ikeda, K. Kohn, E.V. Sampathkumaran and R. Vijayaraghavan, Sol.State Commu. (in press).
[131] A.V. Narlikar- to be published.
[132] T. Den, A. Yamazaki and J. Akimitsu, Jap.J.Appl.Phys., 27 (1988) L1620.
[133] T. Kijima, J. Tanaka and Y. Bando, Jap.J.Appl.Phys., 27 (1988) L1035.
[134] C.N.R. Rao, Sadhana, 13 (1988) 19.
[135] M.A. Subramanian, C.C. Torardi, J. Gopalakrishnan, P.L. Gai, J.C. Calabrese, T.R. Askew, R.B. Flippen and A.W. Sleight, Science, 242 (1988) 249.
[136] T.M. Rice, Superconductivity in Magnetic and Exotic Materials (Ed. T. Matsubara and A. Kotani), Springer Series:Solid State Science (1984), p.181.
[137] U. Welp, W.K. Kwok, G.W. Crabtree, H. Claus, K.G. Vandervoort, B. Dabrowski, A.W. Mitchell, D.R.Richards, D.T.Marx and D.G. Hinks, Physica C, 156 (1988) 27.
[138] D.E. Cox and A.W. Sleight, Sol.State Commu., 19 (1976) 969.
[139] D.E. Cox and A.W. Sleight, Proc.Int.Conf. on Neutron Scattering, Gatlinburg (Ed. R.M. Moon), National Technical Information Service,Springfield,(1976), p.45.
[140] D.E. Cox and A.W. Sleight, Acta Cryst. Sect., B 35 (1979) 1.
[141] M. Oda, Y. Hidaka, A. Katsui and T. Murakami, Sol.State Commu., 60 (1986) 897.
[142] D.G. Hinks, B. Babrowski, J.D. Jorgensen, A.W. Mitchell D.R. Richards, S. Pei and D. Shi, Nature, 333 (1988) 836.
[143] R. M. Fleming, P. Marsh, R.J. Cava and J.J. Krajewski, Phys.Rev. B, 38 (1988) 7026.
[144] D.G. Hinks, .R. Richards, B. Dabrawski, T. Marx and A.W. Mitchell, Nature, 335 (1988) 419.

SUBJECT INDEX